Grün

DOMAIN DECOMPOSITION METHODS

SERIES LIST FOR ALL PROCEEDINGS

Neustadt, L. W., Proceedings of the First International Congress on Programming and Control (1966)

Hull, T. E., Studies in Optimization (1970)

Day, R. H. & Robinson, S. M., Mathematical Topics in Economic Theory and Computation (1972)

Proschan, F. & Serfling, R. J., Reliability and Biometry: Statistical Analysis of Lifelength (1974)

Barlow, R. E., Reliability & Fault Tree Analysis: Theoretical & Applied Aspects of System Reliability & Safety Assessment (1975)

Fussell, J. B. & Burdick, G. R., Nuclear Systems Reliability Engineering and Risk Assessment (1977)

Duff, I. S. & Stewart, G. W., Sparse Matrix Proceedings 1978 (1979)

Holmes, P. J. New Approaches to Nonlinear Problems in Dynamics (1980)

Erisman, A. M., Neves, K. W. & Dwarakanath, M. H., Electric Power Problems: The Mathematical Challenge (1981)

Bednar, J. B., Redner, R., Robinson, E. & Weglein, A., Conference on Inverse Scattering: Theory and Application (1983)

Voigt, R. G., Gottlieb D. & Husaini, M. Yousuff, Spectral Methods for Partial Differential Equations (1984)

Chandra, Jagdish, Chaos in Nonlinear Dynamical Systems (1984)

Santosa, F., Pao, Y-H., Symes, W. W. & Holland, C., Inverse Problems of Acoustic and Elastic Waves (1984)

Gross, Kenneth I., Mathematical Methods in Energy Research (1984)

Babuska, I., Chandra, J. & Flaherty, J., Adaptive Computational Methods for Partial Differential Equations (1984)

Boggs, Paul T., Byrd, Richard H. & Schnabel, Robert B., Numerical Optimization 1984 (1985)

Angrand, F., Dervieux, A., Desideri, J. A. & Glowinski, R., Numerical Methods for Euler Equations of Fluid Dynamics (1985)

Wouk, Arthur, New Computing Environments: Parallel, Vector and Systolic (1986)

Fitzgibbon, William E., Mathematical and Computational Methods in Seismic Exploration and Reservoir Modeling (1986)

Drew, Donald A. & Flaherty, Joseph E., Mathematics Applied to Fluid Mechanics and Stability: Proceedings of a Conference Dedicated to R. C. DiPrima (1986)

Heath, Michael T., Hypercube Multiprocessors 1986 (1986)

Papanicolaou, George, Advances in Multiphase Flow and Related Problems (1987)

Wouk, Arthur, New Computing Environments: Microcomputers in Large-Scale Computing (1987)

Chandra, Jagdish & Srivastav, Ram, Constitutive Models of Deformation (1987)

Heath, Michael T., Hypercube Multiprocessors 1987 (1987)

Glowinski, R., Golub, G. H., Meurant, G. A. & Periaux, J., First International Symposium on Domain Decomposition Methods for Partial Differential Equations (1988)

Salam, Fathi, M. A. & Levi, Mark L., Dynamical Systems Approaches to Nonlinear Problems in Systems and Circuits (1988)

Datta, B., Johnson, C., Kaashoek, M., Plemmons, R. & Sontag, E., Linear Algebra in Signals, Systems and Control (1988)

Ringeisen, Richard D. & Roberts, Fred S., Applications of Discrete Mathematics (1988)

McKenna, James & Temam, Roger, ICIAM '87—Proceedings of the First International Conference on Industrial and Applied Mathematics (1988)

Rodrigue, Garry, Parallel Processing for Scientific Computing (1989)

Chan, Tony F., Glowinski, Roland, Periaux, Jacques & Widlund, Olof B., Domain Decomposition Methods (1989)

DOMAIN DECOMPOSITION METHODS

Edited by Tony F. Chan
Roland Glowinski
Jacques Periaux
Olof B. Widlund

siam
Philadelphia

DOMAIN DECOMPOSITION METHODS

Proceedings of the Second International Symposium on Domain Decomposition Methods, Los Angeles, California, January 14–16, 1988.

This symposium was sponsored by the Office of Naval Research under University Research Initiative Grant N00014-86-K-0691 and by Groupe pour l'Avancement des Méthodes Numériques de l'Ingénieur-Societe de Mathematiques Appliquees et Industrielles.

Library of Congress Cataloging-in-Publication Data

Domain decomposition methods.

Papers presented at the Second International Symposium on Domain Decomposition Methods at the University of California at Los Angeles, Jan. 1988.
1. Decomposition method—Congresses. 2. Differential equations, Partial—Congresses. I. Chan, Tony F. II. International Symposium on Domain Decomposition Methods (2nd: 1988: University of California at Los Angeles)
QA402.2.D66 1989 515.3'53 89-5888
ISBN 0-89871-233-5

PREFACE

Interest in domain decomposition methods has increased tremendously in the past two years, with notable advances taking place in the areas of both theory and application. In January 1988, nearly 100 people attended the Second International Symposium on Domain Decomposition Methods at the University of California at Los Angeles. The proceedings from this conference, including 34 of the 40 presented lectures and a broad range of papers representing worldwide state-of-the-art research, is contained in this volume. For the reader's convenience, this proceedings is organized into four parts: theory, algorithms, parallel implementation and applications. Of course, not all of the papers can be neatly classified into these four categories; in fact, many actually are a successful blend of all four aspects.

The field of domain decomposition methods has matured greatly since the first symposium, held in Paris in January 1987. This is evidenced by the 50% increase in the number of papers contained in this proceedings compared to that of 1987, and is reflected by the growing number of researchers working in this area.

Many aspects of domain decomposition methods are now being explored more thoroughly as a result of the information disseminated and exchanged at the first two meetings. One of the most notable examples is the interest in variants of the Schwarz iteration, which was inspired by a paper by P. L. Lions at the Paris meeting. Researchers are also now beginning to extend their techniques to more realistic applications. For instance, there are now several papers on applications to Stokes and Navier Stokes problems and to problems arising in combustion. Papers are also included on the extension of the domain decomposition principle to time dependent problems and on spectral methods and parallel implementation issues.

Future conferences on domain decomposition methods have already been planned for Houston, Texas in 1989 and Moscow, USSR in 1990.

The authors would like to acknowledge the support of the United States Office of Naval Research and the GAMNI/SMAI of France for their financial support.

Tony F. Chan
University of California at Los Angeles

Roland Glowinski
INRIA and the University of Houston

Jacques Periaux
AMD/BA and INRIA

Olof B. Widlund
New York University

LIST OF CONTRIBUTORS

Christopher R. Anderson, Department of Mathematics, University of California, Los Angeles, Los Angeles, California 90024, USA

Randolph E. Bank, Department of Mathematics, University of California, San Diego, La Jolla, California 92093, USA

Torsten Berglind, FFA, The Aeronautical Research Institute of Sweden, S-161 11, Bromma, Sweden

Petter Bjørstad, Department of Computer Science, Institutt for Informatikk, UiV, Allegaten 55, N-5007 Bergen, Norway

J. F. Bourgat, INRIA, Domaine de Voluceau, 78153 Le Chesnay Cedex, France

Luigi Brochard, IBM Scientific Center, Palo Alto, California 94304, USA, and Ecole Nationale des Ponts et Chaussees, 93194 Noisy le Grand Cedex, France

Tony F. Chan, Department of Mathematics, University of California, Los Angeles, Los Angeles, California 90024-1555, USA

Françoise Coulomb, Service d'Etudes de Réacteurs et de Mathématiques Appliquées, Centre d'Etudes Nucléaires de Saclay, 91191 Gif-sur-Yvette, France, and Laboratoire d'Analyse Numérique, Université Pierre et Marie Curie, 75005 Paris, France

Philippe Destuynder, Office National D'Etudes et de Recherches Aérospatiales, 92322 Chatillon Cedex, France

Milo R. Dorr, Lawrence Livermore National Laboratory, Livermore, California 94550, USA

Maksymilian Dryja, Institute of Informatics, Warsaw University, 00-901 Warsaw, PKiN, p. 850, Poland

D. J. Evans, Loughborough University of Technology, Loughborough, Leicester, LE11 3TU, United Kingdom

Richard E. Ewing, Departments of Mathematics, Petroleum Engineering, and Chemical Engineering, University of Wyoming, Laramie, Wyoming 82071, USA

Roland Glowinski, Department of Mathematics, University of Houston, Houston, Texas 77004, USA

William D. Gropp, Department of Computer Science, Yale University, New Haven, Connecticut 06520, USA

M. Haghoo, Department of Mathematics, University of Southern California, Los Angeles, California 90089-1113, USA

G. W. Hedstrom, Lawrence Livermore National Laboratory, Livermore, California 94550, USA

Richard S. Hirsh, The Johns Hopkins University Applied Physics Laboratory, Laurel, Maryland 20707, USA

F. A. Howes, Department of Mathematics, University of California, Davis, Davis, California 95616, USA

M. Yousuff Hussaini, Institute for Computer Applications in Science and Engineering, NASA Langley Research Center, Hampton, Virginia 23665, USA

Kang Li-shan, Department of Mathematics, Wuhan University, Wuhan, Hubei, People's Republic of China

David E. Keyes, Department of Mechanical Engineering, Yale University, New Haven, Connecticut 06520, USA

David A. Kopriva, Mathematics Department and Supercomputer Computations Research Institute, Florida State University, Tallahassee, Florida 32306, USA

Hwar C. Ku, The Johns Hopkins University Applied Physics Laboratory, Laurel, Maryland 20707, USA

Patrick Le Tallec, Laboratoire Central des Ponts et Chaussées, 75015 Paris, France

C. B. Liem, Department of Mathematical Studies, Hong Kong Polytechnic, Hong Kong

P. L. Lions, Ceremade, Université Paris-Dauphine, 75775 Paris Cedex 16, France, and consultant at Informatique Internationale (CISI)

J. Liou, Department of Aerospace Engineering and Mechanics, and Minnesota Supercomputer Institute, University of Minnesota, Minneapolis, Minnesota 55455, USA

T. Lu, Institute of Mathematical Sciences, Academia Sinica, Chengdu, People's Republic of China

Michele G. Macaraeg, Computational Methods Branch, High-Speed Aerodynamics Division, NASA Langley Research Center, Hampton, Virginia 23665, USA

Yvon Maday, Laboratoire d'Analyse Numerique de l'Université Pierre et Marie Curie, Paris, France, and Massachusetts Institute of Technology, Cambridge, Massachusetts 02139, USA

Jan Mandel, Computational Mathematics Group, University of Colorado, Denver, Denver, Colorado 80202, USA

Cathy Mavriplis, Department of Mechanical Engineering, Massachusetts Institute of Technology, Cambridge, Massachusetts 02139, USA

Steve McCormick, Computational Mathematics Group, University of Colorado, Denver, Denver, Colorado 80202, USA, and Colorado Research and Development Corporation, Fort Collins, Colorado 80525, USA

Gérard Meurant, CEA, Centre d'Etudes de Limeil-Valenton, 94190 Villeneuve St. Georges, France

T. Nguyen, IBM Corporation, Houston, Texas 77056, USA

Joseph E. Pasciak, Brookhaven National Laboratory, Upton, New York 11973, USA

Anthony T. Patera, Department of Mechanical Engineering, Massachusetts Institute of Technology, Cambridge, Massachusetts 02139, USA

S. Poole, IBM Corporation, Houston, Texas 77056, USA

Wlodzimierz Proskurowski, Department of Mathematics, University of Southern California, Los Angeles, California 90089-1113, USA

Alfio Quarteroni, Dipartimento di Matematica, Università Cattolica di Brescia, 25121 Brescia, and Istituto di Analisi Numerica del C.N.R., 27100 Pavia, Italy

Edna Reiter, California State University, Hayward, California 94542, USA

Garry Rodrique, Department of Mathematics, University of California, Davis, and Lawrence Livermore National Laboratory, Livermore, California 94550, USA

Allan P. Rosenberg, The Johns Hopkins University Applied Physics Laboratory, Laurel, Maryland 20707, USA and Sachs/Freeman Associates

François-Xavier Roux, Office National d'Etudes et de Recherches Aérospatiales, 92322 Chatillon Cedex, France

T. M. Shih, Department of Mathematical Studies, Hong Kong Polytechnic, Hong Kong

Craig L. Streett, Theoretical Aerodynamics Branch, Transonic Aerodynamics Division, NASA Langley Research Center, Hampton, Virginia 23665, USA

Wei Pai Tang, Department of Computer Science, University of Waterloo, Waterloo, Ontario, Canada N2L 3G1

Thomas D. Taylor, The Johns Hopkins University Applied Physics Laboratory, Laurel, Maryland 20707, USA

T. E. Tezduyar, Department of Aerospace Engineering and Mechanics, and Minnesota Supercomputer Institute, University of Minnesota, Minneapolis, Minnesota 55455, USA

Marina Vidrascu, INRIA, Domaine de Voluceau, 78150 Le Chesnay, France

Olof B. Widlund, Courant Institute of Mathematical Sciences, New York, New York 10012, USA

Harry Yserentant, Fachbereich Mathematik der Universitat Dortmund, D-4600 Dortmund 50, Federal Republic of Germany

CONTENTS

PART I

Theory

Variational Formulation and Algorithm for Trace Operator in Domain Decomposition Calculations

J. F. Bourgat*
Roland Glowinski†
Patrick Le Tallec††
Marina Vidrascu*

Abstract. A new preconditioning strategy is proposed for solving elliptic problems via domain decomposition techniques. This preconditioner acts on the Steklov-Poincare's operator (represented after discretization by the so-called Schur complement matrix) through the addition of a trace averaging and of the solution of a Neumann problem per subdomain. Such a strategy can operate on arbitrary geometries and unstructured meshes, gives the same role to each subdomain and can be written in a variational form. Two or three-dimensional numerical results are given to illustrate the efficiency of this strategy.

1. Introduction. The idea of reducing the solution of an elliptic problem set on a domain Ω to the parallel solution of problems of same type set on subdomains Ω_i of Ω is ancient ([1],[2],[3]), but it gets new attention with the present development of parallel computers.

These domain decomposition methods are based on simple and intuitive ideas. Nevertheless, their numerical efficiency is very sensitive to the choice of the computational variables, that is to preconditioning. The approach proposed in this paper uses an H^1 norm on each subdomain (§3). On the product of the associated spaces, the problem takes a classical variational form and is associated to an operator which involves on each subdomain the successive solution of

* INRIA, Domaine de Voluceau, 78150 Le Chesnay, FRANCE.
† Dept. of Mathematics, University of Houston, 4800 Cahloun Road, Houston, Texas 77004.
‡ Labo. Central des Ponts et Chaussées, 58 boulevard Lefebvre, 75015 Paris, FRANCE.

a Dirichlet and of a Neumann problem (§4). This formulation is then solved by a conjugate gradient algorithm (§5).

Numerical results will assert the efficiency of our approach (§6): on significative two dimensional situations (irregular mesh, discontinuous coefficients, crossed interfaces), the algorithm has converged in very few iterations. Similarly, convergence occured in 14 iterations in the solution of a Poisson problem set on a complex three-dimensional geometry.

It should be also observed that the present approach is the equivalent in a standard variational form (and using standard discretization techniques) of the mixed variational approach described in [4] within a mixed finite element framework.

2. A Simplified Model Problem. We first describe our approach on the following model problem:

$$\begin{cases} -\Delta u = f \text{ on } \Omega, \\ u = 0 \text{ on } \partial\Omega, \end{cases}$$

the domain Ω being decomposed as indicated in the figure below.

Our goal is to solve the above problem only on the subdomains Ω_i. If we knew the value λ of the solution u on the interface S, then the parallel solution of

$$\begin{cases} -\Delta u_i = f \text{ on } \Omega_i, \\ u_i = \lambda \text{ on } S, \\ u_i = 0 \text{ on } \partial\Omega \cap \partial\Omega_i, \end{cases}$$

on Ω_1 and Ω_2 will achieve this goal. The problem is thus

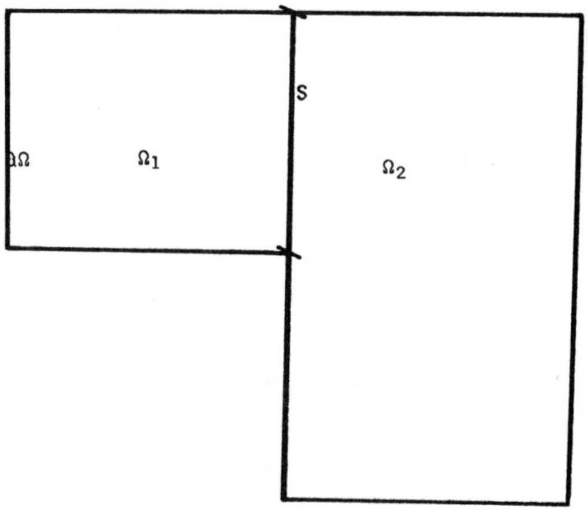

Figure 1: the model problem

reduced to the computation of λ which can be done by the following gradient algorithm operating on the trace space $H_{00}^{1/2}(S)$:

<u>data:</u> λ given in $H_{00}^{1/2}(S)$;

<u>computation of the solution:</u> for any i, solve the Dirichlet problem:

$$\begin{cases} -\Delta u_i = f \text{ on } \Omega_i, \\ u_i = \lambda \text{ on } S, \\ u_i = 0 \text{ on } \partial\Omega \cap \partial\Omega_i; \end{cases}$$

<u>computation of the gradient:</u> for any i, solve the Neumann problem (preconditioner)

$$\begin{cases} -\Delta\psi_i = 0 \text{ on } \Omega_i, \\ \dfrac{\partial\psi_i}{\partial n_i} = \dfrac{1}{2}\left(\dfrac{\partial u_1}{\partial n_1} + \dfrac{\partial u_2}{\partial n_2}\right) \text{ on } S, \\ \psi_i = 0 \text{ on } \partial\Omega \cap \partial\Omega_i; \end{cases}$$

<u>updating:</u> set $\lambda = \lambda - \rho(\psi_1 + \psi_2)$ and reiterate .

The domain decomposition method that we will now introduce is simply a generalization of this algorithm.

3. The Original Problem. Consider the domain Ω, partitioned into subdomains Ω_i as indicated on Figure 2.

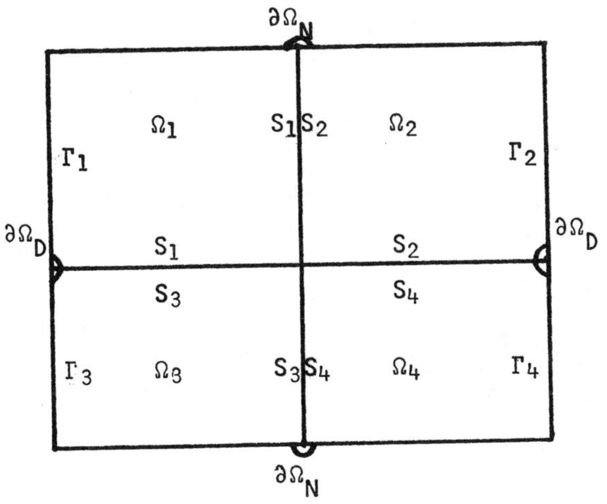

Figure 2: definition of the subdomains and boundaries.

Let us introduce the boundaries (see Figure 2)

$$\partial\Omega = \partial\Omega_N \cup \partial\Omega_D ,$$

$$\Gamma_i = \partial\Omega_D \cap \partial\Omega_i ,$$

$$S_i = \partial\Omega_i - \Gamma_i - \text{interior}(\partial\Omega_N \cap \partial\Omega_i),$$

together with the spaces

$$V = \left\{ v \in H^1 (\Omega ;\mathbb{R}^p),\ v = 0 \text{ on } \partial\Omega_D \right\},$$

$$V_i = \left\{ v \in H^1 (\Omega_i ;\mathbb{R}^p),\ v = 0 \text{ on } \Gamma_i \right\},$$

$$V_{0i} = \left\{ v \in H^1 (\Omega_i ;\mathbb{R}^p),\ v = 0 \text{ on } \Gamma_i \cup S_i \right\}.$$

As in [1], the pairing $\tilde{a}_i (u,v)$ will denote a given scalar product on V_i , X will be the product of the spaces V_i and Y will represent the space of traces on S of functions of V. In addition, $Tr_i^{-1} (\lambda)$ will represent any element z of V_i whose trace on S_i is equal to λ. Finally, we introduce the elliptic form

$$a_i (u,v) = \int_{\Omega_i} A_{mnkl} (x) \frac{\partial u_m}{\partial x_n} \frac{\partial v_k}{\partial x_l} .$$

Under these notations, the problem to solve writes

$$\underline{\text{Find}}\ u \in V\ \underline{\text{such that}}$$
$$\sum_i a_i (u,v) = \langle f,v \rangle,\ \forall\ v \in V .$$

4. Domain Decomposition of the Original Problem.

4.1 Introduction of an operator of Steklov-Poincaré's type.

We first define a trace operator α_i from V_i into Y satisfying

$$\text{(C)} \qquad \sum_i \alpha_i (v) = Tr(v),\ \forall\ v \in V.$$

For example, at the continuous level, we can set $\alpha_i (v) = Tr(v)/2$. A different definition should be used at the finite element level in order for condition (C) to be still satisfied after discretization. A possible choice will be described in §6.

For $w = \{w_i\} \in X = \prod_i V_i$, we then set

$$\lambda = \sum_i \alpha_i (w_i),$$

solve the Dirichlet problems

$$\begin{cases} a_i (z_i, v) = 0 , \ \forall v \in V_{0i}, \\ z_i \in V_i, \ z_i = \lambda \text{ on } S_i , \end{cases}$$

define the usual Steklov-Poincare's operator by

$$< S\lambda , \lambda' > = \sum_j a_j (z_j, Tr_j^{-1}(\lambda')) , \quad \forall \lambda' \in Y,$$

and solve the Neumann problems

$$\begin{cases} \tilde{a}_i (\psi_i, v) = < S\lambda , \alpha_i v > , \ \forall v \in V_i , \\ \psi_i \in V_i . \end{cases}$$

Our final operator is now defined from X into itself by

$$\mathcal{A}(w) = \{\psi_i\} .$$

Remark 4.1: In computing $a_j (z_j, Tr_j^{-1}(\lambda'))$, the choice of the representative element of Tr_j^{-1} is of no importance since, by construction, z_j is orthogonal to any component of this element in $Ker(Tr_j)$.

4.2 Analysis of $\mathcal{A}$. Assuming a_i to be symmetric positive on V_i, we introduce the operator K_i defined on V_i by

$$\tilde{a}_i (K_i u, K_i v) = a_i (u,v) , \ \forall u,v \in V_i ,$$

and the operator $\mathcal{B}$ defined on X by $\mathcal{B}\{w_i\} = \{K_i z_i\}$. By construction, we have:

$$\begin{aligned} (\mathcal{A}(w), w') &= \sum_i \tilde{a}_i (\psi_i, w_i') = \sum_i < S\lambda , \alpha_i w_i' > \\ &= < S\lambda , \lambda' > = \sum_j a_j (z_j, Tr_j^{-1}(\lambda')) \\ &= \sum_j a_j (z_j, z_j') = (\mathcal{B}(w), \mathcal{B}(w')). \end{aligned}$$

Therefore, $\mathcal{A}$ is a self-adjoint positive operator from X into itself, and can be decomposed into the product $\mathcal{A} = \mathcal{B}^T \mathcal{B}$. In particular, standard conjugate gradient algorithms operating on $\mathcal{A}$ will converge in $\mathrm{Im}(\mathcal{B}^T)$ with rate

$$\frac{1 - \mathrm{cond}(\mathcal{B})}{1 + \mathrm{cond}(\mathcal{B})} .$$

Moreover, from the above relations, we have

$$(\mathcal{A}w, w') = < \mathcal{S}\lambda \ , \lambda'>.$$

In other words, $\mathcal{A}$ is a preconditioned version of the Steklov-Poincare's operator $\mathcal{S}$. Such operators have been extensively studied in [5], [6], [7], [8]. Compared to them, our operator $\mathcal{A}$ adds a Neumann problem associated to the boundary condition

$$\frac{\partial \psi_i}{\partial n_i} = \alpha_i \sum_j \frac{\partial z_j}{\partial n_j},$$

which sends the dual of X back in X. The introduction of this preconditioner together with the averaging α_i is the main originality of the present approach.

Remark 4.2: The degrees of freedom in the definition of the operator $\mathcal{A}$ are the choice of the trace averaging α_i and of the scalar product $\tilde{a}_i(u,v)$. The choice of α_i has already been discussed. As for the scalar product, the best possible choice would be:

$$\tilde{a}_i(u,v) = a_i(u,v)$$

($K_i =$ Id), because then the Neumann step corresponds to the exact inverse of the Steklov-Poincaré's operator introduced in the Dirichlet part. With this choice we can expect $\mathcal{A}$ to have a very small condition number. Unfortunately, if Γ_i is empty, $a_i(u,v)$ does not define a scalar product on V_i. In that situation, the best choice is then

$$\tilde{a}_i(u,v) = a_i(u,v) + \sum_{k=1}^{K} <u, v_{ik}> <v, v_{ik}>$$

with $\{v_{ik}\}$ a sequence of elements of V_i adequately chosen. At the discrete level, this means adding positive diagonal terms to the matrix associated to $a_i(u,v)$.

4.3 Variational writing in X. To define the right hand side of our new problem, we solve the Dirichlet problems

$$\begin{cases} a_i(z_i^0, v) = <f,v> , \ \forall \ v \in V_{0i}, \\ z_i^0 \in V_{0i}, \end{cases}$$

and the Neumann problems

$$\left\{\begin{array}{l} \tilde{a}_i(\psi_i^0, v) = \sum_j a_j(z_i^0, Tr_j^{-1}(\alpha_i v)) - \langle f, Tr_j^{-1}(\alpha_i v)\rangle, \quad \forall v \in V_i, \\ \psi_i^0 \in V_i . \end{array}\right.$$

In that framework, the problem to solve can be written under the equivalent form

(P) <u>Solve</u> $\mathcal{A}(w) = -\{\psi_i^0\}$ <u>in</u> $\mathrm{Im}(\mathcal{B}^T)$,

within the identification $u = z_i + z_i^0$ on Ω_i .

<u>4.4 Equivalence proof.</u> Let w be the solution of problem (P), solution which uniquely exists from Lax-Milgram's lemma. Let u be the field given by

$$u = z_i + z_i^0 \text{ on } \Omega_i .$$

By construction the traces of z_i and z_i^0 are compatible on the interfaces S_i and are equal to zero on the boundaries Γ_i. Thus u belongs to V.

Now, let v be any element of V. On each subdomain, v can be decomposed into

$$v = v_j + Tr_j^{-1}(Tr\ v) \text{ with } v_j = v - Tr_j^{-1}(Tr\ v) \in V_{0j} .$$

Therefore, by construction of u, z_i and z_i^0, we have

$$\begin{aligned}
\sum_j a_j(u,v) &= \sum_j a_j(z_j + z_j^0, v_j + Tr_j^{-1}(Tr\ v)) \\
&= \sum_j \langle f, v_j\rangle + \sum_j a_j(z_j + z_j^0, Tr_j^{-1}(Tr\ v)) \\
&= \sum_j \langle f, v_j\rangle + \sum_j a_j(z_j + z_j^0, Tr_j^{-1}(\sum_i \alpha_i v)) \\
&= \sum_j \langle f, v_j\rangle + \sum_i\sum_j a_j(z_j + z_j^0, Tr_j^{-1}(\alpha_i v)) \\
&= \sum_j \langle f, v_j\rangle + \sum_i \tilde{a}_i(\psi_i + \psi_i^0, v) \\
&\quad + \sum_i\sum_j \langle f, Tr_j^{-1}(\alpha_i v)\rangle \\
&= \sum_j \langle f, v_j + Tr_j^{-1}(Tr\ v)\rangle + \sum_i \tilde{a}_i(\psi_i + \psi_i^0, v).
\end{aligned}$$

But, by construction of w, we have $\psi_i = -\psi_i^0$ and thus we finally get

$$\sum_j a_j(u,v) = \sum_j <f, v_j + Tr_j^{-1}(Tr\, v)> = <f,v> , \forall v \in V.$$

In other words, the solution u that we have constructed is the unique solution of our original problem.

5. Solution Algorithm. The solution algorithm that we propose is a standard conjugate gradient method ([9]) applied to the inversion of $\mathcal{A}$ on $Im(\mathcal{B}^T)$. It writes

INITIALIZATION

<u>For λ given in Y, solve the Dirichlet problems:</u>

$$\left\{ \begin{array}{l} a_i(u_{i,0}, v) = <f,v>, \forall v \in V_{0i}, \\ u_{i,0} = \lambda \text{ on } S_i, \\ u_{i,0} \in V_i ; \end{array} \right.$$

<u>compute</u> $L_i(v) = \sum_j \left(a_j(u_j, Tr_j^{-1}(\alpha_i v)) - <f, Tr_j^{-1}(\alpha_i v)> \right)$, $\forall v \in V_i$;

<u>solve the Neumann problems:</u>

$$\left\{ \begin{array}{l} \tilde{a}_i(\varphi_{i,0}, v) = L_i(v), \forall v \in V_i, \\ \varphi_{i,0} \in V_i ; \end{array} \right.$$

<u>compute</u> $d_0 = \sum_i \tilde{a}_i(\varphi_{i,0}, \varphi_{i,0}) = \sum_i L_i(\varphi_{i,0})$;

<u>set</u> $w_{i,0} = \varphi_{i,0}$ <u>and</u> $R_{i,0}(v) = L_i(v)$.

LOOP ON n: for n =0 until satisfied do

<u>computation of $\mathcal{A}w$</u>

<u>compute</u> $\lambda = \sum_j \alpha_j w_{j,n}$;

<u>solve the Dirichlet problems:</u>

$$\left\{ \begin{array}{l} a_i(z_i, v) = 0, \forall v \in V_{0i}, \\ z_i = \lambda \text{ on } S_i, \\ z_i \in V_i ; \end{array} \right.$$

<u>compute</u> $L_i(v) = \sum_j \left(a_j(z_j, Tr_j^{-1}(\alpha_i v)) \right), \ \forall v \in V_i$;

<u>solve the Neumann problems:</u>

$$\begin{cases} \tilde{a}_i(\psi_i, v) = L_i(v), \ \forall v \in V_i, \\ \psi_i \in V_i. \end{cases}$$

<u>descent</u>

<u>set</u> $r_n = \sum_j \tilde{a}_j(w_{j,n}, \psi_j) = \sum_j L_j(w_{j,n})$

<u>set</u> $\rho_n = d_n / r_n$;

<u>set:</u>

$$\begin{cases} u_{i,n+1} = u_{i,n} - \rho_n z_i, \\ \varphi_{i,n+1} = \varphi_{i,n} - \rho_n \psi_i, \\ R_{i,n+1}(v) = R_{i,n}(v) - \rho_n L_i(v). \end{cases}$$

<u>computation of the new descent direction</u>

<u>compute</u> $d_{n+1} = \sum_j \tilde{a}_j(\varphi_{j,n+1}, \varphi_{j,n+1}) = \sum_j R_{j,n+1}(\varphi_{j,n+1})$

or alternatively $d_{n+1} = d_n - \rho_n \sum_j L_j(\varphi_{j,n} + \varphi_{j,n+1})$;

<u>stop if</u> $\sqrt{d_{n+1}/d_0} < 10^{-5}$;

<u>set</u> $w_{i,n+1} = \varphi_{i,n+1} + \dfrac{d_{n+1}}{d_n} w_{i,n}$.

<u>Remark 5.1:</u> Observe that the linear forms $R_i(v)$ and $L_i(v)$ only operate on the traces on S_i of the functions v. This simplifies their computation. Moreover, this means that, in the algorithm, we just have to compute and store the traces on S_i of the functions φ_i and w_i.

<u>Remark 5.2:</u> The above algorithm is modular and, since it does not involve any relaxation parameter, is completely transparent to the user. Moreover, it has a high degree of parallelism: most of the CPU is devoted to the solution on each subdomain of independent Dirichlet and Neumann problems. The only "rendez-vous" are the trace averaging operations which occur twice per iteration.

6. Numerical Results. Our numerical results deal with the Laplace operator for which we have

$$a_i(u,v) = \int_{\Omega_i} \beta_i \ \nabla u . \nabla v \ .$$

The spaces V_i are approximated by finite elements of P1-Lagrange type. The trace operators α_i are defined at each node of the interface by the formula

$$(\alpha_i(v))(M_k) = \left(a_i(\varphi_k,\varphi_k)/(\sum_j a_j(\varphi_k,\varphi_k)) \right) v(M_k)$$

where φ_k denotes the weighting function associated to the node M_k. The three-dimensional coding was done within the MODULEF finite element library in a multielement, multi-problem and multitasking framework.

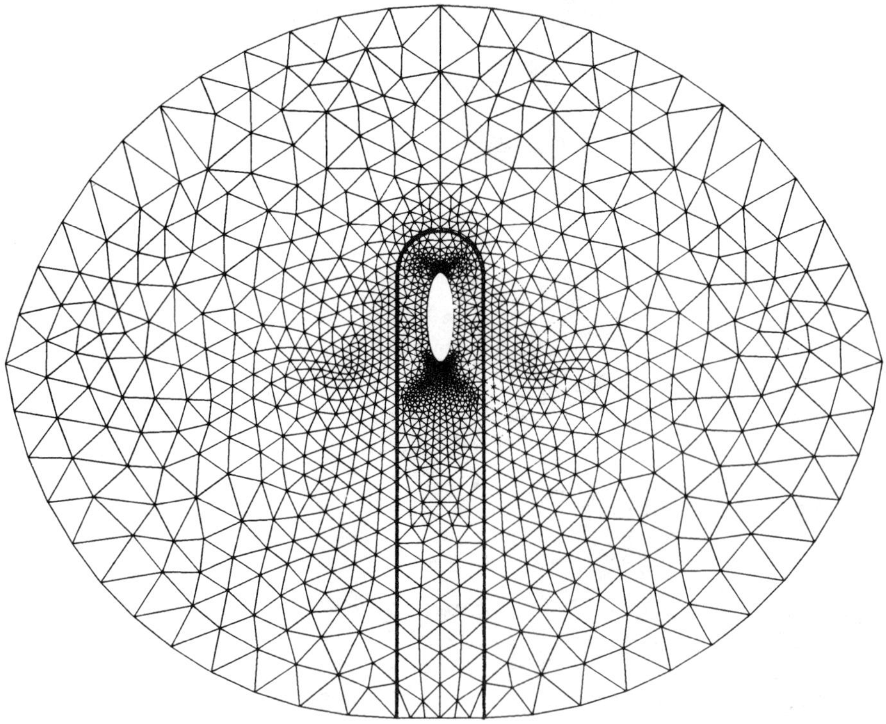

Figure 3

The first computed geometry is described on Figure 3. The internal mesh (on Ω_1) had 1222 triangles and 669 vertices, the external mesh (on Ω_2) had 1440 triangles and 780 vertices. For the pure Dirichlet problem, the algorithm has converged in 4 iterations for $\beta_1=\beta_2=1$ and in 3 iterations for $\beta_1=0.1$, $\beta_2=1$.

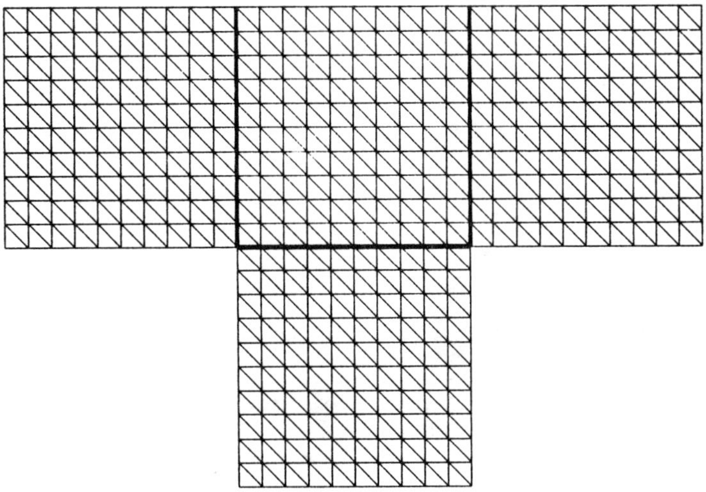

Figure 4: the T shaped domain

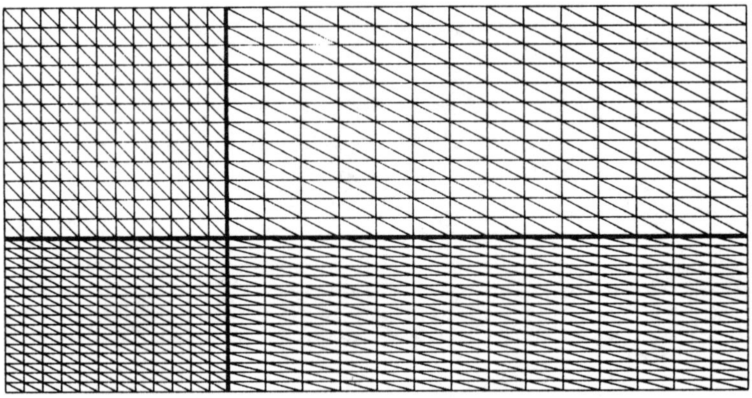

Figure 5: the "square cross" rectangle

The second test considers a pure Dirichlet problem set on the T-shaped domain treated in [10]. On this example, the algorithm has converged in 2 iterations.

The domain for the third example is the rectangle $\Omega=]0,4[\times]0,2[$ which is divided in 4 subdomains with crossing interfaces, as indicated on Figure 5. For both choices $\beta_1=\beta_2=\beta_3=\beta_4=1$ and $\beta_1=1.,\beta_2=10.,\beta_3=0.1,\beta_4=0.01$, the algorithm has converged in 5 iterations in the case of pure Dirichlet boundary conditions.

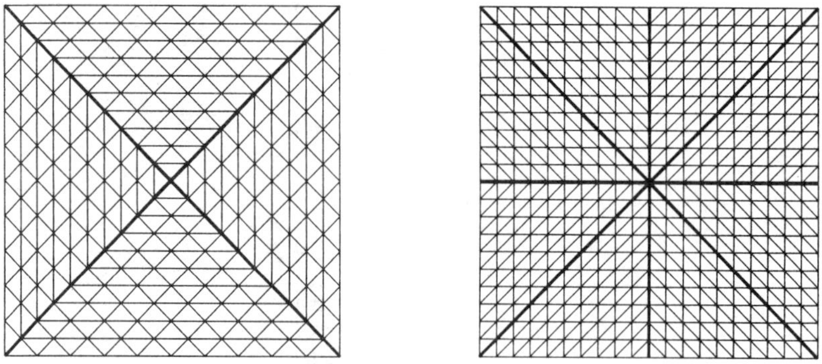

Figure 6: the "oblique cross" rectangle

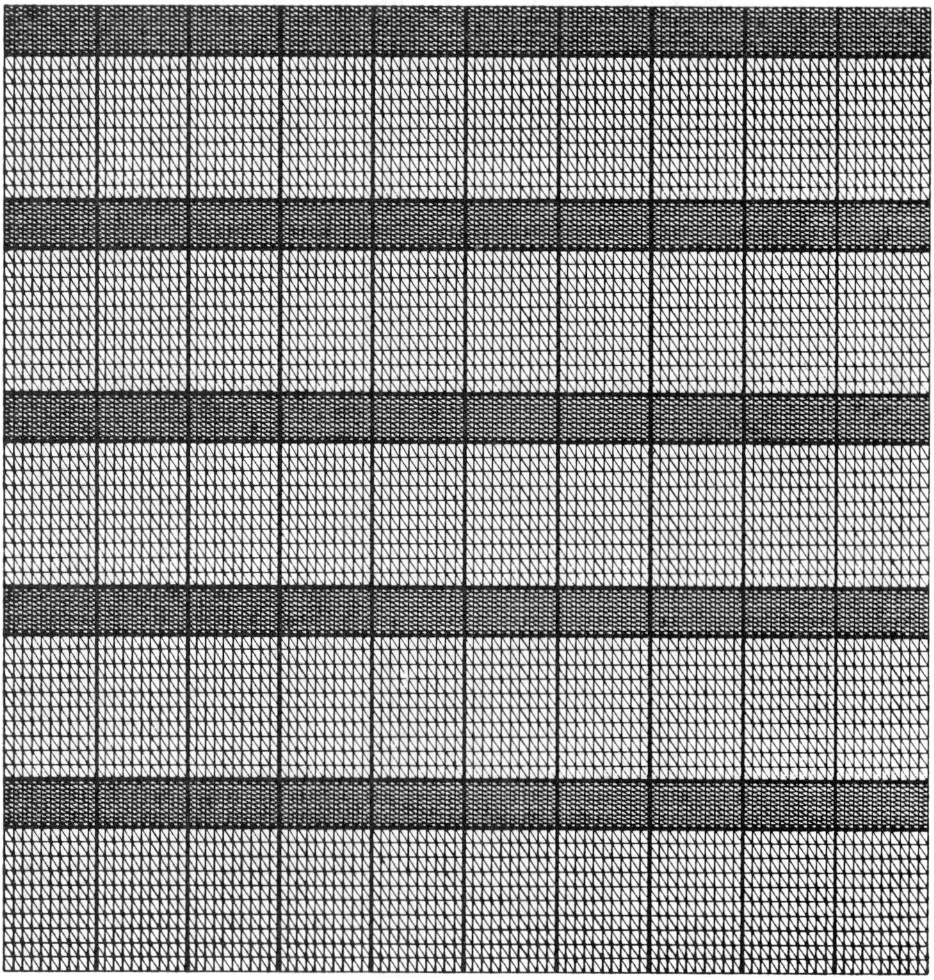

Figure 7: the "checkaboard" domain

The fourth domain that we have treated is the unit square divided as indicated in Figure 6. Here both the pure Dirichlet and the pure Neumann problem were considered, with $\beta_i = 10^{-3}$ if $x_1 - x_2 < 0$ and $\beta_i = 1$ if not . For the Dirichlet (resp. Neumann) problem, convergence was reached after 3 (resp. 8) iterations in the case of 4 subdomains and after 6 (resp. 20) iterations in the case of 8 subdomains. Taking 200 triangles per subdomain instead of 100 did not change the number of required iterations.

The next domain that we have treated has 100 subdomains, 20000 triangles and 10201 nodes. On this domain, represented in Figure 7, we have treated both the pure Dirichlet and the pure Neumann problem. For $\beta_i = 1$ everywhere convergence was reached after 40 iterations for the Dirichlet problem and after 74 iterations for the Neumann problem. For $\beta_i = 1$ in the wide strips and $\beta_i = 10^{-3}$ elsewhere, the Dirichlet (resp. Neumann) problem did require 54 (resp. 173) iterations to converge. Observe that there were 1701 (resp. 1737) nodes on the interface for the Dirichlet (resp. Neumann) problem.

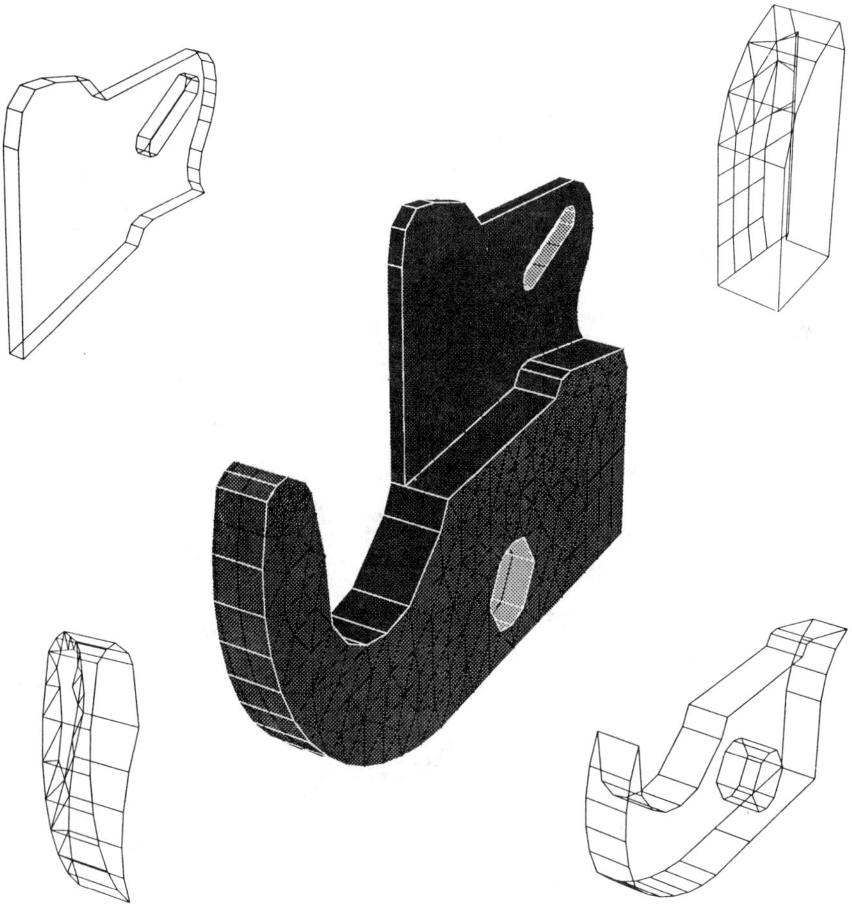

Figure 8: three-dimensional geometry

Our last numerical example deals with the complex three-dimensional geometry of Figure 8. The domain here is divided in 4, contains 937 nodes, with 83 nodes at the interfaces. Dirichlet boundary conditions were imposed only on the internal hole. Convergence occured after 14 iterations.

On all our numerical tests, we have observed that the number of iterations before convergence was independent of:
(i) the initial value of the trace λ,
(ii) the values of the right hand-side,
(iii) the discretization step.

7. Conclusion. In conclusion, all these numerical tests assess the validity of the conjugate gradient algorithm when operating on the product of traces, at least within the framework of a scalar elliptic operator and of a moderate number of subdomains. Corners in the decomposition can be handled easily provided that the averaging trace α_i be properly defined and be present in the Neumann step.

REFERENCES

(1) J.H. BRAMBLE, J.E. PASCIAK et A.H. SCHATZ, The construction of preconditioners for elliptic problems by substructuring,I, Math. Comp. 47 (1986), pp. 103-134.

(2) O. WIDLUND, Iterative methods for elliptic problems on regions partitionned into substructures and the biharmonic Dirichlet problem, Proceedings of the 6th International Conference on Computing Methods In Applied Science and Engineering, Versailles, North-Holland, 1983.

(3) R.GLOWINSKI, G.H. GOLUB, J. PERIAUX, eds.,Proceedings of the 1rst international symposium on decomposition methods for partial differential equations, Paris, SIAM editor, Philadelphia 1988.

(4) R. GLOWINSKI, M. WHEELER, Domain decomposition methods for mixed finite element approximation, in (3).

(5) V.I. AGOSHKOV, Poincare-Steklov's operators and domain decomposition methods in finite dimensional spaces, in (3).

(6) V.I. LEBEDEV, V.I. AGOSHKOV, The variational algorithms of domain decomposition method, Preprint n°54, Department of Numerical Mathematics of the Academy of Sciences, URSS, Moscow, 1983. (In Russian).

(7) V.I. AGOSHKOV, V.I. LEBEDEV, The Poincare-Steklov's operators and the domain decomposition methods in variational problems, in Computational processes and systems, NAUKA, Moscow (1985), pp. 173-227. (In Rusian.)

(8) V.I. LEBEDEV, The decomposition method, Department of Numerical Mathematics of the Academy of Sciences, URSS, Moscow, 1986. (In Russian.)

(9) R. GLOWINSKI, Numerical methods for nonlinear variational problems, Springer Verlag, 1984.

(10) L.D. MARINI, A. QUARTERONI, An iterative procedure for domain decomposition methods: a finite element approach, in (3).

On the Discretization of Interdomain Coupling in Elliptic Boundary-value Problems*

Milo R. Dorr†

Abstract. In this article, we consider the discretization of the interdomain coupling of solutions of elliptic boundary-value problems on decomposed domains without overlaps. Our approach employs Lagrange multipliers to enforce compatibility constraints between adjacent domains. The multipliers explicitly appear as auxiliary dependent variables in continuous and discrete variational formulations of the elliptic problem. As a result, the interdomain coupling is included at a higher level of the discretization than in approaches yielding standard Schur complement systems. Our primary focus is on the convergence of such methods as determined by the choice of multiplier spaces on the interfaces. The main conclusion drawn is that an appropriate selection of the multipliers, based on well-known regularity properties of solutions of elliptic boundary-value problems, yields good resolution of the interdomain coupling in finite-dimensional spaces of very low dimension. This has some important practical advantages, which we demonstrate with some numerical examples.

1. Introduction. In solving an elliptic boundary-value problem on a decomposed domain without overlaps, a typical approach is to apply block elimination to a standard finite-element or finite-difference discretization of the problem. This yields a *Schur complement*, or *reduced*, system to be solved for the unknowns on the interfaces between adjacent subdomains. Much attention has been given (see [18] for a good survey) to the problem of solving these Schur complement systems, which are generally large, dense, and ill-conditioned. Iterative techniques with judiciously-chosen preconditioners are the usual methods of choice.

* This work was supported by the Applied Mathematical Sciences subprogram of the Office of Energy Research, U.S. Department of Energy and by Lawrence Livermore National Laboratory under contract No. W-7405-Eng-48.

† Lawrence Livermore National Laboratory, P. O. Box 808, Livermore, CA 94550.

Since the Schur complement system can be viewed as the linear system resulting from a discretization of a global coupling problem among the subdomains, it is natural to ask the following questions. What are the regularity properties of the global coupling problem? How efficiently does the discretization yielding the Schur complement system resolve the interdomain coupling? What advantage can be taken of any information which may be known *a priori* about the interdomain coupling? The fact that the usual Schur complement system may not represent the best discretization of the coupling problem can be illustrated by simple examples in which very fine meshes are required to resolve the solution on each of the subdomains, yet the coupling between subdomains can be approximated very accurately in a subspace of much smaller dimension than the number of unknowns imposed on the interfaces by the subdomain meshes. The basic problem in such cases is that the need to efficiently discretize the interdomain coupling was not considered early enough in the discretization process. Hence, although the meshes on each subdomain are presumably chosen to sufficiently resolve some desired solution features, the resulting dimension of the Schur complement system may be much greater than is necessary to resolve the components of the solution which are more naturally related to the interdomain coupling.

Rather than obtaining the Schur complement system as a by-product of block elimination applied to a standard finite-difference or finite-element discretization of the problem, one might instead choose to include explicitly in the initial continuous formulation the quantity which will ultimately be approximated by the solution of the interface system. Since this quantity would then be present during the discretization process along with the subdomain variables, a more effective discretization of both the subdomain and interface problems could presumably be obtained. In this paper, we summarize the results of [13], which describe the use of Lagrange multipliers in achieving this goal. The Lagrange multipliers, which are functions defined on the interface between adjacent subdomains, allow the interdomain coupling to be explicitly included in continuous and discrete variational formulations. Since the subdomain basis functions and interface Lagrange multipliers are independently represented in these formulations, there is the opportunity to choose the multipliers in such a way as to resolve the interdomain coupling in finite-dimensional interface spaces of very low dimension.

The utility of Lagrange multipliers specifically for domain decomposition has been recognized by other investigators, e.g., [10]. To our knowledge, in these approaches the Lagrange multipliers are assumed to belong to the reduced space of functions spanned by the traces of the finite-dimensional subdomain basis functions, which represents only one possible choice for the Lagrange multiplier space. In this case, the number of Lagrange multipliers is again equivalent to the number of unknowns L associated with the interface. Since the construction of the dense interface problems to be solved for the Lagrange multipliers requires L linear system solves on each subdomain, iterative procedures are proposed which avoid the otherwise costly construction of the Schur complements. As shown in [13] and the present article, by choosing the Lagrange multiplier space in a more general fashion, the number L of Lagrange multipliers on an interface can be so dramatically reduced that for many problems it is entirely feasible to construct the dense interface problem and solve it directly. Moreover, all of the subdomain solves required to compute the interface matrix can be done in parallel assuming that a sufficient number of processors are available. This provides some important flexibility in achieving load-balanced calculations.

As in [13], the following development will consider only the case of two subdomains separated by a single interface. The results obtained for the two-domain case readily generalize to multidomain cases without crosspoints (e.g., stripwise decompositions). The application of the method described here in the general case with crosspoints will be the subject of a forthcoming paper. Also, although we are considering only self-adjoint problems here, there is nothing preventing the extension of our results to non-self-adjoint problems as well.

2. Lagrange Multipliers and Poincaré-Steklov Operators. In this section, we describe the method of Lagrange multipliers in the context of domain decomposition. We then show how the interface problems obtained by the Lagrange multiplier approach are related to certain dual interface formulations involving Poincaré-Steklov operators. As mentioned in the introduction, we will restrict our attention to the case of only two subdomains. Also, since we intend to focus on the use of Lagrange multipliers to enforce interface rather than boundary conditions, we will assume only homogeneous Dirichlet conditions on the boundary of the given domain.

Let Ω denote a two-dimensional Lipshitz domain whose boundary $\partial\Omega$ is the union of smooth (at least continuously differentiable) arcs. A polygon is an example of such a domain. Consider the following second-order, self-adjoint elliptic problem in Ω:

$$-\sum_{i,j=1}^{2} \frac{\partial}{\partial x_i}\left[a_{ij}(x)\frac{\partial u(x)}{\partial x_j} \right] + c(x)u(x) = f(x) \qquad \text{on } \Omega, \tag{2.1a}$$

with the homogeneous, Dirichlet boundary conditions

$$u = 0 \qquad \text{on } \partial\Omega. \tag{2.1b}$$

Here, the coefficients a_{ij} and c are assumed to belong to $L^\infty(\Omega)$ and satisfy the ellipticity condition

$$\sum_{i,j=1}^{2} a_{ij}\xi_i\xi_j \geq \beta \sum_{i=1}^{2} \xi_i^2, \qquad \text{for all } \xi_1, \xi_2, \quad \text{a.e. on } \Omega,$$

for some constant $\beta > 0$.

Next suppose that Ω is partitioned into two subdomains Ω_1 and Ω_2 by an interface Γ. We assume that Γ is at least continuously differentiable and subdivides Ω in such a way that the boundaries $\partial\Omega_k$ of the resulting subdomains are Lipschitz. For example, Γ might be a line segment partitioning a polygonal Ω.

For each k=1,2, let $H^1(\Omega_k)$ be the usual Sobolev space of functions with square-integrable first derivatives on Ω_k, and let $H_D^1(\Omega_k)$ be the subspace of $H^1(\Omega_k)$ consisting of the those functions which vanish on $\partial\Omega_k \cap \partial\Omega$. Let $H_D^{-1}(\Omega_k)$ be the dual of $H_D^1(\Omega_k)$. We denote by γ_k the operator mapping functions in $H_D^1(\Omega_k)$ to their traces on Γ. Let $H_{00}^{1/2}(\Gamma)$ denote the fractional-order Sobolev space on Γ consisting of all traces on Γ of functions in $H_D^1(\Omega_k)$ equipped with the norm

$$\|g\|_{H_{00}^{1/2}(\Gamma)} \equiv \inf_{u_k \in H_D^1(\Omega_k): g=\gamma_k u_k} \|u_k\|_{H^1(\Omega_k)},$$

and let $H_{00}^{-1/2}(\Gamma)$ denote its dual. With these definitions, it is clear that γ_k maps $H_D^1(\Omega_k)$ onto $H_{00}^{1/2}(\Gamma)$, and it can be shown [14] that γ_k has a continuous right inverse γ_k^{-1}.

Let $H \equiv H_D^1(\Omega_1) \times H_D^1(\Omega_2) \times H_{00}^{-1/2}(\Gamma)$. For each $(u_1,u_2,g^*) \in H$, we define

$$\Phi(u_1,u_2,g^*) \equiv \sum_{k=1}^{2} \int_{\Omega_k} \left[\sum_{i,j=1}^{2} a_{ij} \frac{\partial u_k}{\partial x_i} \frac{\partial u_k}{\partial x_j} + c u_k^2 - 2 f u_k \right] d\Omega \tag{2.2}$$

$$- 2 \int_{\Gamma} g^* (u_1 - u_2)\, d\Gamma.$$

In the last term of (2.2), the Lagrange multiplier g^* is integrated against the difference in the traces of the subdomain solutions u_k on Γ. The constraint that the u_k should agree on Γ is thereby incorporated into the usual quadratic functional to be minimized, which is the first term of (2.2). As a result of adding the Lagrange multiplier term, we have obtained a saddle-point problem, and the critical points (u_1,u_2,g^*) of Φ must satisfy the variational equality

$$B((u_1,u_2,g^*),(v_1,v_2,h^*)) = F((v_1,v_2,h^*)) \qquad \text{for all } (v_1,v_2,h^*) \in H\ , \tag{2.3}$$

where

$$B((u_1,u_2,g^*),(v_1,v_2,h^*)) \equiv \sum_{k=1}^{2} \int_{\Omega_k} \left[\sum_{i,j=1}^{2} a_{ij} \frac{\partial u_k}{\partial x_i} \frac{\partial v_k}{\partial x_j} + c u_k v_k \right] d\Omega$$

$$- \int_{\Gamma} \left[g^*(v_1 - v_2) + h^*(u_1 - u_2) \right] d\Gamma$$

and

$$F(v_1,v_2,h^*) \equiv \sum_{k=1}^{2} \int_{\Omega_k} f v_k \, d\Omega.$$

Let $\vec{n}$ denote the unit normal to Γ pointing from Ω_1 to Ω_2, and let A denote the 2×2 matrix whose entries are the coefficients a_{ij}. For any $u \in H^1(\Omega)$, we define the conormal derivative operator

$$\gamma^* u \equiv \gamma_k^* u_k \equiv \vec{n} \cdot (A \vec{\nabla} u). \tag{2.4}$$

We have the following result [13].

Theorem 2.1. *There exists a unique solution* (u_1,u_2,g^*) *of* (2.3) . *Moreover, if f is such that there exists a classical solution u of* (2.1), *then the element* (u_1,u_2,g^*) *of H given by*

$$u_k \equiv u \quad \text{on } \Omega_k,\ k=1,2 \tag{2.5a}$$

$$g^* \equiv \gamma^* u, \tag{2.5b}$$

is the solution of (2.3).

For each k=1,2, let

$$B_k(u,v) \equiv \int_{\Omega_k} \left[\sum_{i,j=1}^{2} a_{ij} \frac{\partial u}{\partial x_i} \frac{\partial v}{\partial x_j} + cuv \right] d\Omega \tag{2.6}$$

on $H_D^1(\Omega_k) \times H_D^1(\Omega_k)$. Given g^* in $H_{00}^{-1/2}(\Gamma)$ and f_k in $H_D^{-1}(\Omega_k)$, we consider the variational problem of finding u_k in $H_D^1(\Omega_k)$ such that

$$B_k(u_k,v_k) = \int_\Gamma g^* v_k \, d\Gamma + \int_{\Omega_k} f_k v_k d\Omega, \quad \text{for all } v_k \in H_D^1(\Omega_k). \tag{2.7}$$

Lemma 2.1. *For k=1,2, given g^* in $H_{00}^{-1/2}(\Gamma)$ and f_k in $H_D^{-1}(\Omega_k)$, there exists a unique solution u_k of* (2.7) *and*

$$\|\gamma_k u_k\|_{H_{00}^{1/2}(\Gamma)} \le C \left[\|g^*\|_{H_{00}^{-1/2}(\Gamma)} + \|f_k\|_{H_D^{-1}(\Omega_k)} \right]$$

for some constant C.

We now define the *Poincaré-Steklov operator* (see, e.g., [1] and the references therein) $Q_k : H_{00}^{-1/2}(\Gamma) \to H_{00}^{1/2}(\Gamma)$ by

$$Q_k g^* = \gamma_k u_k, \tag{2.8}$$

where, for g^* in $H_{00}^{-1/2}(\Gamma)$, u_k is the solution of (2.7) with $f_k = 0$. Furthermore, let $R_k : H_D^{-1}(\Omega_k) \to H_{00}^{1/2}(\Gamma)$ be given by

$$R_k f_k = \gamma_k u_k \tag{2.9}$$

where, for f_k in $H_D^{-1}(\Omega_k)$, u_k is the solution of (2.7) with $g^* = 0$. From Lemma 2.1, whose simple proof is given in [13], we have that Q_k and R_k are bounded operators.

In terms of Q_k and R_k, the Lagrange multiplier g^* in $H_{00}^{-1/2}(\Gamma)$ obtained by solving (2.3) is the unique solution of

$$(Q_1 + Q_2)g^* = R_2 f_2 - R_1 f_1. \tag{2.10}$$

Equation (2.10) is a statement of the fact that solving the Lagrange multiplier formulation is equivalent to finding interface data g^* in $H_{00}^{-1/2}(\Gamma)$ such that, when (2.7) is solved, the traces on Γ of the resulting solutions u_k, k=1,2, have the same value g in $H_{00}^{1/2}(\Gamma)$. A dual formulation also exists in which one seeks Dirichlet data g on Γ such that when Dirichlet problems are solved on each subdomain Ω_k, the conormal derivatives $\gamma_k^* u_k$ of the resulting solutions have the same value g^* on Γ. In terms of the operators Q_k and R_k, this dual formulation can be written as

$$(Q_1^{-1} + Q_2^{-1})g = Q_1^{-1}R_1f_1 + Q_2^{-1}R_2f_2. \tag{2.11}$$

Equations (2.10) and (2.11) are compact expressions of the interdomain coupling problems to be discretized.

3. Discretization. Discrete analogs of the continuous formulations discussed in the previous section are obtained by the usual procedure of restriction to finite-dimensional subspaces.

For each k=1,2, let $P_{M_k}(\Omega_k)$ be a finite-dimensional subspace of $H_D^1(\Omega_k)$ with basis $\{\phi_{k\mu}, \mu=1,2,...,M_k\}$. We assume that, for some $m_k \geq 1$ and $t_k \geq 2$,

$$P_{M_k}(\Omega_k) \subset H^{m_k}(\Omega_k) , \tag{3.1}$$

and that for every $u_k \in H^{n_k}(\Omega_k)$, $n_k \geq 0$, and $0 \leq s \leq \min(m_k,n_k)$ there exists v_k in $P_{M_k}(\Omega_k)$, independent of s, such that

$$\|u_k - v_k\|_{H^s(\Omega_k)} \leq C M_k^{-\sigma_k/2} \|u_k\|_{H^{n_k}(\Omega_k)} , \qquad \sigma_k = \min(t_k - s, n_k - s) , \tag{3.2}$$

where C is independent of u_k, M_k, and s. An example of such a space is the subspace of $H_D^1(\Omega_k)$ consisting of continuous piecewise linear functions on a quasiuniform triangulation of Ω_k. In this case, $m_k = 1$, $t_k = 2$, and M_k is roughly proportional to h_k^{-2} where h_k is the mesh size.

Next, let $W_L(\Gamma)$ denote a finite-dimensional subspace of $H_{00}^{1/2}(\Gamma)$ with basis $\{w_\lambda , \lambda=1,...,L\}$ such that, for some $m_\gamma \geq -1/2$ and $t_\gamma \geq 2$,

$$W_L(\Gamma) \subset H^{m_\gamma}(\Gamma) , \tag{3.3}$$

and for every $g^* \in H^{n_\gamma}(\Gamma)$, $n_\gamma \geq -1/2$, and $-1/2 \leq s \leq \min(m_\gamma,n_\gamma)$ there exists μ^* in $W_L(\Gamma)$, independent of s, such that

$$\|g^* - \mu^*\|_{H^s(\Gamma)} \leq C L^{-\sigma_\gamma} \|g^*\|_{H^{n_\gamma}(\Gamma)} , \qquad \sigma_\gamma = \min(t_\gamma - s, n_\gamma - s) , \tag{3.4}$$

where C is independent of g^*, L, and s. In addition, we assume that $W_L(\Gamma)$ satisfies the inverse assumption

$$\|g^*\|_{H^s(\Gamma)} \leq C L^{s-q} \|g^*\|_{H^q(\Gamma)} , \qquad \text{for all } g^* \in W_L(\Gamma), \quad -1/2 \leq q \leq s \leq m_\gamma , \tag{3.5}$$

where C is independent of g^*, L, and s. For example, $W_L(\Gamma)$ could be the space of continuous piecewise linear functions on a quasiuniform partitioning of Γ, in which case $m_\gamma = 1$, $t_\gamma = 2$, and L is roughly proportional to the number of partition segments.

Let

$$H_{M,L} \equiv P_{M_1}(\Omega_1) \times P_{M_2}(\Omega_2) \times W_L(\Gamma), \quad M = (M_1, M_2) .$$

Since $H_{M,L}$ is contained in H, we may consider the discrete problem of finding $(\psi_1, \psi_2, \lambda^*)$ in $H_{M,L}$ satisfying

$$B((\psi_1,\psi_2,\lambda^*),(\phi_1,\phi_2,\mu^*)) = F((\phi_1,\phi_2,\mu^*)) \quad \text{for all } (\phi_1,\phi_2,\mu^*) \in H_{M,L} . \tag{3.6}$$

The following result is proved in [13]:

Theorem 3.1. *Assume that the solution u of (2.1) satisfies $u|_{\Omega_k} \in H^{n_k}(\Omega_k)$ for some $n_k \geq 1$, $k=1,2$, and that $\gamma^* u \in H^{n_\gamma}(\Gamma)$ for some $n_\gamma \geq -1/2$. Let K be a constant independent of M_k and L such that*

$$KL \leq M_k^{1/2}, \qquad k=1,2. \tag{3.7}$$

If K is sufficiently large, then there exists a unique $(\psi_1,\psi_2,\lambda^*)$ in $H_{M,L}$ satisfying (3.6). Moreover,

$$\sum_{k=1}^{2} \|u_k - \psi_k\|_{H^1(\Omega_k)} + \|\gamma^* u - \lambda^*\|_{H_{00}^{-1/2}(\Gamma)} \tag{3.8}$$

$$\leq C \left\{ \sum_{k=1}^{2} \inf_{v_k \in P_{M_k}(\Omega_k)} \|u_k - v_k\|_{H^1(\Omega_k)} + \inf_{\mu^* \in W_L(\Gamma)} \|\gamma^* u - \mu^*\|_{H_{00}^{-1/2}(\Gamma)} \right\}$$

$$\leq C \left\{ \sum_{k=1}^{2} M_k^{-\sigma_k/2} \|u\|_{H^{n_k}(\Omega_k)} + L^{-\sigma_\gamma} \|\gamma^* u\|_{H^{n_\gamma}(\Gamma)} \right\},$$

where

$$\sigma_k = \min(t_k, n_k) - 1, \quad k=1,2,$$

$$\sigma_\gamma = \min(t_\gamma, n_\gamma) + 1/2,$$

and C is independent of M_k and L.

Some error estimates obtainable from Theorem 3.1 for specific choices of $P_{M_k}(\Omega_k)$ and $W_L(\Gamma)$ are contained in the following results.

Corollary 3.1. *For $k=1,2$, let $P_{M_k}(\Omega_k)$ denote a subspace of $H_D^1(\Omega_k)$ consisting of continuous piecewise linear polynomials defined on a quasiuniform triangulation of Ω_k, and let $W_L(\Gamma)$ be* *a subspace of* $H_{00}^{1/2}(\Gamma)$ *consisting of continuous piecewise linear polynomials defined on a quasiuniform partitioning of* Γ. *Then, under the hypothesis of Theorem 3.1, there exists a unique* $(\psi_1,\psi_2,\lambda^*)$ *in* $H_{M,L}$ *satisfying* (3.6) *and*

$$\sum_{k=1}^{2} \|u_k-\psi_k\|_{H^1(\Omega_k)} + \|\partial u/\partial \vec{n}-\lambda^*\|_{H_{00}^{-1/2}(\Gamma)} \tag{3.9}$$

$$\le C\left[\sum_{k=1}^{2} M_k^{-[\min(n_k,2)-1]/2} + L^{-\min(n_\gamma,2)-1/2}\right],$$

where $C = C(u, n_k, n_\gamma)$ is independent of M_k and L.

Corollary 3.2. *For* $k=1,2$, *let* $P_{M_k}(\Omega_k)$ *denote a subspace of* $H_D^1(\Omega_k)$ *consisting of continuous piecewise linear polynomials defined on a quasiuniform triangulation of* Ω_k, *and let* $W_L(\Gamma)$ *be the subspace of* $H_{00}^{1/2}(\Gamma)$ *consisting of polynomials of degree* $\le L+1$ *defined on* Γ. *Then, under the hypothesis of Theorem 3.1, there exists a unique* $(\psi_1,\psi_2,\lambda^*)$ *in* $H_{M,L}$ *satisfying* (3.6) *and*

$$\sum_{k=1}^{2} \|u_k-\psi_k\|_{H^1(\Omega_k)} + \|\partial u/\partial \vec{n}-\lambda^*\|_{H_{00}^{-1/2}(\Gamma)} \tag{3.10}$$

$$\le C\left[\sum_{k=1}^{2} M_k^{-[\min(n_k,2)-1]/2} + L^{-n_\gamma-1/2}\right],$$

where $C = C(u, n_k, n_\gamma)$ is independent of M_k and L.

When interpreting the error estimates (3.8-3.10), some care must be taken not to overlook (3.7). This assumed condition must be maintained in the asymptotic limit of increasing M_k and L to ensure that the constants C in (3.8-3.10) are independent of M_k and L. Essentially, (3.7) stipulates that one may not "over-resolve" the interface problem for a given level of subdomain discretization. Thus, in spite of the fact that there are two asymptotic parameters in (3.8), they are not totally independent. A discussion of why (3.7) cannot be removed in general can be found in [2, page 214], and some special cases in which (3.7) can be eliminated are discussed in [4]. For practical reasons which will be discussed later, L will be limited to a finite range of relatively small values, in which case there always exists a constant K such that (3.7) holds for sufficiently large M_k. In the numerical results of the next section, we will see that the need to satisfy (3.7) does not present any difficulties for the problems considered.

In general, the subdomain components ψ_k of the solution of (3.6) will not agree on Γ. If the traces on Γ of all functions in $P_{M_k}(\Omega_k)$, $k=1,2$, span the same space, then continuity of the approximate solution can be recovered by a simple patching procedure. Consider the interface jump function

$$\Delta \equiv \gamma_2\psi_2 - \gamma_1\psi_1 \in H_{00}^{1/2}(\Gamma).$$

For each $k=1,2$, let $\phi_k \in P_{M_k}(\Omega_k)$ be the solution of the Dirichlet problem

$$\int_{\Omega_k} \left[\sum_{i,j=1}^{2} a_{ij} \frac{\partial \phi_k}{\partial x_i} \frac{\partial \chi_k}{\partial x_j} + c\,\phi_k \chi_k \right] d\Omega = 0 \tag{3.11a}$$

for all $\chi_k \in P_{M_k}(\Omega_k) \cap H_0^1(\Omega_k)$,

$$\phi_k = (-1)^{k+1} \frac{\Delta}{2} \qquad \text{on } \Gamma, \tag{3.11b}$$

and define

$$\tilde{\psi}_k = \psi_k + \phi_k \qquad \text{on } \Omega_k . \tag{3.12}$$

The corrected approximations $\tilde{\psi}_1$ and $\tilde{\psi}_2$ then agree on Γ. Moreover, since

$$\|\phi_k\|_{H^1(\Omega_k)} \le C \|\Delta\|_{H_{00}^{1/2}(\Gamma)},$$

it follows that

$$\|u-\tilde{\psi}_k\|_{H^1(\Omega_k)} \le \|u-\psi_k\|_{H^1(\Omega_k)} + C\,\|\psi_2-\psi_1\|_{H_{00}^{1/2}(\Gamma)}$$

$$\le \|u-\psi_k\|_{H^1(\Omega_k)} + C \sum_{k'=1}^{2} \|u-\psi_{k'}\|_{H_{00}^{1/2}(\Gamma)}$$

$$\le C \sum_{k'=1}^{2} \|u-\psi_{k'}\|_{H^1(\Omega_{k'})}.$$

Thus, the patching procedure does not degrade the asymptotic convergence rate of the approximate subdomain solutions.

The algebraic system resulting from (3.6) is

$$S_k \theta_k + (-1)^k U_k \eta^* = T_k, \qquad k=1,2, \tag{3.13a}$$

$$U_2^T \theta_2 - U_1^T \theta_1 = 0, \tag{3.13b}$$

where

$$(S_k)_{\mu,\nu} \equiv \int_{\Omega_k} \left\{ \sum_{i,j=1}^{2} a_{ij} \frac{\partial \phi_{k\mu}}{\partial x_i} \frac{\partial \phi_{k\nu}}{\partial x_j} + c\,\phi_{k\mu}\phi_{k\nu} \right\} d\Omega,$$

$$(T_k)_\mu \equiv \int_{\Omega_k} f\,\phi_{k\mu}\,d\Omega,$$

$$(U_k)_{\mu,\lambda} \equiv \int_\Gamma \phi_{k\mu} w_\lambda\,d\Gamma,$$

and the entries of the unknown vectors θ_k and η^* are the coefficients of the approximate solution components ψ_k and λ^* with respect to the bases $\{\phi_{k\mu}\}$ and $\{w_\lambda\}$, respectively. Solving (3.13a) for θ_k, one obtains, for k=1,2,

$$\theta_k = S_k^{-1}((-1)^{k+1}U_k\eta^* + T_k) \tag{3.14}$$

which, upon substitution in (3.13b) yields the following system to be solved for the discrete Lagrange multiplier η^*:

$$(\sum_{k=1}^{2} U_k^T S_k^{-1} U_k)\eta^* = U_2^T S_2^{-1} T_2 - U_1^T S_1^{-1} T_1. \tag{3.15}$$

Once η^* is determined from (3.15), the θ_k are obtained from (3.14). System (3.15) is the discrete analog of (2.10).

The discrete formulation of (2.11) is

$$[\sum_{k=1}^{2} (U_k^T S_k^{-1} U_k)^{-1}]\eta = \sum_{k=1}^{2} (U_k^T S_k^{-1} U_k)^{-1} U_k^T S_k^{-1} T_k, \tag{3.16}$$

where the entries of η are the coefficients with respect to $\{w_\lambda\}$ of the discrete Dirichlet data to be determined on Γ. After solving (3.16) for η, the subdomain solutions are obtained from

$$\theta_k = S_k^{-1}[U_k(U_k^T S_k^{-1} U_k)^{-1}(\eta - U_k^T S_k^{-1} T_k) + T_k], \qquad k=1,2. \tag{3.17}$$

We refer to (3.16) as a *Generalized Schur Complement (GSC)* system. The motivation for this terminology is the following. Consider the special case in which the finite-dimensional Lagrange multiplier space $W_L(\Gamma)$ is given by

$$W_L(\Gamma) = \left\{ w : w = \gamma_1 v_1 \text{ for some } v_1 \in P_{M_1}(\Omega_1) \right\} \tag{3.18}$$

$$= \left\{ w : w = \gamma_2 v_2 \text{ for some } v_2 \in P_{M_2}(\Omega_2) \right\}$$

where we have implicitly assumed that $P_{M_k}(\Omega_k)$, k=1,2, have the same traces on Γ. We can arrange the unknown vectors θ_k in such a way that the entries $\theta_{k\mu}$ which couple to Γ (i.e., have indices μ such that $(U_k)_{\mu,\lambda} \neq 0$ for some λ) are listed last. Then S_k, T_k, U_k, and θ_k

may be partitioned as

$$S_k = \begin{bmatrix} S_{11}^k & S_{21}^k \\ S_{12}^k & S_{22}^k \end{bmatrix}, \quad T_k = \begin{bmatrix} T_1^k \\ T_2^k \end{bmatrix}, \quad U_k = \begin{bmatrix} 0 \\ \tilde{U} \end{bmatrix}, \quad \theta_k = \begin{bmatrix} \theta_1^k \\ \theta_2^k \end{bmatrix}$$

where $\tilde{U}$ is a nonsingular $L \times L$ matrix. Straightforward matrix algebra yields that (3.16) can then be written as

$$\sum_{k=1}^{2} \left[S_{22}^k - S_{21}^k (S_{11}^k)^{-1} S_{12}^k \right] \tilde{\theta} = \sum_{k=1}^{2} \left[T_2^k - S_{21}(S_{11}^k)^{-1} T_1^k \right] \tag{3.19}$$

where $\tilde{\theta} = \tilde{U}^{-T}\eta = \theta_2^k$, k=1,2. Thus, in the special case defined by (3.18), the system (3.16) is equivalent to the usual Schur complement system, modulo the transformation of the interface unknowns effected by the matrix $\tilde{U}$.

Although it is the GSC system (3.16) which generalizes the well-known and much-studied Schur complement formulation, the presence of the matrices U_k in cases where (3.18) does not hold makes the GSC system more expensive to implement than system (3.15). This is consistent with conclusion that it is better to use the natural interface data represented by the Lagrange multipliers as the unknown quantity to be solved for on the interfaces rather than the Dirichlet data determined by the solution of the GSC system. We note that the estimates of Theorem 3.1 were obtained for system (3.15), since this was the discrete problem directly obtained from the continuous Lagrange multiplier formulation.

It is apparent that the predominant cost in implementing (3.15) is the construction of its coefficient matrix. In each subdomain, it is necessary to solve L linear systems to compute each subdomain's contribution to this matrix with an additional subdomain solve required to obtain the $S_k^{-1}T_k$ contribution to the right-hand side of (3.15). The feasibility of the explicit construction of (3.15) is therefore heavily dependent upon the size of L. In the discretization scheme described above, the space $W_L(\Gamma)$ was chosen independently of the discretizations on the subdomains (i.e., the $P_{M_k}(\Omega_k)$) subject only to the condition (3.7). This implies that we are essentially free to choose $W_L(\Gamma)$ in such a way as to approximate well the natural interface data $\gamma^* u$ on Γ. As can be seen from (3.8), if this is effectively done, then the error on each subdomain Ω_k will be dominated by the usual best-approximation errors on the subdomains. Since we are solving an elliptic partial differential equation, we know that for real problems the natural interface data $\gamma^* u$ will generally be very smooth, except perhaps for singularities of well-known functional form at the endpoints of Γ. Because of this smoothness, it is logical to choose the space $W_L(\Gamma)$ to contain either algebraic or trigonometric polynomials on Γ, since the smooth data $\gamma^* u$ can be approximated well by small numbers of such functions. Even if singularities exist, as the approximation theory [5-8,11,12,15-17] for the p-version of the finite element method demonstrates, on quasiuniform meshes high-order polynomials are superior to the usual low-order piecewise polynomials one is effectively using when the space $W_L(\Gamma)$ is given by (3.18). Therefore, by using such multipliers L should be very small. This means that only a small number of subdomain solves are required to construct the system (3.15), which may even be of sufficiently low dimension for solution via direct methods.

It should be observed that the $M_k \times L$ matrices U_k depend only upon the spaces $P_{M_k}(\Omega_k)$ and $W_L(\Gamma)$. In particular, the U_k are independent of the elliptic operator giving rise to the matrices S_k and therefore could be precomputed. Depending upon the choice of the spaces $P_{M_k}(\Omega_k)$ and $W_L(\Gamma)$, the U_k can also be quite sparse. Such sparsity can therefore be exploited both in the construction of the U_k and in the computations involved in setting up (3.15). For example, if $P_{M_k}(\Omega_k)$ consists of continuous piecewise-linear functions on a fine gridding of Ω_k, only a relatively small number of basis functions $\phi_{k\mu}$ have support meeting the interface Γ. This yields matrices U_k with many zero entries. On the other hand, if $P_{M_k}(\Omega_k)$ consists of globally-defined polynomials, such as in the p-version of the finite element method, then the U_k will necessarily be less sparse, although M_k will generally be much smaller in this case due to the superior approximation properties of such spaces.

We note that in iterative substructuring techniques which use methods such as preconditioned conjugate gradient (PCG) to solve the Schur complement system (3.19), it is necessary to do at least one subdomain solve in each iteration in order to evaluate the action of the Schur complement matrix on a vector. The approach described above in which the system (3.15) is explicitly formed also requires a number of subdomain solves. Here, the subdomain solves are needed to construct the quantities $S_k^{-1}U_k$ and $S_k^{-1}T_k$. Since all of the right-hand sides of the subdomain problems (namely, the columns of U_k and the vector T_k) are available at the same time, these subdomain solves can easily be performed in parallel. This is in contrast to iterative methods where the inherent sequential nature of the PCG iteration precludes exploitation of this extra level of parallelism on each subdomain. Furthermore, if, on a shared-memory multiprocessor, we consider as a pool of tasks the subdomain solves needed in the computation of $S_k^{-1}U_k$ and $S_k^{-1}T_k$, $k=1,2$, then any subdomain solve can be assigned to any available processor with no synchronization required until all subdomain solves have been performed (except for the solves needed to perform the patching procedure). In this way, load-balancing is enhanced and the granularity of the parallel work remains at the level of the subdomain solves.

Another advantage of explicitly forming (3.15) in contrast to iterative approaches is that if many problems involving the same elliptic operator are to be solved (for example in a time-dependent calculation in which S_k and U_k do not change) then the matrix of (3.15) can simply be constructed once, factored and stored.

4. Numerical Results. As our first example, we consider the problem

$$-\Delta u + u = f \qquad \text{in } \Omega, \tag{4.1a}$$

$$u = g \qquad \text{on } \partial\Omega, \tag{4.1b}$$

where Ω is the domain pictured in Figure 4.1, and f and g are such that the solution of (4.1) is

$$u(x, y) = e^{2x+y} \sin(\pi y). \tag{4.2}$$

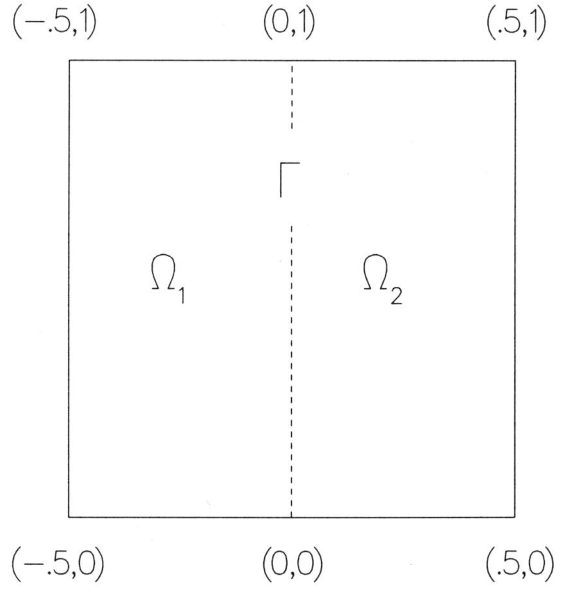

Figure 4.1. *Domain for problem* (4.1).

On the interface Γ separating the two subdomains Ω_1 and Ω_2, the solution conormal derivative, or "flux", is

$$\partial u / \partial \vec{n} = 2e^y \sin(\pi y), \tag{4.3}$$

where $\vec{n}$ is the unit normal to Γ pointing into Ω_2. Let Ω_k, k=1,2, be uniformly triangulated, and let h_k denote the maximum diameter of any mesh triangle. Similarly, let Γ be uniformly partitioned such that $(L+1)^{-1}$ is the length of any mesh segment. Let $P_{M_k}(\Omega_k)$ and $W_L(\Gamma)$ be the corresponding continuous piecewise linear spaces satisfying the appropriate boundary conditions. We see that the restictions u_k, k=1,2, of the exact solution (4.2) belong to $C^\infty(\overline{\Omega}_k)$ and that $\partial u / \partial \vec{n}$ belongs to $C^\infty(\overline{\Gamma})$. If $(\psi_1, \psi_2, \lambda^*)$ is the solution of the discrete problem (3.6) corresponding to (4.1), then, provided that (3.7) holds, it follows from Corollary 3.1 that

$$\sum_{k=1}^{2} \| u_k - \psi_k \|_{H^1(\Omega_k)} + \| \partial u / \partial \vec{n} - \lambda^* \|_{H_{00}^{-1/2}(\Gamma)} \le C(u) \left[\sum_{k=1}^{2} h_k + L^{-5/2} \right], \tag{4.4}$$

where $C(u)$ is independent of h_k and L. In (4.4), we have used the fact that, for a uniform triangulation of Ω_k, $h_k = O(M_k^{-1/2})$.

As a second option for $W_L(\Gamma)$, we consider the space of globally-defined (i.e., not piecewise) polynomials of degree $\le L+1$ on Γ which vanish at the endpoints of Γ. In this case, again assuming (3.7), it follows from Corollary 3.2 that

$$\sum_{k=1}^{2} \|u_k - \psi_k\|_{H^1(\Omega_k)} + \|\partial u / \partial \vec{n} - \lambda^*\|_{H_{00}^{-1/2}(\Gamma)} \qquad (4.5)$$

$$\leq C(u, n_\gamma) \left[\sum_{k=1}^{2} h_k + L^{-(n_\gamma + 1/2)} \right],$$

for arbitrarily large n_γ, where $C(u, n_\gamma)$ is independent of h_k and L.

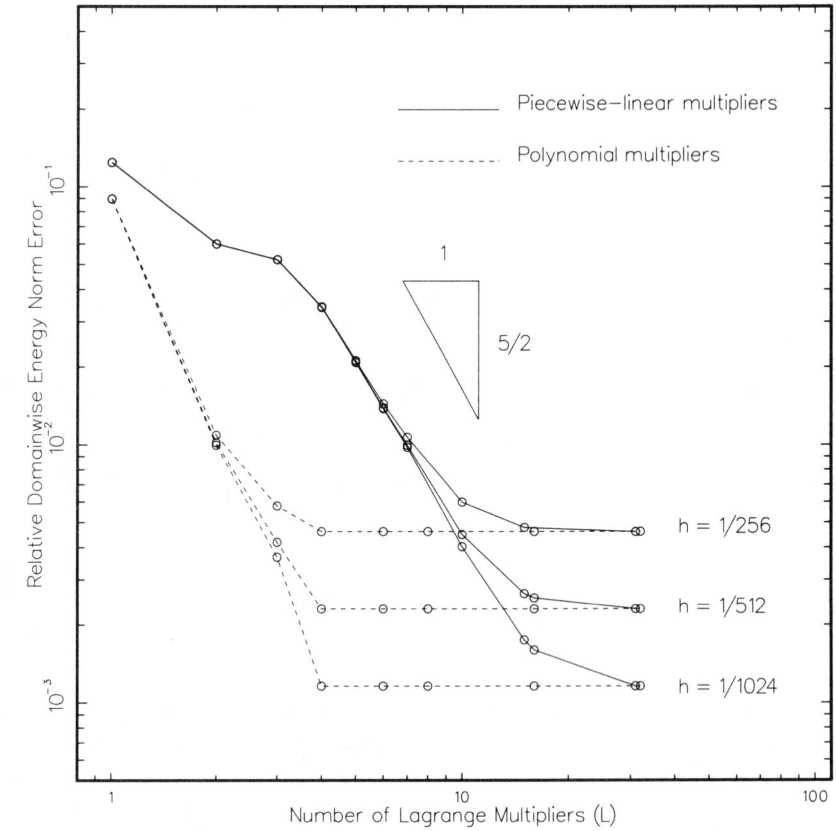

Figure 4.2. *Convergence of the domainwise relative energy norm error with respect to the number of piecewise-linear and polynomial multipliers in the solution of problem* (4.1).

Since the exact solution (4.2) of (4.1) is known, we can calculate the subdomain error terms in the left-hand sides of (4.4) and (4.5) to experimentally observe the convergence rates predicted by the above theory. We cannot directly compute the interface error involving the $H_{00}^{-1/2}(\Gamma)$ norm, but this can be estimated as described in [13]. Figure 4.2 shows some results obtained using the piecewise-linear and polynomial Lagrange multiplier spaces $W_L(\Gamma)$ together with uniform subdomain partitionings corresponding to $h^{-1} = 256, 512,$ and 1024, with $h_1 = h_2 = h$. Plotted against the number of Lagrange multipliers L is the relative domainwise energy norm error

$$\frac{\left[\sum_{k=1}^{2} \|u_k - \psi_k\|^2_{H^1(\Omega_k)} \right]^{1/2}}{\|u\|_{H^1(\Omega)}} . \tag{4.6}$$

Since a log-log scale is used in Figure 4.2, the predicted 5/2 convergence rate with respect to L^{-1} for the case of piecewise-linear multipliers is easy to see. On the other hand, the prediction that the errors using polynomial multipliers are decreasing faster than any polynomial rate with respect to L^{-1} is difficult to verify from the few data points which precede the bottoming out of the error curves. The fact that the errors do not decrease past a certain point for large L is to be expected since the subdomain errors will ultimately predominate as more multipliers are used on Γ. Hence the curve asymptotes represent the errors which would be obtained by a standard finite-element or finite-difference method applied to the original problem (4.1). We note that it was necessary to take a ridiculously fine gridding of the subdomains in order to reduce the subdomain errors enough to even see the convergence rates of the errors introduced by the Lagrange multipliers. This is partially due to the relatively poor approximation properties of the piecewise linear functions used on the subdomains, but it is also an indication of the rapid convergence of the discrete Lagrange multipliers. Using piecewise-linear $W_L(\Gamma)$, 4 multipliers were required to reduce the relative domainwise energy norm error to less than 5%, whereas for polynomial $W_L(\Gamma)$, only 2 multipliers were needed to reach this tolerance. In the latter case, this means that only 2 subdomain solves are required on each subdomain to construct system (3.15), and that the cost of a direct solution of this 2×2 system is negligible.

Since no patching was performed in obtaining the results presented above, the computed solutions of (4.1) are discontinuous at Γ. This discontinuity at Γ can nevertheless be removed using the patching procedure described in Section 3. The quantity (4.6) will then represent the true relative energy norm error on Ω. In Figure 4.3, the error curves obtained using polynomial Lagrange multipliers with patching are plotted against the corresponding unpatched error curves taken from Figure 4.2. Although we showed in Section 3 that the patching procedure will not degrade the rate of convergence of the subdomain solutions with respect to the number of Lagrange multipliers, we cannot observe this in Figure 4.3 since, for all three subdomain mesh sizes, the patching has reduced the interface error below the level of the subdomain error. This means the error obtained using 1, 2, and 3 polynomial Lagrange multipliers in the unpatched version of the algorithm is entirely due to nonsatisfaction of exact continuity at the interface, even though the energy norm error was only computed in a domainwise sense (i.e., discontinuities at the interface do not directly contribute to the errors). It should again be noted that we are using very fine subdomain grids, and from Figure 4.3 we see that if coarser subdomain grids had been used then the difference between the unpatched and patched errors might not be seen. Nevertheless, this example shows that the use of the patching procedure not only recovers the continuity of the computed solutions at the interface, but in fact can even further reduce the errors on the subdomains.

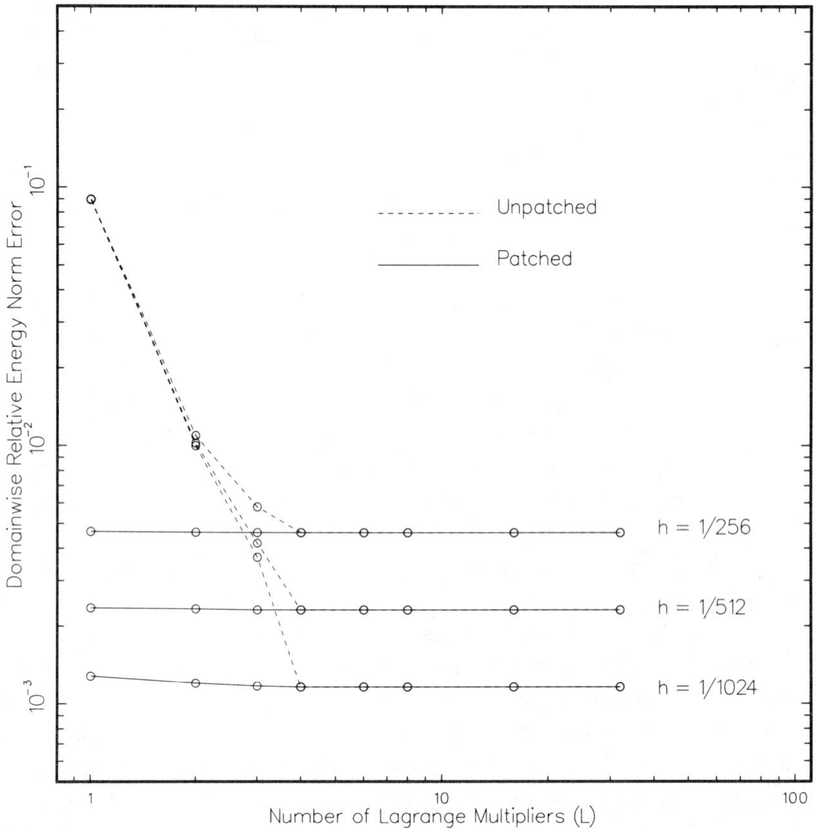

Figure 4.3. *Comparison of the convergence rates of the domainwise relative energy norm error with respect to the number of polynomial Lagrange multipliers before and after application of the patching procedure.*

As an example of a nonsmooth problem, we consider

$$-\Delta u + u = f \qquad \text{in } \Omega, \tag{4.7a}$$

$$u = 0 \qquad \text{on } \partial\Omega, \tag{4.7b}$$

where Ω is the L-shaped domain shown in Figure 4.4. We assume that f is such that the solution of (4.7) is

$$u(x,y) = (1-x^2)(1-y^2)r^{2/3}\sin((2\theta + \pi)/3) \tag{4.8}$$

where r and θ are the usual polar coordinates based at the origin. The singularity in the derivatives of (4.8) at the origin is representative of the singularity existing in all problems with reentrant corner angles of measure $3\pi/2$ and homogeneous Dirichlet conditions on the sides of the domain meeting at such corners [14].

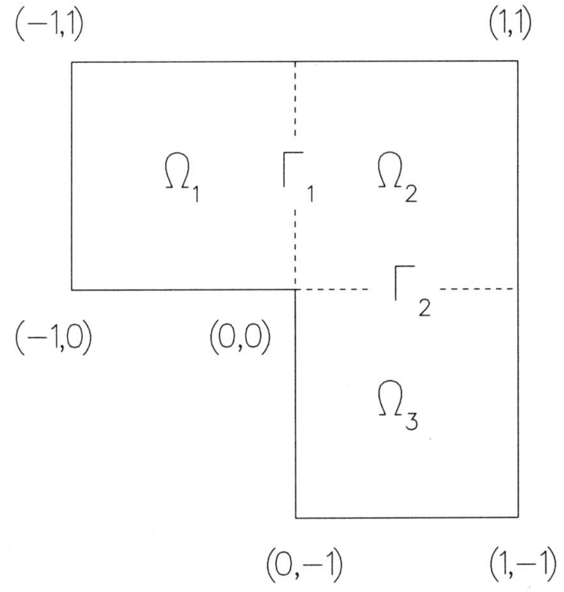

Figure 4.4. *Domain for problem* (4.7).

Figure 4.4 also indicates a very natural domain decomposition of Ω via the introduction of interfaces Γ_1 and Γ_2. Although three subdomains are shown, we observe that the problem is actually symmetric with respect to the line $y = x$ and therefore can be treated as a two-domain problem with one interface. The decomposition shown in Figure 4.4 has the advantage of yielding geometrically simple subdomains, yet both interfaces contain the reentrant corner and therefore also contain the singularity. Letting s denote the arc length along either Γ_1 or Γ_2 as measured from the origin, we have that the solution flux on Γ_j, j=1,2, is given by

$$\partial u / \partial \vec{n}_j = \frac{1}{3}(1-s^2)s^{-1/3} \tag{4.9}$$

modulo a sign depending on the direction of the unit vector $\vec{n}_j$ normal to Γ_j. Letting $P_{M_k}(\Omega_k)$, k=1,2, be the same piecewise linear subdomain space used in the previous example, and taking the Lagrange multiplier spaces $W_L(\Gamma_j)$ to be the spaces of all continuous piecewise linear functions on uniform partitionings of Γ_j, it follows from Corollary 3.1 and the regularity properities of (4.8,4.9) that if (3.7) is satisfied and $(\psi_1,\psi_2,\psi_3,\lambda_1^*,\lambda_2^*)$ is the solution of the discrete problem (3.6) (trivially generalized to the case of three subdomains and two interfaces) corresponding to (4.7), then

$$\sum_{k=1}^{3} \|u_k - \psi_k\|_{H^1(\Omega_k)} + \sum_{j=1}^{2} \|\partial u / \partial \vec{n}_j - \lambda_j^*\|_{H_{00}^{-1/2}(\Gamma_j)} \tag{4.10}$$

$$\le C(u) \left[\sum_{k=1}^{2} h_k^{2/3} + L^{-2/3} \right] .$$

Furthermore, letting $W_L(\Gamma_j)$ be the space of all polynomials of degree $\le L+1$ which vanish at the endpoints of Γ_j, then

$$\sum_{k=1}^{3} \|u_k - \psi_k\|_{H^1(\Omega_k)} + \sum_{j=1}^{2} \|\partial u/\partial \vec{n}_j - \lambda_j^*\|_{H_{00}^{-1/2}(\Gamma_j)} \tag{4.11}$$

$$\le C(u)\left[\sum_{k=1}^{2} h_k^{2/3} + L^{-5/6}\right].$$

Figures 4.5 and 4.6 are the analogs of Figures 4.2 and 4.3 for problem (4.7). Uniform triangulations of the subdomains Ω_k corresponding to mesh sizes of h^{-1}=16, 32 and 64, $h_1=h_2=h_3=h$, were used together with L=1,2,4,8, and 16 multipliers on each of the interfaces Γ_j. Comparing these results with Figures 4.2 and 4.3, the negative effects of the reduced interface flux regularity are evident. Considering Figure 4.5, we see that for h=1/64, at least 10 polynomial multipliers were required to reduce the error to the point at which the subdomain errors are dominant, and the error using 16 piecewise-linear multipliers had not yet

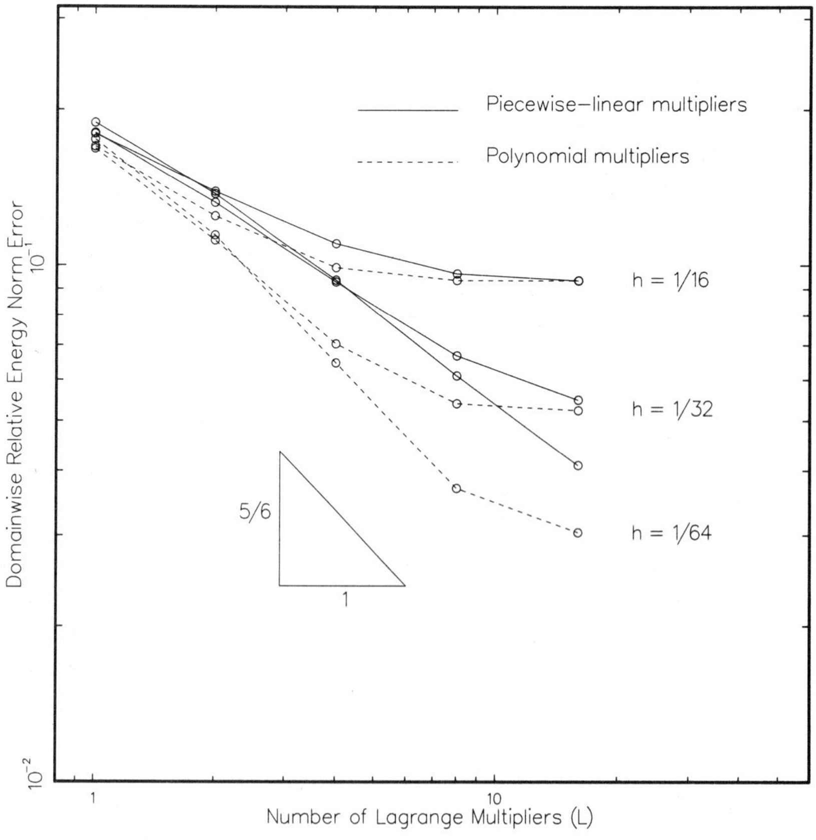

Figure 4.5. *Convergence of the domainwise relative energy norm error with respect to the number of piecewise-linear and polynomial multipliers in the solution of problem* (4.7).

reached this floor. Even 10 multipliers must be considered a rather large number since this would mean that 10 subdomain solves would be required in subdomains Ω_1 and Ω_3 and 20 solves in Ω_2 to construct the system (3.15). As Figure 4.6 shows, by applying the patching procedure the error can nevertheless be reduced to the subdomain error floor using 5 or 6 polynomial multipliers for the case h=1/64 and even fewer for the cases h=1/16 and h=1/32.

In the previous example, our principal concern was the convergence of the Lagrange multipliers. In obtaining our computed results, we therefore used simple uniform subdomain triangulations even though the problem has a singularity at the origin. Indeed, the subdomain approximations can be greatly improved by appropriate mesh refinement near the singularity. However, the same techniques may also be applied on the interface. We could generalize the spaces $W_L(\Gamma_j)$ to consist of arbitrary degree piecewise polynomials on a partitioning of Γ_j as is done in the $h-p$ version of the finite element method [6,7,15-17]. Since we know the form of the singularity at the reentrant corner, we could appropriately grade the interface meshes near the singularity to improve the approximation properties of the Lagrange multiplier spaces $W_L(\Gamma_j)$. It seems clear that, from a practical point of view, it will always be easier to implement effective Lagrange multiplier spaces $W_L(\Gamma_j)$ on the one-dimensional interfaces in order to exploit known solution characteristics than it would be to make an equivalent change in the two-dimensional subdomain spaces $P_{M_k}(\Omega_k)$.

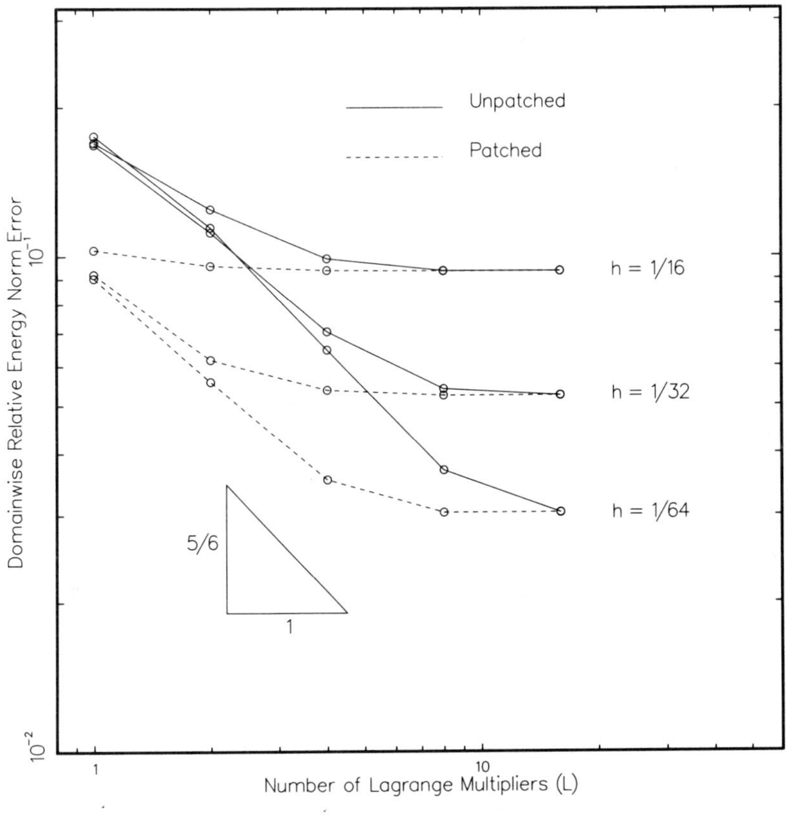

Figure 4.6. *Comparison of the convergence rates of the domainwise relative energy norm error with respect to the number of polynomial Lagrange multipliers before and after application of the patching procedure.*

Based on the results obtained for the smooth model problem (4.1) and the nonsmooth model problem (4.7), we conclude that the set of polynomials of degree $\leq L+1$ on Γ which vanish at the endpoints of Γ are a good general-purpose choice for the Lagrange multiplier space $W_L(\Gamma)$. Considering the fact that the Legendre polynomials mapped to Γ can be used to construct a hierarchical basis for such a space, there are other benefits of this choice as well. Furthermore, we have seen that it is generally a good idea to perform the patching procedure even if the application at hand does not strictly require continuity at the interfaces.

REFERENCES

[1] V. I. Agoshkov, *Poincaré-Steklov operators and domain decomposition methods in finite-dimensional spaces*, in First International Symposium on Domain Decomposition Methods for Partial Differential Equations, R. Glowinski, G. H. Golub, G. A. Meurant, J. Périaux, eds., Society for Industrial and Applied Mathematics, Philadelphia, 1988, pp. 73-112.

[2] I. Babuska and A. K. Aziz, *Survey lectures on the mathematical foundations of the finite element method*, in The Mathematical Foundations of the Finite Element Method with Applications to Partial Differential Equations, A. K. Aziz, ed., Academic Press, New York, 1972, pp. 3-359.

[3] I. Babuska, *The finite element method with Lagrangian multipliers*, Num. Math., Band 20, Heft 3 (1973), pp. 179-192.

[4] I. Babuska, J. T. Oden, and J. K. Lee, *Mixed-hybrid finite element approximations of second-order elliptic boundary-value problems*, Comp. Meth. Appl. Mech. and Eng., 11 (1977), pp. 175-206.

[5] I. Babuska, B. A. Szabo, and I. N. Katz, *The p-version of the finite element method*, SIAM J. Num. Anal., Vol. 18, No. 3 (1981), pp. 515-545.

[6] I. Babuska and M. R. Dorr, *Error estimates for the combined h and p versions of the finite element method*, Num. Math., 37 (1981), pp. 257-277.

[7] I. Babuska and M. Suri, *The h-p version of the finite element method with quasiuniform meshes*, RAIRO Math. Mod. and Num. Anal., Vol. 21, No. 2 (1987).

[8] I. Babuska and M. Suri, *The optimal convergence rate of the p-version of the finite element method*, SIAM J. Num. Anal., 24, No. 4, (1987), pp. 750-776.

[9] J. H. Bramble, *The Lagrange multiplier method for Dirichlet's problem*, Math. Comp., Vol. 37, No. 155 (1981), pp. 1-11.

[10] Q. V. Dihn, R. Glowinski, and J. Periaux, *Solving elliptic problems by domain decomposition methods with applications*, in Elliptic Problem Solvers II, A. Schoenstadt, ed., Academic Press, New York, 1984.

[11] M. R. Dorr, *The approximation theory for the p-version of the finite element method*, SIAM J. Num. Anal., Vol. 21, No. 6 (1984), pp. 1180-1207.

[12] M. R. Dorr, *The approximation of solutions of elliptic boundary-value problems via the p-version of the finite element method*, SIAM J. Num. Anal., Vol. 23, No. 1 (1986), pp. 58-77.

[13] M. R. Dorr, *Domain decomposition via Lagrange multipliers*, submitted to Num. Math.

[14] P. Grisvard, *Boundary Value Problems in Nonsmooth Domains*, Lecture Notes 19, University of Maryland, College Park, 1980.

[15] W. Gui and I. Babuska, *The h, p and h-p versions of the finite element method for one-dimensional problems. Part I: The error analysis of the p-version*, Num. Math., 49 (1986), pp. 577-612.

[16] W. Gui and I. Babuska, *The h, p and h-p versions of the finite element method for one-dimensional problems. Part II: The error analysis of the h and h-p versions*, Num. Math., 49, (1986), pp. 613-657.

[17] W. Gui and I. Babuska, *The h, p and h-p versions of the finite element method for one-dimensional problems. Part III: The adaptive h-p version*, Num. Math., 49 (1986), pp. 659-683.

[18] D. E. Keyes and W. D. Gropp, *A comparison of domain decomposition techniques for elliptic partial differential equations and their parallel implementation*, SIAM J. Sci. Stat. Comput., Vol. 8, No. 2 (1987), pp. s166-s202.

[19] J. Pitkaranta, *Boundary subspaces for the finite element method with Lagrange multipliers*, Num. Math., 33 (1979), pp. 273-289.

A Domain Decomposition Method for a Convection Diffusion Equation with Turning Point[*]

G. W. Hedstrom[†]
F. A. Howes[††]

Abstract. We use asymptotic analysis to determine a domain decomposition method for a singularly perturbed convection-diffusion equation with turning points. The equation

$$-xu_x = \epsilon\Delta u \tag{1}$$

is considered on a square Ω in the plane, with Dirichlet boundary conditions prescribed on $\partial\Omega$. This equation is an idealization of the Navier-Stokes equation with velocity $-x$, parallel to the x-axis. It is known from asymptotic analysis that for small, positive ϵ (large Reynolds number), solutions of (1) very nearly satisfy the reduced equation $u_x = 0$ in subdomains which are at least a distance $C\sqrt{\epsilon}$ from $\partial\Omega$. In the vicinity of $\partial\Omega$ there may be boundary layers, depending on the boundary values. (There is no boundary layer at an inflow boundary.) Our domain decomposition method uses this asymptotic information to determine the partition into subdomains, and it also suggests the basis functions to be used in a finite-element method.

1. Introduction. Several researchers have recently used asymptotic analysis to identify good domain-decomposition strategies for singularly perturbed problems and to suggest efficient numerical algorithms on each subdomain. This work includes both time-dependent problems, such as

$$\partial_t u + a\,\partial_x u + b\,\partial_y u = \epsilon\Delta u, \tag{1.1}$$

and time-independent problems, such as

$$a\,\partial_x u + b\,\partial_y u = \epsilon\Delta u. \tag{1.2}$$

* This work was supported by the Applied Mathematical Sciences subprogram of the office of Energy Research, U. S. Department of Energy, by Lawrence Livermore National Laboratory under contract number W-7405-Eng-48.

† Lawrence Livermore National Laboratory, P. O. Box 808, L–321, Livermore, CA 94550.

‡ Department of Mathematics, University of California Davis, Davis, CA 95616.

Here ∂ denotes partial differentiation, $\partial_x = \partial/\partial x$, and Δ is the Laplace operator. In both cases the singular perturbation aspect of the problem is reflected in the fact that ϵ is a small, positive number. The basic idea of this approach is that in a large portion of the domain it is almost possible to replace, say, (1.2) by the corresponding reduced equation

$$a\,\partial_x u + b\,\partial_y u = 0.$$

There exist subdomains—boundary and internal layers—in which such a reduction is impossible. In many of the layer regions other reductions are possible. Our aim is to identify these layer regions and to use numerical methods there which are suited to the local behavior of the solution.

In this paper we present an algorithm for a problem in which there is an interaction between layer regions of two different types: a boundary layer and a layer generated by a manifold of turning points. Before describing our problem in detail, let us summarize some of the previous work on related problems.

The seminal work using asymptotic analysis to suggest numerical methods was the paper by Chin and Krasny [4] on an algorithm for solving two-point boundary-value problems for equations of the form

$$\epsilon u'' = f(x, u) \tag{1.3}$$

under the condition that $\partial f/\partial u > 0$. These problems are difficult to solve by classical numerical algorithms because the solution may have boundary layers at one or both ends of the interval. The algorithm is based on an approximation of f by a function f_h which is piecewise linear in u. For a given approximate solution u_{n-1}, the knots of a spline u_n are chosen to be the values of x for which $u_{n-1}(x)$ is a point of discontinuity for $\partial f_h/\partial u$. As basis functions for the spline u_n Chin and Krasny use either exact solutions or asymptotic approximations to solutions of

$$\epsilon u'' = f_h(x, u).$$

The method is very efficient because these splines incorporate the boundary-layer behavior of the solutions.

As an example of a numerical algorithm based on asymptotic analysis for a 2-dimensional boundary-value problem, let us summarize the work of Rodrigue and Reiter [9] on the equation

$$\partial_x u = \epsilon \Delta u \tag{1.4}$$

on a square $D = \{(x, y) \mid 0 < x < 1,\ 0 < y < 1\}$, with boundary conditions $u = 0$ on $y = 0$, $u = 1$ on $x = 0$ and on $y = 1$, and $\partial u/\partial x = 0$ on $x = 1$. This problem is a model for steady laminar flow of a fluid over a flat plate. We discuss this work in some detail because ours is an extension of it. The asymptotic behavior of solutions to (1.4) as $\epsilon \downarrow 0$ is well known [5]. The reduced equation for (1.4) is

$$\partial_x U = 0, \tag{1.5}$$

and the solution $U = 1$ of (1.5) fulfills the conditions on three of the four boundaries: the inflow ($x = 0$), the outflow ($x = 1$), and the top ($y = 1$). As a consequence, the solution u to (1.4) is asymptotic to U as $\epsilon \to 0$ except in a boundary-layer region in the vicinity of the bottom boundary ($y = 0$).

One feature of this boundary-layer behavior is that in a neighborhood of the origin of diameter $O(\epsilon)$ (the birth region of the boundary layer), we must keep the full equation (1.4). Elsewhere in the vicinity of the boundary $y = 0$, because the solution u of (1.4)

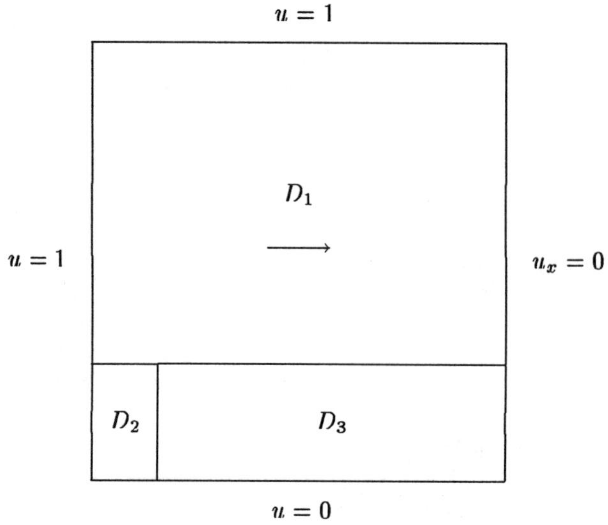

Fig. 1. Domain decomposition.

must have rapid variation in the y-direction, $\partial^2 u/\partial x^2$ is much smaller in magnitude than $\partial^2 u/\partial y^2$. Consequently, near the x-axis and away from the origin, the solution u of (1.4) is asymptotic to the solution of the boundary-layer equation

$$\partial_x v = \epsilon\, \partial_y^2 v. \tag{1.6}$$

It is known that the thickness of this boundary-layer region is $O(\sqrt{\epsilon})$.

On the basis of this asymptotic analysis the method used by Rodrigue and Reiter is as follows. The domain D is partitioned into three subdomains:

$$\begin{aligned} D_1 &= \{(x,y) \mid 0 < x < 1,\ y_0 < y < 1\}, \\ D_2 &= \{(x,y) \mid 0 < x < x_0,\ 0 < y < y_0\}, \\ D_3 &= \{(x,y) \mid x_0 < x < 1,\ 0 < y < y_0\}. \end{aligned}$$

See Fig. 1. On the basis of experience the value of y_0 was taken to be $4\sqrt{\epsilon}$. It is known from the theory that we should take $x_0 = c\epsilon$ for some positive c, but Rodrigue and Reiter determine x_0 dynamically. The following iterative method is used.

1. Take $n = 1$ and set $u_0 = 1$ on D.
2. Solve a discretized version of

$$\partial_x u_n = \epsilon \Delta u_{n-1} \tag{1.7}$$

on D_1. (This iterative scheme provides a mechanism by which the solution on $D_2 \cup D_3$ influences the solution on D_1.)

3. Use invariant embedding (a form of Gaussian elimination) to solve a discrete version of (1.4) for u_n on D_2. The boundary conditions are that $u_n = 1$ on $x = 0$, $u_n = 0$ on $y = 0$, and u_n is as determined by Step 2 on $y = y_0$. On the boundary $x = x_0$ a discrete approximation to (1.6) is used as the boundary condition, and x_0 is selected as the first vertical grid line on which $\epsilon\, \partial_x^2 u_n$ is smaller than a prescribed tolerance.
4. The equation

$$\partial_x u_n = \epsilon \left(\partial_x^2 u_{n-1} + \partial_y^2 u_n\right) \tag{1.8}$$

is solved on D_3 with boundary conditions $u_n = 0$ on $y = 0$, u_n as determined in Step 2 on $y = y_0$, and u_n as determined in Step 3 on $x = x_0$.

5. If the difference $\max_D |u_n - u_{n-1}|$ is sufficiently small, stop. Otherwise, increment n by 1, and return to Step 2.

It may be noted that in this algorithm no attempt is made to fulfill the boundary condition $\partial u/\partial x = 0$ at $x = 1$. This omission is harmless because it overlooks only a weak boundary layer in the vicinity of $x = 1$. In the work of Brown *et al.* [2] the singularly perturbed differential equation

$$a\,\partial_x u + b\,\partial_y u = \epsilon\Delta u + f(x,y) \tag{1.9}$$

is considered on a convex domain Ω with Dirichlet boundary conditions. Here the coefficients a and b may depend on x and y. (We could also permit a and b to depend on u by using an iterative scheme.) It is required, however, that a and b not vanish simultaneously—there are no turning points. The paper [2] has two main thrusts: (1) an implementation of coordinate transformations dictated by asymptotic analysis, and (2) an investigation of the stability of schemes based on (1.7).

The reduced equation for (1.9) is $a\,\partial_x u + b\,\partial_y u = f$, which may be integrated by quadrature of line integrals along the characteristic curves with boundary data specified at inflow boundaries. This process provides the solution over most of Ω, but it must be matched to boundary layers in the vicinity of the remainder of $\partial\Omega$. These boundary layers are obtained by introducing boundary-fitted coordinates and using a finite-element method of Chin-Krasny type.

In an application of this circle of ideas to time-dependent problems Chin *et al.* [3] developed a domain-decomposition method for a time-dependent equation modelling transonic laminar flow in a duct,

$$\partial_t u + u\,\partial_x u + ru = \epsilon\,\partial_x^2 u, \tag{1.10}$$

for $0 < x < L$ and $t > 0$. Here r is a smooth function of x. The boundary and initial data are such that the solution u has a shock layer internal to the domain. Outside of this shock layer one may solve the reduced equation

$$\partial_t u + u\,\partial_x u + ru = 0,$$

while the full equation (1.10) must be solved in the shock layer. An iterative method based on the equation

$$\partial_t u_n + u_{n-1}\,\partial_x u_n + ru_n = \epsilon\,\partial_x^2 u_n$$

is used to locate the shock layer. Scroggs has implemented this algorithm on an 8-processor Alliant computer.

2. Asymptotic Analysis of a problem with turning points.

We consider problems in which there exists a curve of turning points, and as a model we take

$$-x\,\partial_x u = \epsilon\Delta u \tag{2.1}$$

on the unit square $\Omega = \{(x,y) \mid 0 < x < 1,\ 0 < y < 1\}$. For (2.1) on Ω only the portion of the boundary $x = 1$ is an inflow boundary. The direction of flow is from right to left, because the coefficient of $\partial_x u$ is negative in Ω. The characteristics of the reduced equation

$$-x\,\partial_x u = 0 \tag{2.2}$$

are parallel to the parts of the boundary $y = 0$ and $y = 1$, so we would expect to find parabolic boundary layers near the top and bottom of Ω. The boundary $x = 0$ is

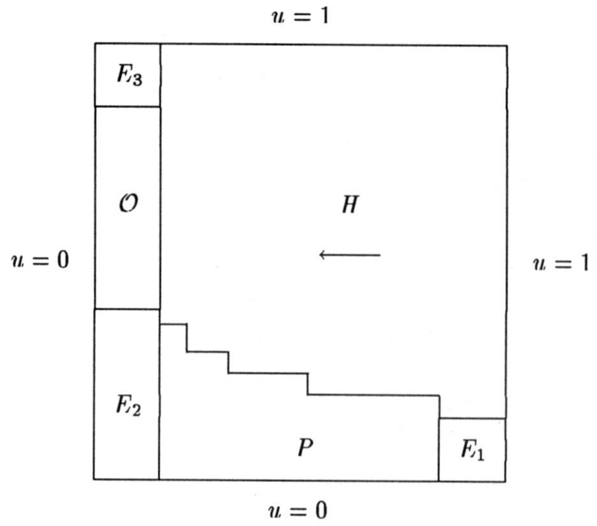

Fig. 2. Domain decomposition with turning points.

special in that it consists of turning points—the coefficient of $\partial_x u$ vanishes there. Let us remark that for this first example we have chosen a case in which the direction of flow is toward the turning points instead of away from them, because the solution may be unstable with respect to perturbation of the boundary data in the other case [1].

If for (2.1) we impose the boundary conditions

$$\begin{aligned} u &= 0 \quad \text{if } x = 0 \text{ or if } y = 0, \\ u &= 1 \quad \text{if } x = 1 \text{ or if } y = 1, \end{aligned} \tag{2.3}$$

then the natural domain decomposition is indicated in Fig. 2. This figure is based on an asymptotic analysis of the behavior of the solution of (2.1) as $\epsilon \to 0$ as given, for example, in [6] and [7]. Let us summarize the principal features of this asymptotic behavior.

There is a parabolic boundary layer region P near the bottom boundary $y = 0$ in which (2.1) may be approximated by the boundary-layer equation

$$-x\,\partial_x u = \epsilon\,\partial_y^2 u. \tag{2.4}$$

In Fig. 2 we have drawn the region P as a sum of rectangles because that is how we implemented the domain decomposition. On the basis of asymptotic analysis it is more natural to select a curve of the family $y^2 = -C\epsilon \log x$ as the upper boundary of P because

$$\eta = \frac{y^2}{\epsilon \log x} \tag{2.5}$$

is a similarity variable for (2.4). Note that the presence of the logarithm in (2.5) indicates that the turning points induce a widening of the parabolic boundary layer. In addition, the boundary-layer approximation (2.4) breaks down near the turning points.

In the vicinity of the corner $(x, y) = (1, 0)$ there is a birth region E_1 for this boundary layer, in which we must keep the full equation (2.1). The diameter of the subdomain E_1 is of order $O(\epsilon)$ as $\epsilon \to 0$.

There is no parabolic boundary layer analogous to P near the top boundary of Ω because the solution $u = 1$ of the reduced equation (2.2) is compatible with the boundary condition there. The subdomain on which the reduced equation (2.2) is a valid approximation to (2.1) is denoted by H in Fig. 2, and it is usually called the outer region.

The region surrounding the turning points $x = 0$ is divided into three parts, an ordinary boundary-layer region $\mathcal{O}$ on which (2.1) may be approximated by the differential equation

$$-x\,\partial_x u = \epsilon\,\partial_x^2 u \tag{2.6}$$

and two regions E_2 and E_3 in which none of the terms in (2.1) may be neglected. Because the change of variable $x = \sqrt{\epsilon}\,\xi$ in (2.6) produces a differential equation with coefficients independent of ϵ, we expect the width of the subdomain $\mathcal{O}$ to be of the order of $O(\sqrt{\epsilon})$ as $\epsilon \to 0$. The special behavior in E_2 arises from the interaction of the parabolic boundary layer in P with the turning points, and the subdomain E_3 is needed because of the incompatibility of the boundary data (2.3) in the vicinity of the corner $(x, y) = (0, 1)$.

3. Iterative Schemes in Individual Subdomains. In this section we describe the different local approximations to (2.1) which are used in the various subdomains. On many subdomains our difference scheme is an iterative method to improve a given approximate solution u_{n-1}. In Section 4 we present our global iterative method, which may be regarded as a form of matched asymptotic expansions without the algebraic manipulations normally associated with matched asymptotics.

Let us first select difference operators, given mesh sizes h_x and h_y in the x- and y-directions, respectively. On all of the subdomains we use a central finite-difference approximation to $\partial_y^2 u$,

$$D_y^2 u(x, y) = \frac{u(x, y + h_y) - 2u(x, y) + u(x, y - h_y)}{h_y^2}. \tag{3.1}$$

On most of the subdomains we use an upstream approximation to $\partial_x u$, namely,

$$D_x^+ u(x, y) = \frac{u(x + h_x, y) - u(x, y)}{h_x} \tag{3.2}$$

and a central-difference approximation to $\partial_x^2 u$,

$$D_x^2 u(x, y) = \frac{u(x + h_x, y) - 2u(x, y) + u(x - h_x, y)}{h_x^2}. \tag{3.3}$$

The exceptional subdomains on which we use different approximations to the partial derivatives in the x-direction are the regions E_2, $\mathcal{O}$, and E_3 adjacent to the turning points. There, in the spirit of Chin and Krasny [4], we approximate the operator $x\,\partial_x + \epsilon\,\partial_x^2$ by a finite-element method based on solutions of

$$-x\,\partial_x u = \epsilon\,\partial_x^2 u.$$

Thus, for a grid $\{x_j\}$ with $j = 0, 1, \ldots, J$ our fundamental basis spline is

$$\begin{aligned}
\phi_j(x) &= \frac{\operatorname{erf}\left\{\frac{x}{\sqrt{2\epsilon}}\right\} - \operatorname{erf}\left\{\frac{x_{j-1}}{\sqrt{2\epsilon}}\right\}}{\operatorname{erf}\left\{\frac{x_j}{\sqrt{2\epsilon}}\right\} - \operatorname{erf}\left\{\frac{x_{j-1}}{\sqrt{2\epsilon}}\right\}} \qquad \text{for } x_{j-1} < x < x_j, \\
\phi_j(x) &= \frac{\operatorname{erf}\left\{\frac{x_{j+1}}{\sqrt{2\epsilon}}\right\} - \operatorname{erf}\left\{\frac{x}{\sqrt{2\epsilon}}\right\}}{\operatorname{erf}\left\{\frac{x_{j+1}}{\sqrt{2\epsilon}}\right\} - \operatorname{erf}\left\{\frac{x_j}{\sqrt{2\epsilon}}\right\}} \qquad \text{for } x_j < x < x_{j+1}, \\
\phi_j(x) &= 0, \qquad \text{elsewhere.}
\end{aligned} \tag{3.4}$$

We define the operator L as the finite-element approximation to $-x\,\partial_x - \epsilon\,\partial_x^2$ obtained from using the basis functions ϕ_j in (3.4).

It might seem reasonable that in the outer region H, where (2.2) is the asymptotic behavior, we use an analogue of (1.7)

$$-x D_x^+ u_n = \epsilon \left(D_x^2 + D_y^2\right) u_{n-1}.$$

It turns out, however, that the stability conditions for this iterative method are quite severe, namely that

$$\epsilon < C h_x^2 \min_H x$$

for some positive constant C. With the notation $Iu = u$ and $T_y u(x,y) = u(x, y + h_y)$, we use in H a scheme which we find to be stable under a wider range of conditions

$$\left(-x D_x^+ + \frac{2\epsilon}{h_y^2} I\right) u_n = \epsilon \left(D_x^2 + \frac{1}{h_y^2}(T_y + T_y^{-1})\right) u_{n-1}. \tag{3.5}$$

In the parabolic layer region P we use

$$\left(-x D_x^+ - \epsilon D_y^2\right) u^n = \epsilon D_x^2 u^{n-1}. \tag{3.6}$$

In the birth region of the boundary layer E_1 we simply discretize (2.1) with a fine grid,

$$-x D_x^+ u_n = \epsilon \left(D_x^2 + D_y^2\right) u_n. \tag{3.7}$$

In the ordinary layer region O we use

$$L u_n = \epsilon D_y^2 u_{n-1}, \tag{3.8}$$

where L is the finite-element approximation to the operator $-x\,\partial_x - \epsilon\,\partial_x^2$ obtained from the splines (3.4). Finally, in regions E_2 and E_3 we use the approximation

$$L u_n = \epsilon D_y^2 u_n. \tag{3.9}$$

4. The Global Iterative Algorithm. Our global iterative method combines the local iterations (3.5–9) as follows.

1. Make an initial domain decomposition based on asymptotic information about the location and size of the layer regions.
2. Set $n = 0$ and provide an initial approximate solution u_0. (One reasonable choice is the outer solution $u_0 = 1$.)
3. Proceed through the subdomains in the following order. Use (3.5) to find u_n on H. Use (3.7) to find u_n on E_1, using (3.6) as the boundary condition on the downstream boundary. Solve (3.6) on P to determine u_n there. Then solve (3.8) on O. Finally use (3.9) to determine u_n on E_2 and E_3.
4. Check for convergence. If we aren't finished, check u_n for consistency with (2.1) across subdomain boundaries, modify the domain decomposition if necessary, increment n, and return to Step 3.

5. Comments. The primary concerns about any numerical method are accuracy and efficiency. Our scheme derives its efficiency from the fact that difficult matrix problems are solved only on the subdomains $\bigcup_{j=1}^3 E_j$. (We use a multigrid method for these problems.) On the largest subdomain H we solve an uncoupled system of ordinary differential equations (3.5). This is not only fast, but it is also highly parallel. The two-point boundary-value problems (3.8) on $\mathcal{O}$ are also highly parallel. Admittedly, the algorithm (3.6) on P is quite sequential, but that is because the inherent nature of the asymptotic behavior there is a parabolic partial differential equation (2.4).

Efficiency also depends on the rate of convergence of the algorithm as $n \to \infty$. On the basis of computational experience and theory for problems with constant coefficients we have found that the rate of convergence of the scheme as a whole is limited by the behavior of (3.5) in H. In fact, for a constant-coefficient version of (3.5)

$$\left(-aD_x^+ + \frac{2\epsilon}{h_y^2}I\right)u_n = \epsilon\left(D_x^2 + \frac{1}{h_y^2}(T_y + T_y^{-1})\right)u_{n-1} \tag{5.1}$$

with $a > 0$ we have the following theorem [2] concerning the discrete l_2-norm of u_n.

Theorem 5.1. *For the iterative scheme (5.1) on the halfspace $-\infty < x < 0$, $-\infty < y < \infty$ with boundary data $u_n(0, y) = 0$ for every positive number κ there exists a continuous function $C(\xi)$ such that*

$$\|u_n\| \le \kappa \|u_{n-1}\|$$

whenever

$$\frac{\epsilon}{ah_x} \le C\left(\frac{h_x\epsilon}{ah_y^2}.\right)$$

Furthermore, there exist functions u_{n-1} such that for any positive δ we have

$$\|u_n\| \ge (1-\delta)\kappa \|u_{n-1}\| \tag{5.2}$$

whenever

$$\frac{\epsilon}{ah_x} \ge C\left(\frac{h_x\epsilon}{ah_y^2}\right).$$

The proof is based on the Godunov-Ryabenky stability theory [8].

The significance of Theorem 5.1 for application to our problem is that the left-hand boundary of H must be sufficiently far from the y-axis for two reasons. On the one hand, we have a variable coefficient $a = x$, so that the theorem can only give heuristic information. Still, we are led by (5.2) to expect instability if x is too small in H. On the other hand, the derivative $|\partial_x^2 u|$ is large in the turning-point region, so that if we are using (3.5), then we need a small step size h_x to resolve the variations in u. This also has a destabilizing effect in (5.2). It should be noted that if the turning-point region $E_2 \cup \mathcal{O} \cup E_3$ is sufficiently wide, then neither of these effects is of concern. The solution u in H is so smooth that the mesh sizes may be taken to be large compared with $\epsilon / \min_H x$.

The criteria for determining the boundaries of P and E_j $(j = 1, 2, 3)$ are not stability, but accuracy and efficiency. If an E_j is larger than necessary, we do more computational work than we need to, but there is no degredation of accuracy. If, on the other hand, an E_j is too small, then we will have to do more global iterations in order to get convergence on the adjoining subdomains. Worse than that, we will also probably loose accuracy because fine meshes are used in the regions E_j in order to resolve the rapid variations of u there. Similar criteria govern the location of the boundary between P and H.

We recognize that classical numerical analysts are likely to regard our stability condition $\epsilon < Ch_x \min_H x$ as quite strange, but it makes sense for singular perturbation problems.

References

(1) K. E. Barrett, *The numerical solution of singular-perturbation boundary-value problems*, Quart. J. Appl. Math. 27 (1974), pp. 57–68.

(2) D. L. Brown, R. C. Y. Chin, G. W. Hedstrom, and T. A. Manteuffel, *A domain decomposition method for singularly perturbed convection-diffusion equations*, to appear.

(3) R. C. Y. Chin, G. W. Hedstrom, J. S. Scroggs, and D. C. Sorensen, *Parallel computation of a domain decomposition method*, Adv. Comp. Meth. Partial Dif. Eq. 6 (1987), pp. 375–381.

(4) R. C. Y. Chin and R. Krasny, *A hybrid asymptotic-finite element method for a stiff two-point boundary value problem*, SIAM J. Sci. Stat. Comput. 4 (1983), pp. 229–243.

(5) W. Eckhaus, *Asymptotic Analysis of Singular Perturbation Problems*, North-Holland, Amsterdam, 1979.

(6) P. P. N. de Groen, *Elliptic singular perturbations of first-order operators with critical points*, Proc. Roy. Soc. Edinburgh 74A (1974/75), pp. 91–113.

(7) E. M. de Jager, *Singular elliptic perturbations of vanishing first order differential operators*, in *Lecture Notes in Mathematics* **280**, Springer Verlag, Berlin, 1972, pp. 73–86.

(8) R. D. Richtmyer and K. W. Morton, *Difference Methods for Initial-Value Problems*, Wiley-Interscience, New York, 1967.

(9) G. Rodrigue and E. Reiter, *A domain decomposition method for boundary layer problems*, in this volume.

On the Schwarz Alternating Method II: Stochastic Interpretation and Order Properties

P. L. Lions*†

Abstract. We continue here a systematic investigation of convergence properties of the Schwarz alternating method and related decomposition methods. Our study here is based upon the maximum principle and the stochastic interpretation of the Schwarz alternating procedure.

Introduction.

This paper is a sequel of [36] and part II of a series of papers devoted to the mathematical study of various decomposition methods (domain decomposition methods) for the solution of various linear or nonlinear partial differential equations. In the recent years, the applications of iterative methods solving subproblems or problems in subdomains to the numerical analysis of boundary value problems have been developed by various authors and a partial list of contributions to this general theme can be found in the bibliography.

Parts I and II of this series of papers are devoted to the study of the classical Schwarz alternating method (that we recall in section I below). In some sense, even if many interesting and important variants have been introduced recently, the Schwarz algorithm remains the prototype of such methods and also presents some properties (like "robustness", or indifference to the type of equations considered...) which do not seem to be enjoyed by other methods. In part I [36], we studied the Schwarz alternating method from a variational view-point (iterated projections in an Hilbert space) and obtained various convergence results. In some sense, with such a variational viewpoint, one is naturally led to variants based upon "control-calculus of variations" considerations (as in [10], [11], [12] ...) which, at least for Laplace equations, are a bit faster for computing applications.

---

* Ceremade, Université Paris-Dauphine, Place de Lattre de Tassigny, 75775 Paris Cedex 16.

† Consultant at Informatique Internationale (CISI).

On the other hand, as it was originally proved by Schwarz [1], the Schwarz alternating method also converges, say for Laplace equations, because of the strong maximum principle for harmonic functions (see for instance the paragraph on Schwarz method in [37]). Let us observe, by the way, that the Schwarz alternating method seems to be the only domain decomposition method converging for two entirely different reasons : variational characterzation of the Schwarz sequence and maximum principle.

This paper is a systematic study of such properties of Schwarz alternating method. First, we recall in section 1 from [38] the stochastic interpretation of Schwarz method in terms of successive exit times from the subdomains of the underlying diffusion processes (Brownian motion in the case of Laplace equation). This interpretation shows that Schwarz method is intimately (even if simply) connected with the deep structure of Laplace's equation.

Next, in section 2, we present a convergence proof for Schwarz method for uniformly elliptic equations in the case of overlapping domains : as we will see the convergence is geometrical and we will indicate an estimate on the rate of convergence. In section 3, we study the same question when we relax the condition of overlapping, allowing the "boundaries of the two subdomains" to touch at the boundary of the original domain". As we will see, if the situation of section 2 is not basically modified for Dirichlat boundary conditions (in this case, our analysis is a minor extension of Schwarz original convergence proof), we will show that drastic changes occur for Neumann boundary conditions. Next, in section 4, we observe that if we start with a subsolution (respectively a supersolution) of the full problem, Schwarz alternating method creates an increasing (respectively decreasing) sequence of subsolutions (respectively supersolutions) and this will allow to prove convergence in some geometrical situations where the condition that the two subdomains overlap is somehow relaxed (including the case of Neumann boundary conditions considered in section 3). Section 5 is devoted to equations which are no more uniformly elliptic like degenerate elliptic equations or parabolic equations (heat equation for instance) possibly with a time discretization : in the cases when the two subdomains strictly overlap, we will show that geometrical convergence is still true, and this will be a consequence of the fact that Schwarz lemma is still true for degenerate equations (and therefore is not always related to the strong maximum principle). Next, we will present in section 6 another convergence proof based upon the stochastic interpretation : this proof will be purely probabilistic and will give some hint on the way to "optimize the domain splitting" in order to obtain the fastest convergence. Section 7 will a brief presentation of the applications of Schwarz alternating methods when the original domain is split into more than two subdomains (section 8).

Let us also mention that we will study in the remaining parts of this series of papers some variants of Schwarz alternating method that we will introduce to take care of the geometrical situation when the domain is split (decomposed) in two (or more) subdomains separated only by an interface (n-1 -dimensional manifold) - in particular, these subdomains do not overlap at all.

Summary

1. Presentation of Schwarz method and stochastic interpretation.
2. Convergence proof via the maximum principle : overlapping domains.
3. Convergence proof via the maximum principle: weakly overlapping domains.
4. Sub and supersolutions.
5. Degenerate equations and time-dependent problems.
6. Convergence proof via the stochastic interpretation.
7. Nonlinear problems.
8. Multidomains decomposition.

1. Presentation of Schwarz method and stochastic interpretation.

We consider a bounded, open domain $\mathcal{O}$ in $\mathbb{R}^N$ and we assume (to simplify) that $\mathcal{O}$ is smooth and connected. We then decompose $\mathcal{O}$ in two subdomains $\mathcal{O}_1$ and $\mathcal{O}_2$ such that

(1) $$\mathcal{O} = \mathcal{O}_1 \cup \mathcal{O}_2$$

and we denote by $\Gamma = \partial\mathcal{O}$, $\Gamma_1 = \partial\mathcal{O}_1$, $\Gamma_2 = \partial\mathcal{O}_2$, $\gamma_1 = \partial\mathcal{O}_1 \cap \mathcal{O}_2$, $\gamma_2 = \partial\mathcal{O}_2 \cap \mathcal{O}_1$, $\mathcal{O}_{12} = \mathcal{O}_1 \cap \mathcal{O}_2$, $\mathcal{O}_{11} = \mathcal{O}_1 \cap \overline{\mathcal{O}}_2^c$, $\mathcal{O}_{22} = \mathcal{O}_2 \cap \overline{\mathcal{O}}_1^c$.

Various decompositions are possible as it can be seen from the following figures

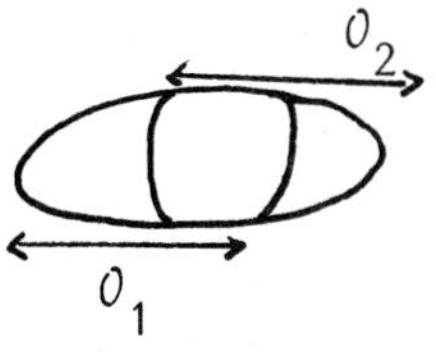

1.a

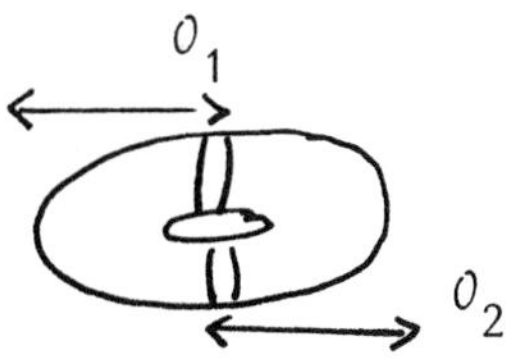

1.b

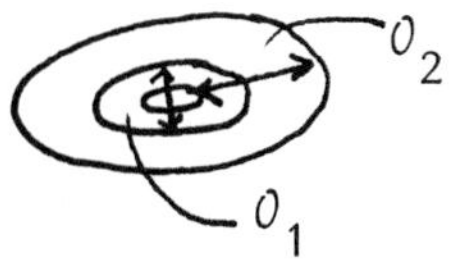

1.c

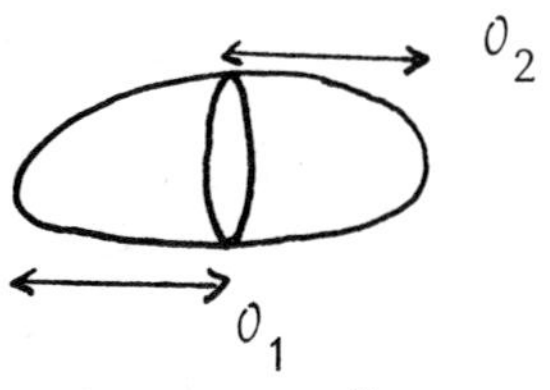

2.a

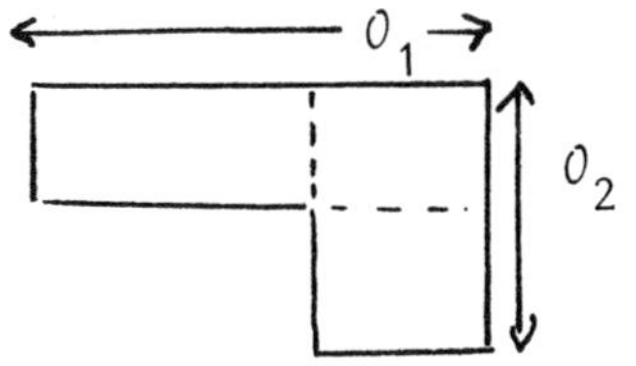

2.b

We will always assume to simplify that γ_1, γ_2 are smooth...

We will also say that $\mathcal{O}_1$ and $\mathcal{O}_2$ overlap if $\overline{\mathcal{O}}_{11}$ and $\overline{\mathcal{O}}_{22}$ do not intersect. Observe that this is the case in figures 1.a, 1.b, 1.c but not in figures 2.a or 2.b.

Next, suppose that we want to solve the following model problem

(2) $$- \Delta u = f \quad \text{in } \mathcal{O} \quad , \quad u = 0 \quad \text{on } \partial\mathcal{O}$$

where f is a given function say in $C(\overline{\mathcal{O}})$ (or in $H^{-1}(\mathcal{O})$ ...) . The Schwarz alternating procedure consists in solving successively the following problems : let u^o be any initialization say in $C_o(\overline{\mathcal{O}})$ (i.e. continuous functions on $\overline{\mathcal{O}}$ vanishing on $\partial\mathcal{O}$), we obtain u^{2n+1} $(n \geqq 0)$ and u^{2n} $(n \geqq 1)$ by solving respectively

(3) $$- \Delta u^{2n+1} = f \quad \text{in } \mathcal{O}_1 \quad , \quad u^{2n+1} = u^{2n} \quad \text{on } \partial\mathcal{O}_1$$

(4) $$- \Delta u^{2n} = f \quad \text{in } \mathcal{O}_2 \quad , \quad u^{2n} = u^{2n-1} \quad \text{on } \partial\mathcal{O}_2 \quad ,$$

and $u^{2n+1} \in C(\overline{\mathcal{O}}_1)$, $u^{2n} \in C(\overline{\mathcal{O}}_2)$, $u^{2n+1} = 0$ on $\Gamma_1 \cap \Gamma$, $u^{2n} = 0$ on $\Gamma_2 \cap \Gamma$. In fact, (3),(4) require that u^{2n}, u^{2n+1} are defined on $\overline{\mathcal{O}}$ and we extend obviously u^{2n+1} and u^{2n} respectively to $\overline{\mathcal{O}}$ by u^{2n} and u^{2n-1} so that $u^{2n+1}, u^{2n} \in C_o(\overline{\mathcal{O}})$ and $u^{2n+1} \equiv u^{2n}$ on $\overline{\mathcal{O}}_{22}$, $u^{2n} \equiv u^{2n+1}$ on $\overline{\mathcal{O}}_{11}$.

In [36] , we explained how u^{2n+1}, u^{2n} correspond to successive projections in subspaces of $H_o^1(\mathcal{O})$. We want to present now a different interpretation of the sequence $(u^n)_n$ in terms of successive exit times from $\overline{\mathcal{O}}_1$ resp. $\overline{\mathcal{O}}_2$. To this end, we consider any standard probability space (Ω, F, F^t, P) equipped with a Brownian motion B_t continuous and F^t adapted and we introduce the following stopping times

(5) $$\tau_x = \inf \{t \geqq 0 \ , \ x+B_t \notin \mathcal{O}\}$$

(6) $$\tau_x^{2n+1} = \inf \{t \geqq \tau_x^{2n} \ , \ x+B_t \notin \mathcal{O}_1\} \quad \text{for } n \geqq 0$$

(7) $$\tau_x^o \equiv 0 \ , \ \tau_x^{2n} = \inf \{t \geqq \tau_x^{2n-1}, x+B_t \notin \mathcal{O}_2\} \quad \text{for } n \geqq 1$$

for all $x \in \overline{\mathcal{O}}$. As it is well-known, the solution u of (2) is given by the formula

(8) $$u(x) = E \int_o^{\tau_x} f(x+B_t)\, dt \quad , \quad \forall\, x \in \overline{\mathcal{O}} \ .$$

At this point, let us mention once for all that B_t is not really a Brownian motion but $B_t = \sqrt{2}\, B'_t$ and B'_t is a Brownian motion. Then, we claim that we have the

<u>Lemma 1.</u> For all $x \in \overline{\mathcal{O}}$ and for all $n \geqq 0$, we have

(9) $$u^n(x) = E \int_o^{\tau_x^n} f(x+B_t)\, dt + u^o\left(x+B_{\tau_x^n}\right)$$

Remarks. i) From the definitions (5),(6),(7), one has immediately

$$\tau_x^n \uparrow_n \ , \quad \tau_x^n \leqq \tau_x \qquad \text{for all} \quad n \geqq 0 \quad \text{and for all} \quad x \in \bar{\mathcal{O}} . \tag{10}$$

ii) Recall also (see for instance [39]) that if λ_1 is the first eigenvalue of $-\Delta$ in $H_0^1(\mathcal{O})$ then we have

$$\sup_{x \in \bar{\mathcal{O}}} E\left[e^{\lambda \tau_x}\right] < \infty \qquad \text{for all} \quad \lambda \in (0,\lambda_1)$$

hence in particular $\tau_x < \infty$ a.e., for all $x \in \bar{\mathcal{O}}$.

Proof of Lemma 1. By induction, suppose that (9) holds for n and let us prove the same formula for $n+1$. Without loss of generality we may assume that n is even. Then, if $x \notin \bar{\mathcal{O}}_1$, $\tau_x^{n+1} = \tau_x^n$ and $u^{n+1}(x) = u^n(x)$ therefore (9) is proved in this case. On the other hand, if $x \in \bar{\mathcal{O}}_1$, we recall from standard facts that u^{n+1} is given by

$$u^{n+1}(x) = E \int_0^{\tau'_x} f(x+B_t)dt + u^n\left(x+B_{\tau'_x}\right) \ , \qquad \forall\, x \in \bar{\mathcal{O}}_1 \tag{11}$$

where $\tau'_x = \inf\,(t \geqq 0 \ , \ x+B_t \notin \mathcal{O}_1)$.

Then, using formula (9) for u^n and the Markov property, (11) immediately yields (3) for u^{n+1} . Δ

2. Convergence proof via the maximum principle : overlapping domains.

All throughout this section, we assume that $\mathcal{O}_1$ and $\mathcal{O}_2$ overlap and that $\partial\mathcal{O}_1 \cap \partial\mathcal{O}$, $\partial\mathcal{O}_2 \cap \partial\mathcal{O}$ are both nonempty. The main convergence result is given by the

Proposition 2. There exist $k_1,k_2 \in (0,1)$ which depend only respectively of $(\mathcal{O}_1,\gamma_2)$ and $(\mathcal{O}_2,\gamma_1)$ such that for all $n \geqq 0$

$$\sup_{\bar{\mathcal{O}}_1} |u-u^{2n+1}| \leqq k_1^n\, k_2^n \sup_{\gamma_1} |u-u^o| \tag{12}$$

$$\sup_{\bar{\mathcal{O}}_2} |u-u^{2n}| \leqq k_1^n\, k_2^{n-1} \sup_{\gamma_2} |u-u^o| \ . \tag{13}$$

Remarks. 1) We will give below some estimates on k_1,k_2 .

2) As we will see from the proof below, it is not necessary to take $u^0 \in C_0(\bar{\mathcal{O}})$: for instance, it is enough to consider $u^0 \in C(\bar{\mathcal{O}})$ and we then define u^1 by (3) replacing the boundary conditions on $\partial\mathcal{O}_1$ as follows

$$u^1 = 0 \quad \text{on} \quad \partial\mathcal{O}_1 \cap \partial\mathcal{O} \ , \qquad u^1 = u^0 \quad \text{on} \quad \partial\mathcal{O}_1 \cap \mathcal{O}_2 \ .$$

Then, $u^2 \in C_0(\bar{\mathcal{O}})$ and the estimates (12),(13) still hold.

3) A similar result with the same proof holds for different boundary conditions on $\partial\mathcal{O}$ or for more general uniformly elliptic second-order operators. More precisely, the same result holds if we replace Dirichlet boundary condition by Neumann or oblique derivative or Robin type or mixed type boundary condition. Furthermore, the operator $-\Delta$ may be replaced by any divergence free operator

$$-\sum_{i,j} \frac{\partial}{\partial x_i}\left(a_{ij}\frac{\partial}{\partial x_j}\right) + \sum_i b_i \frac{\partial}{\partial x_i} + c$$

where $a_{ij}, b_i, c \in L^\infty$, $c \geq 0$ and

$$\sum_{i,j} a_{ij}(x)\xi_i\xi_j \geq \nu|\xi|^2 \quad \text{for all } \xi \in \mathbb{R}^N,\ \text{a.e. } x \in \mathcal{O}, \text{ for some } \nu > 0 \tag{14}$$

or by a nondivergence operator $-\sum_{i,j} a_{ij}(x)\frac{\partial^2}{\partial x_i\,\partial x_j} + \sum_i b_i\frac{\partial}{\partial x_i} + c$ where $a_{ij} \in C(\overline{\mathcal{O}})$ (for instance) satisfy (14), $c, b_i \in L^\infty$, $c \geq 0$.

4) A priori, the rate of convergence given by $(k_1k_2)^{1/2}$ is different from the one obtained by the variational (projections in Hilbert spaces) argument of [36] since the "errors" $u-u^{2n+1}$, $u-u^{2n}$ are estimated in different norms (L^∞ here, instead of H^1 in [36]).

5) It is possible to deduce from Proposition 2 the geometric convergence of u^n to u in H^1 : in fact, more generally, even if u_o is not smooth (say $u_o \in H^1_o(\mathcal{O}) + C_o(\overline{\mathcal{O}})$) one can show easily by interior elliptic regularity - using thus the fact that $\mathcal{O}_1$ and $\mathcal{O}_2$ do overlap - that there exists $n_o \geq 1$ (depending only on the dimension N) such that $u-u^n \in H^1_o(\mathcal{O}) \cap C_o(\overline{\mathcal{O}})$ for $n \geq n_o$. Then, denoting by $\tilde{u}^n = u^{n_o+n}$ for $n \geq 0$, one sees that $\tilde{u}^n$ is a new "Schwarz sequence" corresponding to the new initialization $\tilde{u}^o = u^{n_o}$ hance Proposition 2 applies to $u-\tilde{u}^n$ and one still gets geometric convergence of u^n to u in $L^\infty(\mathcal{O})$. Furthermore, we claim that u^n converges geometrically to u in $H^1_o(\mathcal{O})$: indeed, introducing (as in [36]) $\zeta_1,\zeta_2 \in W^{1,\infty}(\mathcal{O})$ such that $0 \leq \zeta_1,\zeta_2 \leq 1$, $\zeta_1+\zeta_2 \equiv 1$ on $\overline{\mathcal{O}}$, $\zeta_i \equiv 1$ on $\overline{\mathcal{O}}_{ii}$, $\zeta_i \equiv 0$ on $\overline{\mathcal{O}}_{jj}$ for all $i \neq j \in \{1,2\}$, we first obtain multiplying the equation satisfied by $u-u^n$ (for instance) by $\zeta_2^2(u-u^{2n})$

$$\int_{\mathcal{O}} \zeta_2^2|\nabla(u-u^{2n})|^2 = -2\int_{\mathcal{O}} \zeta_2\,\nabla\zeta_2\cdot\nabla(u-u^{2n})(u-u^{2n}) \leq C\left(\int_{\mathcal{O}} \zeta_2^2|\nabla(u-u^{2n})|^2\right)^{1/2} k_1^n k_2^{n-1}\ .$$

Hence,

$$|\zeta_2(u-u^{2n})|_{H^1_o(\mathcal{O})} \leq C\,k_1^n\,k_2^{n-1}\ .$$

Then, observing that $u-u^{2n+1} - \zeta_2(u-u^{2n}) \in H_o^1(\mathcal{O})$, we deduce easily

$$\int_{\mathcal{O}_1} |\nabla(u-u^{2n+1})|^2 = \int_{\mathcal{O}_1} \nabla(u-u^{2n+1}) \cdot \nabla\{\zeta_2(u-u^{2n})\}$$

or

$$\int_{\mathcal{O}_1} |\nabla(u-u^{2n+1})|^2 \leqq \int_{\mathcal{O}_1} |\nabla\{\zeta_2(u-u^{2n})\}|^2 .$$

And since $u-u^{2n+1} = u-u^{2n} = \zeta_2(u-u^{2n})$ on $\mathcal{O}-\mathcal{O}_1$, we deduce finally

$$\int_{\mathcal{O}} |\nabla(u-u^{2n+1})|^2 \leqq \int_{\mathcal{O}} |\nabla\{\zeta_2(u-u^{2n})\}|^2$$

i.e.

$$|u-u^{2n+1}|_{H_o^1(\mathcal{O})} \leqq C\, k_1^n k_2^{n-1} ,$$

proving thus our claim. Notice also that this proof remains valid for general uniformly elliptic second-order operators with straightforward bounds for the first-order terms.

6) In the case when, for example, $\overline{\mathcal{O}}_2 \subset \mathcal{O}$ and thus $\partial\mathcal{O}_2 \cap \partial\mathcal{O} = \emptyset$ then the same result with the same proof holds provided we take $k_2 = 1$.

Δ

We now turn to the proof of Proposition 2 which is an immediate consequence of the following standard lemma.

Lemma 3. Let $w \in L^\infty(\mathcal{O}_1)$ be continuous on $\overline{\mathcal{O}}_1 - \{\overline{\partial\mathcal{O}_1 \cap \mathcal{O} \cap \partial\mathcal{O}}\}$, satisfy

(15) $$-\Delta w = 0 \text{ in } \mathcal{O}_1 ,\quad w = 0 \text{ on } \partial\mathcal{O}_1 - \overline{\partial\mathcal{O}_1 \cap \mathcal{O}} ,\quad w = 1 \text{ on } \partial\mathcal{O}_1 \cap \mathcal{O} .$$

Then,

$$k_1 = \sup \{w(x) \;/\; x \in \partial\mathcal{O}_2 \cap \overline{\mathcal{O}}\} \in (0,1) .$$

Remarks. 1) Of course, a similar lemma holds for $\mathcal{O}_2$.

2) An estimate on k_1 is given below.

3) In one dimension, if $\mathcal{O}_1 = (a,b)$, $\mathcal{O}_2 = (c,d)$ with $a < c < b < d$, then $w(x) = \frac{x-a}{b-a}$ and $k_1 = \frac{c-a}{b-a}$.

Once Lemma 3 is proven, Proposition 2 follows easily from the maximum principle since $|u-u^{2n+1}| \leqq \sup_{\gamma_1} |u-u^{2n}| w$, hence $\sup_{\gamma_2} |u-u^{2n+1}| \leqq k_1 \sup_{\gamma_1} |u-u^{2n}|$ and thus $\sup_{\gamma_2} |u-u^{2n+1}| \leqq$ $\leqq k_1^{n+1} k_2^n \sup_{\gamma_1} |u-u^0|$, $\sup_{\gamma_1} |u-u^{2n}| \leqq k_1^n k_2^n \sup_{\gamma_1} |u-u^0|$. Then, (12) and (13) follows from the maximum principle which yields

$$\sup_{\overline{\mathcal{O}}_1} |u-u^{2n+1}| = \sup_{\gamma_1} |u-u^{2n+1}| = \sup_{\gamma_1} |u-u^{2n}| \text{ for } n \geqq 0,$$

$$\sup_{\mathcal{O}_2} |u-u^{2n}| = \sup_{\gamma_2} |u-u^{2n}| = \sup_{\gamma_2} |u-u^{2n-1}| \text{ for } n \geq 1 .$$

Finally, let us conclude by mentioning that Lemma 3 is an immediate consequence of the strong maximum principle. Δ

It is difficult (apparently) to obtain a sharp estimate on k_1 since k_1 depends very much on the geometries of $\mathcal{O}_1$ and $\mathcal{O}_2$. However, it is possible to estimate k_1 asymptotically when γ_1 "converges" to γ_2. In order to avoid rather unpleasant technicalities, we will consider only the case of figure 1.c (even if a similar analysis can be performed in general situations) i.e. $\gamma_1, \gamma_2 \subset \mathcal{O}$. Then, we just observe that if ν is the unit outward normal to γ_1 by Hopf maximum principle we have $\frac{\partial w}{\partial \nu} \geq \kappa > 0$ on γ_1. Hence, if $\varepsilon = \max_{y \in \gamma_2} d(y,\gamma_1)$, we have up to the second-order terms (ε^2 terms) $k_1 \cong 1-\kappa\varepsilon$. Observe also that if γ_1 "goes to" $\partial\mathcal{O}$, κ behaves like dist $(\gamma_2,\partial\mathcal{O})^{-1}$. We will not push further here this kind of estimates.

3. Convergence proof via the maximum principle : weakly overlapping domains.

We now turn to the case when $\mathcal{O}_1$ and $\mathcal{O}_2$ do not overlap as it is the case in figures 2.a or 2.b. We still assume, of course, that $\gamma_1 \cap \gamma_2 = \emptyset$ and we now assume that $\bar{\gamma}_1$ and $\bar{\gamma}_2$ are not tangent at points of $\partial\mathcal{O}$ (belonging to $\bar{\gamma}_1 \cap \bar{\gamma}_2$). Then, we claim that Proposition 2 still holds in this case. Of course, we just have to explain why lemma 3 is still valid. Since $w < 1$ in $\mathcal{O}$, it is enough to show that

$$\lim\sup \{w(y) \;/\; y \in \gamma_2 \;,\; d(y,\partial\mathcal{O}) \to 0\} < 1 .$$

And this follows easily from potential theory. Observe that in two dimensions it is possible to identify this limit : indeed, if θ_1 is the "angle of $\partial\mathcal{O}_1$" at a point y_0 belonging to $\bar{\gamma}_1 \cap \bar{\gamma}_2$ ($\theta_1 = \pi$ if $\mathcal{O}_1$ is smooth) and θ_2 is the "angle between $\partial\mathcal{O}$ and γ_2 at this point when $w(y) \to \frac{\theta_2}{\theta_1}$ (< 1) as y goes to $y_0, y \in \gamma_2$ (see figure 2 below)

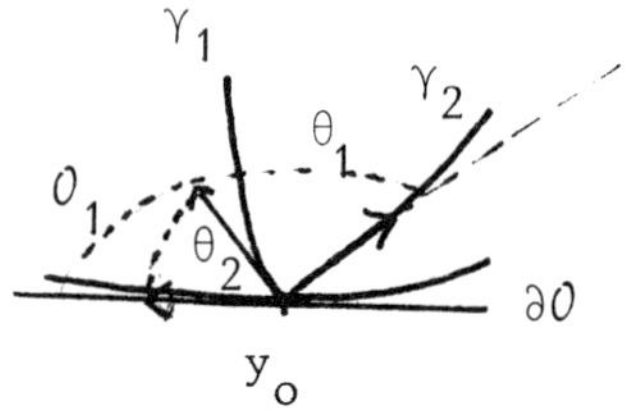

Figure 2.

We now make some comments on the analogues in this case of the remarks following Proposition 2 : first of all, the extension (Remark 3) to more general uniformly elliptic operators is still valid here. Next, if we replace Dirichlet boundary conditions by Neumann boundary conditions on $\partial\mathcal{O}$ (replacing $-\Delta$ by $-\Delta+c$ with $c > 0$), then lemma 3 is no more true and in fact it is not difficult to show that since $u^n = u^o$ at points $y \in \bar{\gamma}_1 \cap \bar{\gamma}_2$, the Schwarz sequence does not converge anymore uniformly on $\bar{\mathcal{O}}$ to the solution u (choose u^o such that $u^o(y_o) \neq u(y_o)$ at some point $y_o \in \bar{\gamma}_1 \cap \bar{\gamma}_2$). This difficulty may also be seen from a variational view-point since (with the notations of [36]) one does not have anymore $V\ (= H^1(\mathcal{O})) = V_1$ $(= \{u \in H^1(\mathcal{O})\ ,\ u \equiv 0 \text{ on } \mathcal{O}_{22}\}) + V_2\ (= \{u \in H^1(\mathcal{O})\ ,\ u \equiv 0 \text{ on } \mathcal{O}_{11}\})$. However, one still has $V = \overline{V_1+V_2}$ (see [36]) and this ensures (cf. [36]) that u^n converges in $H^1(\mathcal{O})$ to u . We will also prove convergence in this case by a different method in section 4.

The analogues of Remarks 4 and 5 are still true here : observe only for Remark 5 that one has to introduce cut-off functions ζ_1,ζ_2 as in [36] involving a singularity at points $y \in \bar{\gamma}_1 \cap \bar{\gamma}_2$.

4. Sub and supersolutions.

In this section, we want to explain a striking property of Schwarz alternating method namely that if u^o is a subsolution of (2) (resp. supersolution) then u^n is also for all $n \geq 0$ a subsolution of (2) (resp. supersolution) and furthermore u^n is an increasing (resp. decreasing) sequence. Next, we will give some applications of this observation to convergence properties even in cases when the usual analysis (either the variational one as in [36] , or the one made in the preceding sections 2-3) fails. We begin with the

Theorem 4. Let $u^o \in C(\bar{\mathcal{O}})$ satisfy

(16) $\qquad -\Delta u^o \leq f$ in $\mathcal{D}'(\mathcal{O})$, $u^o \leq 0$ on $\partial\mathcal{O}$

(resp.

(17) $\qquad -\Delta u^o \geq f$ in $\mathcal{D}'(\mathcal{O})$, $u^o \geq 0$ on $\partial\mathcal{O}$).

Then, for all $n \geq 1$, $u^n \in C_o(\bar{\mathcal{O}} - \partial\mathcal{O} \cap \bar{\gamma}_1 \cap \bar{\gamma}_2)$ is bounded and also satisfies (16) (resp. (17)). Furthermore, we have

(18) $\qquad u^n \leq u^{n+1}$ on $\bar{\mathcal{O}}$ for $n \geq 0$, $u^n \uparrow_n u$ uniformly on $\mathcal{O}$

(resp.

(19) $\qquad u^n \geq u^{n+1}$ on $\bar{\mathcal{O}}$ for $n \geq 0$, $u^n \downarrow_n u$ uniformly on $\mathcal{O}$).

Remarks. 1) Notice that no assumptions on $\mathcal{O}_1,\mathcal{O}_2$ are made except $\gamma_1 \cap \gamma_2 = \emptyset$, $\mathcal{O}_1 \cup \mathcal{O}_2 = \mathcal{O}$. In particular, $\bar{\gamma}_1$ and $\bar{\gamma}_2$ can "meet tangentially" at $\partial\mathcal{O}$, case which was excluded in section 3.

2) It is possible to relax the regularity requirements on u^o as follows : $u^o \in C(\bar{\mathcal{O}}) + H^1(\mathcal{O})$, $(u^o)^+ \in C_o(\bar{\mathcal{O}}) + H_o^1(\mathcal{O})$.

3) The same result holds for Neumann type boundary conditions (and in fact more general ones as well).

Proof. By the maximum principle, we have $u^1 \geq u^o$ in $\bar{\mathcal{O}}_1$ since $u^1 \geq u^o$ on $\partial\mathcal{O}_1$ and u^o satisfies (16). Since $u^1 = u^o$ on $\bar{\mathcal{O}}-\bar{\mathcal{O}}_1$, we also have $u^1 \geq u^o$ in $\bar{\mathcal{O}}$. Next, we have to show

$$- \Delta u^1 \leq f \qquad \text{in } \mathcal{D}'(\mathcal{O}) \quad .$$

This is clearly the case in $\mathcal{O}_1$ by definition of u^1 and it is also the case in $\mathcal{O}-\bar{\mathcal{O}}_1$ since u^1 agrees with u^o there and u^o satisfies (16). Thus, we just have to check that this claim is still true "across γ_1 ". But this follows from a general observation due to H. Berestycki and P.L. Lions [40] since $u^1 \geq u^o$ in $\mathcal{O}$ and $u^1 = u^o$ on γ_1 . We prove the corresponding properties of u^n for $n \geq 2$ similarly.

Hence, $u^n(x)$ is an increasing sequence for each n . Since, by the maximum principle $u^n \leq u$ in $\bar{\mathcal{O}}$ for all $n \geq 0$, $u^n(x)$ converges to some $\tilde{u}$ and $\tilde{u}$ is a bounded function on $\bar{\mathcal{O}}$. Furthermore, since $u^n \leq \tilde{u} \leq u$ in $\bar{\mathcal{O}}$ and $u \in C_o(\bar{\mathcal{O}})$, $\tilde{u}$ vanishes on $\partial\mathcal{O} - \bar{\gamma}_1 \cap \bar{\gamma}_2$ and is continuous at points of $\partial\mathcal{O} - \bar{\gamma}_1 \cap \bar{\gamma}_2$. In addition, by elliptic regularity, u^{2n+1} converges uniformly on compact subsets of $\mathcal{O}_1$ and u^{2n} converges uniformly on compact subsets of $\mathcal{O}_2$. And since u^n is increasing and $\mathcal{O}_1 \cup \mathcal{O}_2 = \mathcal{O}$, this implies that u^n converges uniformly to $\tilde{u}$ on compact subsets of $\mathcal{O}$, therefore $\tilde{u} \in C(\bar{\mathcal{O}})$.

Finally, $\tilde{u}$ satisfies

$$- \Delta\tilde{u} = f \quad \text{in } \mathcal{D}'(\mathcal{O}_1) \quad , \quad - \Delta\tilde{u} = f \quad \text{in } \mathcal{D}'(\mathcal{O}_2)$$

since u^{2n+1}, u^{2n} satisfy respectively these equations. But $\mathcal{O}_1 \cup \mathcal{O}_2 = \mathcal{O}$, therefore we deduce

$$- \Delta\tilde{u} = f \qquad \text{in } \mathcal{D}'(\mathcal{O})$$

hence $\tilde{u} \equiv u$ in $\mathcal{O}$ and we conclude. Δ

Theorem 4 is clearly a convergence result, however it is restricted to some special initial choices of u^o . Nevertheless, we can deduce the general case.

Corollary 5. Let $u^o \in C(\bar{\mathcal{O}})$. Then, u^n converges uniformly to u on $\bar{\mathcal{O}}$.

Proof. It is enough to introduce $\bar{u}^o, \underline{u}^o \in C(\bar{\mathcal{O}})$ satisfying

$$- \Delta \bar{u}^o \geqq f \quad \text{in} \quad \mathcal{D}'(\mathcal{O}) \quad , \quad \bar{u}^o \geqq u^o \quad \text{in} \quad \bar{\mathcal{O}}$$

$$- \Delta \underline{u}^o \leqq f \quad \text{in} \quad \mathcal{D}'(\mathcal{O}) \quad , \quad \underline{u}^o \leqq u^o \quad \text{in} \quad \bar{\mathcal{O}}$$

(the existence of $\bar{u}^o, \underline{u}^o$ is an easy exercise). Then, the Schwarz sequences generated by $\bar{u}^o, \underline{u}^o$ denoted respectively by $\bar{u}^n, \underline{u}_n$ satisfy by the maximum principle

$$\bar{u}^n \geqq u^n \geqq \underline{u}_n \qquad \text{in} \quad \bar{\mathcal{O}}$$

and we conclude applying Theorem 4 to $\bar{u}^n$ and $\underline{u}_n$. Δ

Let us emphasize that Corollary 5 applies to arbitrary decompositions of $\mathcal{O}$ into $\mathcal{O}_1, \mathcal{O}_2$ (such that $\mathcal{O}_1 \cup \mathcal{O}_2 = \mathcal{O}$, $\gamma_1 \cap \gamma_2 = \emptyset$) and that the same result holds for Neumann boundary conditions even if $\bar{\gamma}_1 \cap \bar{\gamma}_2 = \emptyset$, cases which were not always covered by the arguments given in sections 2-3.

In fact, using these arguments, one can even allow some "interior non overlapping", more precisely, in two dimensions for example, one can allow γ_1 and γ_2 to intersect at, say, a finite number of points like in figure 3 below.

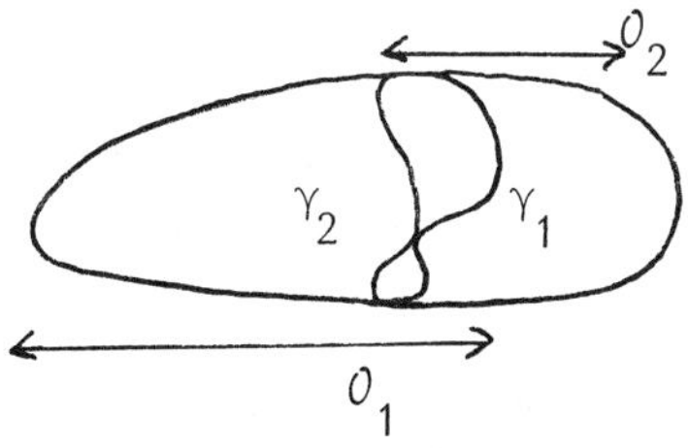

Figure 3.

Indeed, if we repeat the proof of Theorem 4, we find $\tilde{u} \in C(\mathcal{O} - S)$ where S is the finite set $S = \gamma_1 \cap \gamma_2$, $\tilde{u}$ is bounded on $\mathcal{O}$, $\tilde{u}$ vanishes on $\partial\mathcal{O} - \bar{\gamma}_1 \cap \bar{\gamma}_2$ and $- \Delta \tilde{u} = f$ in $\mathcal{D}'(\mathcal{O} - S)$. Then, since $\tilde{u}$ is bounded, the possible singularities at S are easily shown to be removable, hence

$$- \Delta \tilde{u} = f \qquad \text{in} \quad \mathcal{D}'(\mathcal{O})$$

and $\tilde{u} \equiv u$ on $\mathcal{O} - S$. Therefore, the same results as Theorem 4 and Corollary 5 hold in this more general situation provided we replace $\mathcal{O}$ by $\mathcal{O} - S$ in the convergence statements.

5. Degenerate equations and time-dependent problems.

We begin with degenerate second-order elliptic equations. We want to explain in this case how the preceding arguments show that the Schwarz alternating method does converge even for degenerate operators like

$$(20) \qquad A = - \sum_{i,j=1}^{N} a_{ij}\partial_{ij} - \sum_{i=1}^{N} b_i\partial_i + c$$

where $a_{ij} = \sum_{k=1}^{m} \sigma_{ik}\sigma_{jk}$, σ_{ik}, b_i are Lipschitz on $\bar{\mathcal{O}}$, c is continuous on $\mathcal{O}$ $(1 \leq i,j \leq N$, $1 \leq k \leq m)$ and

$$(21) \qquad c \geq c_o > 0 \quad \text{in } \bar{\mathcal{O}} .$$

We will not state precise results since to do so we would need to detail the way boundary conditions are imposed : this technical point can be handled using classical theory (see J.J. Kohn and L. Nirenberg [41] , D.W. Stroock and S.R.S. Varadhan [42]) or the more recent theory of viscosity solutions (see H. Ishii and P.L. Lions [46] for boundary conditions and uniqueness).

We just want to observe here that the method of sub and supersolutions given in section 4 gives the convergence in the general case (Dirichlet or Neumann boundary conditions and $\gamma_1 \cap \gamma_2 = \emptyset$) using the theory of viscosity solutions and in particular the fact that if u^o is a viscosity supersolution then $u^1 \leq u^o$ in $\bar{\mathcal{O}}_1$ and thus u^1 is a viscosity supersolution in $\mathcal{O}$ (see M.G. Crandall and P.L. Lions [43] ; M.G. Crandall, L.C. Evans and P.L. Lions [44] ; P.L. Lions [45] for more details on viscosity solutions). On the other hand, the method of section 3 to prove geometrical convergence fails for general degenerate elliptic equations - it is easy to build first-order operators yielding counter-examples - since it requires some form of strong maximum principle.

Finally, it is possible to adapt the method of section 2 i.e. in the case when $\mathcal{O}_1$ and $\mathcal{O}_2$ overlap. This is due to the following lemma.

Lemma 6. There exists some $\mu > 0$ depending only on c_o , diam $\mathcal{O}_1$ and bounds on b, a such that if $w \in C(\bar{\mathcal{O}}_1)$ solves (in viscosity sense or in classical sense if $u \in C^2(\mathcal{O}_1)$)

$$(22) \qquad Aw = 0 \text{ in } \mathcal{O}_1 , \quad \sup_{\gamma_1} \bar{w} \leq 1 , \quad w|_{\partial\mathcal{O}_1 \cap \partial\mathcal{O}} = 0$$

then we have

$$(23) \qquad \sup_{\gamma_2} w \leq \exp(-\mu\delta^2)$$

where $\delta = \text{dist}(\gamma_1,\gamma_2) = \text{Inf}\{|x-y| \;/\; x \in \gamma_1 , y \in \gamma_2\} > 0$.

Remarks. 1) The same result holds for Neumann boundary conditions.

2) The proof below gives some precise bound on μ .

3) In some sense, going from uniformly elliptic operator A to a degenerate one amounts to replace δ by δ^2 in the convergence rate.

Proof of Lemma 6. Let $x_o \in \gamma_2$. Without loss of generality, we may

assume that $x_o = 0$ and we denote by $\rho = \text{dist}\,(x_o,\gamma_1) = $
$= \text{Inf}\,\{|x_o - y| \;/\; y \in \gamma_1\}$. Then, we introduce

$$\bar{w}(x) = \exp \mu\{|x|^2 - \rho^2\}$$

where $\mu > 0$ will be determined later on. We then compute $A\bar{w}$ to find

$$A\bar{w} = \left(-2\mu \text{ Tr } a - 4\mu^2 \sum_{i,j=1}^{N} a_{ij}x_i x_j + 2\mu \sum_{i=1}^{N} b_i x_i + c\right) \bar{w}$$

while clearly $\bar{w} \geq 0$ on $\partial\mathcal{O}_1$ and $\bar{w} \geq 1$ on γ_1 . Hence, choosing small enough so that

$$c_o \geq 2\mu \|b\|_\infty (\text{diam } \mathcal{O}_1) + 2\mu \text{ Tr } a + 4\mu^2 \|\sigma\|_\infty^2 (\text{diam } \mathcal{O}_1)^2$$

we deduce $A\bar{w} \geq 0$ in $\mathcal{O}_1$. Therefore, by the maximum principle, we have

$$w \leq \bar{w} \quad \text{in } \mathcal{O}_1 \quad .$$

Hence, in particular, $\sup_{\gamma_2} w \leq \sup_{x_o \in \gamma_2} \exp\{-\mu \,\text{dist}(x_o,\gamma_2)^2\}$ and (23) is proven. Δ

A very particular case of the remarks made above is the case of parabolic equations with a spatial domain decomposition in $\mathcal{O}_1,\mathcal{O}_2$ i.e. for example

(24) $$\frac{\partial u}{\partial t} - \Delta u = f \text{ in } \mathcal{O} \;,\; u|_{\partial\mathcal{O} \times [0,T]} = 0 \;,\; u|_{t=0} = u_o \text{ in } \bar{\mathcal{O}}$$

where for instance $u_o \in C_o(\bar{\mathcal{O}})$, $f \in C(\bar{\mathcal{O}} \times [0,T])$. Then, if $u^o \in (\bar{\mathcal{O}} \times [0,T])$ satisfies (and this is not really necessary) $u^o|_{t=0} = u_o$ in $\bar{\mathcal{O}}$, $u^o|_{\partial\mathcal{O} \times [0,T]} = 0$, we define u^n by (for $n \geq 0$)

(25) $$\frac{\partial u^{2n+1}}{\partial t} - \Delta u^{2n+1} = f \text{ in } \mathcal{O}_1 \;,\; u^{2n+1}|_{t=0} = u_o \text{ in } \bar{\mathcal{O}}_1 \;,$$
$$u^{2n+1}|_{\partial\mathcal{O}_1 \times [0,T]} = u^{2n}|_{\partial\mathcal{O}_1 \times [0,T]}$$

(26) $$\frac{\partial u^{2n+2}}{\partial t} - \Delta u^{2n+2} = f \text{ in } \mathcal{O}_2 \;,\; u^{2n+2}|_{t=0} = u_o \text{ in } \bar{\mathcal{O}}_2 \;,$$
$$u^{2n+2}|_{\partial\mathcal{O}_2 \times [0,T]} = u^{2n+1}|_{\partial\mathcal{O}_2 \times [0,T]}$$

and we extend u^{2n+1},u^{2n+2} to $\bar{\mathcal{O}} \times [0,T]$ by u^{2n},u^{2n+1} respectively. Then, all the arguments introduced in sections 2-3-4 apply in this case.

A more interesting situation occurs when we combine Schwarz alternating method with a time discretization. Four possible combinations were proposed in Part I [36] . Let us just mention here that using maximum principle arguments as in the preceding sections, one can prove convergence and error bounds for these methods. We will not pursue this direction here to restrict the length of this paper.

6. Convergence proof via the stochastic interpretation.

Using the notations of section 1, we already observe that since $\tau_x^n \uparrow_n$ and $\tau_x^n \leqq \tau_x < \infty$ a.s. , $\tau_x^n \uparrow_n \sigma \leqq \tau_x$ where σ is a stopping time. Furthermore, since the Brownian paths i.e. the trajectories of $x+B_t$ are continuous (in t) a.s. we have

$$x+B_\sigma \in \partial\mathcal{O}_1 \cap \partial\mathcal{O}_2 \quad \text{a.s.}$$

Therefore, if $\gamma_1 \cap \gamma_2 = \emptyset$, this implies that $x \in B_\sigma \in \partial\mathcal{O}$ a.s. i.e. $\sigma \geqq \tau$ and we conclude

$$\sigma = \tau_x \quad \text{a.s.}$$

And thus we have proven immediately that :

1) for all $x \in \overline{\mathcal{O}}$, $\tau_x^n \uparrow_n \tau_x$ a.s.,

2) hence, by Lebesgue's lemma, $u^n(x) \to_n u(x)$ for all $x \in \overline{\mathcal{O}}$,

3) this convergence is uniform in $\overline{\mathcal{O}}$. For the last claim, we have to show

(27) $$\sup_{x \in \overline{\mathcal{O}}} E\left[\tau_x - \tau_x^n\right] \to_n 0 .$$

But this follows from 2) and Dini's lemma, choosing $u^0 \equiv 0$, $f \equiv 1$ so that $u^n(x) = E\left[\tau_x^n\right]$.

This (striking) proof of convergence clearly adapts to more general situations (general elliptic operators, possibly degenerate, other boundary conditions...). Next, we wish to push these stochastic arguments in order to obtain some estimates on the rate of convergence of u^n to u , that is the rate of convergence of τ_x^n to τ_x . In order to simplify the presentation, we will always assume that $\mathcal{O}_1$ and $\mathcal{O}_2$ overlap i.e. $\bar\gamma_1 \cap \bar\gamma_2 = \emptyset$. We first prove the

Theorem 7. There exist $k_1, k_2 \in (0,1)$ depending only on $\mathcal{O}_1, \mathcal{O}_2$ such that

(28) $$\sup_{x \in \bar\gamma_j} P\left[\sigma_x^i < \tau_x\right] \leqq k_i \quad \text{for } i,j = 1,2 \ , \ i \neq j \ ,$$

where σ_x^i is the first exit time from $\mathcal{O}_i$. In particular, we deduce

(29) $$\sup_{x \in \overline{\mathcal{O}}} P\left[\tau_x > \tau_x^{2n+1}\right] \leqq (k_1 k_2)^n$$

(30) $$\sup_{x \in \overline{\mathcal{O}}} P\left[\tau_x > \tau_x^{2n+2}\right] \leqq k_1^n k_2^{n+1} .$$

Remarks. 1) As usual, the proof given below is valid for general boundary conditions and general uniformly elliptic equations i.e. general nondegenerate diffusion processes.

2) Recalling that $E\left[e^{\lambda\tau_x}\right] \in C(\overline{\mathcal{O}})$ for all $\lambda < \lambda_1$ (where λ_1

is the first eigenvalue of $-\Delta$ in $H_0^1(\mathcal{O})$ - see for instance [39] - we deduce easily some estimate on the convergence of u^n to u since

$$|u^n(x)-u(x)| \leqq C\, E\,[\tau_x-\tau_x^n] + C\, P\,[\tau_x > \tau_x^n]\;,\; \forall\, x \in \bar{\mathcal{O}}$$

and

$$E\,[\tau_x-\tau_x^n] \leqq P\,[\tau_x > \tau_x^n]^{1/\alpha}\, E\,[\tau_x^{\alpha'}]^{1/\alpha'}$$

$$\text{for all } \alpha > 1 \;,\; \alpha' = \frac{\alpha}{\alpha-1} \;.$$

3) As we will see in the proof below, the crucial estimate (28) follows from some knowledge of the support of diffusion processes. Such knowledge - and the available additionnal informations such as "invariant measures"... - should play an important role in an attempt to determine "the optimal decomposition of $\mathcal{O}$ into $\mathcal{O}_1$ and $\mathcal{O}_2$ ". Δ

Proof of Theorem 7. The estimates (29) and (30) follow easily from (28) and the strong Markov property. Next, using the continuity of σ_x^i, τ_x in x, we see that in order to prove (28) we just have to show for all $x \in \bar{\gamma}_j$ $(i = 1,2)$

$$P\left(x + B_{\sigma_x^i} \notin \bar{\gamma}_i\right) > 0 \qquad \text{with} \quad j \neq i \;.$$

To this end, we choose a continuous trajectory (function) ω from $[0,\infty)$ into $\mathbb{R}^N$ such that $\omega(0) = x$, $\omega(1) \in \partial\mathcal{O}_i - \bar{\gamma}_i$, $\omega(t) \notin \bar{\mathcal{O}}_i$ for all $t > 1$, $\omega(t) \in \mathcal{O}_i$ for all $t \in (0,1)$. Then, by a famous result due to D.W. Stroock and S.R.S. Varadhan [47], we have for all $T < \infty$, $\varepsilon > 0$

$$P\left[\sup_{t\,\in\,[0,T]} |x+B_t-\omega(t)| \leqq \varepsilon\right] > 0 \;.$$

We next choose $T = 2$, $\underline{\varepsilon}$ small enough so that $\varepsilon > \text{dist}\,(\omega(2),\bar{\mathcal{O}}_i)$, $\varepsilon < \min_{t\,\in\,[0,T]} \text{dist}\,(\omega(t),\bar{\gamma}_i)$. Then, denoting by

$$\Omega_\varepsilon = \left\{\sup_{t\,\in\,[0,T]} |x + B_t - \omega(t)| \leqq \varepsilon\right\} \;,$$

we see that on Ω_ε, $x+B_T \notin \bar{\mathcal{O}}_i$ hence $\sigma_x^i \leqq T$ and thus $x+B_{\sigma_x^i} \notin \bar{\gamma}_j$. And our claim is proven. Δ

It is also possible to give a pure probabilistic proof of the convergence result implied by Lemma 6 for degenerate elliptic equations. Recall that $\mathcal{O}_1$ and $\mathcal{O}_2$ overlap. It is easy to check that Lemma 6 follows from the following inequality

$$E\,[e^{-c_0\sigma_x^1}\, 1_{\sigma_x^1 < \tau_x}] \leqq e^{-\mu\delta^2} \;, \qquad \forall\, x \in \gamma_2 \;,$$

(where σ_x^1, τ_x are defined as before but correspond now to the relevant diffusion process), inequality which can be proved directly by probabilistic arguments. The mere fact that the right-hand side is strictly less than 1 is obvious intuitively : indeed if $\sigma_x^1 < \tau_x$, this means that the diffusion process starting from $x \in \gamma_2$ exists from $\mathcal{O}_1$ by γ_1 and since $\delta = \text{dist}\,(\gamma_1,\gamma_2) > 0$, it "takes some time" to cross that distance. The phenomenon involved here is much less subtle than in Theorem 7.

7. Nonlinear problems.

First of all, we want to observe that all the convergence arguments given in sections 2-5 are still valid for monotone nonlinear equations of the form

$$(31) \qquad - \Delta u + \beta(u) \ni f \quad \text{in } \mathcal{O} , \qquad u|_{\partial\mathcal{O}} = 0$$

where β is a maximal monotone graph such that $0 \in \text{Dom}\,(\beta)$ - of course, (31) means in particular that $u(x) \in \text{Dom}\,(\beta)$ a.e. in $\mathcal{O}$. A particular an important example is given by $\beta(t) = \emptyset$ if $t < 0$, $\beta(0) = [0,\infty)$, $\beta(t) = 0$ if $t > 0$: it corresponds to variational inequalities or obstacle problems (in this case the obstacle is 0 but more general functions could be considered as well). We then define the Schwarz method exactly as in section 1 replacing only the equations for u^n by the above equation (31)... And all the convergence results are easily adapted to this case. Notice also that obstacle problems have a stochastic interpretation in terms of optimal stopping, more precisely in the above example we have

$$u(x) = \inf_\theta \left\{ E \int_0^\theta f(x+B_t)dt \;/\; \theta \text{ stopping time} \right\} .$$

Then, it is not difficult to show that the Schwarz sequence u^n is then given by

$$u^n(x) = \inf_\theta \left\{ E \int_0^{\theta\wedge\tau_x^n} f(x+B_t)dt + u^0\left(x + B_{\theta\wedge\tau_x^n}\right) \;/\; \theta \text{ stopping time} \right\}$$

Hence, we can also use the probabilistic approach developed in section 6.

The other class of nonlinear second-order equations that can be analysed by our method is the class of fully nonlinear elliptic possibly degenerate second-order equations

$$(32) \qquad F(D^2u,Du,u,x) = 0 \qquad \text{in } \mathcal{O} .$$

This class contains as particular cases the first-order Hamilton-Jacobi equations - F is then independent of D^2u - which are the fundamental partial differential equations for optimal deterministic control and deterministic games (Bellman and Isaacs' equations), the Hamilton-Jacobi-Bellman equations of optimal stochastic control - F convex or concave in D^2u - and the Isaacs' equations of stochastic differential games. For all those problems, we can use the theory of viscosity solutions (as in [46]) to prove (similarly as we did in the preceding sections) convergence results for the Schwarz alternating method.

Notice also that the arguments of section 6 can be used together with the control or games interpretation of solutions to provide other convergence proofs.

We will give only one (extreme) example : we consider the Eikonal equation

(33) $$|\nabla u| = f \quad \text{in } \mathcal{O}\,, \qquad u = 0 \quad \text{on } \partial\mathcal{O}$$

where $f > 0$ in $\bar{\mathcal{O}}$, $f \in C(\bar{\mathcal{O}})$ and u is the unique viscosity solution of (33) - see M.G. Crandall and P.L. Lions [43]. For simplicity, we choose $u^o \equiv 0$. And we build the Schwarz sequence as follows

(34) $$|\nabla u^{2n+1}| = f \quad \text{in } \mathcal{O}_1 \,, \qquad u^{2n+1} = u^{2n} \quad \text{on } \partial\mathcal{O}_1$$

(35) $$|\nabla u^{2n+2}| = f \quad \text{in } \mathcal{O}_2 \,, \qquad u^{2n+2} = u^{2n+1} \quad \text{on } \partial\mathcal{O}_2$$

and we extend as usual u^{2n+1}, u^{2n+2} to $\bar{\mathcal{O}}$ by u^{2n}, u^{2n+1} respectively. Actually, some care is required to build such a sequence. Recall that we assume that $\gamma_1 \cap \gamma_2 = \emptyset$. To actually build (34),(35) we argue inductively and we assume by induction that $0 = u^o \leq u^1 \leq \ldots \leq u^n$ in $\bar{\mathcal{O}}$ and u^n is a viscosity subsolution of (33), then by comparison results on viscosity solutions, we deduce that $u^{n+1} \in C_o(\bar{\mathcal{O}})$, $u^{n+1} \geq u^n$ in $\bar{\mathcal{O}}_1$ if n is even, $\bar{\mathcal{O}}_2$ if n is odd. The fact that u^{n+1} viscosity solution of (34) or (35) exists follows from the results of P.L. Lions [48]. Then, one can check that u^{n+1} is also a viscosity subsolution of (33). Since u^n is bounded in Lipschitz norm and is increasing with respect to n, we deduce that u^n converges uniformly to some $\tilde{u} \in C_o(\bar{\mathcal{O}})$. Furthermore, by the stability results of viscosity solutions, $\tilde{u}$ is a viscosity solution of (33) in $\mathcal{O}_1$ and in $\mathcal{O}_2$ therefore in $\mathcal{O}$. And $\tilde{u} \equiv u$, proving thus the convergence.

8. Multidomains decomposition.

We want now to consider extensions of the classical Schwarz alternating method to the case of a decomposition of $\mathcal{O}$ into more than two subdomains namely

$$\mathcal{O} = \bigcup_{i=1}^{m} \mathcal{O}_i$$

where $m \geq 1$, $\mathcal{O}_1,\ldots,\mathcal{O}_m$ are open sets. We begin with the simple geometrical situation where

(36) $$\bar{\mathcal{O}}_i \cap \bar{\mathcal{O}}_j \cap \bar{\mathcal{O}}_k = \emptyset \quad \text{whenever } 1 \leq i,j,k \leq m\,, \quad i \neq j \neq k \neq i\,.$$

Some examples where (36) holds are given by figures 4.a,4.b below while some examples where (36) does not hold is given by figure 5 below.

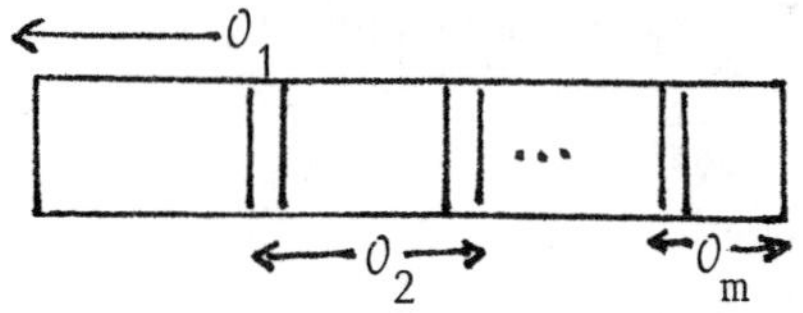

Figure 4.a

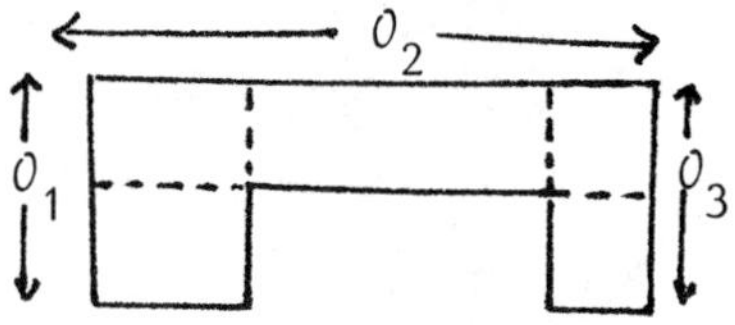

Figure 4.b

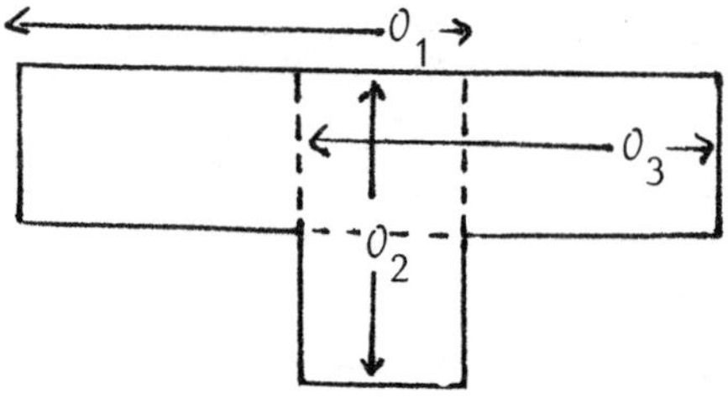

Figure 5.

Then, for all $1 \leqq i \neq j \leqq m$, we denote by $\gamma_{ij} = \partial\mathcal{O}_j \cap \mathcal{O}_i$. We next introduce a (parallel) extension of Schwarz alternating method. For each $i \in \{1,\dots,m\}$, we choose $u_i^o \in C(\bar{\mathcal{O}}_i)$ such that $u_i^o = 0$ on $\partial\mathcal{O}_i \cap \partial\mathcal{O}$ and we build inductively sequences $u_i^n \in C(\bar{\mathcal{O}}_i)$ for $n \geqq 1$, $1 \leqq i \leqq m$, as follows

$$\text{(37)} \qquad \begin{cases} -\Delta u_i^{n+1} = f \quad \text{in } \mathcal{O}_i \,, \quad u_i^{n+1} = 0 \quad \text{on } \partial\mathcal{O} \cap \partial\mathcal{O}_i \,, \\ u_i^{n+1} = u_j^n \quad \text{on } \gamma_{ji} \quad \text{for all } j \neq i \,. \end{cases}$$

We begin with the analysis of the convergence of u_i^n to $u|_{\bar{\mathcal{O}}_i}$ in the case when $\mathcal{O}_i$ and $\mathcal{O}_j$ overlap for all $i \neq j$, $(\bar\gamma_{ij} \cap \bar\gamma_{ji} = \emptyset)$ or when $\mathcal{O}_i$ and $\mathcal{O}_j$ weakly overlap for all $i \neq j$ (if $\bar\gamma_{ij} \cap \bar\gamma_{ji} \neq \emptyset$, $\bar\gamma_{ij}$ and $\bar\gamma_{ji}$ "make a positive angle" at intersection points). In both cases by the arguments of section 2-3, we find some $k \in (0,1)$ such that for all $1 \leqq i \neq j \leqq m$ we have

$$\sup_{\gamma_{ij}} |u - u_i^{n+1}| \leqq k \max_{\ell \neq i} \sup_{\gamma_{\ell i}} |u - u_\ell^n| \,.$$

In particular, this yields

$$\max_{i \neq j} \sup_{\gamma_{ij}} |u - u_i^{n+1}| \leqq k \max_{i \neq j} \sup_{\gamma_{ij}} |u - u_i^n|$$

hence

$$\text{(38)} \qquad \max_{i \neq j} \sup_{\gamma_{ij}} |u - u_i^n| \leqq k^n \max_{i \neq j} \sup_{\gamma_{ij}} |u - u_i^o| \,.$$

And since, by the maximum principle, we have for all i

$$|u-u_i^{n+1}| \leqq \max_{j \neq i} \max_{\gamma_{ji}} |u-u_j^n| \qquad \text{in } \bar{\mathcal{O}}_i$$

we deduce from (38) the following inequality

$$\max_i \sup_{\mathcal{O}_i} |u-u_i^{n+1}| \leqq k^n \max_{i \neq j} \sup_{\gamma_{ij}} |u-u_i^o| \ . \tag{39}$$

It is also worth explaining how it is possible to adapt the arguments of section 4 to this case, proving thus in particular the convergence in the general case when $(\mathcal{O}_i)_{1 \leqq i \leqq m}$ only satisfy $\gamma_{ij} \cap \gamma_{ji} = \emptyset$ for all $1 \leqq i \neq j \leqq m$. Indeed, let $u^o \in C(\bar{\mathcal{O}})$ be a supersolution (resp. subsolution) of (2) and choose $u_i^o = u^o|_{\bar{\mathcal{O}}_i}$ for all $1 \leqq i \leqq m$. Then, we claim that for all $1 \leqq i \neq j \leqq m$, $n \geqq 0$ we have

$$u_i^{n+1} \leqq u_j^n \quad \text{in } \bar{\mathcal{O}}_i \cap \bar{\mathcal{O}}_j \quad , \quad u_i^{n+1} \leqq u_i^n \quad \text{in } \bar{\mathcal{O}}_i \ . \tag{40}$$

This can be show easily by induction since once the second claim is proven, the first one follows remarking that u_i^{n+1} and u_j^n solve the same equation in $\mathcal{O}_i \cap \mathcal{O}_j$, $u_i^{n+1} = u_j^n$ on γ_{ji} and $u_j^n = u_i^{n-1} \geqq u_i^n$ on γ_{ij} . From (40), we deduce easily that u_i^n converges as n goes to $+\infty$ to some u_i and $u_i = u_j$ on $\mathcal{O}_i \cap \mathcal{O}_j$ for all $1 \leqq i \neq j \leqq m$. This allows to define some $\tilde{u}$ which is bounded, continuous on $\bar{\mathcal{O}} - \bigcup_{i \neq j} (\partial\mathcal{O}_i \cap \partial\mathcal{O}_j)$ and satisfies

$$-\Delta\tilde{u} = f \qquad \text{in } \mathcal{O} \ .$$

From this we deduce that $\tilde{u} \equiv u$ and the convergence is proven in this case.

A final remark concerning the situation when (36) holds is the possibility of allowing more flexibility in the iterative method (37) and more precisely of replacing the boundary condition on each γ_{ji} by

$$u_i^{n+1} = u_j^p \quad \text{with } p = p(n,i,j) \text{ , on } \gamma_{ji} \text{ for all } j \neq i \ . \tag{41}$$

We then assume that $p(n,i,j)$ is nondecreasing with respect to n , $p(n,i,j) \leqq n$ for all $n \geqq 0$, $p(n,i,j) \to +\infty$ as $n \to +\infty$, for all $1 \leqq i \neq j \leqq m$. For instance, the first proof we made above is modified as follows

$$\max_{i \neq j} \sup_{\gamma_{ij}} |u-u_i^{n+1}| \leqq k \max_{i \neq j} \sup_{\gamma_{ij}} |u-u_i^{p(n,i,j)}|$$

hence

$$\overline{\lim_n} \max_{i \neq j} \sup_{\gamma_{ij}} |u-u_i^{n+1}| \leqq k \overline{\lim_n} \sup_{\gamma_{i_o j_o}} |u-u_{i_o}^{r(n)}|$$

for some $1 \leq i_o \neq j_o \leq m$, and for some sequence $r(n) \underset{n}{\to} +\infty$ as $n \to +\infty$. And this yields : $\overline{\lim_n} \max_{i \neq j} \sup_{\gamma_{ij}} |u-u_i^{n+1}| = 0$, and the convergence is proven.

We now turn to more general geometrical situations than (36). An example is given by figure 5 and another meaningful example is given by the following.

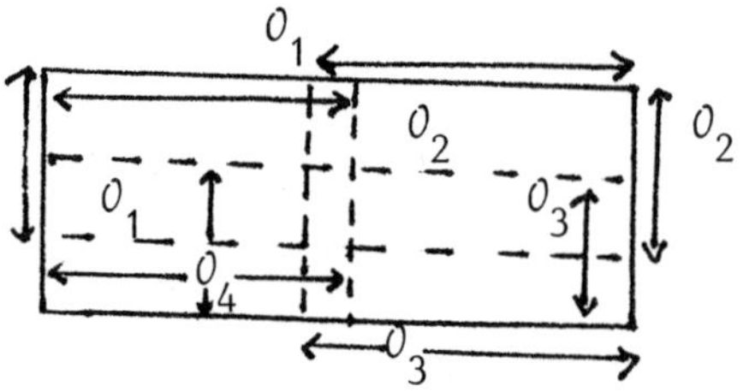

Figure 6.

We will need some restrictions on the decomposition : in two dimensions we will need to assume that $\partial O_i \cap \partial O_j \cap \overline{O}_k$ is a finite set for all i,j,k distinct in $\{1,\ldots,m\}$. In higher dimensions, similar conditions involving the dimensionality of such intersection sets are needed but we will skip them for the sake of simplicity.

We next have to explain how we modify (37) in this general situation : we will use the same equation as in (37) but we modify the boundary condition as follows

$$(42)\quad \begin{cases} u_i^{n+1} = u_{j_1}^n \quad \text{on } \partial O_i \cap O_{j_1} \cap \left[\bigcup_{j \neq i,j_1} O_j^c\right] \text{ for all } j_i \neq i \\ u_i^{n+1} = u_{j_1 j_2}^n \quad \text{on } \partial O_i \cap O_{j_1} \cap O_{j_2} \cap \left[\bigcup_{j \neq i,j_1,j_2} O_j^c\right] \\ \qquad\qquad \text{for all } j_1 \neq j_2 \neq i \neq j_1 \\ u_i^{n+1} = u_{j_1\cdots j_{m-1}}^n \quad \text{on } \partial O_i \cap O_{j_1} \cap O_{j_{m-1}} \\ \qquad\qquad \text{where } \{i,j_1,\ldots,j_{m-1}\} = \{1,\ldots,m\} \end{cases}$$

and $u_{j_1\cdots j_k}^n$ (for all $k \geq 2$) is chosen to satisfy (arbitrarily)

$$(43)\qquad \operatorname{Min}\left(u_{j_1}^n,\ldots,u_{j_k}^n\right) \leq u_{j_1\cdots j_k}^n \leq \operatorname{Max}\left(u_{j_1}^n,\ldots,u_{j_k}^n\right) .$$

To prove the convergence of this method, we first observe that if we choose $\bar{u}_o, \underline{u}_o$ supersolution of (2) (resp. subsolution of (2)) such that $\bar{u}_o \geq u_i^o$ on $\overline{O}_i$, $\underline{u}_o \leq u_i^o$ on $\overline{O}_i$ for all $1 \leq i \leq m$, then it is easy to show

$$(44)\qquad \underline{u}_i^n \leq u_i^n \leq \bar{u}_i^n \quad \text{for all } n \geq 0 ,\ i \in \{1,\ldots,m\}$$

where $\underline{u}_i^n, \bar{u}_i^n$ are the sequences generated by the above iterative method with the special choices

$$(45)\qquad \begin{cases} \bar{u}^n_{j_1\ldots j_k} = \operatorname{Max}\left(\bar{u}^n_{j_1},\ldots,\bar{u}^n_{j_k}\right) \\ \underline{u}^n_{j_1\ldots j_k} = \operatorname{Min}\left(\underline{u}^n_{j_1},\ldots,\underline{u}^n_{j_k}\right) \end{cases}$$

for all $n \geqq 0$, $k \geqq 2$, $j_1\ldots j_k$ distinct in $\{1,\ldots,m\}$. Therefore, it is enough to prove the convergence of $\underline{u}^n_i, \bar{u}^n_i$ and we will do so for $\bar{u}^n_i$ (for instance). Exactly as in section 4, because $\bar{u}^o$ is a supersolution, one checks easily that $(\bar{u}^n_i)_n$ is bounded in $L^\infty(\mathcal{O}_i)$ and

$$(46)\qquad u \leqq \bar{u}^{n+1}_i \leqq \bar{u}^n_i \quad \text{in } \mathcal{O}_i \text{ for all } n \geqq 0,\ i \in \{1,\ldots,m\}.$$

Therefore, $\bar{u}^n_i$ converges uniformly on compact subsets of $\mathcal{O}_i$ to some $\bar{u}_i \in L^\infty(\mathcal{O}_i) \cap C(\mathcal{O}_i)$ which is continuous and vanishes on $\partial\mathcal{O}_i \cap \partial\mathcal{O}$ except maybe at a finite number of points and which satisfies

$$-\Delta u_i = f \quad \text{in } \mathcal{O}_i \quad , \quad u \leqq u_i \quad \text{in } \mathcal{O}_i \quad .$$

Furthermore, there exists a finite set S contained in $\mathcal{O}$ such that for all $i, j_1,\ldots,j_k$ distinct in $\{1,\ldots,m\}$, for all $k \in \{1,\ldots,m-1\}$, $u_i, u_{j_1},\ldots,u_{j_k}$ are continuous on $\partial\mathcal{O}_i \cap \mathcal{O}_{j_1} \cap \ldots \cap \mathcal{O}_{j_k} \cap \left(\bigcup_{j \neq i, j_1,\ldots,j_k}^{i\quad j_k} \mathcal{O}^c_j\right) \cap S^c$ and we have there :

$u_i = \max\left(u_{j_1},\ldots,u_{j_k}\right)$.

We then introduce $w = \max\left(u_{i_1},\ldots,u_{i_k}\right)$ on $\bar{\mathcal{O}}_{i_1} \cap \ldots \cap \bar{\mathcal{O}}_{i_k} \cap \left(\bigcup_{i \neq i_1,\ldots,i_k} \mathcal{O}^c_i\right)$ for all $i_1,\ldots,i_k$ distinct in $\{1,\ldots,m\}$, $k \in \{1,\ldots,m\}$ and we observe that w is bounded on $\mathcal{O}$, continuous on $\partial\mathcal{O}$ except maybe at a finite number of points and satisfies

$$-\Delta w \leqq f \quad \text{in } \mathcal{D}'(\mathcal{O}-S) \quad , \quad u \leqq u_i \leqq w \quad \text{in } \mathcal{O}_i \quad \text{for all } i \in \{1,\ldots,m\} .$$

Because w is bounded and S is finite this yields

$$-\Delta w \leqq f \qquad \text{in } \mathcal{D}'(\mathcal{O})$$

and together with the boundary condition, we deduce $w \leqq u$ in $\mathcal{O}$. Hence $w \equiv u$ in $\mathcal{O}-S$ and in particular $u_i \equiv u$ in $\mathcal{O}_i$, proving thus the convergence.

In some very particular geometrical stituations like the one given by figure 5, we can improve the above convergence proof : indeed, in this case, the same proof as in the case when (36) holds applies and we obtain in this way geometrical convergence.

REFERENCES

[1] H.A. Schwarz : Über einige Abbildungsaufgaben. Ges. Math. Abh., 11 (1869), p. 65-83.

[2] S.L. Sobolev : The Schwarz algorithm in the theory of elasticity. Dokl. Acad. N. USSE, Vol IV (XIII), 1936, p. 236-238 (in Russian).

[3] S.G. Michlin : On the Schwarz algorithm. Dokl. Acad. N. USSR, 77 (1951), p. 569-571 (in Russian).

[4] M. Prager : Schwarzův algoritmus pro polyharmonické funkce. Aplikace matematiky, 3 (1958), 2, p. 106-114.

[5] D. Morgenstern : Begründung des alternierenden Verfahrens durch Orthogonalprojektion. ZAMM, 36 (1956), p. 7-8.

[6] I. Babuska : On the Schwarz algorithm in the theory of differential equations of Mathematical Physics. Tchecosl. Math. J., 8 (83), 1958, p. 328-342 (in Russian).

[7] R. Courant and D. Hilbert : Methoden der matematischen Physik, Vol. 2, Berlin, 1937.

[8] F.E. Browder : On some approximation methods for solutions of the Dirichlet problem for linear elliptic equations of arbitrary order. J. Math. Mech., 7 (1958), p. 69-80.

[9] P.L. Lions : Unpublished notes, 1978.

[10] R. Glowinski, J. Périaux and Q.V. Dinh : Domain decomposition methods for nonlinear problems in fluid dynamics. Rapport INRIA, 147, 1982, to appear in Comp. Meth. Appl. Mech. Eng..

[11] Q.V. Dinh, R. Glowinski and J. Périaux : Solving elliptic problems by decomposition methods with applications. In Elliptic Problem Solvers II, Academic Press, New-York, 1982.

[12] R. Glowinski : Numerical solution of partial differential equation problems by domain decomposition. Implementation on an array processors system. In Proceeding of International Symposium on Applied Mathematics and Information Science, Kyoto University, 1982.

[13] A. Fischler : Résolution du problème de Stokes par une méthode de décomposition de domaines. Application à la simulation numérique d'écoulements de Navier-Stokes de fluides incompressibles en éléments finis. Thèse de 3e cycle, Univ. P. et M. Curie, 1985, Paris.

[14] Q.V. Dinh, A. Fischler, R. Glowinski and J. Périaux : Domain décomposition methods for the Stokes problem. Application to the Navier-Stokes equations. In Numeta 85, Swansea, 1985.

[15] Q.V. Dinh, J. Périaux, G. Terrasson and R. Glowinski : On the coupling of incompressible viscous flows and incompressible potential lows via domain decomposition. In ICNMFD, Pékin, 1986.

[16] J.M. Frailong and J. Pakleza : Resolution of a general partial differential equation on a fixed size SIMD/MIMD large cellular processor. In Proceedings of the IMACS International Congress, Sorente, 1979.

[17] Q.V. Dinh : Simulation numérique en éléments finis d'écoulements de fluides visqueux incompressibles par une méthode de

décomposition de domaines sur processeurs vectoriels. Thèse de 3e cycle, Univ. P. et M. Curie, 1982, Paris.

[18] P.E. Bjorstad and O.B. Widlung : Solving elliptic problems on regions partitioned into substructures. In Elliptic Problem Solvers II, Academic Press, New York, 1982.

[19] P. Lemonnier : Résolution numérique d'équations aux dérivées partielles par décomposition et coordination. Rapport INRIA, 1972.

[20] L. Cambier, W. Ghazzi, J.P. Veillot and H. Viviand : A multi-domain approach for the computation of viscous transonic flows by unsteady methods. In Recent advances in numerical methods in fluids. Vol. 3, Pineridge, Swansea, 1984.

[21] P. Anceaux, G. Gay, R. Glowinski, J. Périaux : Résolution spectrale du problème de Stokes et coordination. Euromech. 159, Spectral methods in computational fluid mechanics, Nice, 1982.

[22] J.P. Benque, J.P. Grégoire, A. Hauguel and M. Maxant : Application des méthodes de décomposition aux calculs numériques en hydraulique industrielle. In 6e Colloque International sur les méthodes de Calcul Scientifique et Technique, Versailles, 1983.

[23] Q.V. Dinh, R. Glowinski, B. Mantel, J. Périaux and P. Perrier : Subdomain solutions of nonlinear problems in fluid dynamics on paralell processors. In 5th International Symposium on computational methods in applied sciences and engineering, Versailles, 1981, North-Holland.

[24] G.I. Marchuk, Yu A. Kuznetsov and A.M. Matsokin : Fictitions domain and domain decomposition methods. Sov. J. Numer. Anal. Math. Modelling, 1 (1986), p. 3-35.

[25] M. Dryja : Domain decompsotion method for solving variational difference systems for elliptic problems. In Variatsionnonaznostnye metody v matematicheskoi fizike, Eds. N.S. Bakhvalov and Yu A. Kuznetsov. Otdel Vichislitel'noi Matematiki Akad. Nauk SSSR, Moscow, 1984 (in Russian).

[26] M. Dryja : A finite element - capacitance matrix method for the elliptic problem. SIAM J. Number. Anal., 20 (1983), p. 671-680.

[27] A.M. Matsokin : Fictitions component method and the modified difference analogue of the Schwarz method. In Vychislitel'nye Metody Lineinoi Algebry, Ed. G.I. Marchuk, Vychisl. Tsentr. Sib. Otdel. Akad. Nauk SSSR, Novosibirsk, 1980 (in Russian).

[28] A.M. Matsokin and S.V. Nepomnyashchikh : On the convergence of the alternating subdomain Schwartz method without intersections. In Metody Interpolyatsii i Approksimatsii, ed. Yu A. Kuznetsov. Vychisl. Tsentr. Sib. Otdel. Akad. Nauk SSSR, Novosibirsk, 1981 (in Russian).

[29] J.B. Baillon and P.L. Lions : Convergence de suites de contractions dans un espace de Hilbert. Rapport Ecole Polytechnique, Palaiseau, 1977.

[30] J. Céa : Optimisation. Théorie et algorithmes. Dunod, Paris, 1971.

[31] Z. Opial : Weak convergence of the sequence of successive approximations for non-expansive mappings. Bull. Amer. Math. Soc. 73 (1967), p. 591-597.

[32] Z. Cai and S. Mac Cormick : Computational complexity of the Schwarz alternating procedure. Preprint.

[33] Z. Cai and B. Cuo : Error estimate of the Schwarz alternating procedure on L -shape domain. Appl. Math. Comp., to appear.

[34] P. Bjorstad and O. Widlund : Iterative methods for the solution of elliptic problems on regions partitioned into substructures. SIAM Num. Anal., 23 (1986), p. 1097-1120.

[35] J. Olger, W. Shamarock and W. Tang : Convergence analysis and acceleration of the Schwarz alternating method. Technical report, Stanford Univ., Center for Large Scale Scientific Computation, 1986.

[36] P.L. Lions : On the Schwarz alternating method. I. In First International Symposium on Domain Decomposition methods for Partial Differential Equations. SIAM, Philadelphia, 1988.

[37] R. Dautray and J.L. Lions : Analyse mathématique et calcul numérique pour les sciences et techniques. Tome 1. Masson, Paris, 1985.

[38] P.L. Lions : Interprétation stochastique de la méthode alternée de Schwarz. C.R. Acad. Sci. Paris, 268 (1978), p. 325-328.

[39] P.L. Lions : Problèmes elliptiques du deuxième ordre non sous forme divergente. Proc. Roy. Soc. Edin., 84 A (1979), p. 263-271.

[40] H. Berestycki and P.L. Lions : Some applications of the method of super and subsolutions. In Bifurcation and Nonlinear Eigenvalue Problems, Springer Lecture Notes in Math., Berlin, 1980;

[41] J.J. Kohn and L. Nirenberg : Noncoercive boundary value problems. Comm. Pure Appl. Math, 18 (1965), p. 443-402.

[42] D.W. Stroock and S.R.S. Varadhan : On degenerate elliptic-parabolic operators od second-order and their associated diffusions. Comm. Pure Appl. Math., 25 (1972), p. 651-674.

[43] M.G. Crandall and P.L. Lions : Viscosity solutions of Hamilton-Jacobi equations. Trans. Amer. Math. Soc., 277 (1983), p. 1-42.

[44] M.G. Crandall, L.C. Evans and P.L. Lions : Some properties of viscosity solutions of Hamilton-Jacobi equations. Trans. Amer. Math. Soc., 282 (1984), p. 487-502.

[45] P.L. Lions : Optimal control of diffusion processes and Hamilton-Jacobi-Bellman equations. Part II. Comm. P.D.E., 8 (1983), p. 1229-1276.

[46] H. Ishii and P.L. Lions : Viscosity solutions of fully nonlinear second-order elliptic partial differential equations. Preprint.

[47] D.W. Stroock and R.S.R. Varadhan : On the support of diffusion processes with applications to the strong maximum principle. Proceedings of Sixth Berkeley Symposium on Mathematical Statistics and Probability, Vol. III, 1970.

[48] P.L. Lions : Generalized solutions of Hamilton-Jacobi equations. Pitman, London, 1982.

Parallel Algorithms for Solving Partial Differential Equations

T. Lu*
T. M. Shih†
C. B. Liem†

Abstract. In §1 to §5, two synchronized parallel algorithms for the solution of boundary value problems of partial differential equations are proved. Algorithm 1 is based on the minimum modulus principle; therefore, it can be applied to nonlinear PDE's. At such cases, a linearization step should be done before each iteration. The proof of the convergence uses the iterative method of groupwise projection. Algorithm 2 is based on the discrete maximum principle. A synchronized parallel algorithm for the Dirichlet problem of linear equations satisfying the uniformly elliptic condition is given in §6.

§1 The parallel algorithm 1

The boundary value problem of a partial differential equation, in general, can be written as

$$\begin{cases} Lu = f, & \text{in } \Omega \\ lu = g, & \text{on } \partial\Omega \end{cases} \tag{1}$$

Ω is a bounded region with boundary $\partial\Omega$. L is a differential operator and l is a boundary operator. We may apply either the finite difference or the finite element method to (1) and obtain an algebraic system

*Institute of Mathematical Sciences, Academia Sinica, Chengdu, China.

† Department of Mathematical Studies, Hong Kong Polytechnic, Hong Kong.

$$L^h u^h(P_j) = \sum_{i \in I} C_{ji}\, u^h(P_i) = f_j, \qquad j \in I \tag{2}$$

where I is the set of indices of all grid points in Ω_h and Ω_h is the set of all grid points inside Ω.

We define the discrete neighbourhood

$$N_j \equiv N(P_j) \equiv \{P_i;\ C_{ji} \neq 0\}.$$

Then (2) becomes

$$L^h u^h(P_j) = \sum_{P_i \in N_j} C_{ji}\, u^h(P_i) = f_j, \qquad j \in I \tag{3}$$

In order to solve (3) by a parallel algorithm, we divide Ω_h into m subsets $\Omega_h^1, \ldots, \Omega_h^m$, $\Omega_h = \bigcup_{i=1}^{m} \Omega_h^i$, where some of the subsets may be overlapping. In order to reduce the waiting time, every Ω_h^i, $i = 1, \ldots, m$ should contain nearly the same number of grid points.

Define the discrete neighbourhood of Ω_h^i as follows

$$N(\Omega_h^i) = \bigcup_{P \in \Omega_h^i} N(P).$$

P_j is called a k-multiple point, denoted by $P_j \in \pi_k$, if there exists at most k subsets $\Omega_h^{i_1}, \ldots, \Omega_h^{i_k}$ such that

$$P_j \in \bigcap_{s=1}^{k} N(\Omega_h^{i_s}).$$

The procedure of the parallel algorithm 1 is as follows:

1° Choose a tolerance $\varepsilon > 0$ and an initial $u_0 = \{u_0(P_j), j \in I\}$. Set $0 \Rightarrow n$.

2° Compute parallelly for each Ω_h^i, $i=1, \ldots, m$ the coefficient C_{js} of the discrete system (for non-linear case, a linearization process is needed) and the residuals

$$\bar{f}_j^i = f_j - \sum_{P_s \in N_j} C_{js}\, u_n^i(P_s), \quad j \in I_i$$

$$F^i = \max_{j \in I_i} |\bar{f}_j^i|, \qquad i = 1, \ldots, m.$$

3° If $F = \max_{1 \le i \le m} F^i \le \varepsilon$, stop the process and output u_n, otherwise proceed to the next step.

4° Set equations for the correction Δu_n^i in each Ω_h^i, $i = 1, \ldots, m$

$$A_i: \sum_{P_s \varepsilon N_j} C_{js} \quad u_n^i(P_s) = \bar{f}_j^i, \qquad j \varepsilon I_i.$$

5° Find the minimum modulus solutions of A_i, $i = 1, \ldots, m$ parallelly

$$\Delta u_n^i = C_i^+ \bar{f}^i$$

where C_i^+ is the Moore-Penrose generalized inverse matrix of the coefficient matrix of A_i.

6° If $P_j \varepsilon \pi_k$, then there exists $i_1, \ldots, i_k$ such that

$$P_j \varepsilon \bigcap_{s=1}^{k} N(\bar{\Omega}_h^{i_s})$$

and define

$$\Delta u_n(P_j) = \frac{1}{k} \sum_{s=1}^{k} \Delta u_n^{i_s}(P_j), \qquad \text{for } j \varepsilon I \tag{4}$$

7° Set $u_n + \omega \Delta u_n \Rightarrow u_n$, $n+1 \Rightarrow n$ and go to 2°, $0 < \omega < 2$.

§2 The parallel algorithm 2

Algorithm 2 is based on the resulting algebraic system is linear and satisfies the discrete maximum principle. For convenience, we consider the following Dirichlet problem

$$\begin{cases} Lu = f, & \text{in } \Omega \\ u = g, & \text{on } \partial\Omega \end{cases} \tag{5}$$

The discrete system in the entire region is

$$\begin{cases} L^h u_h = f^h, & \text{in } \Omega_h \\ u_h = g^h, & \text{on } \partial\Omega_h \end{cases} \tag{6}$$

Divide Ω_h into m subsets: $\Omega_h = \bigcup_{i=1}^{m} \Omega_h^i$, Ω_h^i can be overlapping.

$\partial\Omega_h^i = N(\Omega_h^i)\setminus\Omega_h^i$ is the discrete boundary of Ω_h^i.

Let u* be the unique solution of (6). (6) is equivalent to the system

$$\begin{cases} L^h u_h^i = f^h, \text{ in } \Omega_h^i \\ u_h^i = u^*, \qquad \text{on } \partial\Omega_h^i \end{cases} \tag{7}$$

$i = 1, \ldots, m.$

Since u^* is unknown, (7) can only be solved by iterative methods. The 1st, 2nd and 3rd step in the procedure are the same as those for algorithm 1, the other steps are as follows:

4° Solve

$$\begin{cases} L^h u^i_{n+1} = f^h, & \text{in } \Omega^i_h \\ u^i_{n+1} = u_n, & \text{on } \partial\Omega^i_h \end{cases} \qquad (8)$$

for i = 1, ..., m parallelly.

5° If $P_j \in \pi_k$, then there exist $i_1, \ldots, i_k$ such that

$$P_j \in \bigcap_{s=1}^{k} N(\Omega^{i_s}_h)$$

Set

$$u_{n+1}(P_j) = \frac{1}{k} \sum_{s=1}^{k} u^{i_s}_{n+1}(P_j); \qquad j \in I .$$

6° Set $n+1 \Rightarrow n$ and go to 2°.

§3 The iterative method of groupwise projection for linear systems.

The parallel algorithm 1 introduced in §1 is based on the iterative method of groupwise projection for linear systems. The method was first established by S. Kaczmarz [2]; its further development can be found in [1], [3] and [4]. In those papers the discussion was confined to the case of one equation in one group. To deal with our problem, we shall extend the method to the case of many equations contained in one group.

Consider the linear system

$$(a_j, x) = a_{j1} x_1 + \ldots + a_{j\ell} x_\ell = b_j, \; j = 1, \ldots, \ell \qquad (9)$$

and assume it has a unique solution x^*.

Divide the set of indices $I = \{1, 2, \ldots, \ell\}$ into m subsets:

$I = \bigcup_{i=1}^{m} I_i$, where different subsets can be overlapping. Then the system (9) is divided into m groups:

$$G_i: (a_j, x) = b_j, \qquad j \in I_i \qquad (10)$$

where i = 1, ..., m.

For each i, G_i has at least a solution x^*, in general, the solution is not unique. It is well known that the minimum modulus solution of G_i exists and is unique. This solution is denoted by $E_i x$, where E_i is the projection onto the subspace $H_i = \text{span}\,\{a_j;\ j \in I_i\}$.

The iterative method of groupwise projection is as follows:

Choose a relaxation factor ω $(0 < \omega < 2)$ and an initial approximation $x^0 = \{x^0(P_j),\ j \in I\}$, then the process of getting x^{k+1} from x^k can be proceed as follows:

$$x^k_{(s+1)} = x^k_{(s)} + \omega \Delta x^k_{(s)}, \qquad s = 1, \ldots, 2m$$

$$x^k_{(s)} = \{x^k_{(s)}(P_j),\ j \in I\} \text{ and } x^k_{(1)} = x^k,$$

where the correction $\Delta x^k_{(s)}$ is the minimum modulus solution of the system

$$(a_j, \Delta x^k_{(s)}) = b_j - (a_j, x^k_{(s)}),\ j \in I_i, i = \min(s, 2m+1-s)$$

Obviously, $\Delta x^k_{(s)}(P_j)$ is defined only for $j \in I_i$. We extend it to all $j \in I$ by simply setting $\Delta x^k_{(s)}(P_j) = 0$ for $j \in I \setminus I_i$. Finally, we set

$$x^{k+1} = x^k_{(2m+1)} .$$

Now we are going to prove the convergence. In fact, the exact correction value of $x^k_{(s)}$ is $x^* - x^k_{(s)}$ and the minimum modulus solution $\Delta x^k_{(s)}$ is the projection of $x^* - x^k_{(s)}$ on the subspace H_i, i.e., $\Delta x^k_{(s)} = E_i(x^* - x^k_{(s)})$.

Hence, from $x^k_{(s+1)} = x^k_{(s)} + \omega\, \Delta x^k_{(s)}$, we have

$$x^* - x^k_{(s+1)} = (I - \omega E_i)(x^* - x^k_{(s)})$$

Let $Q_i = I - \omega E_i$. Then

$$x^* - x^{k+1} = Q_1 \cdots Q_m\, Q_m \cdots Q_1(x^* - x^k)$$

$$= (Q\, Q^*)^{k+1}(x^* - x^0) \qquad (11)$$

where $Q = Q_1 \cdots Q_m$. It is known, by direct computation, that $\|Q_i\| \le 1$, $i = 1, \ldots, m$ and hence $\|Q\| \le 1$. The equality holds only if there exists a vector y, $\|y\| = 1$ such that $Qy = y$. This means that y is orthogonal to all a_j, $j \in I$ and hence $y = 0$. The contradiction shows that $\|Q\| = r < 1$ and

$$\|x^* - x^k\| \le \|Q\|^{2k} \|x^* - x^0\| = r^{2k} \|x^* - x^0\|.$$

It follows that $x^k \to x^*$ as $k \to \infty$.

The iterative method of groupwise projection when apply to non-linear systems, the convergence will also follow.

§4 The proof of algorithm 1

In the following, we shall give the proof of linear problems. Non-linear problems can be proved in the similar way.

Let

$$\sum_{P_s \in N_j} C_{js} u_s = f_j, \; j \in I, \; u_s = u^h(P_s) \qquad (12)$$

be the discrete system defined in the entire set Ω_h and

$$A_i: \quad \sum_{P_s \in N_j} C_{js} u_s^i = f_j, \; j \in I_i \qquad (13)$$

$$(i = 1, \ldots, m)$$

be the discrete system defined in the ith subset Ω_h^i.

When we solve A_i by using the parallel algorithm 1, we take in account that $u_s^{i_1}$ and $u_s^{i_2}$ $(i_1 \neq i_2)$ are independent. From this point of view, we have assumed that A_{i_1} and A_{i_2} have no unknowns in common. As a compensation to this assumption, we add the following extra restrictions to A_i,

$$B_s: \quad \begin{array}{l} u_s^{i_1} - u_s^{i_2} = 0 \\ \cdots \quad \cdots \\ u_s^{i_1} - u_s^{i_k} = 0 \end{array} \qquad \text{for } P_s \in \pi_k, \; k \ge 2 \qquad (14)$$

Then, the equations A_i, $i = 1, \ldots, m$ together with the equations $B_s, P_s \in \pi_k$, $k \ge 2$ are equivalent to the equations (12).

We name A_i, $i = 1, \ldots, m$ as group 1 and B_s, $P_s \in \pi_k$, $k \ge 2$ as group 2. Since A_{i_1} and A_{i_2} $(i_1 \neq i_2)$ have no unknowns in common, the minimum modulus solution of group 1 simply is the union of the minimum modulus solutions of A_i, $i = 1, \ldots, m$ and the latter can be found parallelly.

Let $u_0 = \{u_0(P_j), j \in I\}$ be an initial approximation to $u^* = \{u^h(P_j), j \in I\}$, which is the solution of (12). Let $\Delta u_n^i = \{u_n^i(P_j), j \in I_i\}$, $i = 1, \ldots, m$ be the correction of u_n^i

obtained from the minimun modulus solution of group 1 and let $\delta u_n^i(P_s)$, $P_s \in \pi_k$, $k \geq 2$ be the correction of $u_n^i + \Delta u_n^i$ obtained from the minimum modulus solution of group 2. Noting that when $s_1 \neq s_2$, B_{s_1} and B_{s_2} have no unknowns in common, we may find the minimum modulus solutions of $B_s, P_s \in \pi_k$, $k \geq 2$ parallelly. Substituting $u_n^i + \Delta u_n^i + \delta u_n^i$ into B_s, we have

$$C_s: \quad \begin{aligned} &\delta u_n^{i_1}(P_s) - \delta u_n^{i_2}(P_s) = -(\Delta u_n^{i_1}(P_s) - \Delta u_n^{i_2}(P_s)) \\ &\cdot \quad \cdot \quad \cdot \quad \cdot \quad \cdot \quad \cdot \quad \cdot \quad \cdot \quad \cdot \\ &\delta u_n^{i_1}(P_s) - \delta u_n^{i_k}(P_s) = -(\Delta u_n^{i_1}(P_s) - \Delta u_n^{i_k}(P_s)) \end{aligned}$$

The minimum modulus solution of C_s is

$$\delta u_n^{i_j}(P_s) = \frac{1}{k}\left(\sum_{j=1}^{k} \Delta u_n^{i_j}(P_s)\right) - \Delta u_n^{i_j}(P_s)$$

Finally, we have

$$u_{n+1}(P_s) = u_n(P_s) + \frac{1}{k}\left(\sum_{j=1}^{k} \Delta u_n^{i_j}(P_s)\right), \quad P_s \in \pi_k, \quad k \geq 1$$

Let E_1, E_2 be the projections defined by group 1 and group 2 repectively. Let $Q_i = I - \omega E_i$, $i = 1, 2$ and $Q = Q_1 Q_2$. According to the proof given in §3, we conclude that

$$\|u^* - u_n\| = \|(Q Q^*)^n (u^* - u_0)\| \leq r^{2n} \|u^* - u_0\|$$

where $\quad r = \|Q\| = \|(I - \omega E_1)(I - E_2)\| < 1.$

§5 The proof of the algorithm 2

We know, from (7) and (8), that $u_{n+1}^i - u^*$ satisfies

$$\begin{cases} L^h (u_{n+1}^i - u^*) = 0, & \text{in } \Omega_h^i \\ u_{n+1}^i - u^* = u_n - u^*, & \text{on } \partial\Omega_h^i \setminus \partial\Omega_h \\ u_{n+1}^i - u^* = 0, & \text{on } \partial\Omega_h^i \cap \partial\Omega_h \end{cases} \tag{15}$$

where $u_{n+1}^i = \{u_{n+1}^i(P);\ P \in N(\Omega_h^i)\}$, $\partial\Omega_h^i \cap \partial\Omega_h \neq 0$.

Let $u_{n+1} = \{\bar{u}_{n+1}^i(P);\ P \in N(\Omega_h^i),\ i = 1, \ldots, m\}$, where $\bar{u}_{n+1}^i(P)$ is defined as follows: if $P \in \pi_k$, there exist $i_1, \ldots, i_k$ such that $P \in \bigcap_{j=1}^{k} N(\Omega^{i_j})$ and $\bar{u}^i_{n+1}(P) = \frac{1}{k}\sum_{j=1}^{k} u^{i_j}_{n+1}$

It can be shown, by using the discrete maximum principle, that the sequence $\|u^* - u_n\|_\infty$ is strictly monotone decreasing, i.e., if $\|u^* - u_n\|_\infty \neq 0$, then $\|u^* - u_{n+1}\|_\infty < \|u^* - u_n\|_\infty$.

Hence $\{u_{n+1} - u^*\}$ has a convergent subsequence. For convenience, we still write the subsequence as $u_{n+1} - u^*$ and let its limit be $v - u^*$. Let S be an operator such that

$$u_{n+1} - u^* = S(u_n - u^*)$$

then we have

$$v - u^* = S(v - u^*) \tag{16}$$

and

$$\|v - u^*\|_\infty = \|S(v - u^*)\|_\infty < \|v - u^*\|_\infty \tag{17}$$

It proves that $v = u^*$.

We have proved that the sequence $\{\|u_n - u^*\|_\infty\}$ has a limit and there is a subsequence of $\{\|u - u^*\|_\infty\}$ has limit zero, hence $\|u - u^*\|_\infty \to 0$, as $n \to \infty$.

§6 A synchronized parallel algorithm for solving Dirichlet problems of second order elliptic PDE.

Consider the following Dirichlet problem

$$\begin{cases} Lu = f, & \text{in } \Omega \\ u = g, & \text{on } \partial\Omega \end{cases} \tag{18}$$

where $\Omega \in R^2$ is a bounded open set, L is a linear operator satisfying the uniformly elliptic condition.

Let $\Omega = \bigcup_{i=1}^{m} \Omega_i$, $\partial\Omega_i \cap \partial\Omega \neq 0$, and $\partial\pi_k$ either coincide with $\partial\Omega$ or is an arc with endpoints on $\partial\Omega$.

The algorithm is as follows:

1° Choose an initial u^0 with $u^0|\partial\Omega = g$

2° Solve parallelly

$$\begin{cases} Lu^i_{n+1} = f, & \text{in } \Omega i \\ u^i_{n+1} = u^n, & \text{on } \partial\Omega i \setminus \partial\Omega \\ u^i_{n+1} = g, & \text{on } \partial\Omega i \cap \partial\Omega \end{cases} \tag{19}$$

3° $u^{n+1} = \frac{1}{K}\sum_{j=1}^{m} u^{i_j}_{n+1}$

Theorem. If (19) has a bounded solution for each i, i = 1, 2, ...m, then either there exists a $q \in (0,1)$, such that

$$\|u^* - u^{n+1}\|_\infty \le q \; \|u^* - u^n\|_\infty$$

or the above procedure will converge in finite steps. We state the following lemma without proof.

Lemma

$$\begin{cases} Lu = 0, & \text{in } \Omega \\ u = 0, & \text{on } \Gamma_1 \cup \Gamma_2 \\ u = g, & \text{on } \Gamma_3 \end{cases}$$

where $\partial\Omega = \Gamma_1 \cup \Gamma_2 \cup \Gamma_3$.

If $\widehat{MN}$ is a smooth curve in Ω with $M \in \Gamma_1$, $N \in \Gamma_2$, then there exists a constant $q \in (0,1)$ independent of g such that

$$|u(P)| \le qQ, \text{ for } P \in \widehat{MN}.$$

where $Q = \max_{\Gamma_3} |g|$.

Proof of Theorem

Suppose that u* is the solution of (18), then

$$L(u^* - u^i_{n+1}) = 0, \qquad \text{in } \Omega_i$$

$$u^* - u^i_{n+1} = u^* - u^n, \qquad \text{on } \partial\Omega_i \setminus \partial\Omega$$

$$u^* - u^i_{n+1} = 0, \qquad \text{on } \partial\Omega_i \cap \partial\Omega$$

If $\|u^* - u^{n+1}\|_\infty = 0$, the iteration converges in finite steps.

If $\|u^* - u^{n+1}\|_\infty \ne 0$, there exists $P \in \overline{\Omega}$ such that

$$|u^*(P) - u^{n+1}(P)| = \|u^* - u^{n+1}\|_\infty \ne 0 \tag{20}$$

Case I

If $P \in \pi_1$ there exist Ω_{i_0} and $P \in \overline{\Omega}_{i_0}$ such that

$u^{n+1}(P) = u^{i_0}_{n+1}(P)$, hence $u^* - u^{i_0}_{n+1}$ is a constant on $\overline{\Omega}_{i_0}$. Since $\partial\Omega_{i_0} \cap \partial\Omega \ne 0$, we have $\|u^* - u^{n+1}\|_\infty = 0$. This contradicts (20).

Case II

If $P \in \pi_k, k \ge 2$. Since $L(u^* - u^{n+1}) = 0$, for $P \in \mathring{\pi}_k$. From the maximum principle, $P \in \partial\pi_k$. We may assume that $P \in \widehat{MN} \subset \partial\pi_k$ with $M, N \in \partial\Omega$. From our assumption, evidently there is i_1, $\widehat{MN} \subset \Omega_{i_1}$.

From Lemma, there exists $q \in (0,1)$ such that

$$|u^* - u_{n+1}^{i_1}| \leq q \max_{Q \in \partial\Omega_{i_1}} |u^*(Q) - u_{n+1}^{i_1}(Q)|$$

and

$$|u^*(P) - u^{n+1}(P)| \leq \frac{1}{k}|u^*(P) - u_{n+1}^{i_1}(p)| + \frac{1}{k}\sum_{j=2}^{k}|u^*(P) - u_{n+1}^{i_j}(P)|$$

$$\leq \frac{q}{k} \max_{Q \in \partial\Omega_{i_1}} |u^*(Q) - u_{n+1}^{i_1}(Q)| + \frac{1}{k}\sum_{j=2}^{k} \max_{Q \in \partial\Omega_{i_j}} |u^*(Q) - u_{n+1}^{i_j}(Q)|$$

$$\leq \left(\frac{q}{k} + \frac{k-1}{k}\right) \max_{Q \in \partial\Omega_{i_j}} |u^*(Q) - u_{n+1}^{i_j}(Q)|$$

$$= \left(\frac{q}{k} + \frac{k-1}{k}\right) \max_{Q \in \partial\Omega_{i_j}} |u^*(Q) - u^n(Q)|$$

$$\leq \left(\frac{q}{k} + \frac{k-1}{k}\right)\|u^* - u^n\|_\infty < \bar{q}\,\|u^* - u^n\|_\infty, \text{ where } \bar{q} = \frac{q}{k} + \frac{k-1}{k} < 1.$$

REFERENCES

[1] A. BJORCK AND T. ELFVING, Accelerated projection method for computation psendoinverse solution of systems of linear equations, BIT 19 (1979), pp. 143 - 163.

[2] S. KACZMARZ, Angenaherte Auflosung von Systemen Linear Gleichungen, Bull. Internat. Acad. polon. Science et Lettres (1937), pp. 355 - 357.

[3] J. Q. LIU, Convergence of SOR projection method for nonlinear systems, J. Sys. Sci. & Math. Sci., 1(1981), pp. 77 - 79, China (in English).

[4] T. LU AND Q. LIN, A projection iterative method for solving integral equations of the first kind and algebraic systems, Math. Numer. Sinica, 2 (1984), pp. 113 - 120 (in Chinese).

Iterative Solution of Elliptic Equations with Refinement: The Two-Level Case*

Jan Mandel†
Steve McCormick††

Abstract. A two-level theory for FAC [16] and Asynchronous FAC [10] is developed based on the strengthened Cauchy inequality. We obtain convergence bounds that do not depend on regularity of the problem and that can be computed locally. We also establish a relationship between FAC and a closely related preconditioner introduced in [5].

Key Words. Composite Grid, Refinement, Iterative Methods, Orthogonal Projections, Elliptic Partial Differential Equations, Finite Elements, Parallel Computation, FAC

1. Introduction. The Fast Adaptive Composite grid method (FAC [16]) is an iterative method for the solution of composite grid equations using solvers on uniform grids only. It was observed to be a fast, practical method. However, its theoretical foundations have not yet been fully developed.

In brief, the simplest version of FAC can be described as a conforming finite element method using a global coarse grid and one or more (non-overlapping) local fine grids. The spaces of trial and test functions are simply the sum of standard finite element function spaces associated with all grids. To find the minimum of the energy functional, minimization is performed alternatively with respect to the spaces of coarse grid and fine grid finite element functions. Fast solvers can be applied to the resulting linear systems in many cases, especially when the grids are uniform. Several levels of refinement and parallel solution on all levels can be successfully incorporated into this scheme [10].

---

* This research was sponsored by the Air Force Office of Scientific Research under grant AFOSR-86-0126 and by the National Science Foundation under grant DMS-8704169.

† Computational Mathematics Group, University of Colorado at Denver, 1200 Larimer Street, Denver, CO 80202.

‡ Computational Mathematics Group, University of Colorado at Denver, 1200 Larimer Street, Denver, CO 80202, and Colorado Research Development Corporation, P. O. Box 9182, Fort Collins, CO 80525.

In this paper, we are concerned with bounds on the convergence factor of FAC as an iterative method for the solution of the composite equations. The paper is organized as follows. In Section 2, we give some simple algebraical propositions concerning orthogonal projections and the angle between subspaces. In Section 3, we apply these abstract results to obtain a characterization of the convergence factor of FAC in terms of the cosine of the angle between certain subspaces of the space of composite grid functions. We then bound the cosine from locally computable estimates, using a technique known in multigrid literature as the'*strengthened Cauchy inequality*, which makes it possible to use bounds already available in the literature for many classes of finite elements. FAC gives rise to naturally defined preconditioners, studied in Section 4. In particular, we show that the preconditioner defined by Bramble, Ewing, Pasciak, and Schatz in [5] is equivalent to one-and-a-half steps of FAC, so an analysis of FAC applies to the preconditioner from [5] and vice versa. In Section 5, we prove a theorem relating the convergence factor of a parallel version of FAC (AFAC [17]) with the convergence factor of FAC itself.

2. Theoretical preliminaries. We begin with several simple lemmas that establish some algebraical properties of orthogonal projections. Let H be a finite dimensional linear space with inner product $(\cdot,\cdot)$ and norm $||\cdot||$. For $u, v \in H$, $u \neq 0$, $v \neq 0$, we write $\cos(u,v) = (u,v)/||u||\,||v||$. For U, V nontrivial subspaces of H, define the cosine of the angle between U and V by $\cos(U,V) = \sup\{|\cos(u,v)| : u \in U, v \in V\}$. Further, let P_V and $P_{V^\perp}$ denote the respective orthogonal projections from H onto V and $V^\perp$, the orthogonal complement of V. Let $\rho(A)$ denote the spectral radius of a matrix A.

The following lemma is well known and is included only for completeness.

LEMMA 2.1. *Let U, V be nontrivial subspaces of H. Then*

$$\rho(P_U P_V) = ||P_U P_V P_U|| = \cos^2(U,V).$$

Proof. Let $P_U P_V u = \lambda u$, $\lambda \neq 0$. Then u is in the range of P_U, so $P_U P_V P_U u = \lambda u$. This proves that $\rho(P_U P_V) = ||P_U P_V P_U||$ because $P_U P_V P_U$ is symmetric.

To prove the second equality, because the case $\cos(U,V) = 1$ is trivial, we consider only the case $\cos(U,V) < 1$. Because the surface of the unit sphere in H is compact, there exists a $u \in U$ such that $||u|| = 1$ and $||u - P_V u||$ is minimal. Denote $v = P_V u/||P_V u||$ and $c = \cos(u,v)$. Then $|c| = \cos(U,V)$, $P_V u = cv$, and $P_U v = cu$. Consequently, $P_U P_V u = c^2 u$, so $\rho(P_U P_V) \geq \cos^2(U,V)$.

To prove the reverse inequality, note that for any $u \in U$ it holds that $||P_V u||^2 = (P_V u, u) \leq \cos(U,V)||P_V u||||u||$, so $||P_V u|| \leq \cos(U,V)||u||$. Similarly, $||P_U v|| \leq \cos(U,V)||v||$ for any $v \in V$. It follows that $||P_U P_V P_U|| \leq \cos^2(U,V)$. $\square$

The next lemma allows us to estimate $\cos(U,V)$.

LEMMA 2.2. *Let U, V be nontrivial subspaces of H.*

(i) If $U \cap V = \{0\}$, then $\cos(U^\perp, V^\perp) \geq \cos(U,V)$.

(ii) If $H = U \oplus V$, then $\cos(U^\perp, V^\perp) = \cos(U,V)$.

(iii) If $X \subset U$ and $Y \subset V$ and $H = X \oplus Y$, then $\cos(U^\perp, V^\perp) \leq \cos(X,Y)$.

Proof. (i) As in the preceding proof, let $u \in U$ such that $||u|| = 1$ and $||u - P_V u||$ is minimal and denote $v = P_V u / ||P_V u||$ and $c = \cos(u, v)$. Then $|c| = \cos(U, V)$, $P_V u = cv$, and $P_U v = cu$. It follows that $u - P_V u/c \in U^{\perp}$, $v - P_U v/c \in V^{\perp}$, and $\cos(u - P_V u, v - P_U v) = c$, using elementary geometry in the two-dimensional subspace spanned by u an v. This proves (i). Propositions (ii) and (iii) follow immediately from (i). □

The next lemma about localization of the spectrum of the sum of projections will be useful in the study of AFAC.

LEMMA 2.3. *Let* $H = \bigoplus_{j=1}^{n} V_j$. *Then*

$$\lambda_{\min}\left(\sum_{j=1}^{n} P_{V_j}\right) = \min_{v_j \in V_j, ||v_j||=1, j=1,\dots,n} \lambda_{\min}\left(G(v_1, \dots, v_n)\right),$$

$$\lambda_{\max}\left(\sum_{j=1}^{n} P_{V_j}\right) = \max_{v_j \in V_j, ||v_j||=1, j=1,\dots,n} \lambda_{\max}\left(G(v_1, \dots, v_n)\right),$$

where $G(v_1, \dots, v_n) = (g_{ij})$, $g_{ij} = (v_i, v_j)$, *is the Gram matrix of the vectors* $v_1, \cdots v_n$.

Proof. For $j = 1, \dots, n$, let J_j be a matrix whose columns form an orthonormal basis of the space V_j. Then $P_{V_j} = J_j J_j^t$. Define $A = (J_1, \dots, J_n)$. Then A is a square regular matrix and $\sum_{j=1}^{n} J_j J_j^t = AA^t$. Because AA^t is similar to $A^t A = A^{-1}(AA^t)A$, the spectrum of AA^t is same as the spectrum of $A^t A = (J_i^t J_j)_{i,j=1}^{n}$. Now for every $u \in H$ we may write

$$u = \begin{pmatrix} a_1 u_1 \\ \vdots \\ a_n u_n \end{pmatrix},$$

where a_j are scalars, the number of elements in each vector u_j is same as the dimension of V_j, and $u_j^t u_j = 1$. Let $a = (a_j)_{j=1}^{n}$ and $v_j = J_j u_j$. We then have $Au = \sum_{j=1}^{n} a_j v_j$ and the Rayleigh quotient

$$\frac{u^t A^t A u}{u^t u} = \frac{a^t G(v_1, \dots, v_n) a}{a^t a}.$$

The lemma is a direct consequence of these observations. □

In the case $n = 2$, the preceding lemma gives bounds of the spectrum of the sum of two projections in terms of the angle of their ranges.

LEMMA 2.4. *Let* $H = U \oplus V$. *Then*

$$\lambda_{\max}\left(P_U + P_V\right) = 1 + \cos(U, V), \quad \lambda_{\min}\left(P_U + P_V\right) = 1 - \cos(U, V).$$

Proof. Let $u \in U$ and $v \in V$. Then the Gram matrix of u, v is

$$G(u, v) = \begin{pmatrix} 1 & a \\ a & 1 \end{pmatrix}$$

where $|a| \leq \cos(U, V)$, with equality attained for some u and v. The lemma now follows from the observation that the eigenvalues of $G(u, v)$ are $1 + a$ and $1 - a$. □

3. The Fast Adaptive Composite Grid Method (FAC). For simplicity, we restrict ourselves to the case of a diffusion operator with homogeneous Dirichlet boundary conditions. Let Ω_1 be an open polygonal domain in $\mathbf{R}^2$ and $\Omega_2 \subset \Omega_1$ an open polygonal domain or the union of disjoint open polygonal domains. Let $H_{2h} \subset H_0^1(\Omega_1)$ be a conforming Lagrange finite element space associated with Ω_1 and $H_h \subset H_0^1(\Omega_2)$ be a conforming finite element space associated with Ω_2. Define the *composite grid space*

$$H_c = H_{2h} + H_h.$$

Consider the bilinear form

$$a(u,v) = \int_{\Omega_1} a \nabla u \nabla v,$$

with the diffusion coefficient a piecewise smooth and $a > \text{const} > 0$ in Ω_1. Also consider the linear form

$$f(v) = \int_{\Omega_1} fv.$$

The space H_c is equipped with the inner product $a(\cdot,\cdot)$. We are interested in the iterative solution of the *discrete variational equation*

$$\mathbf{u} \in H_c: \quad a(\mathbf{u}, v) = f(v), \quad \forall v \in H_c. \tag{1}$$

The symbol $\mathbf{u}$ will be used to denote the solution of (1).

We consider an iterative method for the solution of (1), whose basic cycle is defined as follows:

ALGORITHM 3.1. (FAC [16]) *Let $u \in H_c$ be the current approximation to* $\mathbf{u}$.

Step 1. *Compute the solution of*

$$u_h \in H_h: \quad a(u + u_h, v_h) = f(v_h), \quad \forall v_h \in H_h, \tag{2}$$

and set $u \leftarrow u + u_h$.

Step 2. *Compute the solution of*

$$u_{2h} \in H_{2h}: \quad a(u + u_{2h}, v_{2h}) = f(v_{2h}), \quad \forall v_{2h} \in H_{2h}, \tag{3}$$

and set $u \leftarrow u + u_{2h}$.

It is easy to see that the *algebraic error*

$$e = \mathbf{u} - u$$

is transformed according to

$$e \leftarrow P_{H_{2h}^{\perp}} P_{H_h^{\perp}} e. \tag{4}$$

An estimate $\rho\left(P_{H_{2h}^{\perp}} P_{H_h^{\perp}}\right) \leq \text{const} < 1$ independent of h was established in [16] using H^2*-regularity* of the elliptic operator, but without restrictions on the *degree* *of refinement* in Ω_2. Here we restrict the degree of refinement — the mesh sizes of H_{2h} and H_h differ by a factor of two — and we obtain a bound without recourse to regularity using a technique well known in the multigrid literature. For the purpose of obtaining local estimates, assume that *the boundaries of the elements associated with H_{2h} contain the boundary of Ω_2 and that every element in Ω_2 associated with H_{2h} is the union of some elements associated with H_h*. We may then define

$$(5) \qquad X = H_{2h}, \quad Y = \{u \in H_h : u = 0 \text{ on nodes of } H_{2h}\}.$$

It is well known, cf., [1], [4], [13], that then a bound on $\cos(X, Y)$ can be evaluated locally: Let $\{K\}$ be the elements associated with the space H_{2h}. For any such K, denote

$$a_K(u,v) = \int_K a \bigtriangledown u \bigtriangledown v.$$

Then if the inequality

$$(6) \qquad |a_K(u,v)| \leq \gamma \sqrt{a_K(u,u)} \sqrt{a_K(v,v)}, \quad \forall u \in X, v \in Y$$

holds for all K with the same constant $\gamma < 1$, it follows that

$$|a(u,v)| = \left| \sum_K a_K(u,v) \right| \leq \gamma \sum_K \sqrt{a_K(u,u)} \sqrt{a_K(v,v)}$$

$$\leq \gamma \sqrt{\sum_K a_K(u,u)} \sqrt{\sum_K a_K(v,v)}.$$

This yields the so-called *strengthened Cauchy inequality*

$$(7) \qquad |a(u,v)| \leq \gamma \sqrt{a(u,u)} \sqrt{a(v,v)}, \quad \forall u \in X, v \in Y.$$

Thus, γ is an upper bound on $\cos(X, Y)$.

REMARK 3.2. It is easy to see from (6) that the value of γ does not change if the diffusion coefficient a is multiplied by a different positive constant in each element. This property is usual for regularity-free estimates; for related — but different — convergence bounds with this property for multigrid methods, see [7] and [11].

REMARK 3.3. Estimates of γ for various elements and a constant diffusion coefficient a are well known, see [1], [4],[13]. Computation of γ on K reduces to the solution of a generalized eigenvalue problem for the local stiffness matrix of the element K. In the two-dimensional case, the H_h elements result from partitioning each H_{2h} element into four elements in the natural way. For this case, the optimal values of γ for triangular linear elements as computed by Maitre and Musy [13] are between $\gamma = \sqrt{3/8}$ for the unilateral triangle and $\gamma \to \sqrt{2/3}$ for the degenerate triangle with one angle close to π. The model problem with Ω_1 a rectangle divided in two rectangles Ω_2 and $\Omega_1 \backslash \Omega_2$ is studied by Fourier analysis in [14]. For triangular linear elements obtained by dividing all squares in a uniform rectangular mesh in two triangles in the same way, it holds that

$$a(u_h, u^{2h\text{-harm}}) = a_{\Omega_2}(u_h, u^{2h\text{-harm}}) \leq \delta\sqrt{a(u_h, u_h)}\sqrt{a_{\Omega_2}(u^{2h\text{-harm}}, u^{2h\text{-harm}})}$$

for all $u_h \in H_h, u^{2h\text{-harm}} \in H_{2h}^{2h\text{-harm}}$, with $\delta \approx 0.669$. This bound is sharp for small h.

Denote

$$H_{2h}^{2h\text{-harm}} = \{u_{2h} \in H_{2h} : a(u_{2h}, v_{2h}) = 0, \quad \forall v_{2h} \in H_{2h} \cap H_h\}.$$

This is the orthogonal complement of $H_{2h} \cap H_h$ in H_{2h}. By analogy with the case when the diffusion coefficient a is constant, the functions from $H_{2h}^{2h\text{-harm}}$ are called $2h$–harmonic functions in Ω_2.

LEMMA 3.4. *It holds that* $P_{H_{2h}^\perp} P_{H_h^\perp} = P_{(H_{2h}^{2h\text{-harm}})^\perp} P_{H_h^\perp}$.

Proof. Let $v \in H_c$ be arbitrary and write $u = P_{H_h^\perp} v$. Define $w_{2h} = P_{H_{2h}^{2h\text{-harm}}} u$, so that

$$w_{2h} \in H_{2h}^{2h\text{-harm}} : \quad a(w_{2h}, z_{2h}) = a(u, z_{2h}), \quad \forall z_{2h} \in H_{2h}^{2h\text{-harm}}.$$

But this must hold for all $z_{2h} \in H_{2h} = H_{2h}^{2h\text{-harm}} \oplus (H_{2h} \cap H_h)$ because, for $z_{2h} \in H_{2h} \cap H_h$, $a(w_{2h}, z_{2h})$ and $a(u, z_{2h})$ are both zero by definition. Consequently, $w_{2h} = P_{H_{2h}} u$, so

$$P_{(H_{2h}^{2h\text{-harm}})^\perp} u = u - w_{2h} = P_{H_{2h}^\perp} u,$$

which proves the lemma. □

We can now exhibit a bound on the convergence factor of Algorithm 3.1.

THEOREM 3.5. *The convergence factor of Algorithm 3.1 is*

$$\rho\left(P_{H_{2h}^\perp} P_{H_h^\perp}\right) = \cos^2(H_{2h}^{2h\text{-harm}}, H_h) \leq \cos^2(X, Y) \leq \gamma^2,$$

where X and Y are given by (5).

Proof. The proof follows immediately from Lemma 3.4, equation (4), Lemmas 2.1 and 2.2, and from the fact that $H = H_{2h}^{2h\text{-harm}} \oplus H_h$. □

REMARK 3.6. There is, in fact, a hidden parallelism in Algorithm 3.1. In the first step, the problem in the space H_h decomposes into independent subproblems if the refinement region Ω_2 consists of several disjoint components, which is often the case in practice. Cf., a similar remark in [5].

REMARK 3.7. The work in [17] uses the regularity-based estimates of [16] to develop a theory that covers the cases of singular equations — e.g., where Neumann replaces Dirichlet boundary conditions — and inexact solvers — i.e., where the solutions of (2) and (3) are only approximated. The present theory extends to such cases in a similar way.

4. FAC as a preconditioner. We can now define a preconditioner in a natural way using one iteration of Algorithm 3.1 with initial value $u = 0$ as an approximate solver. We obtain the following algorithm.

ALGORITHM 4.1. (FAC preconditioner)

Step 1. *Compute u_h from*

$$u_h \in H_h, \quad a(u_h, v_h) = f(v_h), \quad \forall v_h \in H_h. \tag{8}$$

Step 2. *Compute u_{2h} from*

$$u_{2h} \in H_{2h}, \quad a(u_{2h}, v_{2h}) = f(v_{2h}) - a(u_h, v_{2h}), \quad \forall v_{2h} \in H_{2h}. \tag{9}$$

Step 3. *Set $u = u_h + u_{2h}$.*

The value of u obtained by this algorithm is the solution of a variational problem of the form

$$u \in H_c: \quad \tilde{b}(u, v) = f(v), \quad \forall v \in H_c. \tag{10}$$

For the next theorem, we need a simple statement about iterative methods, which we formulate as a lemma for reference.

LEMMA 4.2. *Let $u \leftarrow \mathcal{G}(u, b) = u - B^{-1}(Au - b)$ be an iterative method for the solution of the linear system $A\mathbf{u} = b$. Then $\mathcal{G}(0, b) = B^{-1}b$.*

Because the following result is somewhat easier to formulate in terms of operators rather than bilinear forms, we let $\langle \cdot, \cdot \rangle$ be another inner product on H_c and define the operators $A, \tilde{B} : H_c \to H_c$ by

$$\left. \begin{aligned} \langle Au, v \rangle &= a(u, v) \\ \langle \tilde{B}u, v \rangle &= \tilde{b}(u, v) \end{aligned} \right\} \forall u, v \in H_c. \tag{11}$$

The following theorem is then an immediate consequence of Lemma 4.2.

THEOREM 4.3. *It holds that*

$$I - \tilde{B}^{-1}A = P_{H_{2h}^\perp} P_{H_h^\perp} = P_{(H_{2h}^{2h\text{-harm}})^\perp} P_{H_h^\perp}.$$

Because the product $P_{(H_{2h}^{2h\text{-harm}})^\perp} P_{H_h^\perp}$ is in general not a symmetric operator (with respect to the inner product $a(\cdot, \cdot)$ on H_c), it follows that the bilinear form $\tilde{b}(\cdot, \cdot)$ is in general not symmetric. Therefore, this preconditioner cannot be used directly with the conjugate gradient algorithm. However, an effective use of such a nonsymmetric preconditioner is still possible with more general Krylov space methods, see, for example, [15].

For the application of conjugate gradients, it is therefore natural to apply an additional half-step of the FAC algorithm to obtain a symmetric preconditioner. We then get a preconditioner identical to that of Bramble, Ewing, Pasciak, and Schatz [5]. (Their work is formulated for Neumann rather than Dirichlet boundary conditions, but the present discussion can be easily adapted to that framework.)

Application of Steps 1, 2, and 1 of the FAC Algorithm 3.1 with initial value $u = 0$ yields the following algorithm.

ALGORITHM 4.4. (Symmetric FAC preconditioner)
Step 1. *Compute* u_h *from*

$$u_h \in H_h : \quad a(u_h, v_h) = f(v_h), \quad \forall v_h \in H_h. \tag{12}$$

Step 2. *Compute* u_{2h} *from*

$$u_{2h} \in H_{2h} : \quad a(u_h + u_{2h}, v_{2h}) = f(v_{2h}), \quad \forall v_{2h} \in H_{2h}. \tag{13}$$

Step 3. *Compute* w_h *from*

$$w_h \in H_h : \quad a(u_h + u_{2h} + w_h, v_h) = f(v_h), \quad \forall v_h \in H_h. \tag{14}$$

Step 4. *Set* $u = u_h + u_{2h} + w_h$.

Problem (14) in Step 3 can be rewritten as

$$w_h \in H_h : \quad a(u_{2h} + w_h, v_h) = 0, \quad \forall v_h \in H_h, \tag{15}$$

because of the definition of u_h from (12). Step 3 was introduced in [5] to make sure that the decomposition $u = u_h + (u_{2h} + w_h)$ is orthogonal, which was motivated by the following variational interpretation.

For any $u \in H_c$, we have the orthogonal decomposition

$$u = u_h + u^{h\text{-harm}}, \quad u_h \in H_h, \quad u^{h\text{-harm}} \in H^{h\text{-harm}} = H_h^{\perp}. \tag{16}$$

For $u^{h\text{-harm}}$, we can further define $u^{2h\text{-harm}} \in H_{2h}^{2h\text{-harm}}$ as the unique function in $H_{2h}^{2h\text{-harm}}$ which coincides with $u^{h\text{-harm}}$ on $\Omega_1 \setminus \Omega_2$.

LEMMA 4.5. *The result of Algorithm 4.4 is the solution of the variational problem*

$$u \in H_c : \quad b(u, v) = f(v), \quad \forall v \in H_c, \tag{17}$$

with the bilinear form $b(\cdot,\cdot)$ *defined by*

$$b(u, v) = a(u_h, v_h) + a(u^{2h\text{-harm}}, v^{2h\text{-harm}}). \tag{18}$$

Proof. We adapt the proof from [5] and provide more details. Let $u = u_h + u^{h\text{-harm}}$ be the solution of (17).

First, let $v \in H_h$. Then $v = v_h$ and (17) and (18) show that u_h satisfies (12).

Next, let $v = v_{2h} \in H_{2h} = v_h + v^{h\text{-harm}}$ and note that

$$a(u^{2h\text{-harm}}, v) = a(u^{2h\text{-harm}}, v^{2h\text{-harm}}) \tag{19}$$

because $v - v^{2h\text{-harm}} \in H_{2h} \cap H_h \perp H_{2h}^{2h\text{-harm}}$. Also,

$$a(u_h, v) = a(P_{H_h} u_h, v) = a(u_h, P_{H_h} v) = a(u_h, v_h). \tag{20}$$

Thus, we have from (17) and (18) using (19) and (20) that $u^{2h\text{-harm}}$ satisfies

$$a(u^{2h\text{-harm}}, v_{2h}) = f(v_{2h}) - a(u_h, v_{2h}), \quad \forall v_{2h} \in H_{2h}.$$

Consequently, $u^{2h\text{-harm}}$ is just u_{2h}, the solution of (13).

Finally, $u^{h\text{-harm}} = u^{2h\text{-harm}}$ on $\Omega_1 \setminus \Omega_2$, and we find the values of $u^{h\text{-harm}}$ on Ω_2 from $u^{h\text{-harm}} - u^{2h\text{-harm}} = w_h$, where w_h satisfies (15). □

Spectral equivalence of the forms $a(\cdot,\cdot)$ and $b(\cdot,\cdot)$ was established in [5] using the equality

$$a(u,v) = a(u_h, v_h) + a(u^{h\text{-harm}}, v^{h\text{-harm}})$$

(the decomposition (16) is orthogonal) and the fact that

$$(21) a(u^{h\text{-harm}}, u^{h\text{-harm}}) \le a(u^{2h\text{-harm}}, u^{2h\text{-harm}}) \le C a(u^{h\text{-harm}}, u^{h\text{-harm}}), \quad \forall u \in H_c.$$

The inequality (21) was proved in [6] using the "inverse assumption" and a rather weak form of elliptic regularity. In particular, (21) holds whenever the coefficients of the form $a(\cdot,\cdot)$ are continuous and all components of Ω_2 have Lipschitz boundary.

As above, let $\langle\cdot,\cdot\rangle$ be another inner product on H_c and define the operators $A, B : H_c \to H_c$ by

$$\left.\begin{array}{l}\langle Au, v\rangle = a(u,v)\\ \langle Bu, v\rangle = b(u,v)\end{array}\right\} \forall u, v \in H_c.$$

Now application of Lemmas 4.2 and 3.4 immediately yields the following theorem, which relates our preceding results to properties of the preconditioner $b(\cdot,\cdot)$.

THEOREM 4.6. *It holds that*

$$(22) \qquad I - B^{-1}A = P_{H_h^\perp} P_{H_{2h}^\perp} P_{H_h^\perp} = P_{H_h^\perp} P_{(H_{2h}^{2h\text{-harm}})^\perp} P_{H_h^\perp}.$$

Consequently,

$$(23) \qquad a(u,u) \le b(u,u) \le \kappa a(u,u), \quad \forall u \in H_c,$$

where

$$1 - \frac{1}{\kappa} = \rho\left(P_{H_h^\perp} P_{(H_{2h}^{2h\text{-harm}})^\perp} P_{H_h^\perp}\right) = \rho\left(P_{(H_{2h}^{2h\text{-harm}})^\perp} P_{H_h^\perp})\right) = \cos^2(H_{2h}^{2h\text{-harm}}, H_h) \le \gamma^2.$$

REMARK 4.7. It follows from (22) and from the fact that the product of projections $P_{H_h^\perp} P_{H_{2h}^\perp} P_{H_h^\perp}$ has zero eigenvalues that both inequalities in (23) are sharp, that is, each holds as an equality for some $u \in H_c$. Therefore, the present bound γ, the bound C in (21) from [5] an [6], and the bound from [16] on the convergence factor of FAC, are all in fact equivalent to different bounds on $\cos(H_{2h}^{2h\text{-harm}}, H_h)$.

5. Asynchronous FAC (AFAC). In this section, we present a variation of the FAC algorithm which decomposes into independent processes, and show the relation of its convergence properties to those of FAC. The basic cycle of this algorithm is defined as follows:

ALGORITHM 5.1. (AFAC [10]) *Let* $u \in H_c$ *be the current approximation to* **u**.

Step 1. *Compute u_{2h} from*

$$u_{2h} \in H_{2h}: \ a(u - u_{2h}, v_{2h}) = f(v_{2h}), \ \ \forall v_{2h} \in H_{2h}$$

Step 2. *Compute u_h and w_{2h} from*

$$\begin{aligned} u_h \in H_h: &\quad a(u - u_h, v_h) = f(v_h), &\quad \forall v_h \in H_h, \\ w_{2h} \in H_{2h} \cap H_h: &\quad a(u - w_{2h}, v_{2h}) = f(v_{2h}), &\quad \forall v_{2h} \in H_{2h} \cap H_h \end{aligned}$$

Step 3. *Set $u \leftarrow u - (u_{2h} - w_{2h} + u_h)$.*

REMARK 5.2. The three equations in Steps 1 and 2 can be solved independently of each other, allowing for simultaneous solvers in a multiprocessor computing system. In addition, when the equation for u_h in Step 2 are solved approximately by the full multigrid method, an approximate value of w_{2h} is available at no extra cost in the process of solving for u_h. The algorithm as presented here requires synchronization before Step 3. A completely asynchronous version of the algorithm is obtained by avoiding Step 3 and adding the replacement

$$u \leftarrow u - u_{2h}$$

into Step 1 and the replacement

$$u \leftarrow u - (u_h - w_{2h})$$

into Step 2. Then both steps can be assigned to asynchronously running processes, which requires only locking of memory locations during replacement.

LEMMA 5.3. *The error $e = \mathbf{u} - u$ is transformed by Algorithm 5.1 according to*

$$e \leftarrow e - (P_{H_{2h}^{2h\text{-harm}}} + P_{H_h})e.$$

Proof. Writing $f(v) = a(\mathbf{u}, v)$, we have

$$u_{2h} = -P_{H_{2h}}e, \quad u_h = -P_{H_h}e, \quad w_{2h} = -P_{H_{2h} \cap H_h}e.$$

The proof is concluded by noting that $P_{H_{2h}} - P_{H_{2h} \cap H_h} = P_{H_{2h}^{2h\text{-harm}}}$. □

Using Lemma 2.4, we obtain the following theorem.

THEOREM 5.4. *The convergence factor of AFAC (Algorithm 5.1) is the square root of the convergence factor of FAC (Algorithm 3.1). In particular,*

$$\rho\left(I - P_{H_{2h}^{2h\text{-harm}}} - P_{H_h}\right) = ||I - P_{H_{2h}^{2h\text{-harm}}} - P_{H_h}|| = \cos(H_{2h}^{2h\text{-harm}}, H_h) \le \gamma.$$

Proof. By Lemma 2.4, the extreme eigenvalues of $P_{H_{2h}^{2h\text{-harm}}} + P_{H_h}$ are $1 - \cos(H_{2h}^{2h\text{-harm}}, H_h)$ and $1 + \cos(H_{2h}^{2h\text{-harm}}, H_h)$. It follows that $\rho\left(I - P_{H_{2h}^{2h\text{-harm}}} - P_{H_h}\right) = \cos(H_{2h}^{2h\text{-harm}}, H_h)$. □

REMARK 5.5. This theorem and Lemma 2.4 are related to a result of P. Bjørstad [2], which gives a more complete characterization of the spectrum of the sum of two orthogonal projections using special properties of finite element spaces. A paper containing a generalization of both Theorem 5.4 and some results from [2] is in preparation [3].

REMARK 5.6. Both FAC and AFAC algorithms generalize easily to the case of more refinement levels. Unfortunately, the theory does not carry over immediately. A theory can be developed for the multilevel algorithm in certain model situations [14]. A general multilevel theory for AFAC is the subject of current research. For a convergence bound on a modified multilevel AFAC method, see [9].

REMARK 5.7. The classical Schwarz alternating method is based on geometrical notions of partitioning and overlap, but it can be easily generalized to subspaces [12]. In this way, the refinement methods treated here can be considered as general Schwarz methods. In particular, FAC can be interpreted as the classical "multiplicative" Schwarz process applied to the subspaces H_h and H_{2h}, with "overlap" $H_{2h} \cap H_h$. AFAC corresponds to an "additive" version of the Schwarz method applied to H_{2h} and $H_h \cap (H_{2h} \cap H_h)^{\perp}$, which do not "overlap".

These refinement methods can also be viewed as block relaxation schemes with blocks corresponding to their respective spaces. Thus, FAC and AFAC can be interpreted as block Gauss-Seidel and Jacobi methods, respectively. The relationship between their convergence factors noted in Theorem 5.4 can be obtained as a direct consequence of this viewpoint, cf. [10]. Truly asynchronous FAC corresponds to the well-known method of chaotic relaxation. Finally, this interpretations shows that the symmetrization of FAC as in [5] is analogous to the usual scheme for symmetrizing relaxation that follows each sweep with another in reverse order.

REFERENCES

[1] R. E. BANK AND T. DUPONT, *Analysis of a two-level scheme for solving finite element equations,* Report CNA-159, University of Texas at Austin, Austin, TX, 1980.

[2] P. BJØRSTAD, *Multiplicative and additive Schwarz: Convergence in the 2 domain case,* these proceedings.

[3] P. BJØRSTAD AND JAN MANDEL, *On the spectrum of the sum of orthogonal projections with applications in parallel computing,* in preparation.

[4] D. BRAESS, *The convergence rate of a multigrid method with Gauss-Seidel relaxation for the Poisson equation,* in Multigrid Methods, Proceedings, W. Hackbusch and U. Trottenberg, eds., Lecture Notes in Mathematics 960, Springer-Verlag, Berlin, 1982.

[5] J. H. BRAMBLE, D. E. EWING, J. E. PASCIAK, AND A. H. SCHATZ, *A preconditioning technique for the efficient solution of problems with local grid refinement,* Comp. Meth. in Appl. Mech. Eng., to appear.

[6] J. H. BRAMBLE, D. E. EWING, J. E. PASCIAK, AND A. H. SCHATZ, *The construction of preconditioners for elliptic problems by substructuring. I,* Math. Comp. 47(1986)103-134.

[7] A. BRANDT, *Algebraic multigrid theory: The symmetric case,* Appl. Math. Comput. 19(1986)23-56.

[8] PH. G. CIARLET, *The Finite Element Method for Elliptic Problems,* North-Holland, Amsterdam, 1972.

[9] M. DRYJA AND O. WIDLUND, *Iterative refinement methods,* these proceedings.

[10] L. HART AND S. MCCORMICK, *Asynchronous multilevel adaptive methods for solving partial differential equations on multiprocessors,* manuscript, University of Colorado at Denver, 1987.

[11] M. KOČVARA AND J. MANDEL, *A multigrid method for three-dimensional elasticity and algebraic convergence estimates,* Appl. Math. Comput. 23(1987)121-135.

[12] P. L. LIONS, *On the Schwarz alternating method. I,* First International Symposium on Domain Decomposition Methods, Proceedings, edited by R. Glowinski, G. H. Golub, G. A. Meurant, and J. Périaux, SIAM, Philadelphia, 1988.

[13] J. F. MAITRE AND F. MUSY, *The contraction number of a class of two-level methods; an exact evaluations for some finite element subspaces and model problems,* in Multigrid Methods, Proceedings, W. Hackbusch and U. Trottenberg, eds., Lecture Notes in Mathematics 960, Springer-Verlag, Berlin, 1982.

[14] J. MANDEL AND S. MCCORMICK, *Iterative solution of elliptic equations with refinement: A model multilevel case,* in preparation.

[15] T. MANTEUFFEL, *The Tchebychev iteration for nonsymmetric linear systems,* Numerische Mathematik 28(1977)307-327.

[16] S. F. MCCORMICK, *Fast adaptive composite (FAC) methods: Theory for the variational case,* in Error Asymptotics and Defect Correction, Proceedings, K. Böhmer, H, J. Stetter, eds., Computing Supplementum 5, Springer-Verlag, Wien, 1984.

[17] S. MCCORMICK AND J. THOMAS, *The fast adaptive composite grid method for elliptic boundary value problems,* Math. Comp. 46(1986)439-456.

Iterative Solution of Elliptic Equations with Refinement: The Model Multi-Level Case*

Jan Mandel†
Steve McCormick††

Abstract. A multi-level theory for AFAC [2] in a model problem case is developed that is based on *sectional* energy norm estimates for harmonic functions. This is the companion paper of [3], which treats the two-level case.

Key Words. Composite Grid, Refinement, Iterative Methods, Orthogonal Projections, Elliptic Partial Differential Operators, Finite Elements, Parallel Computation, FAC

1. Introduction. This paper assumes that the reader is familiar with the companion paper [3]. Here we consider the multi-level case for AFAC [2]. We first extend the notation of [2] (Section 2), then develop a theoretical structure for convergence (Section 3), and finally consider a model problem in a specialized geometry (Section 4).

The convergence bounds obtained here involve the relative size of successive refinement regions, that is, the maximum ratio of areas of successive refinement regions. In fact, as this maximum tends to zero, the bounds we obtain tend to the two-grid bounds developed in [3]. The multi-level theory in [1] also obtains bounds that depend on the maximum refinement ratio; this theory applies only to a modified version of AFAC, but unlike ours it does not require that this ratio be sufficiently small.

---

* This research was sponsored by the Air Force Office of Scientific Research under grant AFOSR-86-0126 and by the National Science Foundation under grant DMS-8704169.

† Computational Mathematics Group, University of Colorado at Denver, 1200 Larimer Street, Denver, CO 80204.

‡ Computational Mathematics Group, University of Colorado at Denver, 1200 Larimer Street, Denver, CO 80204, and Colorado Research Development Corporation, P.O. Box 9182, Fort Collins, CO 80525.

2. Notation. For simplicity, we restrict our attention to a 2D diffusion operator with Dirichlet boundary conditions. Let Ω_1 be a polygonal domain in $\mathbf{R}^2$ and $\Omega_1 \supset \Omega_2 \supset \ldots \supset \Omega_k$ a nested sequence of polygonal, or union of disjoint polygonal, domains. Let $H_i \subset H^1_o(\Omega_i)$ be a conforming Lagrange finite element space associated with $\Omega_i, 1 \leq i \leq k$. We assume for simplicity that the element boundaries of Ω_i contain the boundary of Ω_{i+1} and that every element in Ω_{i+1} associated with H_i is the union of some elements associated with $H_{i+1}, 1 \leq i \leq k-1$. We define the ith *composite grid space*

$$H_{c_i} = \sum_{j=i}^{k} H_j$$

and denote $H_c = H_{c_1}$. Consider the bilinear form

$$a(u,v) = \int_\Omega a \nabla u \cdot \nabla v \, d\Omega,$$

with the piecewise smooth diffusion coefficient $a > const > 0$ in Ω. Also consider the linear form

$$f(v) = \int_\Omega f \, v \, d\Omega.$$

The space H_c is equipped with the inner product $a(\cdot,\cdot)$ and its norm $\|\cdot\|$ defined by $\|u\| = (a(u,u))^{\frac{1}{2}}$. In what follows, references to inner products and norms (e.g., 'orthogonality' and 'unit functions') refer to these definitions.

We are interested in the iterative solution of the *discrete variational equation*

$$\mathbf{u} \epsilon H_c : \quad a(\mathbf{u}, v) = f(v) \quad \forall v \epsilon H_c. \tag{1}$$

The symbol $\mathbf{u}$ will be used to denote the solution of (1). The basic cycle of AFAC we consider here for solving (1) is as follows:

Algorithm (AFAC [2]) Let $u \epsilon H_c$ be the current approximation to $\mathbf{u}$. For each $i = 1, 2, \ldots, k$, compute u_i from

$$u_i \epsilon H_i : \quad a(u - u_i, v_i) = f(v_i) \quad \forall v_i \epsilon H_i.$$

For each $i = 1, 2, \ldots, k-1$, compute w_i from

$$w_i \epsilon H_i \cap H_{i+1} : \quad a(u - w_i, v_i) = f(v_i) \quad \forall v_i \epsilon H_i \cap H_{i+1}.$$

With $w_k = 0$, set

$$u \leftarrow u - \sum_{i=1}^{k} (u_i - w_i).$$

Our analysis of AFAC will involve the use of *discrete harmonic functions* and their associated spaces and projections. In particular, for each $i = 1, 2, \ldots, k-1$, define the *i-harmonics*

$$H_i^{i-harm} = \{u_i \epsilon H_i : \quad a(u_i, v_i) = 0 \quad \forall v_i \epsilon H_i \cap H_{i+1}\}.$$

Note that H_i^{i-harm} is just the orthogonal complement of $H_i \cap H_{i+1}$ in H_i. Let P_i^{i-harm} be the orthogonal projector mapping H_c onto H_i^{i-harm}. For each $i = 1, 2, \ldots, k-1$, define the *c_i-harmonics*

$$H_{c_i}^{harm} = \{u_i \epsilon H_{c_i} : \quad a(u_i, v_i) = 0 \quad \forall v_i \epsilon H_{c_{i+1}}\}.$$

Note that

$$H_{c_i} = H_{c_i}^{harm} \oplus H_{c_{i+1}} \tag{2}$$

is an orthogonal decomposition, $1 \leq i < k$. For convenience we define $H_{c_k}^{harm} = H_{c_k}$ and $H_{c_{k+1}} = H_{k+1} = \emptyset$ so that (2) holds for all $i = 1, 2, \ldots, k$. Note that in the decomposition $u_{c_i} = u_{c_i}^{harm} + u_{c_{i+1}}$, $u_{c_i}^{harm}$ is just the c_i–harmonic in $H_{c_i}^{harm}$ that agrees with u_{c_i} in $\Omega_i \backslash \Omega_{i+1}$. Let $P_{c_i}^{harm}$ be the orthogonal projector mapping H_c onto $H_{c_i}^{harm}$.

As in the two-level case [3], the convergence rate for AFAC depends on a measure of how much an i–harmonic deviates from being c_i–harmonic, although now we need to be more precise. In particular, first note that *local* functions with support in $\overline{\Omega}_i \backslash \Omega_{i+1}$ are both i–harmonic and c_i–harmonic. (Overbar denotes set closure.) The space of these functions we denote by H_i^{local}. Note that $H_i^{local} \subset H_i^{i-harm} \cap H_{c_i}^{harm}$. We then define *i-doubly–harmonic* functions by

$$H_i^{i-dharm} = \left\{u_i \epsilon H_i^{i-harm} : a(u_i, v_i) = 0 \quad \forall v_i \epsilon H_i^{local}\right\}.$$

We similarly define *c_i–doubly–harmonic* functions by

$$H_{c_i}^{dharm} = \left\{u_i \epsilon H_{c_i}^{harm} : a(u_i, v_i) = 0 \quad \forall v_i \epsilon H_i^{local}\right\}.$$

Note that

$$H_i^{i-harm} = H_i^{local} \oplus H_i^{i-dharm} \tag{3}$$

is an orthogonal decomposition. (We assume the trivial definitions for these spaces with $i = k$ so that (3) holds for all $i = 1, 2, \ldots, k$.) Define P_i^{local} and $P_i^{i-dharm}$ to be their corresponding orthogonal projectors, that is, P_i^{local} and $P_i^{i-dharm}$ are the respective orthogonal projectors from H_c onto H_i^{local} and $H_i^{i-dharm}$. Then one error measure we will need is given by $\epsilon = \max_{1 \leq i < k} \|(I - P_{c_i}^{harm}) P_i^{i-dharm}\|$. Note that ϵ is just the maximal two-grid estimate defined in [3]:

$$\begin{aligned}
\epsilon = \max_{1 \leq i < k} \{\|e_{i+1}\| : e_{i+1} &= (I - P_{c_i}^{harm}) u_i^{i-dharm}, \\
u_i^{i-dharm} \epsilon H_i^{i-dharm}, &\ \|u_i^{i-dharm}\| = 1\} \\
= \max_{1 \leq i < k} \{\|e_{i+1}\| : e_{i+1} &= (I - P_{c_i}^{harm}) u_i^{i-harm}, \\
u_i^{i-harm} \epsilon H_i^{i-harm}, &\ \|u_i^{i-harm}\| = 1\}.
\end{aligned}$$

We introduce the *sectional energy inner product* $a_{\Omega_j}(\cdot,\cdot) : H_c \times H_c \to \mathbf{R}$ defined for each $j = 1,2,\ldots,k$ by

$$a_{\Omega_j}(u,v) = \int_{\Omega_j} a \nabla u \nabla v \, d\Omega.$$

The *sectional energy norm* is defined by $\|u\|_{\Omega_j} = (a_{\Omega_j}(u,u))^{\frac{1}{2}}$. We also define the *harmonic section measures* for $1 \le i \le k-2$ and $i+2 \le j \le k$ by

$$\delta_{ij} = \max\left\{\|u_i^{i-harm}\|_{\Omega_j} : u_i^{i-harm} \epsilon H_i^{i-harm}, \|u_i^{i-harm}\| = 1\right\}$$

and for $1 \le i \le k-1$ by

$$\delta_{i\,i+1} = \max\left\{|a(u_i^{i-harm}, v_{i+1})| : u_i^{i-harm} \epsilon H_i^{i-harm}, v_{i+1} \epsilon H_{i+1}^{(i+1)-dharm}, \|u_i^{i-harm}\| = \|v_{i+1}\| = 1\right\}.$$

Further, define $\delta_{i\,i} = 0$ and $\delta_{ij} = \delta_{ji}$, $i > j$, and $\delta = \max_{1 \le i \le k} \sum_{j=1}^{k} \delta_{ij}$.

3. Theory. Let $\mathbf{u}$ denote the solution of (1), u the current approximation, and $e = \mathbf{u} - u$ the *algebraic error*. Then one cycle of AFAC applied to u transforms e according to

$$e \leftarrow (I - \sum_{i=1}^{k} P_i^{i-harm})e. \tag{4}$$

Note that if the subspaces H_i^{i-harm} and $H_{c_i}^{harm}$ agree for all i, then $H_c = H_1^{1-harm} \oplus H_2^{2-harm} \oplus \ldots \oplus H_k^{k-harm}$ would be an orthogonal decomposition of H_c. This means that (4) would converge to zero in one cycle, making AFAC a direct solver. In fact, this can happen in practice, namely, for typical one-dimensional problems and higher-dimensional problems with refinement regions Ω_i wholly contained in just one element of the coarser level, H_{i-1}. However, most practical problems have $H_i^{i-harm} \neq H_{c_i}^{harm}$, for which we have the following theorem.

THEOREM 3.1. *Suppose $\gamma = \epsilon + \delta < 1$. Then a bound on the AFAC convergence factor is given by*

$$\rho(I - \sum_{i=1}^{k} P_i^{i-harm}) = \|I - \sum_{i=1}^{k} P_i^{i-harm}\| \le \gamma. \tag{5}$$

PROOF: Let u_i^{i-harm} be unit functions in H_i^{i-harm}. Let $U = (u_{ij})$ be the Gramm matrix for these functions defined by $u_{ij} = a(u_i^{i-harm}, u_j^{j-harm})$. In [2; lemma 2.3] we proved in effect that

$$\rho(I - \sum_{i=1}^{k} P_i^{i-harm}) = \sup\{\rho(I - U)\},$$

where the supremum is taken over all possible choices for u_i^{i-harm}. Thus, to bound the convergence factor for AFAC, we need only bound $\rho(I-U)$ for arbitrary choices of u_i^{i-harm}. To bound $\rho(I-U)$, decomposing by (3) and (2), we have

$$u_i^{i-harm} = \sqrt{1-\beta_i^2}\, x_i^{local} + \beta_i v_i, \; v_i = \sqrt{1-\epsilon_i^2}\, y_{c_i}^{dharm} + \epsilon_i w_{c_{i+1}}$$

where $|\beta_i| \le 1$, $|\epsilon_i| \le \epsilon$, and $x_i^{local} \epsilon H_i^{local}$, $v_i \epsilon H_i^{i-dharm}$, $y_{c_i}^{dharm} \epsilon H_{c_i}^{dharm}$, and $w_{c_{i+1}} \epsilon H_{c_{i+1}}$, which are all unit functions. Define $\theta_i = \frac{\epsilon_i}{\epsilon} a(w_{c_{i+1}}, x_{i+1}^{local})$ and note that $|\theta_i| \le 1$. Now from $a(x_i^{local}, u_{i+1}^{(i+1)-harm}) = 0$ and $a(y_{c_i}^{dharm}, x_{i+1}^{local}) = 0$, we have

$$u_{i\,i+1} = \epsilon\beta_i \sqrt{1-\beta_{i+1}^2}\, \theta_i + \beta_i \beta_{i+1} a(v_i, v_{i+1}).$$

By definition,

$$|a(v_i, v_{i+1})| \le \delta_{i\,i+1},$$

and

$$|u_{ij}| = |a(u_i^{i-harm}, u_j^{j-harm})| \le \delta_{ij}, \quad j > i+1.$$

Now since $u_{ii} = 1$ and $\beta_i \beta_{i+1} \le 1$, we can write $U = I + \epsilon T + E$, where the symmetric matrices $T = (t_{ij})$ and $E = (e_{ij})$ satisfy

$$t_{ij} = \begin{cases} \beta_i \sqrt{1-\beta_{i+1}^2}\, \theta_i & j = i+1 \\ t_{ji} & j = i-1 \\ 0 & otherwise \end{cases}$$

and

$$|e_{ij}| \le \delta_{ij}$$

for $1 \le i, j \le k$. Thus,

$$\rho(I-U) \le \epsilon\rho(T) + \delta.$$

The theorem would now be proved if we could show that $\rho(T) \le 1$. This we do by way of the following lemma.

LEMMA 3.1. *Suppose T is a $k \times k$ tridiagonal matrix of the form $T = tridiag$ $(\beta_{i-1}\sqrt{1-\beta_{i+1}^2}\,\theta_{i-1} \;\; 0 \;\; \beta_i \sqrt{1-\beta_{i+1}^2}\,\theta_i)$, where β_i, θ_i satisfy $|\beta_i| \le 1$, $|\theta_i| \le 1$, $1 \le i \le k$, and $\beta_{k+1} = 0$ (for convenience). Then $\rho(T) < 1$.*

PROOF: If some $\beta_i = 0$ then T reduces to matrices of a smaller size. So, without loss of generality, we assume $\beta_i \ne 0$, $1 \le i \le k$. We shall first prove that $I+T$ is positive definite. Let a_i be the i-th pivot in the LU–decomposition of $I+T$. Then

$$a_1 = 1, \quad a_{i+1} = 1 - \frac{\beta_i^2(1-\beta_{i+1}^2)\theta_i^2}{a_i}.$$

It is easy to see by induction that $a_i \ge \beta_i^2 > 0$. Thus, $I+T$ is positive definite because all pivots are positive. Now by changing the signs of all of the β_i, we can conclude from the above that $I - T$ is also positive definite. Thus, $\rho(T) < 1$ and the lemma is proved.

4. Model Problem. The convergence bounds of the previous section depend on estimates of the "two-grid" factor ϵ and the "total section" δ. While suitable bounds on ϵ have been obtained in a fairly general setting (cf. [1-3]), there appears to be no such theory for δ. Since δ is the sum of integrals of harmonic functions over decreasingly smaller subregions, it is likely that δ can be made as small as desired by requiring that successive refinement regions cover a sufficiently small relative area. We now prove this for a model problem with special geometry.

Our model problem is the Poisson version of (1), with $a = 1$, on the unit square $\Omega = [0,1] \times [0,1]$. Our special geometry is based on uniform grids Ω_i that consist of the respective $m_i - 1$ and $n_i - 1$ interior vertical and horizontal grid lines covering the region $[0, \eta_i] \times [0,1]$, $1 \le i \le k$. We require that every other grid line of Ω_{i+1} coincide with the vertical and horizontal grid lines of Ω_i, and that

$$\frac{m_{i+1}}{\eta_{i+1}} = 2\frac{m_i}{\eta_i} \qquad \text{and} \qquad n_{i+1} = 2n_i, \; 1 \le i < k.$$

To simplify the presentation, we will assume that $m_1 = n_1 \ge 2$ and that the refinement "speed" is constant: $\eta_{i+1} = \eta\eta_i$ for $0 < \eta < 1$, $1 \le i < k$. Finally, we assume that $H_i \subset H_0^1(\Omega_i)$ is the space of continuous piecewise linear functions on a natural triangulation of $\overline{\Omega}_i$: the triangles are formed by connecting the lower left and upper right vertices of each rectangle in $\overline{\Omega}_i$.

In this section, $\|\cdot\|_2$ will be used to denote the Euclidean norm on various "nodal vectors," i.e., vectors of node values of certain functions in $H_0^1(\Omega)$. We will refer to "nodal matrices" associated with certain discrete spaces H. By this we mean the usual stiffness matrix that arises from transforming the bilinear problem (1) on H to an equation involving the nodal values of the discrete solution $\mathbf{u}$ in H. Note that the nodal matrix on the uniform space H_i is just the one that corresponds to the usual 5-point difference stencil

$$\begin{pmatrix} & -1 & \\ -1 & 4 & -1 \\ & -1 & \end{pmatrix}.$$

The following lemmas will be used to bound nodal values of harmonic functions. The first allows us to reduce the two-dimensional problem of obtaining these bounds to a one-dimensional one.

LEMMA 4.1. *Suppose A and B are $m \times m$ symmetric positive definite block tridiagonal matrices consisting of the respective $n \times n$ blocks*

$$A_{ij} = \begin{cases} X & i = j \\ -I & |i-j| = 1 \\ 0 & \textit{otherwise} \end{cases}$$

and

$$B_{ij} = \begin{cases} Y & i = j \\ -I & |i-j| = 1 \\ 0 & otherwise \end{cases} .$$

Suppose $X \geq Y$ (i.e., $X - Y$ is nonnegative definite), X and Y commute, and that $x = (x_1^t \cdots x_m^t)^t$ and $y = (y_1^t \cdots y_m^t)^t$ satisfy

$$Ax = s \quad \text{and} \quad By = t$$

where $s_j = t_j = 0$, $1 \leq j < q \leq m$. Suppose $x_q = y_q$. Then $\|x_j\|_2 \leq \|y_j\|_2$, $1 \leq j \leq q$.

PROOF: By performing elementary block operations on the augmented matrices $(A : s)$ and $(B : t)$, it is easy to see that

$$x_j = \left(\prod_{l=j}^{q-1} S_l^{-1} \right) x_q$$

and

$$y_j = \left(\prod_{l=j}^{q-1} T_l^{-1} \right) x_q$$

where $S_1 = X$, $S_{l+1} = X - S_l^{-1}$, $T_1 = Y$, and $T_{l+1} = Y - T_i^{-1}$, $1 \leq l < q$. By induction it is easy to see that $S_l \geq T_l$ and, hence, $S_l^{-1} \leq T_l^{-1}$. The lemma now follows from using the fact that S_l^{-1} and T_l^{-1} are rational functions of X and Y, respectively.

In the following, A will be the nodal matrix of the Laplace equation on H_i, that is, $X = diag[-1 \quad 4 \quad -1]$; B will be given by $Y = 2I$.

LEMMA 4.2. *Suppose A in Lemma 4.1 is the nodal matrix for H_i. Then $\|x_j\|_2 \leq \frac{j}{q}\|x_q\|_2$. This also holds for the nodal matrix for H_{c_i} provided the indices j and q are interpreted with respect to grid Ω_i lines.*

PROOF: Using Lemma 4.1 with $Y = 2I$, we have

$$\|x_j\|_2 \leq \gamma_j \|x_q\|_2$$

with equality for the case $n = 1$, where

$$\gamma_j = \rho \left(\prod_{l=j}^{q-1} T_l^{-1} \right)$$

is independent of n. We can thus examine the case $n = 1$ which corresponds to a simple one–dimensional case where the scalars x_l satisfy $x_j = \frac{j}{q} x_q$. This proves the first assertion. The final assertion follows from a similar argument using the fact that the presence of the interfaces $x = \eta_l$ with $l > i + 1$ cannot increase $\|x_j\|_2$ for any j (which we assume corresponds to a grid Ω_i line). This proves the lemma.

Our final lemma will be used to bound nodal values of harmonics at grid interfaces.

LEMMA 4.3. *Let $u_i^{i-dharm}$ be a unit function in $H_i^{i-dharm}$ and let w_j be the n_i–dimensional vector of its values at the nodes on the vertical grid line $x = \eta_i(1-\frac{j}{m_i})$, $1 \le j \le m_i$. Then*

$$\|w_{r_i}\|_2 \le \left(\frac{r_i l_i}{r_i + l_i}\right)^{1/2} \tag{8}$$

and

$$\|w_1\|_2 \le \left(\frac{l_i}{r_i(r_i + l_i)}\right)^{1/2} \tag{9}$$

where $r_i = m_i - l_i$ and $l_i = m_{i+1}/2$, the number of vertical grid lines of $\overline{\Omega}_i$ to the respective right and left of $x = \eta_{i+1}$. Moreover, (9) also holds if we replace the function $u_i^{i-dharm}$ by a unit function $u_{c_i}^{dharm}$ in $H_{c_i}^{dharm}$.

PROOF: By Lemma 4.2 and the fact that $u_i^{i-dharm}$ is i–doubly harmonic, we have

$$\|w_j\|_2 \le \frac{j}{r_i}\,\|w_{r_i}\|_2\,, \quad 1 \le j \le r_i, \tag{10}$$

and

$$\|w_{r_i+j}\|_2 \le \frac{l_i - j}{l_i}\,\|w_{r_i}\|_2. \tag{11}$$

Since $u_i^{i-dharm}$ is of unit length, so is its vector of nodal values, w, in the sense that $(Aw)^t w = 1$. Since $u_i^{i-dharm}$ is i–doubly harmonic, all but the $r_i th$ entry of Aw is nonzero. Using $j = r_i - 1$ in (10) and $j = 1$ in (11), we thus have

$$\begin{aligned}
1 &= \left(Xw_{r_i} - w_{r_i-1} - w_{r_i+1}\right)^t w_{r_i} \\
&\ge 2\|w_{r_i}\|_2^2 - \|w_{r_i-1}\|_2\|w_{r_i}\|_2 - \|w_{r_i+1}\|_2\|w_{r_i}\|_2 \\
&\ge \left(2 - \frac{r_i - 1}{r_i} - \frac{l_i - 1}{l_i}\right)\|w_{r_i}\|_2^2 \\
&= \left(\frac{1}{r_i} + \frac{1}{l_i}\right)\|w_{r_i}\|_2^2.
\end{aligned}$$

This proves (8). Now using Lemma 4.2 with $j = 1$ and $q = r_i$, we have

$$\|w_1\|_2 \le \frac{1}{r_i}\left(\frac{r_i l_i}{r_i + l_i}\right)^{1/2}$$

which proves (9). The final assertion follows from Lemma 4.2, so the proof is complete.

THEOREM 4.1. *For the model problem and special geometry considered here, if $\eta < \frac{1}{2}$ then*

$$\delta < 2\left(\eta + \frac{\sqrt{2\eta}}{1 - \sqrt{\eta}}\right).$$

Thus, $\delta = 0(\sqrt{\eta})$.

PROOF: Let v_{i+1} and v_i be arbitrary unit functions in $H_{i+1}^{(i+1)-dharm}$ and $H_i^{i-dharm}$, respectively. Then

$$|a(v_i, v_{i+1})| \le |a_{\Omega_i \setminus \Omega_{i+2}}(v_i, v_{i+1})| + |a_{\Omega_{i+2}}(v_i, v_{i+1})|.$$

Define $\overline{v}_i \epsilon H_i$ by

$$\overline{v}(x,y) = \begin{cases} 0 & x \le \eta_{i+1} - \frac{1}{n_{i+1}} \\ v_i & x \ge \eta_{i+1} \end{cases}$$

and $\tilde{v}_i \epsilon H_i$ by

$$\tilde{v}_i(x,y) = \begin{cases} v_i(x,y) & x \le \eta_{i+2} \\ v_i(2\eta_{i+2} - x, y) & \eta_{i+2} \le x \le 2\eta_{i+2} \\ 0 & x \ge 2\eta_{i+2} \end{cases}.$$

Then $v_i - \overline{v}_i - \tilde{v}_i \epsilon H_{i+1}^{local}$ and $\|\overline{v}_i\| = \sqrt{2}\|v_i\|_{\Omega_{i+2}}$, so

$$\begin{aligned} |a(v_i,\, v_{i+1})| &= |a(\overline{v}_i + \tilde{v}_i, v_{i+1})| \\ &\le |(Jz)^t w| + \sqrt{2}\delta_{i\,i+2}, \end{aligned}$$

where z and w are the vectors of node values of v_i and v_{i+1} on $x = \eta_{i+1}$ and $x = \eta_{i+1} - \frac{1}{n_{i+1}}$, respectively, and where J is the linear interpolation operator mapping Ω_i nodal vectors on $x = \eta_{i+1}$ to Ω_{i+1} nodal vectors on $x = \eta_{i+1}$.

Note that $\|J\|_2 \le \sqrt{2}$ so that

$$|(Jz)^t w| \le \sqrt{2}\ \|z\|_2\ \|w\|_2.$$

By (8) and (9), and the relations $r_{i+1} = 2r_i$ and $\frac{l_i}{r_i + l_i} = \eta$, we have

$$\begin{aligned} \delta_{i\,i+1} &\le \sqrt{2}\left(\left(\frac{r_i l_i}{r_i + l_i}\right)\left(\frac{l_{i+1}}{r_{i+1}(r_{i+1} + l_{i+1})}\right)\right)^{1/2} + \sqrt{2}\,\delta_{i\,i+2} \\ &\le \eta + \sqrt{2}\delta_{i\,i+2}. \end{aligned} \tag{12}$$

Now let $j \ge i+2$ and define z_q as the n_i–dimensional vector of nodal values of v_i on $x = \frac{q}{m_i}\eta_i$, $0 \le q \le m_i$. Assume for the moment that $x = \eta_j$ coincides with a grid line of Ω_i, say the pth one. Let A and X be as in Lemma 4.2. It holds that

$$Xz_q - z_{q-1} - z_{q+1} = 0,\ q = 1, \ldots, l_i - 1. \tag{13}$$

Let

$$z_{l_i} = \sum_k c_k u_k,$$

where u_k are eigenvectors of X, $Xu_k = \lambda_k u_k$, $\|u_k\|_2 = 1$. Note that $\lambda_k \epsilon (2,6)$. Then by (13) we have

$$z_q = \sum_k c_k t_k(q) u_k,$$

where

$$t_k(q) = \frac{\mu_k^q - \mu_k^{-q}}{\mu_k^{l_i} - \mu_k^{-l_i}} = \frac{\sinh(q ln \mu_k)}{\sinh(l_i ln \mu_k)}$$

and where $\mu_k = \frac{\lambda_k + \sqrt{\lambda_k^2 - 4}}{2} > 1$ and $\mu_k^{-1} < 1$ are the roots of the characteristic equation $\lambda_k - \frac{1}{\mu_k} - \mu_k = 0$.

Writing $e_q = ((X - I)z_q - z_{q-1})^t z_q$, we have from the fact that z_q are values of a discrete harmonic function that

$$a_{\Omega_j}(v_i, v_i) = e_p, \; a_{\Omega_{i-1}}(v_i, v_i) = e_{l_i} \le \|v_i\|^2 = 1.$$

Substituting for z_q, we have

$$e_q = \sum_k c_k^2 (\lambda_k t_k^2(q) + (t_k(q) - t_k(q-1)) t_k(q)).$$

From the properties of hyperbolic functions, it follows that

$$\frac{t_k(p)}{t_k(l_i)} = \frac{\sinh(p ln \mu_k)}{\sinh(l_i ln \mu_k)} < \frac{p}{r}$$

and

$$\frac{(t_k(p) - t_k(p-1)) t_k(p)}{(t_k(l_i) - t_k(l_i - 1)) t_k(l_i)} = \frac{\cosh(\frac{2p-1}{2} ln \mu_k)}{\cosh(\frac{2l_i - 1}{2} ln \mu_k)} \frac{\sinh(p ln \mu_k)}{\sinh(l_i ln \mu_k)} < \frac{p}{l_i}.$$

Consequently,

$$\|v_i\|_{\Omega_j}^2 = a_{\Omega_j}(v_i, v_i) \le \frac{e_p}{e_{l_i}} < \frac{p}{l_i} = \mu^{j-i-1}.$$

Hence,

$$\delta_{ij} \le \eta^{\frac{j-i-1}{2}}, \; j > i + 1. \tag{14}$$

To see now that (14) also holds for the case that $x = \eta_j$ is not a grid line of Ω_i, it is enough to recognize that $\nabla u \cdot \nabla u$ for u in H_i is constant on triangles of Ω_i.

Now from (12) and (14) we have that

$$\delta < 2\left(\eta + \sqrt{2} \sum_{q=1}^{\infty} \eta^{q\backslash 2}\right) = 2\left(\eta + \frac{\sqrt{2\eta}}{1 - \sqrt{\eta}}\right),$$

which proves the theorem.

References

[1] M. Dryja and O. Widlund, *Iterative refinement methods*, these proceedings.

[2] L. Hart and S. McCormick, *Asynchronous multilevel adaptive methods for solving partial differential equations on multiprocessors*, manuscript, University of Colorado at Denver, 1987.

[3] J. Mandel and S. McCormick, *Iterative solution of elliptic equations with refinement: the two-level case*, these proceedings.

Remarks on Spectral Equivalence of Certain Discrete Operators

Wlodzimierz Proskurowski*

Abstract We consider the Neumann-Dirichlet preconditioner for the discrete Laplacian in the unit square. We show that the capacitance matrix C is equal to the identity even for problems with discontinuous coefficients. In the case of many Neumann-Dirichlet strips this is no longer true if the strips are extremely thin. The conditioning of C in this case is significantly better when the Neumann strips correspond to regions with larger coefficients.

1. Introduction It is well known that all uniformly elliptic operators L defined on Ω with the same boundary conditions are *spectrally equivalent* [5]:

$$c_1(L_1x,x) \le (L_2x,x) \le c_2(L_1x,x),$$

where c_1, c_2 are positive constants, and (x,y) is a proper inner product. Similar relations hold for the descretized form of these operators:

$$a_1x^TAx \le x^TBx \le a_2x^TAx, \quad \forall x \in \Re^n, \quad \forall n,$$

where A, B are nxn symmetric positive definite matrices, and a_1, a_2 are positive constants independent of n. These inequalities imply that the ratio of extreme eigenvalues of AB^{-1} is bounded by a_2/a_1, called the *spectral equivalence bound* [1]. Thus, we could use one discrete elliptic operator as an efficient preconditioner for another one. In particular, the discrete Laplacian would be an excellent candidate for such a preconditioner. Let us have a closer look at some questions that can be posed: How large is the bound for $\kappa(AB^{-1})$? What is the effect of changing the boundary conditions? And finally, how all this relates to domain decomposition?

*Dept. of Mathematics, University of Southern California, Los Angeles, CA 90089-1113.

1.1 Changing boundary conditions Let us consider two nxn matrices, A and B, that represent the one dimensional -Laplacian on $\Omega = [0,1]$ with the Dirichlet and Neumann boundary condition at the left end of the interval, and with the Dirichlet condition at the other end, respectively. They differ only in the (1,1) element:

$$A = \begin{pmatrix} 2 & -1 & & & \\ -1 & 2 & -1 & & \\ & \cdot & \cdot & \cdot & \\ & & -1 & 2 & -1 \\ & & & -1 & 2 \end{pmatrix}, \qquad B = \begin{pmatrix} 1 & -1 & & & \\ -1 & 2 & -1 & & \\ & \cdot & \cdot & \cdot & \\ & & -1 & 2 & -1 \\ & & & -1 & 2 \end{pmatrix}.$$

It has been shown by Hald, see [7, p.457], that AB^{-1} is a rank one modification of an identity, $AB^{-1} = I + uv^T$, where $u^T=(1,0,...,0)$ and $v^T=(n,n-1,...,1)$. Moreover, only two singular values of AB^{-1} differ from 1. These two coincide with the eigenvalues of $I_2 + (v+\alpha u,u)^T(u,v+ \alpha u)$, where $\alpha = v^Tv/2 \approx n^3/6$. Its characteristic polynomial is $\lambda^2-2(1+\alpha+\gamma)\lambda+(1+ \gamma^2+2\gamma)=0$, where $\gamma=u^Tv=n$. Thus, the smallest singular value of AB^{-1} is equal to about $\sigma_{min} \approx 3/n$, the largest $\sigma_{max} \approx n^3/3$, and $\kappa(AB^{-1}) = \sqrt{\sigma_{max} / \sigma_{min}} \approx n^2/3$. Consequently, A and B are far from being spectrally equivalent, although they represent the same operator (-Laplacian), albeit with different boundary conditions.

2. Neumann-Dirichlet preconditioner (two subdomains) Let us now consider the same - Laplacian on $\Omega = [0,1]$ with the Dirichlet boundary conditions at both end points, represented by the nxn matrix A. Let us impose artificial inner boundary conditions at x=0.5 such that we have the Dirichlet condition to the left of it, and the Neumann condition to the right. The resulting matrix B has the form:

$$B = \begin{pmatrix} 2 & -1 & & & & & \\ \cdot & \cdot & \cdot & & & & \\ & -1 & 2 & -1 & & & \\ & & 0 & 2 & -2 & & \\ & & & -1 & 2 & -1 & \\ & & & & \cdot & \cdot & \cdot \\ & & & & & -1 & 2 \end{pmatrix}$$

Using the same technique as above, we can show that $AB^{-1} = I + uv^T$, where $u^T=(0,...,0,1,0,...,0)$, i.e., $u_i = \delta_{ik}$, $v^T = 1/k\ (-1,...-(k-1),0,k-1,...,1)$, and $k = (n+1)/2$. Since now $\gamma = u^Tv = 0$, the characteristic polynomial of $I_2 + (v+\alpha u,u)^T(u,v+ \alpha u)$ is simplified to $\lambda^2 - 2(1+\alpha)\lambda + 1 = 0$. Here, the new $\alpha = v^Tv/2 \approx n/3$. Thus, $\sigma_{min} \approx 3/n$, similarly as before, but $\sigma_{max} \approx n/3$. As a result, we obtain a much more favorable ratio, and $\kappa(AB^{-1}) = \sqrt{\sigma_{max} / \sigma_{min}} \approx 2(\alpha+1) \approx n/3$. Yet again, A and B fail to be spectrally equivalent. Numerical experiments confirm this analysis, see Table 1.

# points	5	15	25	35	45
$\kappa(AB^{-1})$ in **1.1**	11.08	79.49	214.50	416.16	684.50
$\kappa(AB^{-1})$ in **2.**	2.75	6.21	9.59	12.94	16.29

Table 1. Condition numbers $\kappa(AB^{-1})$ as a function of the number of grid points.

In a two dimensional analog of our example, n^2x n^2 matrix $\mathbb{A}$, represents the - Laplacian on a unit square, $\Omega=[0,1]\times[0,1]$ with the Dirichlet boundary conditions at $\partial\Omega$, and $\mathbb{B}$ represents the same operator with added Neumann-Dirichlet conditions at the artificial interface (x;y=0.5):

$$\mathbb{A} = \begin{pmatrix} T & -I & & & \\ -I & T & -I & & \\ & \cdot & \cdot & \cdot & \\ & & -I & T & -I \\ & & & -I & T \end{pmatrix}, \qquad T = \begin{pmatrix} 4 & -1 & & & \\ -1 & 4 & -1 & & \\ & \cdot & \cdot & \cdot & \\ & & -1 & 4 & -1 \\ & & & -1 & 4 \end{pmatrix} \tag{1'}$$

$$\mathbb{B} = \begin{pmatrix} T & -I & & & & & \\ \cdot & \cdot & \cdot & & & & \\ & -I & T & -I & & & \\ & & 0 & T & -2I & & \\ & & & -I & T & -I & \\ & & & & \cdot & \cdot & \cdot \\ & & & & & -I & T \end{pmatrix} \tag{1''}$$

Here, $\mathbb{A}\mathbb{B}^{-1} = I + UV^T$, where the rectangular n^2xn matrices U and V have the form $U^T=(0,I_n,0)$, $V^T=(-X^T,0_n,X^T)$, and mxn matrix X will be defined further, m=n(n-1)/2. As an example, for the simplest case, n=3, we have

$$X = 1/56 \begin{pmatrix} 15 & 4 & 1 \\ 4 & 16 & 4 \\ 1 & 4 & 15 \end{pmatrix}$$

As before, we can conclude that there are only 2n singular values of $\mathbb{A}\mathbb{B}^{-1}$ that differ from 1.

$N=n^2$	λ_{min}	λ_{max}	cond #
9	.76	1.31	1.72
25	.66	1.51	2.28
49	.60	1.68	2.82
81	.55	1.83	3.35

Table 2. Extreme eigenvalues and the condition number of $A\mathbb{B}^{-1}$ in 2D.

Table 2 demonstrates the results of numerical experiments. Here, the condition number grows linearly with $\sqrt{N}$ (approximately as $\sqrt{N}/4+1$), yet, once more, $\mathbb{A}$ and $\mathbb{B}$ fail to be spectrally equivalent.

It should be noted, that application of a special symmetrizer to $\mathbb{B}$, see [8], results in a matrix B that *is* spectrally equivalent to $\mathbb{A}$. More precisely, only n eigenvalues of $\mathbb{A}\mathsf{B}^{-1}$ are equal to 2, while all the rest are 1.

3. Capacitance matrix The method of domain decomposition often can be considered as a process in a *subspace*, see [6]. This amounts to performing the main iteration with the capacitance matrix C of the form $C = S^TAB^{-1}S$, where S^T is a restriction operator $S^T=(I_p,0)$, and p is the number of grid points on the separator, $p<<n$. Note, that for our one dimensional examples in sections 1.1 and 2., p equals to 1 and the capacitance $C = S^T(I + uv^T)S$ is equal to $n+1$ and 1, respectively.

In general, we can write matrices $\mathbb{A}$ and $\mathbb{B}$ in a 2 by 2 block form:

$$\mathbb{A} = \begin{pmatrix} A_{11} & A_{12} \\ A_{21} & A_{22} \end{pmatrix} \qquad \mathbb{B} = \begin{pmatrix} A_{11} & A_{12} \\ A_{21} & A_{22} \end{pmatrix} \qquad (2)$$

It has been shown, see [3] and also [4], that if A_{11} is invertible then $\mathbb{A}\mathbb{B}^{-1}$ has the form:

$$\mathbb{A}\mathbb{B}^{-1} = \begin{pmatrix} I & 0 \\ Z & C \end{pmatrix} \qquad (3)$$

where $C = C_1C_2^{-1}$, $C_1 \equiv (S^T\mathbb{A}S)^{-1} = A_{22} - A_{21}A_{11}^{-1}A_{12}$, $C_2 \equiv (S^T\mathbb{B}S)^{-1} = B_{22} - B_{21}A_{11}^{-1}A_{12}$, and $Z = (A_{21} - CB_{21})A_{11}^{-1}$.

C_1 and C_2 are called Schur complements of $\mathbb{A}$ and $\mathbb{B}$, respectively, and thus it is appropriate to call C the Schur complement of $\mathbb{A}\mathbb{B}^{-1}$.

Matrix $\mathbb{B}$ defined by (1") can be used efficiently as the so-called Neumann-Dirichlet preconditioner for $\mathbb{A}$, see [2]. We want to show not only that $\mathbb{A}$ and $\mathbb{B}$ are spectrally equivalent in a *subspace* :

$$a_1 x^T S^T \mathbb{A}\, Sx \le x^T S^T \mathbb{B}\, Sx \le a_2 x^T S^T \mathbb{A}\, Sx, \qquad \forall x \in \Re^n, \ \forall n$$

but also find the proportionality constants a_1, a_2.

When we reorder $\mathbb{A}$ and $\mathbb{B}$ from (1) into the block form (2) we obtain:

$$\mathbb{A} = \begin{pmatrix} T & -I & & & & & & & & & 0 \\ -I & T & -I & & & & & & & & 0 \\ & \cdot & \cdot & \cdot & & & & & & & \cdot \\ & & -I & T & -I & & & & & & 0 \\ & & & -I & T & & & & & & -I \\ & & & & & T & -I & & & & -I \\ & & & & & -I & T & -I & & & 0 \\ & & & & & & \cdot & \cdot & \cdot & & \cdot \\ & & & & & & & -I & T & -I & 0 \\ & & & & & & & & -I & T & 0 \\ 0 & 0 & 0 & 0 & -I & -I & 0 & 0 & 0 & 0 & T \end{pmatrix}$$

where matrix A_{11} is $(n^2\text{-}n)x(n^2\text{-}n)$, $A_{22}=T$ is nxn, and $A_{12}={A_{21}}^T$ is $(n^2n)xn$.

Matrix $\mathbb{B}$ differs from $\mathbb{A}$ only in the (21) block as $B_{22} = A_{22} = T$:

$$B_{21} = (0 \quad 0 \quad 0 \quad 0 \quad 0 \quad -2I \quad 0 \quad 0 \quad 0 \quad 0).$$

A_{11} is a 2x2 block diagonal matrix with two identical diagonal blocks. As a consequence, so is its inverse. Let us denote ${A_{11}}^{-1} = \mathrm{diag}(G,G)$, where the mxm matrix G represents the discrete Green's function for Δ_n on a half of the unit square with Dirichlet boundary conditions. The action of A_{21} on ${A_{11}}^{-1}A_{12}$ is identical to that of B_{21} on ${A_{11}}^{-1}A_{12}$ and is equal to $2S^TGS$, where the mxn matrix S is defined by $S^T = (0,\dots,0,I)^T$. Therefore, $C_1 = C_2 = T - 2S^TGS$, and finally we get $C = C_1{C_2}^{-1} = I$.

We have thus shown that C, the Schur complement of $AB^{-1}$for the case of the Neumann-Dirichlet preconditioner, is equal to the identity. Consequently, in a subspace, $\mathbb{B}$ is an excellent preconditioner for $\mathbb{A}$.

Additional item. Matrix $Z = (A_{21} - CB_{21}){A_{11}}^{-1} = (0,\dots,0,-I,I,0,\dots,0)\ \mathrm{diag}(G,G) = (X,-X)$, where the nxm matrix $X = -S^TG$. Thus, $\mathbb{A}\mathbb{B}^{-1} = I + SZ^T$.

4. Many Neumann-Dirichlet strips In the case of many Neumann-Dirichlet strips the situation is somewhat different: the capacitance C is not equal to the unity matrix any more. Actually, it is not even symmetric. Nevertheless, since C_1 and C_2 remain symmetric, all eigenvalues of C are real, as the following argument shows.

$$C\phi = C_1{C_2}^{-1}\phi = \lambda\phi, \quad {C_2}^{-1/2}C_1{C_2}^{-1/2}\varphi = \lambda\varphi, \qquad \text{where } {C_2}^{-1/2}\phi = \varphi.$$

Table 3 demonstrates the dependence of the number of grid points across and along each strip on the condition number of C in the case of four strips. For very narrow strips, $\kappa(C)$ is not independent of the grid size h. In the extreme case of only one inner grid point across the strip, $\kappa(C)$ grows roughly linearly with the number of grids in the other direction, i.e., with the inverse of the grid size, 1/h. On the other hand, when the number of grid points across the strip grows even slightly, $\kappa(C)$ rapidly decays to one.

	k=1	k=2	k=3	k=4	k=5	k=7	k=9	k=11
m=1	1.226	1.520	1.926	2.431	3.030	4.510	6.375	8.631
m=2	1.056	1.171	1.353	1.590	1.874	2.570	-	-
m=3	1.015	1.062	1.153	1.283	1.446	-		
m=4	1.004	1.023	1.070	1.145	-			

Table 3. Condition number of the capacitance matrix in the case of four strips, d=4. Here, k and m are the number of inner grid points along and across the strip, respectively.

In the simplest example with four strips, when m=k=1, we have

$$C_1 = 1/4 \begin{pmatrix} 14 & -1 & 0 \\ -1 & 14 & -1 \\ 0 & -1 & 14 \end{pmatrix} \qquad C_2 = 1/4 \begin{pmatrix} 14 & -2 & 0 \\ -2 & 14 & 0 \\ 0 & 0 & 14 \end{pmatrix}$$

$$C = C_1 C_2^{-1} = 1/96 \begin{pmatrix} 97 & 7 & 0 \\ 7 & 97 & -48/7 \\ -1 & -7 & 96 \end{pmatrix}$$

with the eigenvalues equal to 1.1130, 1.0000 and 0.9078, which results in $\kappa(C)$=1.226. We note that out of total k(d-1) eigenvalues of C only k of them are exactly equal to 1, the rest are unequally clustered around the value of 1. For example, for k=5 and m=3 we have λ_{max}=1.22, λ_{min}=0.85, and the remaining 13 eigenvalues are between 0.97 and 1.03.

	d=8					d=16	
	k=1	k=2	k=3	k=4	k=5	k=1	k=2
m=1	1.305	1.723	2.331	3.117	4.076	1.326	1.780
m=2	1.074	1.229	1.483	-	-	1.078	-
m=3	1.019	1.082	-	-	-	-	-
m=4	1.005	1.031	-	-	-	-	-

Table 4. Condition number of the capacitance matrix in the case of d=8 and d=16 strips. Here, k and m are the number of inner grid points along and across the strip, respectively.

In the case of eight strips, see Table 4, the values of $\kappa(C)$ do not change much. A further subdivision to 16 strips produces an increase in $\kappa(C)$ of only a few percent even for the extreme case of only one inner grid point across the strip.

5. Discontinuous coefficients Let us now, instead of the Laplace equation consider problems with self-adjoint operators $\mathrm{div}(k(x)\mathrm{grad}u)$ in Ω and Dirichlet boundary conditions at $\partial\Omega$. Here, the diffusion function $k(x)$ is discontinuous at the interface across Ω, and constant in each of the subdomains. As before, we impose the Neumann-Dirichlet condition at this interface.

Let us consider one dimensional problems with self-adjoint operators:

$$-\mathrm{div}(k(x)\mathrm{grad}u) = f(x) \quad \text{on} \quad \Omega = [0,1],$$

with Dirichlet boundary conditions at $\partial\Omega$. We shall use a standard uniform *staggered grid* approximation that gives rise to a symmetric matrix representation of A:

$$-k_{i-1/2}u_{i-1} + (k_{i-1/2} + k_{i+1/2})u_i - k_{i+1/2}u_{i+1} = h^2 f_i \qquad \text{for } i=1,\dots,n.$$

For $k_1(x)=1$ in the Dirichlet strip and $k_2(x)=10$ in the Neumann strip this approximation results in the following matrices A and B:

$$A = \begin{pmatrix} 2 & -1 & & & & & \\ \cdot & \cdot & \cdot & & & & \\ & -1 & 2 & -1 & & & \\ & & -1 & 11 & -10 & & \\ & & & -10 & 20 & -10 & \\ & & & & \cdot & \cdot & \cdot \\ & & & & & -10 & 20 \end{pmatrix}, \qquad B = \begin{pmatrix} 2 & -1 & & & & & \\ \cdot & \cdot & \cdot & & & & \\ & -1 & 2 & -1 & & & \\ & & 0 & 11 & -11 & & \\ & & & -10 & 20 & -10 & \\ & & & & \cdot & \cdot & \cdot \\ & & & & & -10 & 20 \end{pmatrix}.$$

The resulting matrix AB^{-1} is better conditioned than in the case $k_1(x)=k_2(x)$ reported in section **2** as the numerical data in Table 5 indicate. Here, $\kappa(AB^{-1})$ is about n/6.

# points	5	15	25	35	45
$\kappa(AB^{-1})$ in **2.**	2.75	6.21	9.59	12.94	16.29
$\kappa(AB^{-1})$ in **5.**	2.08	3.96	5.71	7.43	9.14

Table 5. Condition numbers $\kappa(AB^{-1})$ as a function of the number of grid points.

In a two dimensional analog of our example, we use the following staggered grid scheme:

$$-(k_{i+1/2,j-1/2}+k_{i+1/2,j+1/2})u_{i+1,j}-(k_{i-1/2,j-1/2}+k_{i+1/2,j-1/2})u_{i,j-1}-(k_{i-1/2,j+1/2}+k_{i+1/2,j+1/2})u_{i,j+1}$$

$$-(k_{i-1/2,j-1/2}+k_{i-1/2,j+1/2})u_{i-1,j}+2(k_{i-1/2,j-1/2}+k_{i-1/2,j+1/2}+k_{i+1/2,j-1/2}+k_{i+1/2,j+1/2})u_{i,j}=2h^2 f_{i,j}$$

$$\text{for } i,j=1,\dots,n$$

The matrix $\mathbb{A}$ that arises from this discretization has a form similar to that in section **3.**:

$$
\mathbb{A} = \begin{pmatrix}
T & -I & & & & & & & & & 0 \\
-I & T & -I & & & & & & & & 0 \\
 & \cdot & \cdot & \cdot & & & & & & & \cdot \\
 & & -I & T & -I & & & & & & 0 \\
 & & & -I & T & & & & & & -I \\
 & & & & & cT & -cI & & & & -cI \\
 & & & & & -cI & cT & -cI & & & 0 \\
 & & & & & & \cdot & \cdot & \cdot & & \cdot \\
 & & & & & & & -cI & cT & -cI & 0 \\
 & & & & & & & & -cI & cT & 0 \\
0 & 0 & 0 & 0 & -I & -cI & 0 & 0 & 0 & 0 & sT
\end{pmatrix},
$$

where the constant $c=k_2(x)/\,k_1(x)$ and $s=(c+1)/2$.

As before, matrix $\mathbb{B}$ differs from $\mathbb{A}$ only in the (21) block :

$$B_{21} = (0 \quad 0 \quad 0 \quad 0 \quad 0 \quad -2sI \quad 0 \quad 0 \quad 0 \quad 0).$$

Theorem Let -div(k(x)gradu)=f in $\Omega=(0,1)^2$ with Dirichlet boundary conditions at $\partial\Omega$. Let the diffusion function k(x) be discontinuous at the interface across that halves Ω, and constant in each of the subdomains. Let the Neumann-Dirichlet conditions be imposed at this interface. Then the capacitance matrix C is equal to the identity.

Proof We have $A_{11}^{-1} = \mathrm{diag}(G,1/cG)$, where the mxm matrix G represents, as before, the discrete Green's function for Δ_n on the half of the unit square with Dirichlet boundary conditions. Here, as before, S is defined by $S^T = (0,\ldots,0,I)^T$. Then

$$A_{11}^{-1}A_{12} = \begin{pmatrix} G & 0 \\ 0 & 1/cG \end{pmatrix}\begin{pmatrix} S \\ cS \end{pmatrix} = \begin{pmatrix} GS \\ GS \end{pmatrix}$$

$$A_{21}A_{11}^{-1}A_{12} = (S^T, cS^T)\begin{pmatrix} GS \\ GS \end{pmatrix} = (c+1)S^TGS.$$

$$B_{21}A_{11}^{-1}A_{12} = (0,(c+1)S^T)\begin{pmatrix} GS \\ GS \end{pmatrix} = (c+1)S^TGS.$$

Thus the action of A_{21} on $A_{11}^{-1}A_{12}$ is identical to that of B_{21} on $A_{11}^{-1}A_{12}$ and is equal to $(c+1)S^TGS$. Therefore, $C_1 = C_2 = sT - 2S^TGS$, and finally we get $C = C_1C_2^{-1} = I$. Thus, C is equal to the identity, even in the case of discontinuous coefficients.

In the case of many Neumann-Dirichlet strips the results are anti-intuitive: the conditioning of the capacitance matrix C is significantly better for problems with discontinuous coefficients than with continuous ones, as the following table indicates.

	k=1	k=2	k=3	k=4	k=5	k=7	k=9	k=11
m=1	1.038	1.081	1.134	1.194	1.261	1.420	1.612	1.842
m=2	1.010	1.029	1.057	1.090	1.127	1.210	-	-
m=3	1.003	1.011	1.026	1.047	1.071	-		
m=4	1.001	1.004	1.012	1.025	-			

Table 6. Condition number of the capacitance matrix for the problem with discontinuous coefficients ($c=k_2/k_1=10$) in the case of four strips, d=4. Here, k is the number of inner grid points along the strip, and m is the number of grid points across the strip.

To gain some insight, let us examine the simplest example with four strips, when m=k=1.

$$B = \left(\begin{array}{cccc|ccc} 4 & & & & -1 & & \\ & 40 & & & -10 & -10 & \\ & & 4 & & & -1 & \\ & & & 40 & & & -10 \\ \hline & -11 & & & 22 & & \\ & -11 & & & & 22 & \\ & & & -11 & & & \cdot 22 \end{array}\right)$$

while A is symmetric ($A_{21}=A_{12}^T$). In accordance with (3), we obtain

$$C_1 = 1/4 \begin{pmatrix} 77 & -10 & 0 \\ -10 & 77 & -1 \\ 0 & -1 & 77 \end{pmatrix}, \qquad C_2 = 1/4 \begin{pmatrix} 77 & -11 & 0 \\ -11 & 77 & 0 \\ 0 & 0 & 77 \end{pmatrix},$$

$$C = C_1 C_2^{-1} = 1/528 \begin{pmatrix} 529 & 7 & 0 \\ 7 & 529 & -48/7 \\ -1 & -7 & 528 \end{pmatrix}$$

with the $\lambda_i(C)$ equal to 1.0205, 1.0000 and 0.9832, which results in $\kappa(C)=1.038$. Note that matrix C_2 is the same as the one for continuous coefficients, scaled by $(c+1)/2=11/2$. Increasing the number of strips above 4 does not change much the conditioning of C:

	k=1	k=2	k=3	k=4	k=5	k=1	k=2
m=1	1.050	1.108	1.182	1.271	1.374	1.053	1.115
m=2	1.013	1.038	1.076	-	-	-	-

Table 7. Condition number of the capacitance matrix for the problem with discontinuous coefficients ($c=k_2/k_1=10$) in the case of d=8 strips. Here, k and m are the number of inner grid points along tand across the strip, respectively.

Let us now investigate the influence of the ratio of the diffusion coefficients on the condition number of matrix C.

$c=k_2/k_1$	0.001	0.01	0.1	0.3	1.0	3.	10.	100.	1000.
$\kappa(C)$ (m=k=1)	1.502	1.497	1.448	1.368	1.226	1.107	1.038	1.004	1.0004
$\kappa(C)$ (m=k=3)	1.329	1.326	1.295	1.245	1.153	1.074	1.026	1.003	1.0003

Table 8. Condition number of the capacitance matrix for the problem with discontinuous coefficients k_2 and k_1 in the case of d=4 strips. Here, k and m are the number of inner grid points along tand across the strip, respectively.

Results in Table 8 strongly indicate that as c, the ratio of the coefficients, grows, the condition number approaches the value of 1.0, and $\kappa(C)$ approaches a somewhat larger value as c decreases, see Fig.1. Coefficient k_1 corresponds here to the Dirichlet and k_2 to the Neumann strips. Let us investigate the problem for the simplest case when k=m=1. Using the formulas in (3) we arrive at the following representation

$$C_1 = \frac{1}{4}\begin{pmatrix} 7(c+1) & -c & 0 \\ -c & 7(c+1) & -1 \\ 0 & -1 & 7(c+1) \end{pmatrix}, \qquad C_2 = \frac{1}{4}\begin{pmatrix} 7(c+1) & -(c+1) & 0 \\ -(c+1) & 7(c+1) & 0 \\ 0 & -1 & 7(c+1) \end{pmatrix},$$

$$C = C_1 C_2^{-1} = I + 1/48(c+1)R, \qquad \text{where } R = \begin{pmatrix} 1 & 7 & 0 \\ 7 & 1 & -48/7 \\ -1 & -7 & 0 \end{pmatrix}$$

Fig.1

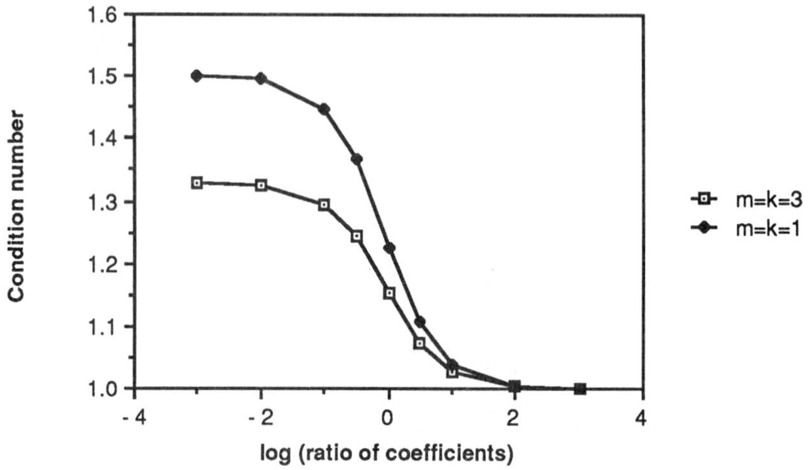

Thus the eigenvalues of C are $\lambda(C) = 1 + 1/48(c+1)\ \lambda(R)$. It is easy to verify that the eigenvalues of R are 0 (R is singular), and $1 \pm \sqrt{97}$. Consequently, the condition number of C is $\kappa(C) = (48(c+1) + 1 + \sqrt{97})\ /\ (48(c+1) + 1 - \sqrt{97})$. This formula gives us $\kappa(C)=1.226$ for c=1, and $\kappa(C)=1.503$ for c=0 (the limiting case).
Examining C_1 and C_2 we see that for large values of c these matrices are "similar", in contrast to the situation when c are small (in the limit, for c=0, the zeros appear in the "wrong" positions). This strongly suggests that *in the preconditioner $\mathbb{B}$ the Neumann strips should correspond to the regions with larger diffusion coefficients.*
Finally, we investigate the influence of the number of inner grid points along the strip on the condition number when the strips are *wrongly* placed, i.e., c<1.

k	1	2	3	4	5	7	9	11
$\kappa(C)$	1.448	2.137	3.273	4.950	7.278	14.33	25.21	40.44

Table 9. Condition number of the capacitance matrix for the problem with discontinuous coefficients ($c=k_2/k_1=0.1$) in the case of d=4 strips and m=1. Here, the k and m are the number of inner grid points along tand across the strip, respectively.

Clearly, for c<1, in this worst case of extremely thin strips (m=1), $\kappa(C)$ grows almost quadratically with k, the number of inner grid points along the strip.

References

[1] O.Axelsson and V.A.Barker, *Finite Element Solution of Boundary Value Problems,* Acadamic Press, 1984.

[2] P.Bjørstad and O.Widlund, SIAM J. Num.Anal, v.23, 1987, 1097.

[3] M.Dryja, SIAM J.Num.Anal, v.20, 1983, 671.

[4] M.Dryja and W.Proskurowski, CRI-86-12 Report, May 1986.

[5] V.Faber, T.Manteuffel and S.Parter, "On the equivalence of operators and the implications to preconditioned iterative methods for elliptic problems", 1987, preprint.

[6] G.Marchuk and Yu.Kuznetsov, in *Problems of Computational Mathematics and Mathematical Modelling,* G.Marchuk and V.Dymnikov (eds.), Mir Publ., 1985.

[7] W.Proskurowski and O.Widlund, Comp. Math., v.30, 1976, 433.

[8] O.Widlund, in *Proc. 1st Symposium on Domain Decomposition*, SIAM, 1988, 113.

Optimal Iterative Refinement Methods*

Olof B. Widlund†

Abstract. We consider the solution of the linear systems of algebraic equations which arise from elliptic finite element problems defined on composite meshes. Such problems can systematically be built up by introducing a basic finite element approximation on the entire region and then repeatedly selecting subregions, and subregions of subregions, where the finite element model is further refined in order to gain higher accuracy. We consider conjugate gradient algorithms, and other acceleration procedures, where, in each iteration, problems representing finite element models on the original region and the subregions prior to further refinement are solved. We can therefore use solvers for problems with uniform or relatively uniform mesh sizes, while the composite mesh can be strongly graded.

In this contribution to the theory, we report on new results recently obtained in joint work with Maksymilian Dryja. We use a basic mathematical frame work recently introduced in a study of a variant of Schwarz' alternating algorithm. We establish that several fast methods can be devised which are optimal in the sense that the number of iterations required to reach a certain tolerance is independent of the mesh size as well as the number of refinement levels. This work is also technically quite closely related to previous work on iterative substructuring methods, which are domain decomposition algorithms using non-overlapping subregions.

1. Introduction. In this paper, we consider the solution of the large linear systems of algebraic equations which arise when working with elliptic finite element approximations on composite meshes. In this contribution to the theory, we report on new results recently obtained in joint work with Maksymilian Dryja.

Finite element models on composite meshes can systematically be built up inside a frame work of conforming finite elements; cf. Ciarlet [3]. We do so to be able to use

* This work was supported in part by the National Science Foundation under Grant NSF-CCR-8703768 and, in part, by the U. S. Department of Energy under contract DE-AC02-76ER03077-V at the Courant Mathematics and Computing Laboratory.

† Courant Institute of Mathematical Sciences, 251 Mercer Street, New York, N.Y. 10012.

a number of technical tools which are available primarily in the conforming case. We begin by introducing a basic finite element approximation on the entire region and we then repeatedly select subregions, and subregions of subregions, where the finite element model is further refined. We solve the resulting linear system by using an iterative method such as the conjugate gradient, Richardson or Chebyshev method. We accelerate the convergence by solving so called standard problems. These correspond to the finite element models on the original region, prior to any refinement, and those on the subregions prior to further refinement. We can thus use solvers for problems with uniform or relatively uniform mesh sizes, while the composite mesh can be strongly graded.

This approach offers a number of advantages. An existing code can be upgraded, in order to increase the accuracy locally, without a radical redesign of the data structures etc., since we can use the old code to solve one or several of the standard problems. Issues of data structures and geometry are generally simpler if we design programs for composite mesh problems in terms of simpler standard problems. The use of simple standard problems also tends to improve the performance of the programs on vector machines. Finally we note that if each of the standard problems has appreciatively fewer degrees of freedom than the composite model, then we might benefit from solving a number of smaller problems rather than one large one. In other words, we might view our approach as a divide-and-conquer strategy.

We study two families of algorithms which we call multiplicative and additive respectively. The so called additive algorithms are particularly well suited for parallel computing in that in each iteration step we can simultaneously solve all the standard problems on the the different levels of refinement. Synchronization between the processors is required once in each iteration, namely when the residual is computed and assembled. One or a group of processors can therefore be assigned in a straightforward way to each of the standard problems. As we will see, we can view the multiplicative algorithms as preconditioned conjugate gradient methods, while the additive variant involves the solution of a transformed equation with the same solution as the original one using no further preconditioning.

The systematic study of the methods under consideration goes back at least to 1983, when the Fast Adaptive Composite (FAC) method was introduced by McCormick [12]. Issues related to the implementation of this method on parallel computers led to the introduction of Asynchronous FAC (AFAC) methods [8] a few years later. Numerical experiments have now been reported for model cases and recently Richard Ewing, Steve McCormick and others have begun to test the algorithms for more difficult problems arising in industry. The convergence of the FAC method is discussed in McCormick et al. [12], [13] under a certain additional regularity assumption. An important contribution to the theory is given by Bramble, Ewing, Pasciak and Schatz [2], who outlined a proof of optimality for a two level FAC algorithm. They require no additional regularity in their proof. In recent papers Bjørstad [1] and Mandel and McCormick [11], develop a theory for both multiplicative and additive algorithms, using two levels. In particular, they obtain interesting results concerning the relationship between the spectra of the two methods. In a recent paper, Mandel and McCormick [10] establish optimality for a multi-level AFAC algorithm for a special model problem.

Our study began with the discovery that FAC and AFAC methods have a structure quite similar to that of the classical Schwarz procedure, see Schwarz [14], and an additive variant thereof recently considered in Dryja [4] and Dryja and Widlund [5]. Our earlier work was in turn inspired by a recent paper by P.-L. Lions [9] in which a variational frame work for the classical, multiplicative Schwarz' method is developed for continuous elliptic problems. We were able to establish a rate of convergence which is independent of the number of degrees of freedom as well as the number of subregions for a special kind of additive Schwarz algorithm, see Dryja [4] and Dryja and Widlund [5]. In our view, the central theoretical issue of the iterative refinement algorithms, which are discussed in this paper, is similar to that of Schwarz type algorithms namely the design and study of algorithms for which the rate of convergence is independent of the number of subproblems, i.e. the number of refinement levels as well as and the mesh sizes.

In section 2, we introduce the finite element problems on composite meshes, certain projections and our algorithms in their basic form.

In Section 3, we consider a multiplicative algorithm for the multi level case and show that its rate of convergence is optimal.

In Section 4, we introduce several multilevel additive algorithms and provide a number of bounds. The principal result is that one of these methods has a rate of convergence which is independent of the number of refinement levels, as well as the mesh sizes.

2. Composite Finite Element Problems and Basic Iterative Methods. We consider linear, self adjoint, elliptic problems discretized by finite element methods on a bounded Lipschitz region Ω in R^n. To simplify the presentation, we assume that the differential operator is the Laplacian and that we use continuous, piecewise linear finite elements. However, almost all of our results can be extended immediately to general conforming finite element approximations of any self adjoint elliptic problem which can be formulated as a minimization problem. The continuous and discrete problems are of the form

$$a(u,v) = f(v), \ \forall\ v \in V,$$

and

$$a(u_h, v_h) = f(v_h), \ \forall\ v_h \in V^h, \tag{1}$$

respectively. The spaces V and V^h are defined in the next few paragraphs. The bilinear form is defined by

$$a(u,v) = \int_\Omega \nabla u \cdot \nabla v \ dx \ .$$

This form defines a semi-norm $|u|_{H^1} = (a(u,u))^{1/2}$ in $H^1(\Omega)$.

To simplify the presentation, we assume that we have a zero Dirichlet boundary condition on $\partial\Omega$, the boundary of Ω. The space V is thus $H_0^1(\Omega)$. We note that we equally well could have considered an inhomogeneous Dirichlet problem. The space V^h is defined on a composite triangulation, which is possibly the result of a large number of successive refinements. The triangulation of Ω is given in the following way.

We first introduce a relatively coarse triangulation of Ω, also denoted by Ω_1, and denote the corresponding space of finite element functions by V^{h_1}. We can think of this space as having a relatively uniform (or uniform) mesh size h_1. Let Ω_2 be an area where we wish to increase the resolution. We do so by subdividing the elements and introducing an additional finite element space V^{h_2}. We assure that the resulting composite space $V^{h_1} + V^{h_2}$ is conforming by having the functions of V^{h_2} vanish on $\partial\Omega_2$. We repeat this process by selecting a subregion Ω_3 of Ω_2 and introducing a further refinement of the mesh and finite element space etc.. We denote the the resulting nested subregions and subspaces by Ω_i and V^{h_i} respectively. Throughout, we have $\Omega_i \subset \Omega_{i-1}$ and $V^{h_{i-1}} \cap H_0^1(\Omega_i) \subset V^{h_i} \subset H_0^1(\Omega_i)$, $i = 2, \ldots, k$. The composite finite element space on the repeatedly refined mesh, is

$$V^h = V^{h_1} + V^{h_2} + \ldots + V^{h_k} \ .$$

We assume that all the elements are shape regular in the sense that there is a uniform bound on h_K/ρ_K. Here h_K and ρ_K are the diameter and the radius of the largest inscribed sphere of any element K, respectively. Our bounds in the theory developed below also depend on the shape of the subregions Ω_i. Thus in order for our proofs to work, we cannot allow the sets $\Omega_{i-1} \setminus \Omega_i$ to become arbitrarily thin in comparison with the diameter of Ω_{i-1}. We also assume that the area of any triangle on level i can be bounded by *const.* q^{j-i}, $j < i$ times the area of the triangle on level j of which it is a part. Here q is a constant < 1.

The finite element problem is defined by equation (1) and the corresponding stiffness matrix can conveniently be computed by using a process of subassembly. Introducing subscripts to indicate the domain of integration, we write

$$a(u,v) = a_{\Omega_1 \setminus \Omega_2}(u,v) + a_{\Omega_2 \setminus \Omega_3}(u,v) + \ldots + a_{\Omega_k}(u,v) .$$

The stiffness matrices corresponding to the regions $\Omega_i \setminus \Omega_{i+1}$, $i \leq k-1$, and Ω_k are computed by working with basis functions related to the mesh size h_i . The quadratic form corresponding to the composite stiffness matrix is the sum of the quadratic forms corresponding to $\Omega_i \setminus \Omega_{i+1}$ and Ω_k. In our algorithms, we use solvers for the same elliptic problem on the subregions Ω_i , $i = 1, \ldots, k$, and the relatively uniform meshes corresponding to h_i . The corresponding stiffness matrices can in fact be obtained at a small extra cost during the assembly of the composite mesh model. When we refine a finite element model locally, the modified stiffness matrix is obtained by replacing the quadratic form associated with the subregion in question by the one corresponding to the refined model on the same subregion. It is therefore relatively easy to design a method which systematically generates the stiffness matrices for all the standard problems necessary while the stiffness matrix of the composite model is computed.

The fundamental building blocks of our algorithms are the P_j^i, $i \leq j$, the projections onto the spaces $V^{h_i} \bigcap H_0^1(\Omega_j)$. We note that if $j > i$, then we solve a problem on Ω_j with a coarser mesh than if V^{h_j} were used. The projection P_j^i, $i \leq j$, is defined, in terms of the unique element of $V^{h_i} \cap H_0^1(\Omega_j)$, which satisfies

$$a(P_j^i v_h, \phi_h) = a(v_h, \phi_h), \ \forall \phi_h \in V^{h_i}. \tag{2}$$

We now introduce the multiplicative and additive algorithms. Since no further effort is involved, we develop a framework which is also useful in other contexts such as the study of algorithms of Schwarz type; cf. Dryja [4], Dryja and Widlund [5] and Lions [9]. We begin by considering the case of two subspaces V_1 and V_2 and the multiplicative (sequential) algorithms. In a first fractional step, we find a correction $\delta_1 u^n \in V_1$ of the current approximation u^n by solving

$$a(\delta_1 u^n, v) = f(v) - a(u^n, v) = a(u^* - u^n, v), \ \forall \ v \in V_1.$$

Here $u^* \in V_1 + V_2$ is the solution of the given problem. The calculation of a second correction $\delta_2 u^n \in V_2$ completes the $(n+1)$th step.

$$a(\delta_2 u^n, v) = f(v) - a(u^n + \delta_1 u^n, v) = a(u^* - (u^n + \delta_1 u^n), v), \ \forall v \in V_2 .$$

As shown in Lions [9], it is easy to see that

$$\begin{aligned} \delta_1 u^n &= P_1(u^* - u^n) \\ \delta_2 u^n &= P_2(u^* - u^n - \delta_1 u^n) = P_2(I - P_1)(u^* - u^n) \end{aligned}$$

and thus the error propagates as

$$u^{n+1} - u^* = (I - P_2)(I - P_1)(u^n - u^*).$$

Here P_1 and P_2 are the orthogonal projections associated with the bilinear form $a(\cdot,\cdot)$ and V_1 and V_2, respectively.

We can thus view this algorithm as a simple iterative method for solving

$$(P_1 + P_2 - P_2 P_1) u_h = g_h,$$

with an appropriate right hand side g_h. We note that this operator is a polynomial of degree two and thus not ideal for parallel computing, since two sequential steps are involved. If we use more than two subspaces and therefore more projections this effect is further pronounced. The basic idea behind the additive form of the algorithm is to work with a simplest possible polynomial in the projections. With k subspaces we thus solve the equation

$$Pu_h = (P_1 + P_2 + \ldots + P_k)u_h = g_h' , \tag{3}$$

by an iterative method. If we can show that the operator P is symmetric and positive definite, the iterative method of choice is the conjugate gradient method at least on a computer with conventional architecture. We must also make sure that this equation has the same solution as equation (1), i.e. we must find the correct right hand side. Since by equation (1), we have

$$a(u_h, \phi_h) = f(\phi_h) ,$$

we can construct the right-hand side g_h' by solving this equation restricted to all the different subspaces and adding the results. It is similarly possible to apply the operator P of equation (3) to any given element of V^h by applying each projection P_i once. Most of the work , in particular that which involves the individual projections, can be carried out in parallel.

It is well known that the number of steps required to decrease an appropriate norm of the error of a conjugate gradient iteration by a fixed factor is proportional to $\sqrt{\kappa}$, where κ is the condition number of P; see e.g. Golub and Van Loan [7]. We therefore need to establish that the operator P of equation (3) is not only invertible but that satisfactory upper and lower bounds on its eigenvalues can be obtained.

We end this section by discussing a symmetric version of the multiplicative algorithm. The projections generally do not commute and if we wish to have a symmetric expression in the operators P_i, which allows us to accelerate the convergence by the standard conjugate gradient method, we have to use additional fractional steps. In the case of two subspaces, we can solve the first problem again. Since $(I - P_2)^2 = (I - P_2)$ we can write the resulting operator as

$$I - (I - P_1)(I - P_2)(I - P_1) = P_1 + P_2 - P_1P_2 - P_2P_1 + P_1P_2P_1 = I - T_2T_2^*,$$

where $T_2 = (I - P_1)(I - P_2)$. The corresponding operator for k subspaces involves $2k - 1$ fractional steps and has the form

$$I - T_kT_k^*, \text{ where } T_k = (I - P_1)(I - P_2)\cdots(I - P_k).$$

The error propagation operator for the basic multiplicative algorithm is T_k^*, while the convergence rate of the symmetrized algorithm can be bounded in terms of the condition number of $I - T_kT_k^*$. We note that a bound on the spectral radius of T_k is obtained immediately from the spectral bounds on the symmetric operator.

3. The Multiplicative Algorithm. In this section, we establish the following result. We note that our analysis resembles that of Bramble et al. [2] in the case of $k = 2$.

Theorem 1. *The symmetrized, multiplicative iterative refinement algorithm based on the projections P_i^i, $i \leq k$, has a condition number which is independent of k and the number of degrees of freedom. The spectral radius of the basic multiplicative algorithm is bounded by a constant which is uniformly less than one.*

We begin by designing a preconditioner $b(u, v)$ for the finite element problem on the space

$$V^h = V^{h_1} + \cdots + V^{h_k}.$$

We then show that this quadratic form can be bounded uniformly from above and below by the quadratic form of the composite finite element problem. Finally we establish that we can work practically with this preconditioner and that the error propagation matrix is the same as that of the symmetrized multiplicative algorithm, introduced in the previous section. The algorithms are thus the same.

In addition to the projections P_i^i, we also use another family of operators, H_j^i, $i = j-1, j$, defined by

$$H_j^i v_h(x) \in V^{h_1} + \cdots + V^{h_i},$$
$$H_j^i v_h(x) = v_h(x), \quad x \in \Omega \setminus \Omega_j,$$
$$a(H_j^i v_h, w_h) = 0, \quad \forall w_h \in V^{h_i} \cap H_0^1(\Omega_j).$$

We can call $H_j^i v_h$ the h_i–harmonic extension of v_h to Ω_j, since it is the solution in V^{h_i} of a discrete Dirichlet problem with zero right hand side and with boundary data on $\partial\Omega_j$ given by v_h.

In order to prepare for the work that remains, we formulate three lemmas.

Lemma 1. $H_{i-1}^{i-2} H_i^{i-1} = H_{i-1}^{i-2}$, $3 \le i \le k$.

The proof follows immediately from the definition.

Lemma 2. $a(H_i^{i-1} u_h, H_i^{i-1} v_h) = a(H_i^{i-1} u_h, v_h)$, $\forall u_h \in V^h$, $\forall v_h \in V^{h_{i-1}}$, $2 \le i \le k$.

Proof: From the definition of H_i^{i-1} follows that $H_i^{i-1} v_h - v_h \in V^{h_{i-1}} \cap H_0^1(\Omega_i)$. Since $H_i^{i-1} u_h$ is h_{i-1}–harmonic on Ω_i the result follows by the orthogonality inherent in the definition of h_{i-1}–harmonic functions.

The following lemma is a consequence of the extension theorem given in Widlund [15].

Lemma 3. *There exists a constant C which is independent of u_h and the mesh sizes, such that for $i = j-1, j$,*

$$a_{\Omega_j}(H_j^i u_h, H_j^i u_h) \le C a_{\Omega_{j-1} \setminus \Omega_j}(u_h, u_h), \quad \forall u_h \in V^h.$$

We will not discuss the proof of this lemma. We note that the constant C necessarily blows up if we let the area of $\Omega_{j-1} \setminus \Omega_j$ shrink to zero, keeping Ω_{j-1} fixed. Such situations are not of particular interest in our applications.

Trivially, we can write $u_h = P_k^k u_h + (I - P_k^k) u_h$. It is easy to see that the second term equals $H_k^k u_h$. The two terms are orthogonal in the sense of the bilinear form and thus

$$\begin{aligned} a(u_h, v_h) &= a(P_k^k u_h, P_k^k v_h) + a((I - P_k^k) u_h, (I - P_k^k) v_h) \\ &= a(P_k^k u_h, P_k^k v_h) + a(H_k^k u_h, H_k^k v_h)\,. \end{aligned}$$

We now introduce a preconditioner, which in the case k=2 is the final one but which in the general case is only a first step of our construction.

$$a(P_k^k u_h, P_k^k v_h) + a(H_k^{k-1} u_h, H_k^{k-1} v_h)\,.$$

The h_k–harmonic function on Ω_k has thus been replaced by the h_{k-1}–harmonic function with the same boundary values. Since the latter space of discrete harmonic functions is smaller, it is easy to see that the preconditioner is bounded from below by the original quadratic form

$$a(u_h, u_h) \le a(P_k^k u_h, P_k^k u_h) + a(H_k^{k-1} u_h, H_k^{k-1} u_h)\,.$$

To find a bound from above, we have to show that the energy attributable to the h_{k-1}-harmonic function can be bounded by that of the h_k–harmonic function. This follows directly from Lemma 3.

When we now turn to the case of $k > 2$, we repeatedly replace certain discrete harmonic functions by others. In each step, we replace the current last term of the preconditioner by two. We begin by considering the identity

$$H_k^{k-1}u_h = P_{k-1}^{k-1}H_k^{k-1}u_h + (I - P_{k-1}^{k-1})H_k^{k-1}u_h \; .$$

It is easy to see that the second term equals $H_{k-1}^{k-1}H_k^{k-1}u_h$. As in the first step, we replace this last term by another, namely $H_{k-1}^{k-2}H_k^{k-1}u_h$, and we obtain a preconditioner with three terms:

$$a(P_k^k u_h, P_k^k v_h) + a(P_{k-1}^{k-1}H_k^{k-1}u_h, P_{k-1}^{k-1}H_k^{k-1}v_h) + \\ a(H_{k-1}^{k-2}H_k^{k-1}u_h, H_{k-1}^{k-2}H_k^{k-1}v_h)$$

We simplify this expression, by using Lemma 1 and replace the third term of the preconditioner by $a(H_{k-1}^{k-2}u_h, H_{k-1}^{k-2}v_h)$. By repeating the process just outlined, we arrive at the following preconditioner:

$$\begin{aligned} b(u_h, v_h) = {} & a(P_k^k u_h, P_k^k v_h) + a(P_{k-1}^{k-1}H_k^{k-1}u_h, P_{k-1}^{k-1}H_k^{k-1}v_h) \\ & + \cdots a(P_2^2 H_3^2 u_h, P_2^2 H_3^2 v_h) + a(H_2^1 u_h, H_2^1 v_h). \end{aligned} \tag{4}$$

We can now complete our proof of the optimality of the preconditioner. By using the definitions of the operators P_i^i and H_j^i, we find that

$$\begin{aligned} a(P_k^k u_h, P_k^k u_h) &= a_{\Omega_k}(P_k^k u_h, P_k^k u_h) \\ &= a_{\Omega_k}(u_h, u_h) - a_{\Omega_k}(H_k^k u_h, H_k^k u_h), \end{aligned}$$

and, for $3 \le i \le k$,

$$\begin{aligned} a(P_{i-1}^{i-1}H_i^{i-1}u_h, P_{i-1}^{i-1}H_i^{i-1}u_h) &= a_{\Omega_{i-1}}(P_{i-1}^{i-1}H_i^{i-1}u_h, P_{i-1}^{i-1}H_i^{i-1}u_h) \\ = a_{\Omega_{i-1}}(H_i^{i-1}u_h, H_i^{i-1}u_h) &- a_{\Omega_{i-1}}(H_{i-1}^{i-1}H_i^{i-1}u_h, H_{i-1}^{i-1}H_i^{i-1}u_h) \\ = a_{\Omega_{i-1}}(H_i^{i-1}u_h, H_i^{i-1}u_h) &- a_{\Omega_{i-1}}(H_{i-1}^{i-1}u_h, H_{i-1}^{i-1}u_h). \end{aligned}$$

In the last step, we have used Lemma 1.

By using the definition of the H_j^i operators, we can rewrite the preconditioner as

$$\begin{aligned} b(u_h, v_h) = {} & a_{\Omega_k}(u_h, u_h) - a_{\Omega_k}(H_k^k u_h, H_k^k u_h) \\ & + a_{\Omega_{k-1}\backslash\Omega_k}(u_h, u_h) + a_{\Omega_k}(H_k^{k-1}u_h, H_k^{k-1}u_h) - a_{\Omega_{k-1}}(H_{k-1}^{k-1}u_h, H_{k-1}^{k-1}u_h) \\ & + a_{\Omega_{k-2}\backslash\Omega_{k-1}}(u_h, u_h) + a_{\Omega_{k-1}}(H_{k-1}^{k-2}u_h, H_{k-1}^{k-2}u_h) - a_{\Omega_{k-2}}(H_{k-2}^{k-2}u_h, H_{k-2}^{k-2}u_h) \\ & + \cdots a_{\Omega_1\backslash\Omega_2}(u_h, u_h) + a_{\Omega_2}(H_2^1 u_h, H_2^1 u_h) \end{aligned}$$

This quadratic form can be written as the sum of $a_\Omega(u_h, u_h)$ and a number of terms of the form

$$a_{\Omega_i}(H_i^{i-1}u_h, H_i^{i-1}u_h) - a_{\Omega_i}(H_i^i u_h, H_i^i u_h)$$

Since, on Ω_i, $H_i^{i-1}u_h$ and $H_i^i u_h$ are h_{i-1}–harmonic and h_i–harmonic functions, respectively, with the same boundary values on $\partial\Omega_i$, it is easy to see that these terms are positive. This shows that the preconditioner is bounded from below by $a_\Omega(u_h, u_h)$. To get an upper bound, we only have to estimate the positive terms. By Lemma 3, the energy attributable to the subregion Ω_i, of the h_{i-1}–harmonic function $H_i^{i-1}u_h$ can be estimated by a constant times $a_{\Omega_{i-1}\backslash\Omega_i}(u_h, u_h)$. The proof of the upper bound is completed by summing over i.

What remains is to establish that the use of this preconditioner leads to a series of problems which are directly related to the projections P_i^i in particular the symmetric multiplicative algorithm given in section 2. For the unaccelerated algorithm, the equation that we have to solve is of the form

$$b(\delta u_h, v_h) = a(e_h, v_h) ,$$

where e_h is the error before the current step and δu_h the next correction. We first choose test functions v_h in the subspace V^{h_k}. All terms of the preconditioner, except the first, vanish for such test functions. Therefore the result of this first fractional step equals $P_k^k \delta u_h = P_k^k e_h$. We choose $V^{h_{k-1}}$ as our second space of test functions. All the terms of the preconditioner except the first two vanish and the first is already known and can therefore be moved over to the right hand side. A straight forward calculation and Lemma 2 show that the new right hand side is equal to $a((1-P_k^k)e_h, v_h)$. In this second fractional step, we obtain $P_{k-1}^{k-1} H_k^{k-1} \delta u_h = P_{k-1}^{k-1}(1-P_k^k)e_h$. Proceeding in this manner, we compute, in the k–th fractional step, $H_2^1 \delta u_h = P_1^1(I-P_2^2)\cdots(I-P_k^k)e_h$. At this time, the values of δu_h are available on $\Omega_1 \setminus \Omega_2$. We can use its boundary values on $\partial\Omega_2$ as Dirichlet values and solve a problem in V^{h_2} to obtain the values of δu_h in $\Omega_2 \setminus \Omega_3$. Proceeding in this manner, using a total of $(2k-1)$ fractional steps, we finally obtain δu_h everywhere. A calculation shows that the error propagation operator corresponding to the whole step is equal to

$$(I - P_k^k)(I - P_{k-1}^{k-1}) \cdots (I - P_1^1)(I - P_2^2) \cdots (I - P_k^k) .$$

This shows that the method defined by the preconditioner indeed is the same as that of the symmetrized multiplicative algorithm defined in section 2.

4. Some Additive Algorithms. We recall, that in the so called additive algorithms, we solve equation (3) by the conjugate gradient or some other standard iterative algorithm. The different algorithms can simply be defined by specifying the subspaces V_i or alternatively the projections P_i. Before we discuss specific algorithms, we make some remarks on estimating the eigenvalues of P from above and below.

The upper bound on the spectrum is obtained by bounding

$$a(Pv_h, v_h) = a(P_1 v_h, v_h) + a(P_2 v_h, v_h) + \cdots + a(P_k v_h, v_h) ,$$

from above in terms of $a(v_h, v_h)$. We can use Schwarz' inequality and the fact that P_i is a projection to prove that each term is bounded by $a(v_h, v_h)$ and thus the spectrum of P is bounded from above by k. Our goal, however, is to establish a uniform bound on the condition number. This can be done if the terms are orthogonal or almost orthogonal; cf. discussion below.

A lemma from Lions [9] provides a method for obtaining lower bounds. Since the proof of his result is quite short, we include it in this paper.

Lemma 4. *Let $u_h = \sum_{i=1}^k u_{h,i}$, where $u_{h,i} \in V_i$, be a partition of an element of V^h and assume further that $\sum_{i=1}^k a(u_{h,i}, u_{h,i}) \le C_0^2 a(u_h, u_h)$, $\forall u_h \in V^h$. Then $\lambda_{min}(P) \ge C_0^{-2}$.*

Proof: By elementary properties of symmetric projections and the representation of u_h as a sum, we find that

$$a(u_h, u_h) = \sum_{i=1}^{k} a(u_h, u_{h,i}) = \sum_{i=1}^{k} a(u_h, P_i u_{h,i}) = \sum_{i=1}^{k} a(P_i u_h, u_{h,i}) .$$

Therefore,

$$a(u_h, u_h) \le (\sum_{i=1}^{k} a(P_i u_h, P_i u_h))^{1/2} (\sum_{i=1}^{k} a(u_{h,i}, u_{h,i}))^{1/2} .$$

By the assumption of the lemma

$$a(u_h, u_h) \le C_0^2 \sum_{i=1}^{k} a(P_i u_h, P_i u_h) = C_0^2 \sum_{i=1}^{k} a(P_i u_h, u_h) = C_0^2 a(P u_h, u_h),$$

and the lemma is established.

We consider three different algorithms and distinguish between them by using a superscript 1, 2 or 3. The most natural algorithm amounts to using the projections $P_i^{(1)} = P_i^i$, where the projection operators P_i^i have been defined in equation (2). The condition number of this algorithm grows linearly with k. We have already remarked that the eigenvalues of P always are bounded from above by k. This bound is attained if $V^{h_1} \cap H_0^1(\Omega_k)$ is not empty, i.e. when the mesh size is fine enough. Any such function belongs to V^{h_i}, $i = 1, 2, \ldots, k$, and is exactly reproduced by each of the projection operators. It is thus an eigenfunction with eigenvalue k. Similarly, any function which belongs to $V^{h_1} \cap H_0^1(\Omega_1 \setminus \Omega_2)$ is an eigenfunction with eigenvalue 1. We will show later that all the eigenvalues are bounded from below by a constant, which completes the proof that the condition number of $P^{(1)}$ is of order k.

A very promising method, for which to our knowledge the optimality has only been established in a quite special model case, cf. [10], is based on using $P_i^{(2)} = P_i^i - P_{i+1}^i$, $i \le k-1$ and $P_k^{(2)} = P_k^k$ as the basic projections in equation (3). It easy to show that these differences of projections are projections and that the composite finite element space V^h is the direct sum of the corresponding subspaces. A difficulty experienced when trying to establish that the eigenvalues of $P^{(2)}$ are bounded uniformly from below, by using Lions' lemma, is related to this complete lack of flexibility in representing a given element of V^h as a sum of elements of the subspaces. So far we have only been able to prove that $C_0^2 \le const.\ \ell$.

We can, however, prove the following result.

Theorem 2. *The eigenvalues of $P^{(2)}$ are uniformly bounded from above by a constant.*

Before we turn to the proof proper, we observe that the unbalance of the first method, which resulted in the amplification of certain functions by a factor k, is no longer possible. By regrouping terms, we also see that

$$P^{(2)} = P_1^1 + (P_2^2 - P_2^1) + \cdots + (P_k^k - P_k^{k-1}) . \tag{5}$$

All the terms, except the first, are similar to multigrid corrections since they each represent the difference between two solutions on the same subregion, using two different mesh sizes.

By using the representation of $P^{(2)}$, given in equation (5), we obtain

$$\begin{aligned} P^{(2)} u_h &= P_1^1 u_h + (P_2^2 - P_2^1) u_h + \cdots + (P_k^k - P_k^{k-1}) u_h \\ &= u_1 + u_2 + \cdots + u_k. \end{aligned}$$

We show that these terms are increasingly orthogonal:

$$a(u_\ell, u_m) \le const.\ q_1^{|\ell - m|} (a(u_\ell, u_\ell))^{1/2} (a(u_m, u_m))^{1/2},$$

where $q_1 < 1$, uniformly. This is a so-called strengthened Cauchy inequality. Without loss of generality, we assume that $\ell < m$. It is easy to show that $a(u_m, v_h) = 0$, $\forall v_h \in$

$V^{h_{m-1}} \cap H_0^1(\Omega_m)$, i.e. u_m is h_{m-1}–harmonic on Ω_m. Let $u_\ell = \sum \alpha_i \phi_i^{h_\ell}$ where the $\phi_i^{h_\ell}$ are basis functions in V^{h_ℓ}. Any such basis function is orthogonal to u_m if its support does not intersect $\partial\Omega_m$ since then either its support belongs to $\Omega_\ell \setminus \Omega_m$ or $\phi_i^{h_\ell} \in V^{h_{m-1}} \cap H_0^1(\Omega_m)$. Thus the only contributions to $a(u_m, u_\ell)$ originate from the triangles of V^{h_ℓ} which intersect $\partial\Omega_m$. Consider one such triangle T. The restriction of u_ℓ to T is a linear function and thus has a constant gradient. It can also be written as a linear combination of basis functions in $V^{h_{m-1}}$. The support of most of these do not intersect $\partial\Omega_m$ and therefore

$$\begin{aligned} a_T(u_m, u_\ell) &= a_{T\cap S}(u_m, u_\ell) \leq (a_T(u_m, u_m))^{1/2}(a_{T\cap S}(u_\ell, u_\ell))^{1/2} \\ &\leq Const.\ q_1^{|m-\ell|} a_T(u_{m,m})^{1/2} a_T(u_\ell, u_\ell)^{1/2} . \end{aligned}$$

Here S denotes the union of the small triangles that are next to $\partial\Omega_m$. We also use our assumptions on the triangulations, given in section 2, and the fact that the gradient of u_ℓ is constant on T and that therefore its contribution to the quadratic form is directly proportional to the area of integration. The proof of the strengthened Cauchy inequality is completed by summing over the triangles of V^{h_ℓ}.

We now note that

$$a(P^{(2)}u_h, P^{(2)}u_h) = \sum_{i,j=1}^{k} a(u_i, u_j) \leq U^T A U.$$

Here

$$A = const. \begin{pmatrix} 1 & q_1 & q_1^2 & \cdots & q_1^{k-1} \\ q_1 & 1 & q_1 & \cdots & q_1^{k-2} \\ \cdot & \cdot & \cdot & \cdots & \cdot \\ q_1^{k-1} & q_1^{k-2} & q_1^{k-3} & \cdots & 1 \end{pmatrix}$$

and

$$U^T = (a(u_1, u_1)^{1/2}, (a(u_2, u_2)^{1/2}, \ldots, a(u_k, u_k)^{1/2}).$$

It is easy to show that the Euclidean norm of A is uniformly bounded. Thus

$$a(P^{(2)}u, P^{(2)}u) \leq |A|_{\ell_2} \sum_{i=1}^{k} a(u_i, u_i) \leq const. \sum_{i=1}^{k} a(u_i, u_i).$$

To complete the proof, we note that

$$\begin{aligned} a(u_i, u_i) &= a(P_i^i u_h - P_i^{i-1} u_h, P_i^i u_h - P_i^{i-1} u_h) \\ &= a(P_i^i u_h - P_i^{i-1} u_h, P_i^i u_h), \end{aligned}$$

since $P_i^i u_h - P_i^{i-1} u_h$ is h_{i-1}–harmonic. By using elementary properties of the projections, we find that $a(u_i, u_i) = a(u_i, u_h)$. Therefore

$$\sum_{i=1}^{k} a(u_i, u_i) = \sum_{i=1}^{k} a(u_i, u_h) = a(P^{(2)}u_h, u_h).$$

Thus

$$a(P^{(2)}u_h, P^{(2)}u_h) \leq const.\ a(P^{(2)}u_h, u_h),$$

from which the upper bound on $P^{(2)}$ follows.

If we modify our projections, we arrive at an algorithm for which we have a proof of optimality.

Theorem 3. *The additive method defined by the projections* $P_i^{(3)} = P_i^i - P_{i+2}^i$, $i \leq k-2$, $P_{k-1}^{(3)} = P_{k-1}^{k-1}$ *and* $P_k^{(3)} = P_k^k$ *has a condition number which is independent of* k *and the number of degrees of freedom.*

A uniform upper bound is obtained by using Theorem 2 and the following formula.

$$\begin{aligned} P^{(3)} &= P_1^1 - P_3^1 + P_2^2 - P_4^2 + \cdots + P_{k-1}^{k-1} + P_k^k \\ &= P_1^1 - P_2^1 + P_2^2 - P_3^2 + \cdots + P_k^k \\ &\quad + P_2^1 - P_3^1 + P_3^2 - P_4^2 + \cdots + P_{k-1}^{k-1} \\ &= P^{(2)} + \widetilde{P}^{(2)} \end{aligned}$$

By Theorem 2, there is an upper bound for $P^{(2)}$ and the same result also provides a uniform upper bound for $\widetilde{P}^{(2)}$, which is an operator corresponding to a composite finite element problem using $k-1$ levels.

From this argument and the fact that $\widetilde{P}^{(2)}$ is positive definite, it follows that the condition number of $P^{(2)}$ cannot be much better than that of $P^{(3)}$. A lower bound of the spectrum of $P^{(3)}$ is given by a partitioning formula and Lions' Lemma. Before we give the details, we note that this argument also provides a lower bound for the spectrum of $P^{(1)}$ since the subspaces related to the projections of that method include those of $P^{(3)}$.

We partition an arbitrary $u_h \in V^h$ as

$$u_h = \sum_{i=1}^{k} u_{h,i}, \qquad u_{h,i} \in V_i^{(3)} \ ,$$

where $V_i^{(3)}$ is the range of $P_i^{(3)}$. The formulas for the $u_{h,i}$ are given by

$$u_{h,1} = \begin{cases} u_h & \text{on} \quad \Omega_1 \setminus \Omega_2 \\ h_1 - harmonic & \text{on} \quad \Omega_2 \setminus \Omega_3 \\ 0 & \text{on} \quad \Omega_3 \end{cases}$$

and for $2 \leq i \leq k-2$, and

$$u_{h,i} = \begin{cases} u_h - u_{h,i-1} & \text{on} \quad \Omega_i \setminus \Omega_{i+1} \\ h_i - harmonic & \text{on} \quad \Omega_{i+1} \setminus \Omega_{i+2} \\ 0 & \text{on} \quad \Omega_{i+2} \end{cases}$$

Since the construction of $u_{h,h-1}$ and $u_{h,h}$ is straightforward, we do not provide details. We note that on each region $\Omega_i \setminus \Omega_{i+1}$ only two of the terms differ from zero and that $u_{h,i} = u_h$ on $\partial\Omega_{i+1}$. By applying the projection $P_i^{(3)}$ to $u_{h,i}$, we find that $u_{h,i} \in V_i^{(3)}$. The proof is completed by showing that

$$a(u_{h,i}, u_{h,i}) \leq const.\ a_{\Omega_i \setminus \Omega_{i+1}}(u_h, u_h).$$

This follows from a variant of Lemma 3 and elementary considerations. The proof of the lower bound is completed by summing over i.

REFERENCES

[1] PETTER E. BJØRSTAD, *Multiplicative and Additive Schwarz: Convergence in the 2 domain case,* Technical Report, Department of Computer Science, University of Bergen, Allégatan 55, Bergen 500, Norway, 1988. In these Proceedings.

[2] JAMES H. BRAMBLE, RICHARD E. EWING, JOSEPH E. PASCIAK, AND ALFRED H. SCHATZ, *A preconditioning technique for the efficient solution of problems with local grid refinement.* Computer Methods in Applied Mechanics and Engineering, 1988, to appear.

[3] PHILIPPE G. CIARLET, *The Finite Element Method for Elliptic Problems,* North-Holland, 1978.

[4] MAKSYMILIAN DRYJA, *An Additive Schwarz Algorithm for Two- and Three- Dimensional Finite Element Elliptic Problems,* Technical Report, Institute of Informatics, University of Warsaw, 00-901 Warsaw, PKiN, p. 850, Poland, 1988. In these Proceedings.

[5] MAKSYMILIAN DRYJA and OLOF B. WIDLUND, *An Additive Variant of the Schwarz Alternating Method for the Case of Many Subregions,* Technical Report 339; also Ultracomputer Note 131, Dept. of Computer Science, Courant Institute, 1987.

[6] RICHARD E. EWING, *Domain Decomposition Techniques for Efficient Adaptive Local Grid Refinement,* Technical Report, Department of Mathematics, University of Wyoming, Laramie, Wyoming, 1988. In these Proceedings.

[7] GENE H. GOLUB AND CHARLES F. VAN LOAN, *Matrix Computations,* Johns Hopkins Univ. Press, 1983.

[8] LESLIE HART and STEVE McCORMICK, *Asynchronous Multilevel Adaptive Methods for Solving Partial Differential Equations on Multiprocessors: Computational Analysis,* Technical Report, Comp. Math. Group, Univ. of Colorado at Denver, 1987. Submitted to Parallel Computing.

[9] PIERRE-LOUIS LIONS, *On the Schwarz alternating method, I,* in: Domain Decomposition Methods for Partial Differential Equations, Roland Glowinski, Gene H. Golub, Gérard A. Meurant, and Jacques Périaux, eds., SIAM, Philadelphia, 1988.

[10] JAN MANDEL and STEVE McCORMICK, *Iterative Solution of Elliptic Equations with Refinement: The Model Multi-Level Case,* Technical Report, Comp. Math. Group, Univ. of Colorado at Denver, 1988. In these Proceedings.

[11] JAN MANDEL and STEVE McCORMICK, *Iterative Solution of Elliptic Equations with Refinement: The Two-Level Case,* Technical Report, Comp. Math. Group, Univ. of Colorado at Denver, 1988. In these Proceedings.

[12] STEVE McCORMICK, *Fast adaptive composite grid (FAC) methods,* in: Defect Correction Methods: Theory and Applications, K. Böhmer and H. J. Stetter, eds., Computing Supplementum 5, Springer-Verlag, Wien, 1984, pp. 115-121.

[13] STEVE McCORMICK and JIM THOMAS, *The fast adaptive composite grid (FAC) method for elliptic equations,* Math. Comp. 46 (174) (1986), pp. 439-456.

[14] H. A. SCHWARZ, *Gesammelete Mathematische Abhandlungen,* Volume 2, Springer, Berlin (1890). First published in Veirteljahrsschrift der Naturforschenden Gesellschaft in Zürich,Vol. 15 (1870),pp. 272-286.

[15] OLOF B. WIDLUND, *An extension theorem for finite element spaces with three applications,* in: Numerical Techniques in Continuum Mechanics, (Notes on Numerical Fluid Mechanics, v. 16), Wolfgang Hackbusch and Kristian Witsch, eds., Friedr. Vieweg und Sohn, Braunschweig Wiesbaden (1987); Proceedings of the Second GAMM-Seminar, Kiel, January 1986.

PART II

Algorithms

Domain Decomposition Techniques and the Solution of Poisson's Equation in Infinite Domains*

Christopher R. Anderson†

Abstract. We discuss how domain decomposition ideas can be used to construct efficient methods for solving Poisson's equation in domains of infinite extent. We give the details for the construction of methods to solve Poisson's equation in the region external to a cylinder and for an infinite backstep. Computational results are presented.

1. Introduction. In some recent work on the numerical solution of incompressible fluid motion in two dimensions [1] it was necessary to construct a solver which would compute the solution to Poisson's equation in a domain of infinite extent. The particular problem was that associated with flow around a circular cylinder. The problem domain was the infinite region external to the cylinder and the computational domain, the region in which the flow quantities were being tabulated, was an annulus about the cylinder. The outer radius of this annulus was not sufficiently far so that setting the value of the solution of Poisson's equation or it s normal derivative equal to zero on this boundary was an acceptable boundary condition for a finite difference solution in the annulus. The purpose of these proceedings is to discuss how domain decomposition techniques can be used to construct a solver for this particular infinite domain problem. The basic idea is to treat the annulus and the infinite component exterior to the annulus as two domains in a domain decomposition procedure. We shall discuss the ideas behind domain decomposition in such a way that the method of construction of a solver is straight forward. The resulting algorithm which we have obtained is not really "new", for example, similar results could be obtained employing the work of Bramble and Pasciak [3], Hariharan [5], or Kang and Dehao [6] and presumably many others. However, the implementation we give here is rather easy to carry out and is efficient since it is just a combination of a fast Fourier transform routine (to solve the interface problem) and a fast Poisson solver (for the solution in the interior of the annulus). The approach which is presented here can also be applied to other infinite domains. In order to demonstrate this, we discuss the implementation of a Poisson solver when the domain is an infinite backstep - a domain which is often used in

*Research Supported by ONR Contracts #N00014-86-K-0691, #N00014-86-K-0727 and NSF Grant DM586-57663

†Department of Mathematics, UCLA, Los Angeles, California, 90024

testing of two-dimensional incompressible fluids codes. To our knowledge the method has not been presented before. Whereas, the procedure for the region external to a cylinder is a direct method, the procedure for the backstep uses the conjugate gradient method to solve the equations which arise in the domain decomposition implementation. This iterative technique is certainly well known to those who apply domain decomposition techniques to bounded domains, but has not necessarily been used by those who are in the business of implementing "infinite" boundary conditions - which is what we are essentially doing here. Thus, for those who are familiar with domain decomposition techniques what follows is a description of how those techniques can be extended to problems in which the domain may be of infinite extent. For those who are familiar with solution procedures on infinite domains, the following shows how domain decomposition ideas can be used to construct efficient implementations of infinite domain boundary conditions.

2. Basic Strategy. In this section we discuss the ideas of domain decomposition in such a way that the application to the problem of calculating the solution to Poisson's equation in an infinite domain is clearly revealed.

Consider the problem associated with finding the solution of Poisson's equation for a rectangle. The problem to be solved is

$$\begin{aligned} \Delta u &= f \quad \text{in} \quad \Omega \\ u &= g \quad \text{on} \quad \partial\Omega. \end{aligned} \tag{1}$$

The basic strategy of the domain decomposition procedure is to decompose the region into two pieces Ω_1 and Ω_2 and construct a solution to (1) by taking the union of solutions to (1) on sub-domains. If the domain is split up into two rectangular pieces then the sub-domain problems are specified by

$$\begin{aligned} \Delta u_i &= f_i \quad \text{in} \quad \Omega_i \\ u_i &= g \quad \text{on} \quad \partial\Omega_i \backslash \Gamma \\ u_i &= u_\Gamma \quad \text{on} \quad \partial\Gamma \end{aligned} \tag{2}$$

for $i = 1, 2$ and where Γ is the interface between the two regions. (See Figure 1) The difficulty in implementing this technique is the determination of the boundary values u_Γ. What is done is to find equations which the u_Γ satisfy and then solve these equations. With the u_Γ determined, the complete solution over the domain is then obtained by just solving (2) for u_1 and u_2.

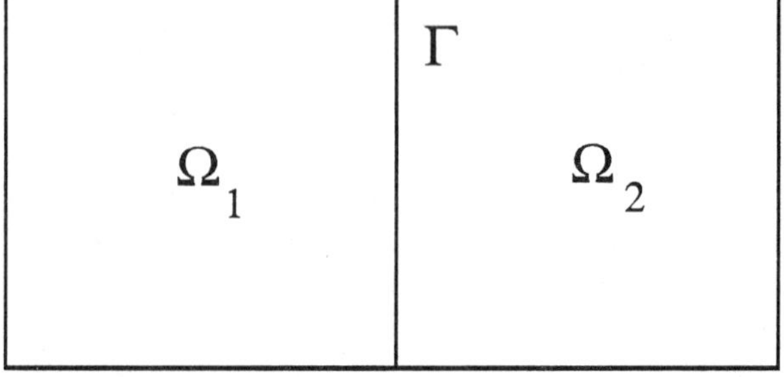

Figure 1

The equations which determine u_Γ are precisely the equations which ensure that if they are solved, and these values are used in the two boundary value problems (2), then the resulting solutions combine to form a solution on the whole domain. Simply; the equations you solve are those which will guarantee that the domain decomposition procedure works.

We now formally derive the equations which the boundary values u_Γ satisfy. In order for the two solutions of (2) , u_1 and u_2, to combine to form a solution of the problem on the whole domain they must be continuous and their normal derivatives must also be continuous across the interface Γ. Under sufficient regularity assumptions this guarantees that the combined solution forms a weak solution of the equations. Under further regularity assumptions this also guarantees that u_1 and u_2 combine to form a strong solution of the equations. If we take the normal derivative along Γ to be outward, then these two conditions can be expressed as

$$u_1 = u_2 \quad \text{on} \quad \Gamma \tag{3}$$

$$\frac{\partial u_1}{\partial n} = -\frac{\partial u_2}{\partial n} \quad \text{on} \quad \Gamma. \tag{4}$$

We are assuming that $u_1 = u_2 = u_\Gamma$ along Γ so relation (3) is satisfied. It is the second relation, the so called "transmission" or "flux" boundary conditions, which determine the equations for u_Γ.

To understand the manner in which the relation (4) determines u_Γ we express it as

$$\frac{\partial u_1}{\partial n}[u_b, f_1, g_1] + \frac{\partial u_2}{\partial n}[u_b, f_2, g_2] = 0. \tag{5}$$

This form reveals the dependence of these normal derivatives on the data in the sub-domains. Here $g_1 = g$ on $\partial\Omega_1\backslash\Gamma$ and $g_2 = g$ on $\partial\Omega_2\backslash\Gamma$. Since the problem is linear, we can separate out the contribution to the normal derivatives in (5) from u_Γ and that from the other data. We are lead to the following equation,

$$\frac{\partial \bar{u}_1}{\partial n}[u_\Gamma] + \frac{\partial \bar{u}_2}{\partial n}[u_\Gamma] = -(\frac{\partial \tilde{u}_1}{\partial n}[f_1, g_1] + \frac{\partial \tilde{u}_2}{\partial n}[f_2, g_2]) \tag{6}$$

with $u_1 = \bar{u}_1 + \tilde{u}_1$ and $u_2 = \bar{u}_2 + \tilde{u}_2$. The equations which determine u_Γ are normal derivative (or flux) balance equations – equation (6) expresses the fact that the flux jump of $\bar{u}_1$ and $\bar{u}_2$ induced by the values u_Γ must be equal to the flux jump in $\tilde{u}_1$ and $\tilde{u}_2$ which is induced by the other boundary values and the forcing function.

The terms in (6) which determine u_Γ can be evaluated operationally. Specifically, given u_Γ to evaluate $\frac{\partial \bar{u}_1}{\partial n}[u_\Gamma]$, one first solves

$$\begin{aligned} \Delta \bar{u}_1 &= 0 && \text{in} \quad \Omega_1 \\ \bar{u}_1 &= 0 && \text{on} \quad \partial\Omega_1\backslash\Gamma \\ \bar{u}_1 &= u_\Gamma && \text{on} \quad \partial\Gamma \end{aligned}$$

and then takes the normal derivative of this function. Similarly, to evaluate the first term on the right hand side of the equation (6), we first solve a Poisson problem of the form

$$\begin{aligned} \Delta \tilde{u}_1 &= f_1 && \text{in} \quad \Omega_1 \\ \tilde{u}_1 &= g_1 && \text{on} \quad \partial\Omega_1\backslash\Gamma \\ \tilde{u}_1 &= 0 && \text{on} \quad \partial\Gamma \end{aligned}$$

and evaluates the normal derivative of this solution. It is worth noting that in this process of evaluating these quantities, we have not depended on any specific discretization of the equations. It is this fact which allows one to have one or both of the domains of infinite extent - all that is needed is some technique for evaluating the terms of (6).

The basic strategy, as will be exemplified in the next two sections, will be to use (6) in the context of different discretizations and domains to define a system of equations which determine the interfacial values of a solution of Poisson's equation. This system will be solved, and the resulting values will be used as boundary data for Poisson solves on subdomains to construct the remaining parts of the solution.

3. A Solver For The Region External to a Circular Cylinder. The domain under consideration is composed of two pieces, Ω_1 and Ω_2. Ω_1 is the annulus between the circle of radius r_a and a circle of radius r_b. Ω_2 is the region external to the circle of radius r_b. Our goal is to find the restriction of the solution to

$$\Delta u = f \qquad \text{for} \quad r_a < r \leq \infty \tag{7}$$

$$u = g \qquad \text{on } r = r_a$$

on the set of grid points of a polar grid in the annular region Ω_1. (See Figure 2.)

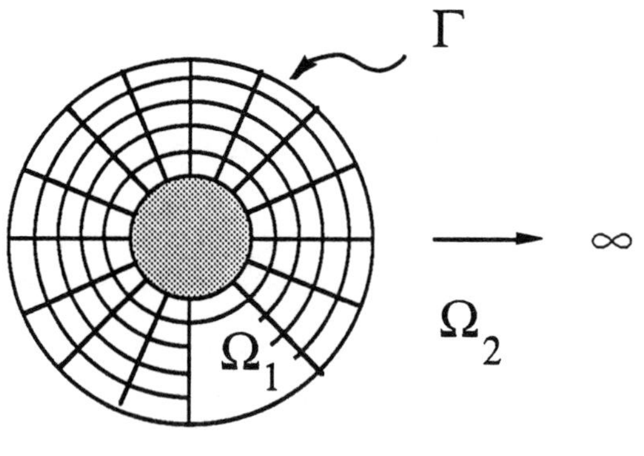

Figure 2

In order for our procedure to work as it is described here we assume that the support of the forcing function f is contained in Ω_1 and the logarithmic behavior of the solution to (7) at infinity is known. (In fluids problems in which the solution is a stream function, this is equivalent to specifying the circulation of the velocity field.)

Following the basic ideas of domain decomposition, the technique here will be to first find the values of the solution to (7) on the boundary Γ located at $r = r_b$. We shall refer to these values as u_Γ. We then employ a fast Poisson solver to calculate the remainder of the solution in Ω_1 by solving a problem with forcing function f and boundary values g on $r = r_a$ and u_Γ on $r = r_b$.

The determination of the solution values u_Γ is done by solving a finite dimensional analog of the equations described by (6) in the previous section. Taking note of our assumptions on the data of the problem, the equations which determine u_Γ. are

$$\frac{\partial \tilde{u}_1}{\partial n}[u_\Gamma] + \frac{\partial \tilde{u}_2}{\partial n}[u_\Gamma] = -\frac{\partial \tilde{u}_1}{\partial n}[f_1, g] \tag{8}$$

In these equations $\bar{u}_1$ represents an approximate solution of Poisson's equation in Ω_1 with homogeneous forcing function and boundary data non-zero only along $r = r_b$. $\bar{u}_2$ is defined similarly. $\tilde{u}$ is the solution of Poisson's equation in Ω_1 with forcing function f and boundary data $\tilde{u}_1 = g$ for $r = r_a$ and $\tilde{u}_1 = 0$ for $r = r_b$. In order to reduce the problem to a finite dimensional one,we require that a relation of the form (8) hold at N equally spaced points on the boundary $r = r_b$. We take N to be the number of panels in the θ direction associated with the underlying finite difference approximation.

In this problem, Fourier analysis can be used effectively. The technique for constructing and solving the system of equations on the left hand side of (8) is done using the Fourier series solutions of the boundary value problems which implicitly define these operators. Given values of u_Γ at $N = 2m+1$ equally spaced points on the circle $r = r_b$ we first form the trigonometric interpolant -

$$u_\Gamma(\theta) = a_0 + \sum_{k=1}^{m} a_k \cos(k\theta) + \sum_{k=1}^{m} b_k \sin(k\theta)$$

The value of $\frac{\partial \bar{u}_1}{\partial n}[u_\Gamma]$, the normal derivative of the function obtained by solving a Laplace equation with this interpolated boundary data, is given by

$$\frac{\partial \bar{u}_1}{\partial n} = \alpha_0 a_0 + \sum_{k=1}^{m} \alpha_k a_k \cos(k\theta) + \sum_{k=1}^{m} \alpha_k b_k \sin(k\theta). \tag{9}$$

Here,

$$\alpha_0 = \frac{1}{r_b} \frac{1}{log(r_b) - log(r_a)} \tag{10}$$

$$\alpha_k = \frac{1}{r_b} \frac{(1 + (\frac{r_b}{r_a})^{2k})}{(1 - (\frac{r_b}{r_a})^{2k})}$$

The same approach allows us to evaluate $\frac{\partial \bar{u}_2}{\partial n}$, and we find

$$\frac{\partial \bar{u}_2}{\partial n} = \frac{\gamma}{r_b} + \sum_{k=1}^{m} \beta_k a_k \cos(k\theta) + \sum_{k=1}^{m} \beta_k b_k \sin(k\theta). \tag{11}$$

With

$$\beta_k = \frac{k}{r_b}. \tag{12}$$

γ is the coefficient of the log term in the solution. (If the solution is a stream function, then $2\pi\gamma =$ the circulation of the velocity field.)

The right hand side of equation (8) can be calculated by using a second order difference approximation to $\frac{\partial \tilde{u}_1}{\partial n}$, where $\tilde{u}_1$ is the solution of the discrete Poisson problem associated with the equation

$$\begin{aligned} \Delta\tilde{u}_1 &= f \quad \text{in } \Omega_1 \\ \tilde{u}_1 &= g \quad \text{on } r = r_a \\ \tilde{u}_1 &= u_\Gamma \quad \text{on } r = r_b. \end{aligned} \tag{13}$$

We denote this normal derivative by f_Γ.

What we have just described is the procedure for evaluating both sides of equation (8) at the points $r = r_b$ and $\theta_j = j(\frac{2\pi}{N})$ $j = 1 \ldots N$. Usually in domain decomposition procedures an iterative approach is taken to solve (8), and we have sufficient information to begin the process of creating an iterative scheme for solving the equations. However, the problem at hand is such that the matrix which represents our approximation to the operator on the right hand side of (8) diagonalized by the discrete fourier transform. Specifically, if we designate $\hat{f_\Gamma}$ as the discrete transform of f_Γ, then we have the following matrix equation for the coefficients of the transform of u_Γ,

$$(14) \qquad \begin{bmatrix} \\ & A & \\ \\ \end{bmatrix} \begin{bmatrix} a_0 \\ a_1 \\ \vdots \\ a_m \\ b_1 \\ \vdots \\ b_m \end{bmatrix} = \begin{bmatrix} \hat{f_\Gamma}_0 \\ \hat{f_\Gamma}_1 \\ \\ \vdots \\ \\ \hat{f_\Gamma}_N \end{bmatrix} + \begin{bmatrix} \frac{\gamma}{r_b} \\ 0 \\ 0 \\ 0 \\ \vdots \\ 0 \end{bmatrix}$$

Where A is the matrix

$$\begin{bmatrix} \alpha_0 & & & & & & \\ & \alpha_1 + \beta_1 & & & & & \\ & & \ddots & & & & \\ & & & \alpha_m + \beta_m & & & \\ & & & & \alpha_1 + \beta_1 & & \\ & & & & & \ddots & \\ & & & & & & \alpha_m + \beta_m \end{bmatrix}$$

If we put all the pieces together, the algorithm is as follows:

(i) Solve (13) for $\tilde{u}_1$ and evaluate f_Γ by differencing the result.

(ii) Solve (14) for the coefficients of the transform of u_Γ. Evaluate u_Γ using the inverse Fourier transform.

(iii) Find the restriction to Ω_1 of the solution to (7) by solving

$$\Delta u = f \qquad \text{in } \Omega_1$$

$$u = g \quad \text{on} \quad r = r_a \qquad u = u_\Gamma \quad \text{on} \quad r = r_b$$

We have implemented the above algorithm, and performed some tests on the method. We were primarily interested in demonstrating that the effect of truncating the domain gives errors which are always on the order of the truncation error of the finite difference method used in Ω_1 (i.e. that the solution on the finite difference mesh is second order in δr and $\delta\theta$ independent of the mesh size). We expect that this should be the case since there is no error in the boundary conditions -we are using the exact analytic boundary conditions for the first m modes of the solution. In Table 1 we present the results of a problem for which $r_a = 1$ and $r_b = 2$. The solution computed was that induced by a unit charge located inside the cylinder and offset from the origin (so all modes of the boundary conditions would be tested). These results show that second order accuracy is preserved.

We were also interested in observing the change in the solution when the outer boundary of Ω_1 at $r = r_b$ was made larger, but the mesh size was kept fixed. Ideally there should be no change in the solution, but in light of our implementation, it can be expected to be on the order of the truncation error of the finite difference scheme. This behavior is reflected by the results presented in Table 2. Here the grid size was 16×64 for an outer radius of $r_b = 2$, and was changed to 32×64, 64×64 and 128×64 for the radii used to construct the table. The L^2 error of the 16×64 grid results was 1.580×10^{-3} while the L^∞ error was 8.437×10^{-4}

Mesh Size (r, θ)	L^2 Err.	L^∞ Err.
32×64	1.580×10^{-4}	8.437×10^{-4}
64×128	3.646×10^{-5}	1.972×10^{-4}
128×256	8.446×10^{-6}	4.928×10^{-5}

Table 1
Error With Change in Mesh Size

Outer Radius	L^2 Err.	L^∞ Err.
$r_b = 3$	1.883×10^{-3}	1.686×10^{-3}
$r_b = 5$	2.419×10^{-3}	2.183×10^{-3}
$r_b = 9$	2.646×10^{-3}	2.389×10^{-3}

Table 2
Error With Change in External Radius

4. Solver For An Infinite Step. In this section we use the ideas of domain decomposition to construct a solution to Poisson's equation in a domain which is an infinite step. Figure 3 shows the region which we are considering. The problem to be solved is

$$\Delta u = f \quad \text{in} \quad \Omega \tag{15}$$

$$u = 0 \quad \text{on } \partial\Omega$$

where $\Omega = \cup_{i=1}^{4} \Omega_i$. Ω_1 and Ω_4 are the unbounded components on the left and right ends of the step while Ω_2 is the region in which $x_a \leq x \leq x_b$, while Ω_3 is the adjoining region in which $x_b \leq x \leq x_c$. We designate the boundaries at $x = x_a$, $x = x_b$ and $x = x_c$ by Γ_a, Γ_b and Γ_c and the values of the solution to (15) on these boundaries as u_a, u_b and u_c. The idea is to compute u_a, u_b and u_c first and then obtain the restriction of the solution in the domains Ω_2 and Ω_3 by solving the appropriate Poisson problem in these domains. As before we assume that the support of f is compact, and contained in a region $x_a \leq x \leq x_c$.

The key differences between this problem and that for the cylinder is that the outer computational boundary (the one connecting the computational region to the infinite components of the domain) is not connected and the interior computational domain is not regular. This fact precludes the use of analytic procedures to evaluate the analog of equations (6) directly. However, as will be seen below, the construction of a solution procedure for this domain is only a minor modification of that for a region consisting of bounded rectangles. Moreover, the solution we find will be the restriction of the *exact* solution (up to roundoff) of the solution of the finite difference equations over the whole domain Ω.

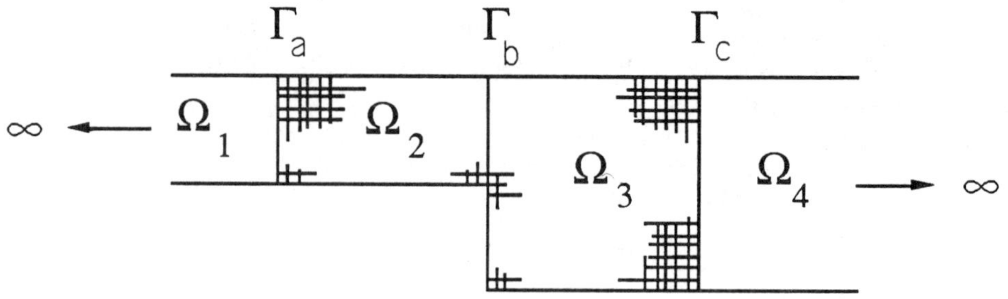

Figure 3

In the notation of the previous two sections, the conditions which determine the values of the solution along the interfaces are the following

(16) $$\frac{\partial \bar{u}_1}{\partial n}[u_a] + \frac{\partial \bar{u}_2}{\partial n}[u_a, u_b] = -\frac{\partial \tilde{u}_2}{\partial n}[f_2] \quad \text{along} \quad \Gamma_a$$

(17) $$\frac{\partial \bar{u}_2}{\partial n}[u_a, u_b] + \frac{\partial \bar{u}_3}{\partial n}[u_b, u_c] = -(\frac{\partial \tilde{u}_2}{\partial n}[f_2] + \frac{\partial \tilde{u}_3}{\partial n}[f_3]) \quad \text{along} \quad \Gamma_b$$

(18) $$\frac{\partial \bar{u}_3}{\partial n}[u_b, u_c] + \frac{\partial \bar{u}_4}{\partial n}[u_c] = -\frac{\partial \tilde{u}_3}{\partial n}[f_3] \quad \text{along} \quad \Gamma_c.$$

As before, we interpret the terms in these equations operationally, that is

$$\frac{\partial \bar{u}_2}{\partial n}[u_a, u_b]$$

is the normal derivative of $\bar{u}_2$ the solution of

$$\Delta \bar{u}_2 = 0 \quad \text{in} \quad \Omega_2$$

$$\bar{u}_2 = u_a \quad \text{at} \quad \Gamma_a \qquad \text{and} \qquad \bar{u}_2 = u_b \quad \text{at} \quad \Gamma_b.$$

What is in brackets indicates what data is used for the Poisson problem and the subscript indicates in which domain the solution is computed.

We construct a finite dimensional analog of (16) - (18) by using finite difference approximations. The finite difference approximations are those which are inherited by the underlying discretization which we are attempting to solve rather than just difference approximations to the operators in (16) - (18). We assume that the underlaying finite difference mesh is a uniform rectangular one, with mesh widths δx and δy in the x and y directions respectively. If we consider the (i,j)th node of the finite difference grid along Γ_a, then the equations used to approximate (16) have the form

$$
\begin{aligned}
(19)\qquad & \frac{\bar{u}_1(i,j)-\bar{u}_1(i-1,j)}{\delta x}+(\frac{\delta x}{2\delta y})\frac{\bar{u}_1(i,j+1)-2\bar{u}_1(i,j)+\bar{u}_1(i,j-1)}{\delta y}+\\
& \frac{\bar{u}_2(i+1,j)-\bar{u}_2(i,j)}{\delta x}+(\frac{\delta x}{2\delta y})\frac{\bar{u}_2(i,j+1)-2\bar{u}_2(i,j)+\bar{u}_2(i,j-1)}{\delta y}\\
&=\frac{\tilde{u}_2(i+1,j)-\tilde{u}_2(i,j)}{\delta x}+(\frac{\delta x}{\delta y})\frac{\tilde{u}_2(i,j+1)-2\tilde{u}_2(i,j)+\tilde{u}_2(i,j-1)}{\delta y}
\end{aligned}
$$

The equations for the other interfaces are of similar form. In these difference equations there are differences in the direction tangent to the interface, whereas there are only normal derivatives in the flux balance equations (16) - (18). These terms arise in the finite difference implementation because tangential differences contribute to the discrete flux balance equations which are the analogs of (16) - (18). The evaluation of the terms in (19) and of the equations for the other interfaces is accomplished by solving the appropriate Poisson problem and then differencing the result. This procedure works well for the problems which are associated with the bounded components, but for the infinite components, this poses some difficulty. In particular, we must evaluate the difference approximations to $\frac{\partial \bar{u}_1}{\partial n}$ and $\frac{\partial \bar{u}_4}{\partial n}$ in which the functions $\bar{u}_1$ and $\bar{u}_4$ are solutions of a Poisson problem in half infinite tubes.

Fortunately, one can solve such problems exactly using discrete Fourier analysis. Specifically, if we have discrete data along the interface Γ_a of the form

$$(20)\qquad u_a(y_j)=\sum_{k=1}^{m} b_k \sin(\frac{2\pi y_j}{2\mathrm{L}})$$

where $y_j = y_a + (j-1)\delta y$ and L is the length of Γ_a, then a solution of a discrete Laplace equation in the half-tube Ω_1 is given by

$$(21)\qquad \bar{u}_1(x_i,y_j)=\sum_{k=1}^{m}\lambda_k^{i-1} b_k \sin(\frac{2\pi y_j}{2\mathrm{L}})$$

where the values λ_k are given by

$$\lambda_k=\frac{2+4(\frac{\delta x}{\delta y})^2\sin^2(\frac{\pi k\delta y}{2\mathrm{L}})-\sqrt{\left[(2+4(\frac{\delta x}{\delta y})^2\sin^2(\frac{\pi k\delta y}{2\mathrm{L}})\right]^2-4}}{2}$$

The explicit form of (21) allows us to compute the differences needed in the evaluation of (19). The evaluation of the term involving $\bar{u}_4$ is accomplished in a similar fashion.

In light of the specific form of the solutions which define $\bar{u}_1$ and $\bar{u}_4$, we see that the evaluation of the appropriate differences can be accomplished with fast sin transforms. It is also the case that the other operators in the difference approximations to (16) - (18) can be evaluated using discrete sine transforms. This follows because all of the boundary value problems which define these operators can be solved explicitly using discrete Fourier analysis. (See [2] for more details.) Thus, the forward application of the operators which determine the boundary values u_a, u_b, and u_c can be carried out using fast sine transforms.

There is a natural block structure to the equations when one groups the unknowns together according to which interface they occur on. With this blocking the discrete equations for the values u_a, u_b and u_c take the form,

$$
\begin{bmatrix} A_{11}^1 + A_{11}^2 & B_{12} & 0 \\ B_{21} & A_{22}^1 + A_{22}^2 & B_{23} \\ 0 & B_{32} & A_{33}^1 + A_{33}^2 \end{bmatrix} \begin{bmatrix} u_a \\ u_b \\ u_c \end{bmatrix} = \begin{bmatrix} \tilde{f}_a \\ \tilde{f}_b \\ \tilde{f}_c \end{bmatrix} \tag{22}
$$

Here $\frac{\partial \bar{u}_1}{\partial n}$ is approximated by A_{11}^1, $\frac{\partial \bar{u}_2}{\partial n}$ is approximated by A_{11}^2 etc. The right hand side of the equations, designated by $(\tilde{f}_a, \tilde{f}_b, \tilde{f}_c)^t$ are determined by differencing solutions of the appropriate finite difference Poisson problem.

This equation is not completely diagonalized by the discrete sin transform, for the number of points which define the transform used to evaluate the equations for Γ_b are different for the terms involving u_2 and u_3. However, the matrices in the first and third rows are simultaneously diagonalized by the fast sine transform.

Unlike the cylinder case, the equations are not solved directly but iteratively. As is typical with domain decomposition techniques we used pre-conditioned conjugate gradients. The preconditioner used was the inverse of the matrix

$$
\begin{bmatrix} A_{11}^1 + A_{11}^2 & 0 & 0 \\ 0 & A_{22}^1 & 0 \\ 0 & 0 & A_{33}^1 + A_{33}^2 \end{bmatrix}.
$$

Each of these matrices is diagonalized by the discrete sin transform, so the application of the preconditioner can be accomplished using fast sin transforms. This preconditioner worked well and converged rapidly. (This is to be expected by analogy with results for the bounded domain case [4])

In sum the algorithm is as follows:

(I) Solve appropriate Dirichlet problems in Ω_2 and Ω_3 and evaluate $(\tilde{f}_a, \tilde{f}_b, \tilde{f}_c)^t$ by differencing the result.

(ii) Solve (22) using the method of pre-conditioned conjugate gradients. (Use the fact that the the foward action of the operator in (22) can be computed using fast sin transforms.)

(iii) Find the restriction to Ω_2 and Ω_3 of the solution to (15) by solving Dirichlet problems in each of these domains.

In the first test of this method, a fixed uniform mesh (with $\delta x = \delta y = 0.1$) was used and the upstream and downstream boundaries were changed. The domain was taken to be an infinite tube with width 0.8 on the left and 1.0 on the right. A unit source was placed at a fixed location in Ω_2 and the values in Ω_2 were monitored when the computational domains Ω_2 and Ω_3 were extended from a length of 1.0 to 2.0 and from 2.0 to 4.0. The maximum change which occurred was 9.357×10^{-5} and 9.359×10^{-5} - a quantity on the order of the accuracy with which the fast Poisson solver can calculate the solution to the discrete Laplacian. This degree of accuracy is to be expected since we are constructing an exact solution to the finite difference equations in the whole infinite region. In the second test we measured the number of conjugate-gradient iterations which were needed to reach a given residual size. In each case the domain was fixed to be of width 0.8 on the left and 1.0 on the right. The length of Ω_2 and Ω_3 was fixed at 1.0. In Table 3 we present the L^2 norm of the residual for different size grids. As can be seen from the table the solution converges rapidly demonstrating the efficiency of the conjugate-gradient method with our choice of pre-conditioner.

Iterations	$\delta x = \delta y = 0.1$	$\delta x = \delta y = 0.05$	$\delta x = \delta y = 0.025$
1	3.813×10^{-2}	8.693×10^{-2}	1.923×10^{-1}
2	6.494×10^{-4}	1.747×10^{-3}	6.317×10^{-3}
3	1.332×10^{-5}	1.343×10^{-5}	8.397×10^{-5}

Table 3
Residual Error

5. Conclusion. We have discussed the manner in which domain decomposition techniques can be used to construct solvers for infinite domains. The central idea is to observe that the conditions which determine the interface values (values which are necessary to implement the domain decomposition technique) are derived from a condition of an equality of normal derivative's or a flux balance condition. With this observation, such conditions can be used to derive equations for the interfacial values even if one or more of the domains is of infinite extent. The two applications, one for a domain external to a cylinder and one for an infinite step exhibit the potential of the approach.

REFERENCES

[1] C. R. Anderson, *Vorticity Boundary Conditions and Boundary Generation of Vorticity*, J. Comp. Phys. *to appear.*

[2] C. R. Anderson, *The Application of Domain Decomposition to the Solution of Laplace's Equation in Infinite Domains*, CAM Report 87-19, Dept. of Mathematics, UCLA, 1987.

[3] J. H. Bramble and J. E. Pasciak, *A Boundary Parametric Approximation To The Linearized Scaler Potential Magnetostatic Field Problem*, App. Num. Math. 1, 1985, pp. 493-514.

[4] T.F. Chan and D.C. Resasco, *A Survey of Preconditioners for Domain Decomposition*, Research Report YALEU/DCS/RR-414, September 1985.

[5] S. I. Hariharan, *Absorbing Boundary Conditions for Exterior Problems*, ICASE Report #95-33, NASA Langley Research Center, Virginia, 1985.

[6] F. Kang and Y. De-hao, *Canonical Integral Equations of Elliptic Boundary-value Problems and Their Numerical Solutions*, Proceedings of the China-France Symposium on Finite Element Methods, Ed. F. Kang and J.L. Lions, 1983, pp. 211-252.

CHAPTER 11

Some Remarks on the Hierarchical Basis Multigrid Method

Randolph E. Bank*
Harry Yserentant†

1. Introduction. Originally, multigrid methods were developed for elliptic partial differential equations discretized by difference methods on sequences of uniform and uniformly refined meshes. For most of these problems, multigrid methods are extremely fast solvers for the resulting discrete equations. On the other hand, for many problems of practical interest, uniform meshes are far from optimal. In such cases, finite element methods based on adaptively refined, strongly nonuniform grids are much more appropriate.

It is not obvious how the multigrid method can be effectively applied to problems discretized on nonuniform grids of this type. For many simple decompositions of the final mesh into meshes of different levels, the number of operations per iteration may not be proportional to the number of unknowns in the finest mesh. Indeed, to keep the operation count of optimal order, it is necessary that the number of nodes increase by a constant factor larger than one from one level to the next. As the experience with adaptive local mesh refinement packages like PLTMG [1] [2] shows, this condition is not satisfied for many examples. Often only a few nodes are added at each refinement level. It is therefore necessary to reduce the number of levels artificially in such situations. This can be justified by practical experience [1]. Nevertheless, a complete and sufficiently general theoretical analysis of the convergence behavior in such situations has not yet been given.

The hierarchical basis multigrid method [4] has been developed to overcome these difficulties. In section 2, we formulate the method for general second order boundary value problems in two space dimensions, which are not necessarily self-adjoint and positive definite. We survey some of the properties of the method in sections 3 to 5. In section 6, we present some numerical illustrations using the new version of the package PLTMG [2], which employs the hierarchical basis multigrid method as its linear equation solver.

2. The Hierarchical Basis Multigrid Method. Mathematically, the hierarchical basis multigrid method is a classical block symmetric Gauss-Seidel iteration

* Department of Mathematics, University of California at San Diego, La Jolla, California 92093.
The work of this author was supported by the Office of Naval Research under contract N00014-82K-0197.
† Fachbereich Mathematik der Universität Dortmund, D-4600 Dortmund 50, Federal Republic of Germany.

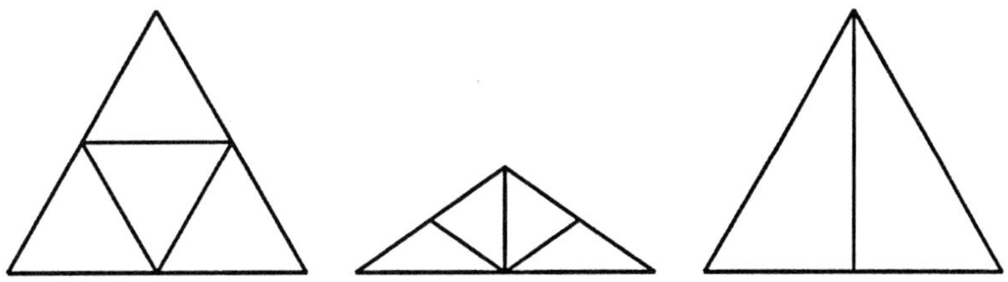

FIG. 1. *Regular refinement (left, center) and irregular refinement(right)*

for solving the discrete boundary value problem, not with respect to the usual nodal basis of the finite element space, but rather with respect to its hierarchical basis. It is closely related to the two-level methods of [3] [7], and the Jacobi-like hierarchical basis method of [9]

In this section we describe the hierarchical basis multigrid method applied to the finite element spaces used in PLTMG [2], but it is straightforward to extend the method to other finite elements, other refinement schemes, and to higher degree polynomial spaces.

Assume that $\mathcal{T}_1$ is an intentionally coarse initial triangulation of a given polygonal domain $\Omega \subset \mathcal{R}^2$. This triangulation is refined several times, yielding a family of nested triangulations $\mathcal{T}_1, \mathcal{T}_2, \mathcal{T}_3 \ldots$. A triangle of $\mathcal{T}_{k+1}$ is either a triangle of $\mathcal{T}_k$ or is generated by dividing a triangle in $\mathcal{T}_k$ into two or four triangles as illustrated in Figure 1.

Refinement into four triangles is called *regular refinement.* If the triangle in question has an obtuse angle, it is refined as in the figure above center; otherwise, it is refined as in the figure above left. Refinement into two triangles is called *irregular refinement.* Irregularly refined elements may not be refined further. This insures that the interior angles of all refined triangles remain bounded away from zero.

The triangles of the initial triangulation $\mathcal{T}_1$ are called level 1 elements and the triangles obtained by the refinement of a level $k-1$ element are called level k elements. In creating the $\mathcal{T}_{k+1}$ from $\mathcal{T}_k$, we allow only the refinement of triangles of level k; this uniquely defines the sequence $\mathcal{T}_k$, $1 \leq k \leq j$. This sequence of meshes does not usually correspond to the actual sequence of meshes generated by an adaptive refinement process like that used in PLTMG, where unrefined triangles of all levels are considered for refinement in every adaptive step. Nonetheless, each fine mesh $\mathcal{T}_j$ generated by such a dynamic process can be uniquely decomposed *a posteriori* into a sequence of triangulations of the type described here. The vertices of the level 1 elements are called level 1 nodes, and the vertices created by the refinement of level $k-1$ elements are called level k nodes. With the definition of $\mathcal{T}_k$ given above, the level k vertices are those generated by the refinement of $\mathcal{T}_{k-1}$.

The triangulations $\mathcal{T}_1, \mathcal{T}_2, \mathcal{T}_3, \ldots$ correspond to a sequence $\mathcal{S}_1 \subseteq \mathcal{S}_2 \subseteq \mathcal{S}_3 \subseteq \ldots$ of finite element spaces consisting of piecewise linear polynomials. The space $\mathcal{S}_{k+1}$ can be written as the direct sum

$$\mathcal{S}_{k+1} = \mathcal{S}_k \oplus \mathcal{V}_{k+1} \tag{1}$$

where $\mathcal{V}_{k+1}$ is the subspace of $\mathcal{S}_{k+1}$ containing those functions which are zero at all of the nodes of $\mathcal{T}_k$; that is, the vertices of level ℓ for $1 \leq \ell \leq k$. This decomposition leads to the definition of the hierarchical basis for the space $\mathcal{S}_k$. The hierarchical basis for the space $\mathcal{S}_1$ is just the usual nodal basis for $\mathcal{S}_1$. The hierarchical basis for $\mathcal{S}_{k+1}$ consists of the union of the hierarchical basis for $\mathcal{S}_k$ and the nodal basis for $\mathcal{V}_{k+1}$ (the nodal basis functions associated with the level $k+1$ nodes).

For convenience, we fix a finite element space $\mathcal{S} = \mathcal{S}_j$. Formally, the discrete boundary value problem corresponding to $\mathcal{S}$ can be formulated with respect to the hierarchical basis of $\mathcal{S}$. Assume that the hierarchical basis functions are ordered according to level; that is, corresponding to the decomposition

$$\mathcal{S} = \mathcal{S}_1 \oplus \mathcal{V}_2 \oplus \cdots \oplus \mathcal{V}_j \tag{2}$$

Then the stiffness matrix A corresponding to the hierarchical basis has a block structure given by

$$A = \begin{bmatrix} A_{11} & A_{12} & \cdots & A_{1j} \\ A_{21} & A_{22} & & A_{2j} \\ \vdots & & \ddots & \vdots \\ A_{j1} & A_{j2} & \cdots & A_{jj} \end{bmatrix} \tag{3}$$

where A_{kk} corresponds to the inner products of basis functions in $\mathcal{V}_k$.

We split A into the sum

$$A = L + D + U \tag{4}$$

where L is block lower triangular, D is block diagonal, and U is block upper triangular. Our basic scheme for the solution of the discrete boundary value problem

$$Ax = b \tag{5}$$

is the block symmetric Gauss-Seidel iteration given by

$$x^{(i+1/2)} = x^{(i)} + (D+U)^{-1}(b - Ax^{(i)}) \tag{6}$$

$$x^{(i+1)} = x^{(i+1/2)} + (L+D)^{-1}(b - Ax^{(i+1/2)}) \tag{7}$$

To avoid the expensive direct solution of the $2j$ linear systems involving the diagonal blocks of A, we use inner iterations, and modify (6)-(7) to

$$x^{(i+1/2)} = x^{(i)} + (\tilde{D}+U)^{-1}(b - Ax^{(i)}) \tag{8}$$

$$x^{(i+1)} = x^{(i+1/2)} + (L+\bar{D})^{-1}(b - Ax^{(i+1/2)}) \tag{9}$$

The block diagonal matrices $\bar{D}$ and $\tilde{D}$ correspond to the situation where only the systems involving A_{11} are solved directly, whereas the systems involving the coefficient matrices A_{kk} for $k > 1$ are solved approximately by a fixed number of inner iteration steps. The theory developed in [4] allows:

- a single backward Gauss-Seidel step for each diagonal system in (8) and a single forward Gauss-Seidel step for each system in (9) (except those involving A_{11}).
- an arbitrary fixed number of symmetric Gauss-Seidel steps for all systems (except those involving A_{11}) arising in both half steps (8)-(9).

In PLTMG [2], one symmetric Gauss-Seidel is used for the inner iteration for self-adjoint problems. For strongly nonsymmetric problems, it is desirable to damp the inner symmetric Gauss-Seidel iteration. The resulting iteration is like SSOR except that the relaxation parameter ω is less than one.

3. Convergence Results. A complete convergence theory for this process in the case of self-adjoint, positive definite, second order elliptic boundary value problems in two space dimensions has been given in [4]. In this case, $U = L^t$, $D = D^t$, $\bar{D} = \tilde{D}^t$, and the iteration (8)-(9) can be written as

$$x^{(i+1)} = x^{(i)} + B^{-1}(b - Ax^{(i)}) \tag{10}$$

with a symmetric and positive definite matrix B given by

$$B = (L + \tilde{D})^t(\tilde{D} + \tilde{D}^t - D)^{-1}(L + \tilde{D}) \tag{11}$$

The optimal estimate for the speed of convergence of this iteration with respect to the energy norm induced by the matrix A is

$$||x^{(i+1)} - x|| \leq (1 - \kappa^{-1})||x^{(i)} - x|| \tag{12}$$

In (12), x denotes the exact solution of the linear system (5), $||x||^2 = x^tAx$ is the energy norm, and κ is the spectral condition number of the preconditioned matrix

$$B^{-1/2}AB^{-1/2} \tag{13}$$

Using only very weak assumptions, which are almost always satisfied in practice, we were able to show that

$$\kappa = O(j^2) \tag{14}$$

where j is the number of refinement levels. In particular, we assume shape regularity of the finite elements, but do not require that the global mesh be quasiuniform. No global regularity of the problem is used (beyond the $\mathcal{H}^1$ regularity necessary to define the weak form), and only local ellipticity constants enter into the proofs.

This result means that only $O(j^2|\log\epsilon|)$ iterations are required to reduce the energy norm of the error by a factor of ϵ. This can be improved to

$$O(j\log\epsilon) \tag{15}$$

iteration steps by using a conjugate gradient or minimum residual acceleration scheme in conjunction with the basic iteration.

4. Implementation. Algorithmically, the hierarchical basis stiffness matrix (3) should not be assembled and stored explicitly. As we have shown in [4], and have implemented in PLTMG [2], the iteration (8)-(9) can be realized in a fashion similar to a standard multigrid V-cycle [5] using a point symmetric Gauss-Seidel (or SSOR) smoother. The difference is that, on a given level k, for the hierarchical basis method only those unknowns corrresponding to the space $\mathcal{V}_k$ are smoothed, as opposed to the unknowns associated with all the vertices in $\mathcal{T}_k$, as in the standard multigrid method.

Therefore only the (sparse) diagonal blocks A_{kk} for $1 \leq k \leq j$ need to be stored, along with certain entries allowing one to generate the product $A_{ik}y$ for $i \neq k$. The details of this scheme are described in [4]. In any event, the total number of floating point numbers required for this implicit representation of the stiffness matrix is $9N + O(1)$ for the nonsymmetric case, and $5N + O(1)$ for the symmetric case, where N is the dimension of $\mathcal{S}$. This should be compared with $7N + O(1)$ and $4N + O(1)$, respectively, for the standard nodal stiffness matrix.

In the same fashion, one can show that the number of operations necessary to perform one iteration is $O(N)$, regardless of the distribution of the nodes among the levels. In view of the results cited in section 3, at least for symmetric, positive definite problems, a total of

$$O(jN\log\epsilon) \tag{16}$$

operations are required to reduce the energy norm of the error by a factor of ϵ. Practically, this represents a logarithmic-like growth in the number of operations as a function of the number of unknowns. This estimate is slightly suboptimal when compared to standard multigrid methods applied to sufficiently regular problems with geometrically increasing subspace dimensions, where the corresponding work estimate would be

$$O(N\log\epsilon) \tag{17}$$

Unlike the usual multigrid methods, however, (16) requires only the weak assumptions of sections 2 and 3 for the continuous problem and for the sequence of meshes.

5. A Counterexample. Both a standard multigrid V-cycle, using a point Gauss-Seidel smoother, and a cycle of the hierarchical basis multigrid method can be interpreted as a sequence of one-dimensional corrections. Compared with the standard multigrid method, in the hierarchical basis multigrid method certain directions are skipped. As a curiousity, we present a simple example that this does not necessarily mean slower convergence.

Consider the one-dimensional boundary value problem

$$-u'' = f \qquad 0 < x < 1 \tag{18}$$

$$u(0) = u(1) = 0 \tag{19}$$

discretized by piecewise linear finite elements on the grids

$$x_i = i2^{-k} \qquad i = 0, \ldots, 2^k \tag{20}$$

for $k = 1, 2, \ldots, j$. If we first smooth the unknowns associated with odd indices, and then those with even indices, the standard V-cycle is not an exact solver. On the other hand, the discretization matrix for this boundary value problem with respect to the hierarchical basis is a diagonal matrix. Therefore, the hierarchical basis multigrid method, smoothing only the unknowns associated with the odd indices on every level, is an exact solver.

6. Numerical Results. As mentioned above, the hierarchical basis multigrid method is the linear solver used in the new version of PLTMG [2]. This package was used to solve the problems presented in this section.

The first example is the Helmholtz equation

$$-\Delta u - 100u = 1 \tag{21}$$

on the unit square $\Omega = [0,1]^2$. The boundary conditions

$$u = 0 \tag{22}$$

are imposed on $\partial\Omega$. The eigenvalues of the differential operator associated with this boundary value problem are

$$\lambda_{k\ell} = (k^2 + \ell^2)\pi^2 - 100 \tag{23}$$

for $k, \ell = 1, 2, \ldots$. The corresponding eigenfunctions are

$$u_{k\ell}(x, y) = \sin(k\pi x)\sin(\ell\pi y) \tag{24}$$

This means that four eigenvalues are negative and the corresponding eigenfunctions form a six-dimensional space. To obtain meaningful results, the triangulation associated with the first level in the hierarchical basis multigrid method must be fine enough to represent these eigenfunctions. We began with a uniform 9×9 grid. PLTMG was used to adaptively refine this mesh, obtaining a fine mesh with 517 points distributed among three levels. Both grids, together with the final solution, are shown in Figure 2.

In the table below we summarize the numerical results obtained using the hierarchical basis multigrid method in conjunction with conjugate gradient acceleration.

iteration	1	2	3	4	5	6	7	8	9
digits	-0.013	0.91	2.20	2.85	3.33	3.44	3.50	3.60	4.74

The number of correct digits is defined by

$$digits = -\log_{10}(||x^{(i)} - x||/||x||) \tag{25}$$

where the norm is the continuous $\mathcal{H}^1$ norm of the functions represented by the given coefficient vectors, and where x is the exact solution of the discrete equation. The initial guess $x^{(0)}$ was zero.

The average rate of convergence is approximately 0.3, which is not very different from the rate of convergence observed in [4] for a boundary value problem involving the

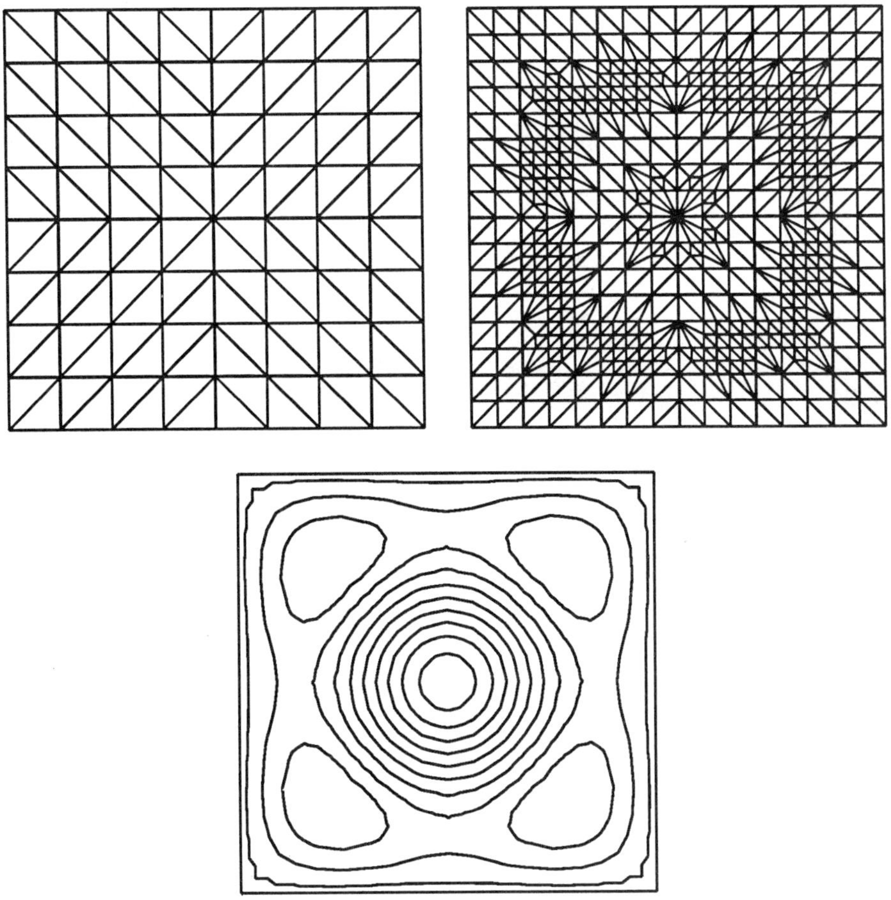

FIG. 2. *The coarse mesh (top left), refined mesh (top right), and the solution (bottom) for (21)*

Laplace operator. With coarser initial triangulations, the convergence rate decreases, at least in the initial steps. For example, using the same fine grid, but using four levels and a coarse 5×5 grid, the convergence rate is 0.5. For five levels, using a 3×3 coarse mesh, the convergence rate is 0.84. This behavior is typical of many multigrid procedures, and is explained by poor approximation properties of the coarsest mesh. The 3×3 mesh, for example, had only one interior point (one unknown), and therefore could not hope to give a satisfactory approximation of the six-dimensional space corresponding to the negative eigenvalues.

Our second example is the convection-diffusion problem

$$-\Delta u + \beta \cdot \nabla u = 1 \tag{26}$$

with $\beta^t = (100, 100)$. The region Ω is the unit square, and the boundary conditions are the same as in the first example. This is a moderately difficult convection dominated problem with sharp boundary layers. For problems like this, PLTMG uses a stabilized version of the finite element method, the streamline diffusion method [6]. We started with the same coarse grid as in the first example, and ended with the highly nonuniform final grid shown in Figure 3.

This grid has 502 vertices distributed among five levels. The discrete equations have been solved using a generalized minimum residual acceleration procedure similar

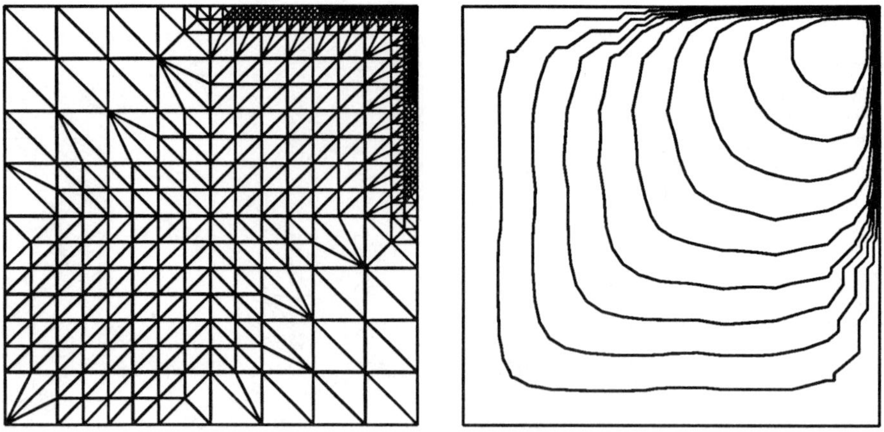

FIG. 3. *The refined mesh (left) and the solution (right) for (26)*

to orthomin [8], starting from an initial guess of zero. The results are summarized below:

iteration	1	2	3	4	5	6	7	8	9	10
digits	0.43	0.79	0.97	1.62	2.00	2.49	2.78	3.27	3.73	4.17

The average rate of convergence was approximately 0.38. These results demonstrate that the hierarchical basis multigrid method is a reasonable solver for this class of problems too. As in the first example, there is sensitivity to the level of approximation of the coarsest mesh. Solving the same problem using a 5×5 coarse mesh and six levels led to a convergence rate of 0.76; the same rate was observed using a 3×3 coarse grid with seven levels.

The solver in PLTMG adaptively determines how many previous directions should be used for orthogonalization in the acceleration procedure. The criterion is based on the observed rate of reduction in the ℓ^2 norm of the residual. In solving the problem using the 9×9 coarse grid, the minimum residual procedure orthogonalized the current direction with respect to only one previous direction. The same was true for the 5×5 case, except for one iteration where two previous directions were used. For the 3×3 case, up to seven previous directions were used; four to six directions were typical. Thus, although the convergence rate was nearly the same for the 5×5 and 3×3 cases, the acceleration procedure worked much harder to achieve that rate in the latter case.

REFERENCES

[1] R. E. BANK, *PLTMG Users' Guide, Edition 4.0*, Tech. Rep., Department of Mathematics, University of California, San Diego, 1985.

[2] R. E. BANK, *PLTMG Users' Guide, Edition 5.0*, Tech. Rep., Department of Mathematics, University of California, San Diego, 1988.

[3] R. E. BANK AND T. F. DUPONT, *Analysis of a Two Level Scheme for Solving Finite Element Equations*, Tech. Rep. CNA-159, Center for Numerical Analysis, University of Texas at Austin, 1980.

[4] R. E. BANK, T. F. DUPONT, AND H. YSERENTANT, *The hierarchical basis multigrid method*, Numer. Math., (1988).

[5] W. HACKBUSCH, *Multigrid Methods and Applications*, Springer-Verlag, Berlin, 1985.

[6] C. JOHNSON, *Finite element methods for convection-diffusion problems*, in Computing Methods in Applied Sciences and Engineering, North Holland, Amsterdam, 1982, pp. 311–324.

[7] J. F. MAITRE AND F. MUSY, *The contraction number of a class of two-level methods; an exact evaluation for some finite element spaces and model problems*, in Multigrid Methods (Lecture Notes in Mathematics 960), Springer-Verlag, Berlin, 1982, pp. 535–544.

[8] Y. SAAD AND M. H. SCHULTZ, *GMRES: a generalized minimun residual algorithm for solving nonsymmetric linear systems*, SIAM J. Sci. Stat. Comput., 7 (1986), pp. 856–869.

[9] H. YSERENTANT, *On the multi-level splitting of finite element spaces for indefinite elliptic boundary value problems*, SIAM J. Numer. Anal., 23 (1986), pp. 581–595.

Multiplicative and Additive Schwarz Methods: Convergence in the Two-Domain Case*

Petter E. Bjørstad†

Abstract. We consider the classical Schwarz alternating algorithm and an additive version more suitable for parallel processing. The two methods are compared and analyzed in the case of two domains. We show that the rate of convergence for both methods, can be directly related to a generalized eigenvalue problem, derived from subdomain contributions to the global stiffness matrix. Analytical expressions are given for a model case.

Key Words. Schwarz' Method, Orthogonal Projections, Elliptic Equations, Domain Decomposition, Finite Elements, Parallel Computation.

1. Introduction. Recently there has been a strong revival of the interest in domain decomposition algorithms for the solution of elliptic problems; cf. e.g. Glowinski et al [12], and Chan et al [7]. This is to a large extent due to their potential in parallel computing environments.

Substructuring methods, with a long history from the structural analysis community [20,1], are methods where the global domain is partitioned into disjoint (non-overlapping) pieces. More recently, the use of iterative methods for this class of problems, has been investigated, see [3] and [6] and the references given in those papers.

In a famous paper more than hundred years ago, Schwarz [21] developed his domain decomposition algorithm, based on computing the solution of two overlapping subproblems in an alternating fashion. This algorithm is therefore quite sequential in its original form, and not necessarily the best candidate for a parallel implementation. More recently P. L. Lions [14,15] has obtained a number of interesting results on Schwarz' method and inspired new work on this type of algorithms. His reformulation

* This work was supported in part by the Norwegian Research Council for Science and the Humanities under grant D.01.08.054 and by The Royal Norwegian Council for Scientific and Industrial Research under grant IT2.28.28484.

† Institutt for Informatikk, University of Bergen, Allegaten 55, N-5007 Bergen, Norway.

of the classical method in terms of orthogonal projections has motivated the development of an alternative additive algorithm, see [11]. This particular formulation was apparently first described by Matsokin and Nepomnyashchikh [17].

Grid refinement algorithms can also be formulated and analyzed using orthogonal projections see the papers by Dryja and Widlund [24] and Mandel and McCormick [16]. The close relationship between Schwarz type methods, iterative substructuring methods [3] and promising iterative grid refinement algorithms; cf. e.g. McCormick et al [18,19,13] and Bramble et al [5], has only recently been realized.

As a beginning contribution to the analysis of these algorithms, we will derive precise relations between the classical (multiplicative) and the newly proposed additive Schwarz' methods in the case of two domains. We will do this by expressing the relevant orthogonal projection operators in terms of elementary contributions to the global stiffness matrix from the subdomains.

A model problem and the necessary notation is described in section 2. We review the classical Schwarz algorithm, including its variational form in section 3. The related additive form, is introduced in section 4. Since this algorithm is relatively new, we also describe some implementation details. In section 5, we explicitly compute the projection operators introduced in section 2, in terms of stiffness contributions from the individual subdomains. It is important to note that this derivation is quite general, and valid for any geometrical shape of the domains. This leads to a decomposition of the spectrum giving a precise relationship between the multiplicative and the additive algorithm. We also note that the classical Schwarz' algorithm is identical to a method recently proposed by Chan and Resasco [8,9]. A closed form expression for the eigenvalues as a function of the geometry of the subdomains can only be given for special cases. We give the eigenvalues as a function of the aspect ratio, for the case where Ω is rectangular in section 6. Section 7 contains a few numerical experiments, confirming the theoretical results.

2. Notations and Preliminaries. To simplify our presentation, we assume that the elliptic operator is the Laplacian and that we have a zero Dirichlet condition. Thus,

$$\begin{aligned} -\Delta u &= f \qquad in\ \Omega, \\ u &= 0 \qquad on\ \partial\Omega. \end{aligned} \tag{1}$$

The region Ω is bounded, two- or three-dimensional, with a Lipschitz continuous boundary. Our algorithms and results can be extended immediately to linear, self-adjoint elliptic problems. We use a variational formulation of the problem, which, as shown by Sobolev [22] makes the maximum principle superfluous.

In variational form equation 1 is written as

$$a_\Omega(u,v) = \int_\Omega \nabla u \cdot \nabla v \, dx = \int_\Omega fv \, dx = f(v) \,, \qquad \forall v \in H_0^1(\Omega) \,,$$

where the solution $u \in H_0^1(\Omega)$, the closure in the Sobolev space $H^1(\Omega)$ of the space of smooth functions which vanish in a neighborhood of $\partial\Omega$. As always, the space $H^1(\Omega)$ is the subspace of $L_2(\Omega)$ for which $| u |^2_{H^1(\Omega)} = a(u,u)$ is finite. In our analysis we work

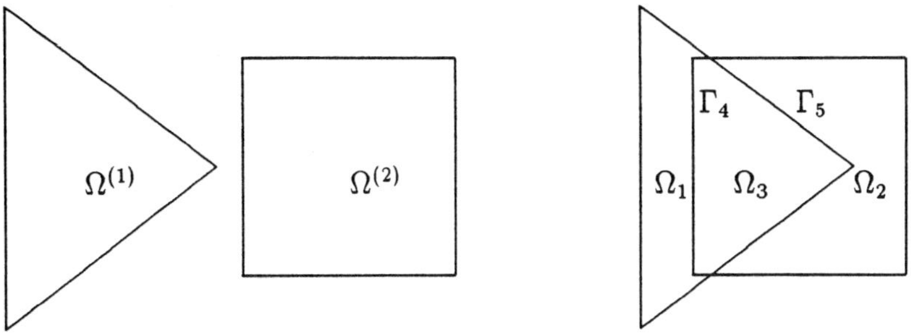

Fig. 1. *The partition of the domain.*

exclusively with the inner product defined by $a(\cdot,\cdot)$. Thus in this paper orthogonality and symmetry always refer to this inner product.

In this paper we consider domain decomposition algorithms with two overlapping subregions $\Omega^{(1)}$ and $\Omega^{(2)}$. We also use the notations $\Omega_1 = \Omega \setminus \overline{\Omega}^{(2)}$, where $\overline{\Omega}^{(2)}$ is the closure of $\Omega^{(2)}$, $\Omega_2 = \Omega \setminus \overline{\Omega}^{(1)}$ and $\Omega_3 = \Omega^{(1)} \cap \Omega^{(2)}$. The region Ω is thus also divided into three nonoverlapping subregions Ω_1, Ω_2, and Ω_3 which are separated from each other by the curves (or surfaces) $\Gamma_4 = \overline{\Omega}_1 \cap \overline{\Omega}_3$ and $\Gamma_5 = \overline{\Omega}_2 \cap \overline{\Omega}_3$. In figure 1 on the left, we display separately, the two subregions from which Ω is built and on the right the partitioning of Ω into the five subsets just defined. We assume that Γ_4 and Γ_5 follow element boundaries, that they are Lipschitz, and that they intersect only in at most a few points (or along a few curves).

The problem is discretized by finite elements in the customary fashion; cf. Ciarlet [10]. The region Ω is triangulated and a conforming finite element space $V^h \subset H_0^1(\Omega)$ is introduced. We assume that all the elements are shape regular, i.e. there is a bound on h_K/ρ_K which is independent of the number of degrees of freedom and of K. Here h_K is the diameter of the element K and ρ_K the diameter of the largest sphere that can be inscribed in K. The approximate solution $u_h \in V^h$ is defined by

$$a_\Omega(u_h, v_h) = f(v_h)\,, \qquad \forall\, v_h \in V^h. \tag{2}$$

We note that the exact solution to the discrete problem is given as $u_h = P_{V^h}u$, the orthogonal projection of the continuous solution $u \in H_0^1(\Omega)$, from $H_0^1(\Omega)$ into V^h.

3. The multiplicative Schwarz' algorithm. We return to the continuous case and write down the Schwarz algorithm in its traditional form. There are two fractional steps. Let u^n be the n-th iterate. Then the updated solution u^{n+1} is determined by,

$$\begin{aligned} -\Delta u^{n+1/2} &= f && \text{in } \Omega^{(1)},\\ u^{n+1/2} &= 0 && \text{on } \partial\Omega^{(1)} \cap \partial\Omega,\\ u^{n+1/2} &= u^n && \text{on } \Gamma_5, \end{aligned}$$

and,

$$\begin{aligned} -\Delta u^{n+1} &= f && \text{in } \Omega^{(2)},\\ u^{n+1} &= 0 && \text{on } \partial\Omega^{(2)} \cap \partial\Omega,\\ u^{n+1} &= u^{n+1/2} && \text{on } \Gamma_4. \end{aligned}$$

Following P. L. Lions [14,15], we write these equations in variational form. This formulation is also valid in the discrete case.

$$\begin{aligned} a_\Omega(u_h^{n+1/2} - u_h^n, v_h) &= f(v_h) - a_\Omega(u_h^n, v_h) \\ &= a_\Omega(u_h - u_h^n, v_h), \qquad \forall\, v_h \in V^h \cap H_0^1(\Omega^{(1)}), \\ a_\Omega(u_h^{n+1} - u_h^{n+1/2}, v_h) &= f(v_h) - a_\Omega(u_h^{n+1/2}, v_h) \\ &= a_\Omega(u_h - u_h^{n+1/2}, v_h), \qquad \forall\, v_h \in V^h \cap H_0^1(\Omega^{(2)}). \end{aligned}$$

Here $u_h^{n+1/2} - u_h^n \in H_0^1(\Omega^{(1)})$ and $u_h^{n+1} - u_h^{n+1/2} \in H_0^1(\Omega^{(2)})$, since the boundary values do not change from one fractional step to the next. The equations can now be restated using orthogonal projections,

$$\begin{aligned} u_h^{n+1/2} - u_h^n &= P_1(u_h - u_h^n), \\ u_h^{n+1} - u_h^{n+1/2} &= P_2(u_h - u_h^{n+1/2}), \end{aligned}$$

where P_i, $i = 1, 2$, is the orthogonal projection of V^h into $V_i = V^h \cap H_0^1(\Omega^{(i)})$. The error $e_h^n = u_h^n - u_h$ satisfies the relationship

$$e_h^{n+1} = (I - P_2)(I - P_1)e_h^n = (I - (P_1 + P_2 - P_2P_1))e_h^n.$$

In a certain sense Schwarz' method is therefore a straightforward iterative method of solving the equation

$$(P_1 + P_2 - P_2P_1)u_h = g_m, \tag{3}$$

for a certain right hand side g_m.

4. The Additive Schwarz' Algorithm. The product term P_2P_1 in 3 prevents a straight forward parallel implementation, although many subdomain computations can be carried out simultaneously in the case where the classical algorithm is extended to many subdomains. Assuming that we can solve linear systems involving polynomials in P_1 and P_2, it is natural to consider alternatives to equation 3.

The simplest, additive form of Schwarz more suitable for parallel processing, appears by simply removing the product term in 3. We are then faced with solving the system

$$(P_1 + P_2)u_h = g_a. \tag{4}$$

In order to retain the same solution u_h, the right hand side g_m in 3 has changed to g_a. In order to use 4 for the computation of u_h, we must first find this new right hand side g_a. Fortunately it is computable from solving subproblems on the individual subdomains. We compute g_a by writing

$$g_a = \sum_i g_i$$

where $g_i \in V_i$ and each g_i solves the subproblem

$$a(g_i, v_h) = a(u_h, v_h) = f(v_h) \quad \forall v_h \in V_i.$$

Once the appropriate right hand side g_a has been computed, we can apply a conjugate gradient iteration (working with the K-inner product) to solve 4 with no further preconditioning. This strategy requires a procedure for computing the products $v_h = P_i w_h$ for a given function w_h. Let us consider this problem in some detail for the subdomain $\Omega^{(1)}$ corresponding to $i = 1$. The orthogonal projection $P_1 w_h$ of w_h on V_1, is defined by

$$a_\Omega(P_1 w_h, \hat{v}_h) = a_\Omega(w_h, \hat{v}_h)\ , \quad \forall\ \hat{v}_h \in V_1,$$

and $P_1 w_h \in H_0^1(\Omega^{(1)})$. If we write $w_h = P_1 w_h + \hat{w}_h$, we conclude that $\hat{w}_h$ must be harmonic in $\Omega^{(1)}$ and have the boundary value w_h on Γ_5. We must take $\hat{w}_h = 0$ on the rest of the boundary $\partial\Omega \cap \partial\Omega^{(1)}$. Clearly, $\hat{w}_h = w_h$ in $\Omega \setminus \Omega_1$, making $v_h = 0$ outside Ω_1. The conjugate gradient iteration for solving 4 therefore requires the computation of a harmonic function on each subdomain in every iteration. All the subdomains can be processed in parallel. A representation of the projection operator in terms of the discrete matrix operator of the appropriate subdomain can also be given. Let K be the global stiffness matrix for the problem on Ω and let $K^{(i)}$ be the corresponding matrix for a problem on $\Omega^{(i)}$. We define the correspondence between the discrete finite element functions v_h and w_h and the vectors of nodal values y and x in the usual fashion. If the nodal unknowns are numbered appropriately, we can write:

$$y = P_i x = \begin{pmatrix} K^{(i)^{-1}} & 0 \\ 0 & 0 \end{pmatrix} Kx.$$

5. Representation of the Projection Operators in terms of Block Stiffness Matrices. The computer program implementation of the finite element method most often uses the so called subassembly method when building the matrix corresponding to the discrete, global problem. This principle is based on the observation that the bilinear form in 2 can be written as a sum of bilinear forms over disjoint subregions as expressed in equation 6 below. In this way contributions from the elementary elements can be added to form substructures of the full model, the substructures can again be added to form the full matrix.

The same concept is very useful in the study of certain domain decomposition methods. Many properties of these algorithms can be understood and interpreted in terms of the stiffness submatrices corresponding to the subdomains into which the domain has been decomposed [3].

With subvectors and subscripts corresponding to the degrees of freedom associated with the open sets Ω_1, Ω_2 and Ω_3 and the curves (surfaces) Γ_4 and Γ_5, the entire discrete problem can be written as

$$(5) \qquad Kx = \begin{pmatrix} K_{11} & 0 & 0 & K_{14} & 0 \\ 0 & K_{22} & 0 & 0 & K_{25} \\ 0 & 0 & K_{33} & K_{34} & K_{35} \\ K_{14}^T & 0 & K_{34}^T & K_{44} & K_{45} \\ 0 & K_{25}^T & K_{35}^T & K_{45}^T & K_{55} \end{pmatrix} \begin{pmatrix} x_1 \\ x_2 \\ x_3 \\ x_4 \\ x_5 \end{pmatrix} = \begin{pmatrix} b_1 \\ b_2 \\ b_3 \\ b_4 \\ b_5 \end{pmatrix}.$$

As always the elements of K are given by $a(\phi_i, \phi_j)$ where ϕ_i and ϕ_j are finite element basis functions. The zero blocks are a consequence of the fact that, when using standard finite element basis functions, there is no direct coupling between Ω_1 and Ω_2 etc..

Since the bilinear form is defined in terms of an integral,

$$(6) \qquad a_\Omega(\phi_i, \phi_j) = a_{\tilde{\Omega}}(\phi_i, \phi_j) + a_{\Omega\setminus\tilde{\Omega}}(\phi_i, \phi_j),$$

for any subset $\tilde{\Omega} \subset \Omega$. We note that for a pair of basis functions associated with $\partial\tilde{\Omega}$, we get contributions from both terms. Thus, for example the submatrix K_{44} in 5 can be written $K_{44} = K_{44}^{(1)} + K_{44}^{(3)}$ where $K_{44}^{(i)}$ is the contribution from triangles in Ω_i , $i = 1, 3$.

Using the procedure to compute $v_h = P_i w_h$, described in the previous section, we can easily derive explicit expressions for the operator P_i itself. We first introduce some notations for Schur complements that frequently occur in the calculations to follow. Schur complements arise quite naturally in block Gaussian elimination and play an important role in the analysis of iterative substructuring methods as well [3]. We define the Schur complement $S_j^{(i)}$ as the Schur complement corresponding to the domain Ω_i with respect to the interior subdomain boundary Γ_j, as follows:

$$S_j^{(i)} = K_{jj}^{(i)} - K_{ij}^T K_{ii}^{-1} K_{i,j}, \quad j = 4,\ i = 1, 3 \ \ and \ \ j = 5,\ i = 2, 3.$$

In a similar way, we define $S_4^{(2)}$ to be the Schur complement corresponding to the entire region $\Omega^{(2)} = \Omega_2 \cup \Omega_3 \cup \Gamma_5$ with respect to Γ_4. It is of the form

$$S_4^{(2)} = K_{44}^{(3)} - \begin{pmatrix} 0 & K_{34}^T & K_{45} \end{pmatrix} \begin{pmatrix} K_{22} & 0 & K_{25} \\ 0 & K_{33} & K_{35} \\ K_{25}^T & K_{35}^T & K_{55} \end{pmatrix}^{-1} \begin{pmatrix} 0 \\ K_{34} \\ K_{45}^T \end{pmatrix}.$$

We also define

$$S_{45} = K_{45} - K_{34}^T K_{33}^{-1} K_{35}$$

and write $S_4 = S_4^{(1)} + S_4^{(3)}$ and $S_5 = S_5^{(2)} + S_5^{(3)}$. We can now write down an explicit, block matrix representation for the projection operators.

$$P_1 x = \begin{pmatrix} I & 0 & 0 & 0 & C_4 \\ 0 & 0 & 0 & 0 & 0 \\ 0 & 0 & I & 0 & B_4 \\ 0 & 0 & 0 & I & A_4 \\ 0 & 0 & 0 & 0 & 0 \end{pmatrix} \begin{pmatrix} x_1 \\ x_2 \\ x_3 \\ x_4 \\ x_5 \end{pmatrix}$$

and

$$P_2x = \begin{pmatrix} 0 & 0 & 0 & 0 & 0 \\ 0 & I & 0 & C_5 & 0 \\ 0 & 0 & I & B_5 & 0 \\ 0 & 0 & 0 & 0 & 0 \\ 0 & 0 & 0 & A_5 & I \end{pmatrix} \begin{pmatrix} x_1 \\ x_2 \\ x_3 \\ x_4 \\ x_5 \end{pmatrix}.$$

We have introduced the quantities A_i , B_i and C_i , $i = 4, 5$ in order to simplify the notation. These quantities are defined as:

$$\begin{aligned} A_4 &= S_4^{-1} S_{45} \\ A_5 &= S_5^{-1} S_{45}^T \\ B_4 &= K_{33}{}^{-1}(K_{35} - K_{34}A_4) \\ B_5 &= K_{33}{}^{-1}(K_{34} - K_{35}A_5) \\ C_4 &= -K_{11}{}^{-1}K_{14}A_4 \\ C_5 &= -K_{22}{}^{-1}K_{25}A_5. \end{aligned}$$

We can now find the matrix representation of the operators defined in 3 and 4,

(7)
$$P_1 + P_2 = \begin{pmatrix} I & 0 & 0 & 0 & C_4 \\ 0 & I & 0 & C_5 & 0 \\ 0 & 0 & 2I & B_5 & B_4 \\ 0 & 0 & 0 & I & A_4 \\ 0 & 0 & 0 & A_5 & I \end{pmatrix}$$

and

(8)
$$P_1 + P_2 - P_2P_1 = \begin{pmatrix} I & 0 & 0 & 0 & C_4 \\ 0 & I & 0 & 0 & -A_4C_5 \\ 0 & 0 & I & 0 & -A_4B_5 \\ 0 & 0 & 0 & I & A_4 \\ 0 & 0 & 0 & 0 & I - A_5A_4 \end{pmatrix}.$$

We observe that 7 is 2 by 2 block upper triangular with an invariant subspace corresponding to the unknowns on the two interior interfaces Γ_i $i = 4, 5$, while 8 is block upper triangular. The spectra of the operators are readily available and we observe a very simple relationship between the two. Let the number of unknowns on Ω_i or Γ_i be N_i $i = 1, 2, 3, 4, 5$.

In the additive case, we have $N_1 + N_2$ eigenvalues equal to 1 and N_3 eigenvalues equal to 2. The remaining eigenvalues have the form $1 \pm \mu_i$ where μ^2 is an eigenvalue of

$$A_5A_4y_5 = \mu^2 y_5$$

or in terms of the Schur complements

$$S_{45}^T S_4^{-1} S_{45} y_5 = \mu^2 S_5 y_5. \tag{9}$$

One should note that if $N_4 \neq N_5$, then there will be $|N_5 - N_4|$ additional eigenvalues equal to one. This observation may be important in cases where there is a significant difference between the number of points on Γ_4 and Γ_5. Only the smallest number of unknowns (shortest boundary) on the two interfaces will contribute nontrivial eigenvalues to the operator.

The algorithms presented in this paper and the finite element framework provided the motivation for this work. We have obtained a complete characterization of the spectra of the two operators from 3 and 4. It is possible to obtain a similar characterization for projection operators in general, we refer to a joint paper with J. Mandel [2] on this topic.

The matrix 7 is only symmetric in the K-inner product, for completeness we therefore include the explicit representations of 7 and 8 multiplied by the global stiffness matrix K.

$$K(P_1 + P_2) = \begin{pmatrix} K_{11} & 0 & 0 & K_{14} & 0 \\ 0 & K_{22} & 0 & 0 & K_{25} \\ 0 & 0 & 2K_{33} & 2K_{34} & 2K_{35} \\ K_{14}^T & 0 & 2K_{34}^T & K_{44} + D_4 & 2K_{45} \\ 0 & K_{25}^T & 2K_{35}^T & 2K_{45}^T & K_{55} + D_5 \end{pmatrix} \tag{10}$$

where

$$D_4 = K_{34}^T K_{33}^{-1} K_{34} + S_{45} S_5^{-1} S_{45}^T$$

and

$$D_5 = K_{35}^T K_{33}^{-1} K_{35} + S_{45}^T S_4^{-1} S_{45}.$$

For the multiplicative case,

$$K(P_1 + P_2 - P_2 P_1) = \begin{pmatrix} K_{11} & 0 & 0 & K_{14} & 0 \\ 0 & K_{22} & 0 & 0 & K_{25} \\ 0 & 0 & K_{33} & K_{34} & K_{35} \\ K_{14}^T & 0 & K_{34}^T & K_{44} & K_{45} + E_{45} \\ 0 & K_{25}^T & K_{35}^T & K_{45}^T & K_{55} \end{pmatrix} \tag{11}$$

with

$$E_{45} = S_{45}(I - A_5 A_4).$$

We observe that the matrix 10 is symmetric and that the entries corresponding to Ω_3 have been multiplied by two as should be expected. Except for this, the matrix is different from K only by the addition of a symmetric, positive definite term to each of the block diagonal positions representing the two interior interfaces. The matrix 11 differs from K only in the (4,5) block position. The matrix is therefore unsymmetric. Notice that the correction term is closely related to 9.

It remains to show that μ in 9 is bounded away from one, independent of the discretization. Consider the generalized eigenvalue problem

$$(S_4^{(1)} + S_4^{(2)})x_4 = \lambda(S_4^{(1)} + S_4^{(3)})x_4. \tag{12}$$

This problem arises if we consider the (non-overlapping) domain decomposition of Ω into $\Omega^{(2)}$ and Ω_1 with the common interface Γ_4. If 5 is reduced by block Gaussian elimination, to a system on Γ_4 only, and this system is preconditioned by the Schur complement resulting from a similar reduction of the equations on $\Omega^{(1)}$ to Γ_4, then $\lambda_{max}/\lambda_{min}$ is the relevant condition number of the iteration operator. We can establish uniform lower and upper bounds on the corresponding generalized Rayleigh quotient by using:

$$x_4{}^T S_4^{(i)} x_4 \leq C x_4{}^T S_4^{(j)} x_4\,, \qquad \forall x_4,\ \ i,j = 1,2,3.$$

Proofs of this so called extension theorem are given in [3] and [23]. The constants in these inequalities depend only on Ω_1, Ω_2 and Ω_3.

It should be noted that this iterative method has recently been proposed by Chan and Resasco [8,9]. Their results show that this method is quite robust, when applied to certain model problems, having subdomains with large aspect ratios, in the sense that the rate of convergence seems to be independent of the aspect ratios. As opposed to this, it is shown in [3] that the so called Neumann-Dirichlet algorithm do have a weak aspect ratio dependent rate of convergence.

We remark that the Neumann-Dirichlet algorithm [3], results when the preconditioning in 12 contains the term $S_4^{(1)}$ only. A simple calculation will show that

$$S_4^{(2)} = S_4^{(3)} - S_{45}S_5^{-1}S_{45}^T.$$

Substituting this into 12 gives

$$\lambda = 1 - \mu^2$$

where μ^2 solves the eigenvalue problem 9.

We have therefore shown that $1-\mu^2$ is uniformly bounded from below and above independent of the discretization. This observation shows that the classical Schwarz' algorithm can be reduced to a symmetric form on the subspace corresponding to the interface Γ_4, and accelerated by conjugate gradients. At the same time this argument shows that the method proposed by Chan and Resasco is nothing but the classical Schwarz' method accelerated by conjugate gradients. We refer to a joint paper with Widlund [4] for a more detailed discussion of this point.

The condition number of the iteration operator providing an upper bound on the number of conjugate gradient iterations in order to achieve a given tolerance, is $(1+\mu_1)/(1-\mu_1)$ for the additive method. From 8, it follows that the appropriate eigenvalues are $1-\mu_i^2$ in the multiplicative case. The condition number of the iteration operator is therefore $1/(1-\mu_1^2)$ in this case. It follows that the condition number of the additive method is bounded by 4 times the condition number of the multiplicative method. We conclude that the number of conjugate gradient iterations needed in the additive method in order to achieve a given tolerance, cannot be more than twice the number of iterations when using the multiplicative algorithm.

This result is somewhat negative, showing that in the worst case we would solve a problem using the additive algorithm and two processors in the same time as one processor would run the multiplicative algorithm. Domain decomposition clearly depends on having (many) more than two subdomains in order to be efficient on parallel processor systems. Unfortunately, the analysis presented in this paper does not extend easily to the case of many subdomains.

6. Exact Results for a Model Problem. The results of the previous section are quite general, they do not depend on the shape of the various domains, the particular discretization, or the specific form of the elliptic equation. The actual numerical values of the spectra and in particular, the condition number of the iteration operators can, if necessary, be computed from equation 9.

More details can be given in model cases only. We will in this section derive the exact values for the rectangular model problem. Study of model problems can lead to a better understanding of the influence of the relative sizes of the subdomains on the rate of convergence. We consider the rectangle shown in figure 2. The domain has q interior nodes in the vertical direction, while there are r , $n-l-1$, and l interior nodes in the horizontal direction of Ω_1, Ω_2 and Ω_3 respectively. We use linear triangular elements, this discretization is equivalent with the well known 5-point finite difference stencil. The notation in this section will be consistent with the paper [3]. Let

$$\lambda_{iq} = 4sin(\frac{i\pi}{2(q+1)})^2 , \quad i = 1, 2, .. \ q ,$$

and

$$a_{iq} = 1 + \lambda_{iq}/2 - (\lambda_{iq} + (\lambda_{iq}/2)^2)^{1/2}.$$

The quantity a_{iq} appears frequently in the analysis of model problems using Fourier analysis. It is bounded by

$$3 - 2\sqrt{2} < a_{iq} < 1.$$

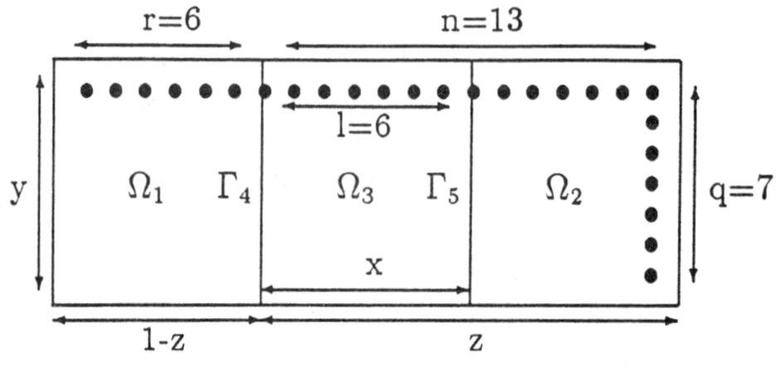

Fig. 2. *The model problem geometry.*

Using a derivation very similar to the one in section 4 of [3], we proceed to diagonalize the eigenvalue problem 9. After a rather tedious calculation one obtains the following exact formula for the eigenvalues μ_i:

$$\mu_i^2 = \frac{a_{iq}^{2r}(a_{iq}^{2n+4} - a_{iq}^{2l+4}) - a_{iq}^{2n+2} + a_{iq}^{2l+2}}{a_{iq}^{2r}(a_{iq}^{2n+2l+6} - a_{iq}^{2l+4}) - a_{iq}^{2n+2} + 1}. \tag{13}$$

We observe that $l = n$ ie. full overlap gives $\mu_i = 0$ and one can derive the upper bound

$$\mu_i \leq \frac{2a_{iq}^2}{(l+1)(1+a_{iq})(1-a_{iq})}.$$

The bound therefore decreases with the size of the overlap as $1/(l+1)$, where l is the number of interior nodes in the x-direction, inside the overlap subdomain Ω_3.

We can also compute the eigenvalues μ_i^2 in the continuous limit as the discretization parameter $h = 0$. We scale the rectangle to unit length and introduce the relative sizes x, y, z as indicated in figure 2. The eigenvalues are

$$\mu_i^2 = \frac{(e^{\alpha_i z} - e^{\alpha_i})(e^{\alpha_i z} + e^{\alpha_i})(e^{\alpha_i z} - e^{\alpha_i x})(e^{\alpha_i z} + e^{\alpha_i x})}{(e^{\alpha_i z} - 1)(e^{\alpha_i z} + 1)(e^{\alpha_i z} - e^{\alpha_i (x+1)})(e^{\alpha_i z} + e^{\alpha_i (x+1)})} \tag{14}$$

where we write $\alpha_i = i\pi/y$. It turns out that 14 is a very good approximation to 13 for almost all interesting cases. The example in figure 2 has $\mu_1 = 0.06$, while an overlap corresponding to only one grid line in the figure, increases μ_1 to 0.67 .

7. Numerical Examples. We give a few numerical results to confirm the theoretical results and to give the reader an idea of the performance of the algorithms. For simplicity, Ω is the union of two rectangles.

We use a uniform mesh spacing $h = 1/256$ with the size of Ω_1 being .75 by .5, and $\Omega^{(2)}$ extending 3/16 above and 5/16 below Ω_1, see figure 2. This gives a total of 40449 unknowns, with 127 along Γ_4. We consider the same problem as in [3] with the exact solution $u(x, y) = x^2 + y^2 - xe^x \cos y$. The number of iterations required to

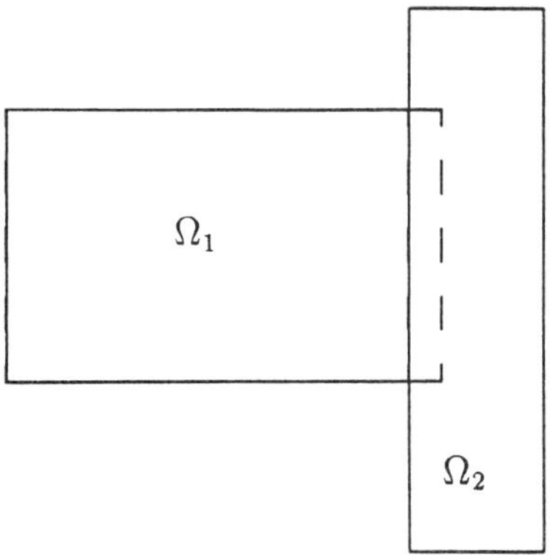

Fig. 3. *The computational domain.*

reduce the initial error to the truncation error level (a factor of $3 \cdot 10^{-5}$) are reported both for the classical Schwarz' method and for the additive version. The conjugate gradient method is used in both cases. We create the overlap Ω_3 by extending the rectangle Ω_1 a distance x into Ω_2. Table 1 gives results for six different sizes of Ω_3. The first column is the overlap length x. Column 2 shows the largest eigenvalue μ_1, the next two columns give the condition number and the number of iterations N_m, for the multiplicative method. Column 5 and 6 have the same information for the additive method. The last column lists μ_1^R in the continuous limit, for the rectangular model case just studied, with $y = .5$ and $\Omega^{(2)}$ being an extension of Ω_1 of length .25.

x	μ_1	$\frac{1}{1-\mu_1^2}$	N_m	$\frac{1+\mu_1}{1-\mu_1}$	N_a	μ_1^R
1/128	0.95	10.70	17	40.75	33	0.95
1/64	0.91	5.61	13	20.41	24	0.90
1/32	0.82	3.08	9	10.24	17	0.81
1/16	0.68	1.85	7	5.19	11	0.66
1/8	0.47	1.29	6	2.80	9	0.41
1/4	0.37	1.16	5	2.19	7	0.00

Table 1. *Multiplicative and additive Schwarz.*

We note that the estimated eigenvalues from our conjugate gradient iteration, behaves exactly as predicted by the theory. The numerical results demonstrate that the condition number of the additive method approaches four times the size of the multiplicative condition number, as the overlap tends to zero. The actual number of iterations required by the additive method, is also two times what is needed for the multiplicative method when the overlap is very small. The iteration counts of the two methods are closer in the case of a little larger, perhaps more realistic overlap. Finally, observe that the exact eigenvalues for the rectangular model case in the continuous limit, are quite close to the actual eigenvalues for this T-shaped region, except for the special case when the model problem has complete overlap.

8. Acknowledgments. The author thanks professor Olof Widlund for many valuable discussions related to the present work.

REFERENCES

[1] K. Bell, B. Hatlestad, O. E. Hansteen, and P. O. Araldsen, *NORSAM, a programming system for the finite element method. users manual, part 1, general description.*, NTH, Trondheim, 1973.

[2] P. E. Bjørstad and J. Mandel, *On the Spectra of Sums of Orthogonal Projections with Applications to Parallel Computing*, Tech. Rep., Department of Computer Science, University of Bergen, Allégaten 55, N-5007 Bergen, Norway., 1988.

[3] P. E. Bjørstad and O. B. Widlund, *Iterative methods for the solution of elliptic problems on regions partitioned into substructures*, SIAM J. Numer. Anal., 23 (1986), pp. 1093–1120.

[4] ——, *To overlap or not to overlap; A note on a domain decomposition method for elliptic problems*, Tech. Rep., Department of Computer Science, University of Bergen, Allégaten 55, N-5007 Bergen, Norway., 1988.

[5] J. H. Bramble, R. E. Ewing, J. E. Pasciak, and A. H. Schatz, *A preconditioning technique for the efficient solution of problems with local grid refinement*, Computer Methods in Applied Mechanics and Engineering, (1988). To appear.

[6] J. H. BRAMBLE, J. E. PASCIAK, AND A. H. SCHATZ, *The construction of preconditioners for elliptic problems by substructuring, I*, Math. Comp., 47 (1986), pp. 103–134.

[7] T. CHAN, R. GLOWINSKI, G. A. MEURANT, J. PÉRIAUX, AND O. WIDLUND, eds., *Domain Decomposition Methods for Partial Differential Equations II*, SIAM, Philadelphia, 1989. Proceedings of the second international symposium on domain decomposition methods for partial differential equations, UCLA, January 1988.

[8] T. F. CHAN AND D. C. RESASCO, *A domain-decomposed fast Poisson solver on a rectangle*, SIAM J. Sci. Stat. Comput., 8 (1987), pp. s14–s26.

[9] ——, *A survey of preconditioners for domain decomposition*, Tech. Rep. /DCS/RR-414, Yale Univeristy, 1985.

[10] P. G. CIARLET, *The finite element method for elliptic problems.*, North-Holland, Amsterdam, 1978.

[11] M. DRYJA AND O. B. WIDLUND, *An Additive Variant of the Schwarz Alternating Method for the Case of Many Subregions*, Tech. Rep. 339, also Ultracomputer Note 131, Department of Computer Science, Courant Institute, 1987.

[12] R. GLOWINSKI, G. H. GOLUB, G. A. MEURANT, AND J. PÉRIAUX, eds., *Domain Decomposition Methods for Partial Differential Equations*, SIAM, Philadelphia, 1988. Proceedings of the first international symposium on domain decomposition methods for partial differential equations, Paris, January 1987.

[13] L. HART AND S. MCCORMICK, *Asynchronous Multilevel Adaptive Methods for Solving Partial Differential Equations on Multiprocessors: Computational Analysis*, Tech. Rep., Comp. Math. Group, Univ. of Colorado at Denver, 1987. Submitted to Parallel Computing.

[14] P. L. LIONS, *On the Schwarz alternating method. I.*, in Domain Decomposition Methods for Partial Differential Equations, R. Glowinski, G. H. Golub, G. A. Meurant, and J. Périaux, eds., SIAM, Philadelphia, 1988.

[15] ——, *On the Schwarz alternating method. II.*, in Domain Decomposition Methods for Partial Differential Equations II, T. Chan, R. Glowinski, G. A. Meurant, J. Périaux, and O. Widlund, eds., SIAM, Philadelphia, 1989. (Proceedings of the Second International Symposium on Domain Decomposition Methods for Partial Differential Equations, UCLA, January 1988.).

[16] J. MANDEL AND S. MCCORMICK, *Iterative solution of elliptic equations with refinement: the two-level case*, in Domain Decomposition Methods for Partial Differential Equations II, T. Chan, R. Glowinski, G. A. Meurant, J. Périaux, and O. Widlund, eds., 1989. (Proceedings of the Second International Symposium on Domain Decomposition Methods for Partial Differential Equations, UCLA, January 1988.).

[17] A. M. MATSOKIN AND S. V. NEPOMNYASHCHIKH, *A Schwarz alternating method in subspaces*, Tech. Rep. 114, Computing Center of the Siberian Branch of the USSR Academy of Sciences, Novosibirsk, 1985. In Russian.

[18] S. MCCORMICK, *Fast adaptive composite grid (FAC) methods*, in Defect Correction Methods: Theory and Applications, K. Böhmer and H. J. Stetter, eds., Computing Supplementum 5, Springer-Verlag, Wien, 1984, pp. 115–121.

[19] S. MCCORMICK AND J. THOMAS, *The fast adaptive composite grid (FAC) method for elliptic equations*, Math. Comp., 46 (1986), pp. 439–456.

[20] J. S. PRZEMIENIECKI, *Matrix structural analysis of substructures.*, Am. Inst. Aero. Astro. J., 1 (1963), pp. 138–147.

[21] H. A. SCHWARZ, *Gesammelete Mathematische Abhandlungen*, vol. 2, Springer, Berlin, 1890, pp. 133–143. First published in Vierteljahrsschrift der Naturforschenden Gesellschaft in Zürich, volume 15, 1870, pp.272–286.

[22] S. SOBOLEV, *The Schwarz algorithm in the theory of elasticity*, Dokl. Acad. Nauk. USSR, IV (1936), pp. 236–238. In Russian.

[23] O. B. WIDLUND, *An extension theorem for finite element spaces with three applications*, in Numerical Techniques in Continuum Mechanics, W. Hackbusch and K. Witsch, eds., Notes on Numerical Fluid Mechanics, v. 16, Friedr. Vieweg und Sohn, Braunschweig/Wiesbaden, 1987, pp. 110–122. Proceedings of the Second GAMM-Seminar, Kiel, January , 1986.

[24] ——, *Optimal iterative refinement methods*, in Domain Decomposition Methods for Partial Differential Equations II, T. Chan, R. Glowinski, G. A. Meurant, J. Périaux, and O. Widlund, eds., 1989. (Proceedings of the Second International Symposium on Domain Decomposition Methods for Partial Differential Equations, UCLA, January 1988.).

CHAPTER 13

Boundary Probe Domain Decomposition Preconditioners for Fourth Order Problems*

Tony F. Chan†

Abstract. The boundary probing technique is a class of methods for the construction of efficient interface preconditioners in domain decomposition algorithms. The main idea is to capture the strong local coupling of the interface operator through subdomain solves with a few appropriately chosen "probing" boundary conditions. For second order elliptic problems, this technique has proven to be very successful and frequently performs better than other preconditioners. In this paper, we show how this technique can be extended to derive efficient domain decomposition preconditioners for fourth order problems. The main modifications are that the interface between subdomains now consists of two grid lines and a new set of probing boundary conditions is used. Numerical results for the biharmonic equation are presented.

Key words. Parallel algorithms, domain decomposition, fourth order partial differential equations, biharmonic equation, preconditioned conjugate gradient.

1. Introduction. In this paper, domain decomposition refers to a class of algorithms for solving boundary value problems for elliptic partial differential equations. The main idea is to decompose the original domain into smaller subdomains, solve the original problem on the subdomains, and somehow "patch" the subdomain solutions to form the solution to the original problem. In general, the above process has to be repeated through an iterative process until some convergence criteria is satisfied. There are two main approaches, characterized by the way the subdomains are constructed, namely overlapping and nonoverlapping. In this paper, we shall consider only the nonoverlapping approach.

There are several reasons why such a procedure would be useful. First, irregular domains can be decomposed into regular subdomains on which more efficient solvers can be used. Second, it is a natural way to design parallel algorithms for elliptic boundary value problems. Third, it allows large problems to be solved on computers with relatively small core memory. Finally, different mathematical models and different grid resolutions can be adaptively used in the different subdomains.

In the nonoverlapping approach of domain decomposition, the original problem

* This version was last modified August 30, 1988. The research of the author is supported in part by The Department of Energy under contract DE-FG03-87ER25037, by the National Science Foundation under contract NSF-DMS87-14612 and by the Army Research Office under contract No. DAAL03-88-K-0085.

† Department of Mathematics, UCLA, Los Angeles, CA 90024. E-mail: chan@math.ucla.edu or na.tchan@na-net.stanford.edu.

is reduced to an equivalent one defined on the interfaces separating the subdomains. This problem (sometimes referred to as the Schur Complement system or capacitance system) is then solved iteratively, each iteration requiring a solve on each subdomain. To improve the rate of convergence, preconditioners (which can be thought of as easily invertible approximations to the interface operator) are often used. Due to its critical impact on the overall efficiency of the domain decomposition algorithm, the design and analysis of such preconditioners has been a main topic of research in this area. For a survey of this activity, see [4,9,6,2].

In this paper, we shall study in particular a class of preconditioners that we shall call *boundary probe preconditioners.* The main motivation behind these preconditioners is the observation that for many elliptic problems, the reduced interface operator has strong spatial local coupling and weak global coupling. In the discrete case, the corresponding interface matrix has elements whose magnitude decay rapidly away from the main diagonal [7]. The boundary probe preconditioners are designed to capture the strong coupling (i.e. the main diagonals) via a few subdomain solves with appropriately chosen *probing* boundary conditions. They were first proposed in [3] for second order elliptic problems and have proven to be quite successful for a large class of problems [9], including convection diffusion problems [8]. Unlike some of the other domain decomposition preconditioners, its performance has also been shown to be relatively insensitive to the mesh size, the shapes of the subdomains and the variations in the coefficients of the differential equation.

The main purpose of this paper is to extend the boundary probe preconditioners to fourth order boundary value problems discretized with standard compact finite difference approximations. There are two main modifications needed. First, the interface between any two subdomains must now consist of two grid lines instead of one in order to competely decouple the subdomain problems. Second, slightly different probing boundary conditions must be used. These will be discussed in more details in Section 2. Results from numerical experiments for the biharmonic problem will be presented in Section 3. They show that the boundary probing technique does produce effective preconditioners.

We note that several preconditioners designed specifically for second order elliptic operators cannot be extended directly to fourth order problems, short of employing them in an inner iteration of an algorithm which solves second order problems at each step. For the biharmonic operator arising from the stream function formulation for the Stoke's problem in incompressible flows, some of these ideas are used in [10,11] to derive domain decomposition preconditioners through the alternate velocity-pressure formulation. On the other hand, the boundary probing technique makes it possible in a direct way to construct efficient domain decomposition preconditioners for problems with more complicated operators. For applications of the techniques developed here to the Navier Stoke's equations, see [1,5].

2. Formulation. In this case, we consider only the simplest case of a domain Ω split into two subdomains Ω_1 and Ω_2 sharing the interface Γ. Consider the problem $Lu = f$ on Ω where L is a fourth order partial differential operator with appropriate boundary conditions on u and its derivatives on $\partial\Omega$. We consider the use of a compact general 25-point discrete approximation of L on a finite difference grid, i.e. the discrete approximation at a point (i,j) involves only values at grid points (k,l) with $|i-k| \leq 2$ and $|j-l| \leq 2$. Further, we assume that the interface Γ consists of two adjacent grid lines, which we shall denote by Γ_1 and Γ_2. If we order the unknowns for the internal points of the subdomains (which we denote by u_1 and u_2) first and those on the interface Γ (which we denote by u_3) last, then the discrete solution vector $u = (u_1, u_2, u_3)^T$ satisfies the linear system :

$$Au \equiv \begin{pmatrix} A_{11} & & A_{13} \\ & A_{22} & A_{23} \\ A_{31} & A_{32} & A_{33} \end{pmatrix} \begin{pmatrix} u_1 \\ u_2 \\ u_3 \end{pmatrix} = \begin{pmatrix} f_1 \\ f_2 \\ f_3 \end{pmatrix} , \tag{2.1}$$

where the discrete vector $f = (f_1, f_2, f_3)^T$ contains the contribution of the right hand side f of the differential equation and of the boundary conditions. Note that in (2.1) the subdomain problems (the blocks A_{11} and A_{22}) are not coupled together.

System (2.1) can be solved by block Gaussian elimination which gives the equations for the interface variables u_3 :

$$Su_3 = \hat{f}_3 \quad , \tag{2.2}$$

with

$$S = A_{33} - A_{31}A_{11}^{-1}A_{13} - A_{32}A_{22}^{-1}A_{23}$$

and

$$\hat{f}_3 = f_3 - A_{31}A_{11}^{-1}f_1 - A_{32}A_{22}^{-1}f_2 \quad .$$

The matrix S is the Schur complement of A_{33} in the matrix A. It corresponds to the reduction of the operator L on Ω to an operator on the internal boundary Γ. Constructing the Schur complement would require the solution of n_Γ elliptic problems on each subdomain, where n_Γ is the number of internal points on Γ. Furthermore it is dense, so that factoring would be expensive.

Instead of solving the system (2.2) directly, iterative methods such as preconditioned conjugate gradient (PCG) can be applied in which only matrix vector product Sy are required. This product can be computed by one solve on each subdomain with boundary condition on Γ determined by y. Since each iteration is rather expensive, it is important to precondition this iteration with a good preconditioner in order to keep the number of iterations small.

We now consider using the boundary probing technique to construct efficient preconditioners for S. The main motivation for this approach is the observation that, for many elliptic operators, the reduced interfacial operator S exhibits a strong spatial local coupling and weak global coupling among the interfacial unknowns. To show this effect, consider the special case of the biharmonic operator $L \equiv \Delta^2$, on the unit square Ω with u and its normal derivatives given on $\partial\Omega$ and the usual 13-point second order central difference approximation on a uniform n by n grid. Let the interface Γ consist of the two vertical grid lines closes to $x = 0.5$, which we shall denote by w_1 and w_2. We shall order the unknowns first on one of the grid lines starting from $y = 0$ to $y = 1$ and then similarly on the other grid line. The Schur complement system corresponding to (2.2) can be written as a block 2 by 2 system:

$$S\begin{pmatrix} w_1 \\ w_2 \end{pmatrix} \equiv \begin{pmatrix} S_{11} & S_{12} \\ S_{21} & S_{22} \end{pmatrix} \begin{pmatrix} w_1 \\ w_2 \end{pmatrix} = \begin{pmatrix} \hat{f}_{31} \\ \hat{f}_{32} \end{pmatrix} . \tag{2.3}$$

The blocks S_{11} and S_{22} account for the coupling of the unknowns on Γ_1 and Γ_2 respectively among themselves and the blocks S_{12} and S_{21} account for the coupling between the unknowns on the two interfaces. Figure 1 shows a plot of the elements of the matrix S for the case $n = 20$. The decay property can be seen clearly in the figure: the magnitude of the elements of the individual subblocks of S decays rapidly away from their respectively main diagonals, reflecting the strong local coupling and weak global coupling. Figure 2 shows the elements of the eighth row of S, showing that the elements of the main diagonal blocks S_{11} and S_{22} are negligible except for the 5 main diagonals and that the off diagonal blocks S_{12} and S_{21} have only 3 non-negligible main diagonals. This shows that the interfacial unknowns are most strongly coupled to its four nearest neighbors on its own grid line and to the three nearest neighbors on the adjacent grid line.

A natural way to construct an interface preconditioner is therefore to capture efficiently the effects of these main diagonals of the individual sublocks of S. However, it would not be efficient to calculate all the elements of S in order to do this, for

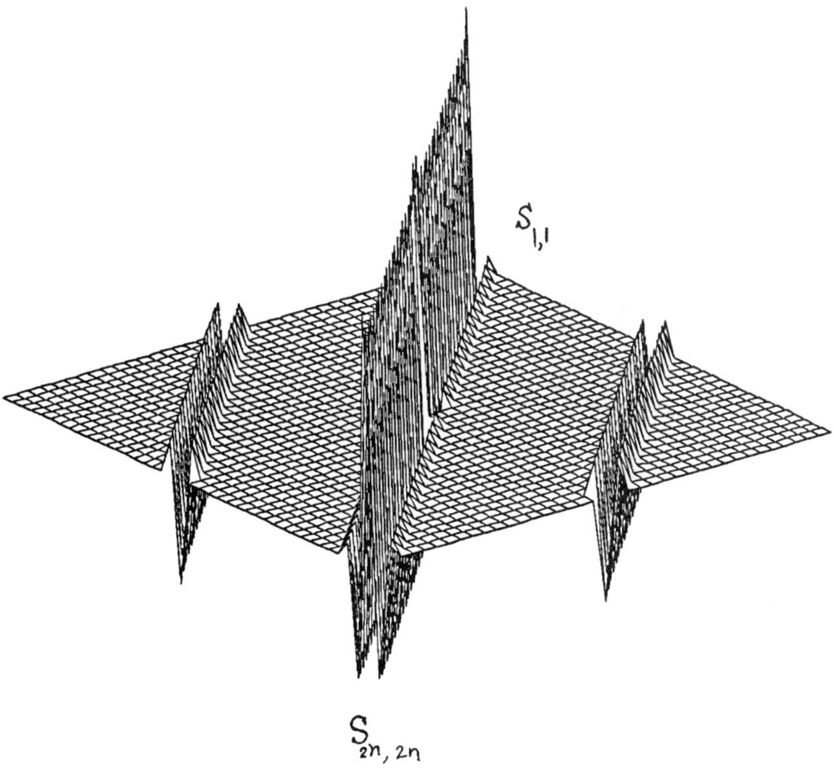

Fig. 1 Plot of elements of S, n=20

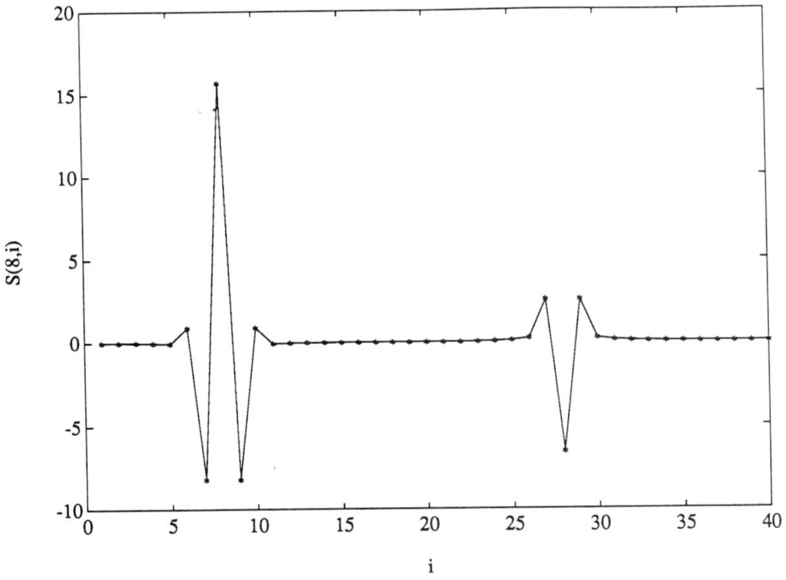

Fig. 2 elements of the Eighth row of S, n=20

this would require $2n$ subdomains solves. Instead, we shall do this by probing the individual subblocks via a few matrix vector products $S_{ij}v_l$ with appropriately chosen probing vectors v_l's. As proposed in [3], a $2k+1$ diagonal approximation to subblock S_{ij} can be constructed by computing the action of S_{ij} on $2k+1$ "probing" vectors $v_l, l = 1, \ldots, 2k+1$ which for the cases $k = 0$ and $k = 1$ are given by:

$$\begin{aligned} k=0: \quad & v_1 = (1,1,1,1,1,1,1,\ldots)^T \\ k=1: \quad & v_1 = (1,0,0,1,0,0,1,\ldots)^T \\ & v_2 = (0,1,0,0,1,0,0,\ldots)^T \\ & v_3 = (0,0,1,0,0,1,0,\ldots)^T. \end{aligned}$$

The case $k = 0$ corresponds to a scaling of each row of the matrix S_{ij} by the sum of the elements of the row. For $k = 1$, if S_{ij} were indeed tridiagonal, all of its elements would be recovered in the vectors $S_{ij}v_l$, $l = 1, 2, 3$. Generalization to cases with $k > 1$ is straightforward [8].

Suppose we want to compute a preconditioner M_{kl} for S consisting of k-diagonal approximations for the diagonal blocks S_{11} and S_{22}, and l-diagonal approximations for the off-diagonal blocks S_{12} and S_{21}. Let V_k be a n by k matrix consisting of k probing vectors for any one of the subblocks described above. Then M_{kl} can be obtained by probing S by the columns of the following matrix:

$$\begin{pmatrix} V_k & V_l & 0 & 0 \\ 0 & 0 & V_k & V_l \end{pmatrix}.$$

This requires solving $2(k+l)$ subdomain problems with boundary conditions consisting of probing vectors from V_k or V_l on one grid line and zero on the other grid line. More efficient probing techniques, with fewer probing vectors and hence fewer subdomain solves for given values of k and l, can be constructed [5] but since our main concern in this paper is on the convergence rates of the preconditioned interface system, for simplicity we shall not present them here.

Finally, the block matrix M_{kl} can be permuted into a narrowly banded matrix by reordering the unknowns to preserve their physical proximity. For example, if we start from $y = 0$ and alternatively order the unknowns on the two grid lines, then M_{kl} is reordered into a banded matrix with bandwidth $2k - 1$, assuming $l < k$ for simplicity. Therefore, the product $M_{kl}^{-1}w$ for a given interfacial vector w can be computed efficiently by banded Gaussian elimination.

3. Numerical Results. We now present some numerical results for the performance of the above boundary probing techniques on the biharmonic problem described in the last section. Figures 3a and 3b show the eigenvalue distribution of the unpreconditioned interface matrix S and the preconditioned matrix $M_{kl}^{-1}S$ for several values of (k, l). The figures show that the preconditioners produce a dramatic improvement in the conditioning of the interface operator. As a sample measure, the eigenvalues of S lie in the interval $(.02, 45)$ while those of the preconditioned system $M_{53}^{-1}S$ lie in $(0.3, 1.2)$. Moreover, many eigenvalues of the preconditioned system are clustered around unity. Figure 4 shows the condition number $M_{kl}^{-1}S$ in the spectral norm as a function of n for several values of k and l. These results show that not only are the condition numbers of the preconditioned matrix much lower than S itself, but also that they grow at a slower rate (approximately $O(n)$ for the preconditioned cases versus $O(n^{2.8})$ for the unpreconditioned case). The plots also show that M_{53} is in some sense optimal because the more expensive M_{75} produces negligible improvement in the condition numbers. Finally, to show that the improvement in the condition number and the eigenvalue distribution of the preconditioned matrix does improve the performance in an iterative solution of the interfacial unknowns, we solve the interfacial system by the preconditioned conjugate gradient algorithm. Figure 5 shows the history of an iteration, with the norm of the residual plotted against the iteration step. It is clear that M_{53} produces a much faster convergence rate.

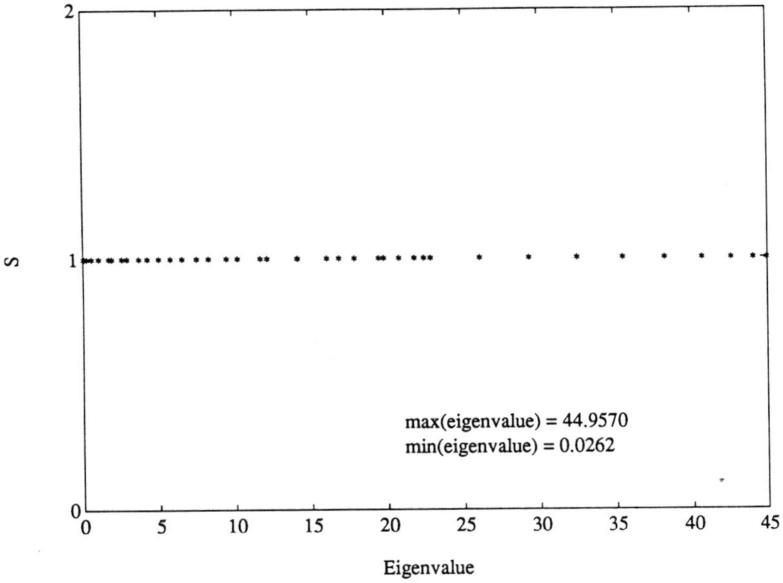

Fig.3a Eigenvalue Distribution of S, n=20

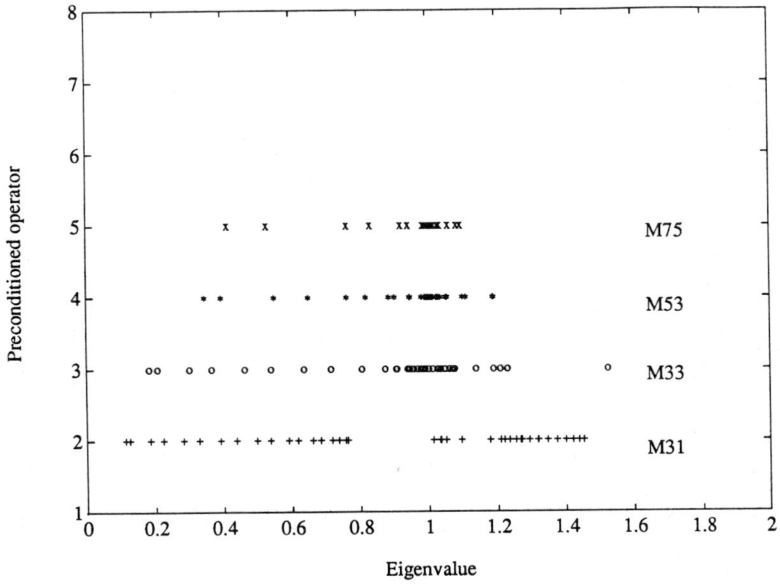

Fig.3b Eigenvalue Distribution of inv(m)*S, n=20

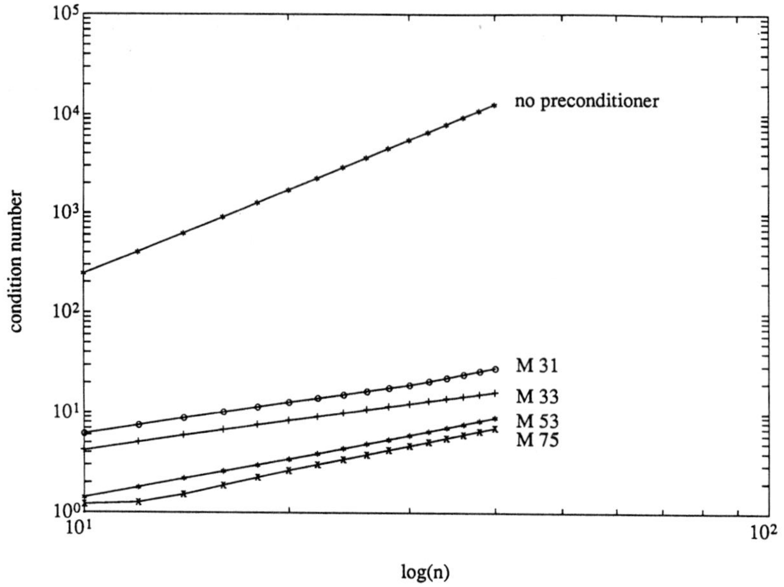

Fig. 4 Condition number of inv(m)*S vs n

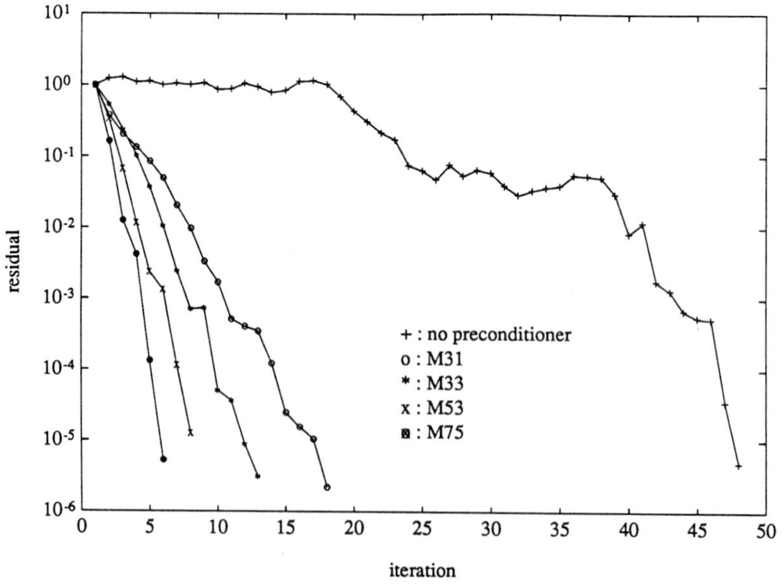

Fig. 5 Convergence history of PCG iteration

We have also performed experiments with the biharmonic operator on nonrectangular domains, such as the L-shaped domain resulting from cutting away a quarter of the unit square. The results are completely similar. More recently, we have successfully applied the above boundary probing technique to the driven cavity problem of incompressible flow [1,5]. However, these results are only preliminary evidence that the boundary probing technique can be applied successfully to fourth order problems. Much further work needs to be done, especially concerning the decay rates of the elements of S and the properties of the preconditioners derived from it.

Acknowledgement. The author would like to acknowledge the assistance of Mr. Wai-Kin Tsui who carried out the numerical experiments in this paper.

REFERENCES

[1] Tony F. Chan, *Domain Decomposition Algorithms and Computational Fluid Dynamics*, UCLA Comp. and App. Math. Report No. 88-25, August, 1988. To appear in the Proceedings of the Second International Conference on Parallel and Vector Processing, Tromso, Norway, June 1988, a special issue of the journal "Parallel Computing".

[2] Tony F. Chan, Roland Glowinski, Jacques Periaux and Olof Widlund (editors), *Domain Decomposition Algorithms*, Proceedings of the 2nd International Symposium on Domain De composition Methods, Jan 14-16, 1988, UCLA, SIAM Publ., to appear.

[3] Tony F. Chan and Diana Resasco, *A Survey of Preconditioners for Domain Decomposition*, in Proceedings of the IV Coloquio de Matematicas, TAller de Analisis Numerico y sus Aplicaciones, Taxco, Guerrero, Mexico, August, 1985, to be published by Springer Verlag. Also appeared as Yale Univ. Dept. of Comp. Sci. Report YALEU/DCS/RR-414, July, 1985.

[4] Tony F. Chan and Diana Resasco, *A Framework for the Analysis and Construction of Domain Decomposition Preconditioners*, in "Domain Decomposition Methods for Partial Differential Equations", Proceedings of First Int'l Symposium on Domain Decomposition Methods, Jan, 1987, Paris, France, R. Glowinski et al (editors), SIAM Publ., 1988, pp. 217-230.

[5] Tony F. Chan and Wai-Kin Tsui, *Boundary Probe Domain Decomposition Preconditioners for Incompressible Flows*, UCLA Comp. and App. Math. Report, to appear.

[6] Roland Glowinski, G.H. Golub, G.A. Meurant and J. Periaux, *Domain Decomposition Methods for Partial Differential Equations*, Proceedings of First Int'l Symposium on Domain Decomposition Methods, Jan, 1987, Paris, France, R. Glowinski et al (editors), SIAM Publ., 1988.

[7] G.H. Golub and D. Mayers, *The Use of Preconditioning Over Irregular Regions*, in "Computing Methods in Applied Sciences and Engineering VI", R. Glowinski and J.L. Lions (Editors), North Holland (1984), pp. 3-14.

[8] D. Keyes and W. Gropp, *Domain Decomposition Techniques for Nonsymmetric Systems of Elliptic Boundary Value Problems : Examples from CFD*, in Proceedings of Second Int'l Symposium on Domain Decomposition Methods, Jan, 1988, Los Angeles, Calif., T.F. Chan et al (editors), SIAM Publ., to appear.

[9] D. Keyes and W. Gropp, *A Comparison of Domain Decomposition Techniques for Elliptic Partial Differential Equations*, SIAM J. Sci. Stat. Comp., 8/2 March (1987), pp. s166-s202.

[10] J. Pasciak, *Two Domain Decomposition Techniques for the Stokes Problems*, in Proceedings of Second Int'l Symposium on Domain Decomposition Methods, Jan, 1988, Los Angeles, Calif., T.F. Chan et al (editors), SIAM Publ., to appear.

[11] Alfio Quarteroni, *Domain Decomposition Algorithms for the Stokes Equations*, in Proceedings of Second Int'l Symposium on Domain Decomposition Methods, Jan, 1988, Los Angeles, Calif., T.F. Chan et al (editors), SIAM Publ., to appear.

CHAPTER 14

An Additive Schwarz Algorithm for Two- and Three-Dimensional Finite Element Elliptic Problems*

Maksymilian Dryja†

Abstract. In this paper, we present an additive variant of the Schwarz alternating method for finite element approximations of two and three dimensional elliptic problems with mixed boundary conditions. We establish that the rate of convergence of this method is optimal even when the number of subregions is large. In our analysis techniques previously developed for iterative substructuring methods are used. All these methods show great promise for parallel computers. The reported results have been recently obtained in joint work with Olof Widlund.

1. Introduction. In this paper, we will consider conforming finite element discretizations of linear, self adjoint two and three dimensional second order elliptic equations with mixed boundary conditions. Our aim is to develop and study an additive variant of the classical Schwarz alternating method, see [8], and in particular to discuss the case of many subregions. The method is described in terms of projections from the finite element space onto subspaces related to overlapping subregions which cover the region. These projections are defined by using the symmetric bilinear form associated with the original elliptic problem. In addition, a special subspace defined on a coarse triangulation is introduced. It provides a mechanism for the global transportation of information which is necessary to obtain fast convergence; cf. [9] for a discussion of this matter in the context of iterative substructuring methods. The additive form of the Schwarz algorithm is thus an iterative method for solving an equation with a symmetric and positive definite operator which is a sum of the projections. This permits the use of the conjugate gradient method. We have proved, see the theorem in section 3, that the rate of convergence is optimal

* This work was supported in part by the National Science Foundation under Grant NSF-CCR-870368.

† Institute of Informatics, Warsaw University, 00-901 Warsaw, PKiN, p. 850, Poland.

even when the region is divided into many subregions. In the last section, we discuss the question of inexact solvers in additive Schwarz algorithms.

The methods described are interesting for parallel computations particularly for computers with many processors. The subproblems can be solved separately and at the same time. The methods also offer real benefits even when one processor is used, particularly when the subregions have simple geometry. We note that compared with iterative substructuring methods, see [1], [2], [3] and literature therein, the additive Schwarz algorithms satisfy somewhat better asymptotic bounds.

In this paper, we report on certain extensions of our report [4]. We note that our work was inspired by the paper [6] in which a variational framework for the classical multiplicative Schwarz method for continuous elliptic problems is discussed. As for earlier work on additive algorithms of this kind see [7] in which certain iterative substructuring methods on regions divided into a fixed number of subregions are discussed.

2. Statement of problems. We consider the following weak form of a linear, self-adjoint second order elliptic equation with mixed boundary conditions. Let Ω be a bounded Lipschitz region in R^n, $n = 2$ or 3. We assume that $\partial\Omega$, the boundary of Ω, is the union of two nonoverlapping sets Γ_D and Γ_N on which zero Dirichlet and arbitrary Neumann boundary conditions are given. We introduce the Sobolev space $V(\Omega) \subset H^1(\Omega)$:

$$V(\Omega) := \{v \in H^1(\Omega) : \gamma_D v = 0\}$$

where $\gamma_D v$ is the trace of v on Γ_D.

The problem takes the form: Find $u \in V(\Omega)$ such that

$$a(u, v) = \ell(v), \quad v \in V(\Omega) \tag{2.1}$$

where

$$a(u, v) = \int_\Omega \sum_{i,j} a_{ij}(x) D_i u D_j v \, dx$$

and $\ell(v)$ is a linear continuous functional defined on $V(\Omega)$. We assume that the bilinear form $a(u, v)$ is symmetric and positive definite, i.e.

$$\begin{aligned} &a(v, u) = a(u, v), \\ &c\|u\|^2_{H^1(\Omega} \leq a(u, u), \quad u \in V \end{aligned} \tag{2.2}$$

where c is a positive constant.

The problem (2.1) is approximated by a conforming finite element method. To simplify the presentation we assume that Ω is a polyhedral (polygonal) region and that it is divided into tetrahedral (triangular) elements e_i. We assume that the curves where the boundary conditions change consist of edges of elements. We also assume that the triangulation of Ω is regular, see [2]. Let V^h denote a conforming finite element space defined on the given triangulation with zero values on Γ_D. The discrete problem is of the form:

Find $u_h \in V^h$ such that

$$a(u_h, v_h) = \ell(v_h), \quad v_h \in V^h(\Omega) \tag{2.3}$$

Our aim is to solve (2.3) by an additive version of the Schwarz alternating method.

3. The Additive Schwarz Algorithm. In this section we give a description of the additive Schwarz algorithm for solving (2.3). We assume that triangulation of Ω can be subdivided into nonoverlapping tetrahedral (triangular) subregions Ω_i, $i = 1, \ldots, N$, called substructures, in such a way that the boundary of Ω_i follows element boundaries. We assume that the substructures form a regular triangulation of Ω, with a parameter H in the sense of finite element theory. We next extend each substructures Ω_i to a larger region Ω_i'. In this way we get the partitioning of Ω into overlapping subregions Ω_i' needed in Schwarz-type domain decomposition methods. We assume that the distance between the boundaries $\partial\Omega_i$ and $\partial\Omega_i'$ is bounded from below by a fixed fraction of H, and that $\partial\Omega_i'$ does not cut through any element. In the case when part of Ω_i' is outside of Ω we cut off that part and denote the resulting subregion by Ω_i'.

We represent the finite element space V^h as the sum of $N+1$ subspaces

$$V^h = V_0^h + V_1^h + \ldots + V_N^h \tag{3.1}$$

The first subspace V_0^h is the space of continuous, piecewise linear functions defined on the coarse triangulation with parameter H. It is also called V^H and plays a special role in the algorithm. The remaining subspaces V_i^h are associated with the subregions Ω_i' and are defined as follows. For interior subregions Ω_i'

$$V_i^h(\Omega) = \{v \in V^h \cap H_0^1(\Omega_i') : v(x) = 0, \ x \in C\Omega_i'\}$$

where $C\Omega_i'$ is the complement of Ω_i' with respect to the region Ω. In the case when $\partial\Omega_i'$ intersects $\partial\Omega$, we use the original boundary condition on $\partial\Omega_i \cap \partial\Omega$ and zero Dirichlet conditions at all other points of $\partial\Omega_i'$. We see that $V_i^h(\Omega) \subset V^h$ and that the representation (3.1) of V^h is possible.

We now introduce the projections $P_i : V^h \to V_i^h$ with respect to the bilinear form $a(u,v)$. By definition $P_i v_h \in V_i^h$ is the unique element of V^h satisfying the equation

$$a(P_i v_h, \phi_h) = a(v_h, \phi_h), \quad \phi_h \in V_i^h. \tag{3.2}$$

We note that $P_i v_h$ can be computed, for any $v_h \in V^h$, at the expense of solving the finite element problem in the subspace V_i^h. For u_h, the solution of (2.3) can be computed by solving

$$a(P_i u_h, \phi_h) = \ell(\phi_h), \quad \phi_h \in V_i^h,$$

since

$$a(u_h, u_h) = \ell(\phi_h).$$

We can therefore replace the problem (2.3) by the equation

$$Pu_h \equiv (P_0 + P_1 + \cdots + P_N)u_h = g_h, \tag{3.3}$$

where the right hand side g_h can be computed as $g_h = \sum_{i=0}^{N} g_{h,i}$, $g_{h,i} = P_i u_h$. Inorder to establish that equations (2.3) and (3.3) have the same solutions, we need only show that P is invertible. This follows from our theorem given below.

The additive form of the Schwarz algorithm for solving (2.3) is simply an iterative method for solving equation (3.3). A natural choice is the conjugate gradient method; cf. [5]. Our main result, cf. [4], is that P is uniformly well conditioned.

Theorem. *For any $v_h \in V^h$ the following inequalities hold*

$$C_0 |v_h|^2_{H^1(\Omega} \leq |Pv_h|^2_{H^1(\Omega)} \leq C_1 |v_h|^2_{H^1(\Omega)} \tag{3.4}$$

where C_0 and C_1 are positive constant independent of H and h.

The derivation of a conjugate gradient method for equation (3.3) is a relatively routine matter. Details will therefore not be given here. We only note that equation (3.3) requires no further preconditioning but that it is important to work with the inner product defined by the bilinear form $a(u, v)$ since the operator P is symmetric with respect to that bilinear form but not the Euclidean inner product. In this respect this iterative method differs from the standard preconditioned conjugate gradient method, where the quadratic form associated with the preconditioner as well as the Euclidean inner product are used.

4. The Use of Preconditioners. In the algorithm described, the subproblems on Ω_i' are solved directly, see (3.2) and (3.3). These subproblems can also be solved inexactly using another bilinear form $b(u, v)$. If this problem satisfies

$$C_2 b(u, v) \leq a(u, v) \leq C_3 b(u, v), \quad u, v \in V_i^h(\Omega) \tag{3.7}$$

with positive constants C_2 and C_3 independent of H and h, then the condition number of the resulting method is bounded by $\kappa(P) C_3 / C_2$. We can, for example, use

$$b(u, v) = \int_\Omega \nabla u \cdot \nabla v \, dx$$

which satisfies (3.7). We use the bilinear form $b(u, v)$, when we solve the equations in the subspaces, and the original form $a(u, v)$ when we evaluate the residuals in the iterative method. Otherwise the algorithm is unchanged.

REFERENCES

[1] J. H. BRAMBLE, J. E. PASCIAK and A. H. SCHATZ, *The construction of preconditioners for elliptic problems by substructuring, I*, Math. Comp., 47 (1986), pp. 103-134.

[2] P. G. CIARLET, *The Finite Element Method for Elliptic Problems*, North-Holland, Amsterdam, 1978.

[3] M. DRYJA, *A method of domain decomposition for three-dimensional finite element problems,* in Domain Decomposition Methods for Partial Differential Equations, R. Glowinski, G. H. Golub, G. A. Meurant and J. Périaux, eds., SIAM, Philadelphia, 1988, pp. 43-61.

[4] M. DRYJA and O. WIDLUND, *An additive variant of the Schwarz alternating method for the case of many subregions,* TR#339, Courant Institute, Dept. of Computer Science, December 1987.

[5] G. H. GOLUB and C. F. Van LOAN, *Matrix Computations,* Johns Hopkins University Press, 1983.

[6] P.-L. LIONS, *On the Schwarz alternating method, I,* in Domain Decomposition Methods for Partial Differential Equations, R. Glowinski, G. H. Golub, G. A. Meurant and J. Périaux, eds., SIAM, Philadelphia, 1988, pp. 1-42.

[7] A. M. MATSOKIN and S. V. NEPOMYASHCHIKH, *Schwarz alternating method in subspace,* TR#114, Computing Center of the Siberian Branch of the USSR Academy of Sciences, Novosibirsk, 1985, in Russian.

[8] H. A. SCHWARZ, *Gesammelte Mathematische Abhandlungen,* v. 2, Springer, Berlin, 1890.

[9] O. WIDLUND, *Iterative substructuring methods: algorithms and theory for problems in the plane,* in Domain Decomposition Methods for Partial Differential Equations, R. Glowinski, G. H. Golub, G. A. Meurant and J. Périaux, eds., SIAM, Philadelphia, 1988, pp. 113-128.

New Domain Decomposition Strategies for Elliptic Partial Differential Equations

D. J. Evans*
Kang Li-shan†

Abstract. In this paper some new strategies for numerically solving elliptic partial differential equations by domain decomposition are presented which are applicable to non-overlapped and overlapped regions of the domain.

1. Introduction. The solution of elliptic partial differential equations over a 2 dimensional region can be shown to be accomplished efficiently by a range of *Domain Decomposition* techniques in which the solution of smaller problems on subdomains can be grouped together to produce the overall solution for the whole domain. Such techniques seem ideally suited for the solution of elliptic problems on irregular domains and on the present day multiprocessor systems.

The paper considers *firstly* the case of non-overlapping regions in which the vital factor is the efficient solution of the linear systems of equations governing the variables on the interfaces between the subdomains. Here we propose a *preconditioned iterative method* called P^2CG which involves the application of 2 preconditioned stages to the well known conjugate gradient method resulting in an improved convergence rate for the method.

Secondly, the case of overlapping regions is discussed and convergence factors and acceleration strategies for the Schwarz Alternating Procedure (SAP) are presented which again leads to computationally efficient algorithms for solving problems involving partial differential equations.

2. Problem Formulation for Non-Overlapping Domains. We will first formulate our approach in the simplest case of a domain split into two

*Loughborough University of Technology, Loughborough, Leics., LE11 3TU, England.
† Wuhan University, Wuhan, Hubei, China.

subdomains with one interface. Consider the problem:

$$Lu = f \text{ on } \Omega \,, \tag{2.1}$$

with boundary condition $u = u_b$ on $\partial\Omega$,

where L is a linear elliptic operator and the domain Ω is as illustrated in Fig.2.1. We will call the interface between Ω_1 and Ω_2, Γ.

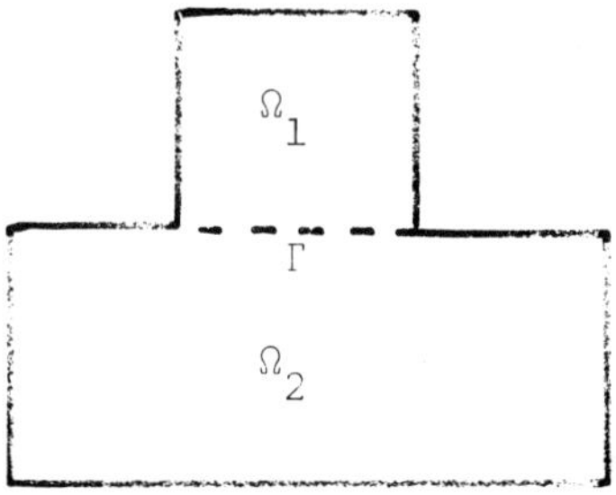

FIG. 2.1. The domain Ω and its partition

If we order the unknowns for the internal points of the subdomains first and those in the interface Γ last, then the discrete solution vector $u=(u_1,u_2,u_3)$ satisfies the linear system,

$$Au = b \,, \tag{2.2}$$

which can be expressed in block form as:

$$\begin{pmatrix} A_{11} & & A_{13} \\ & A_{22} & A_{23} \\ A_{13}^T & A_{23}^T & A_{33} \end{pmatrix} \begin{pmatrix} u_1 \\ u_2 \\ u_3 \end{pmatrix} = \begin{pmatrix} b_1 \\ b_2 \\ b_3 \end{pmatrix} . \tag{2.3}$$

The solution of system (2.3) can now proceed as follows:

<u>Step 1</u>: Compute

$$C = A_{33}-A_{13}^TA_{11}^{-1}A_{13}-A_{23}^TA_{22}^{-1}A_{23} \,, \tag{2.4}$$

$$w_1 = A_{11}^{-1}b_1 \,, \tag{2.5}$$

$$w_2 = A_{22}^{-1}b_2 \,, \tag{2.6}$$

and solve,

$$Cu_3 = b_3-A_{13}^Tw_1-A_{23}^Tw_2 \,, \tag{2.7}$$

by a suitable iterative procedure to be discussed later. Then,

<u>Step 2</u>: Compute

$$u_1 = w_1-A_{11}^{-1}A_{13}u_3 \,, \tag{2.8}$$

and $$u_2 = w_2 - A_{22}^{-1} A_{23} u_3 \ . \tag{2.9}$$

Note, that except for (2.7), the algorithm only requires the solution of problems with A_{11} and A_{22}, which corresponds to solving independent problems on the subdomains. The matrix C (2.4) is the Schur complement of A_{33} in A and it is sometimes called the capacitance matrix in this context. It corresponds to the reduction of the operator L on Ω to an operator on the boundary Γ.

The basic idea of the solution technique is to guess an initial solution on the boundary Γ and then to iterate to the solution of (2.7) using the preconditioned conjugate gradient method. Several alternative strategies have been proposed in which a variant of the capacitance matrix method becomes an iterative solution for the capacitance system.

In general, each iteration of the capacitance/PCG algorithm requires one solution of the smaller subproblems on domains Ω_1 and Ω_2 to form the product of the capacitance matrix C and righthand side vector. In addition, the initialisation step requires a solution of the Dirichlet problem on each of the subdomains.

3. Explicit Diagonal Block Preconditioning. We now reconsider the approach used by Evans (1984), i.e. of using a small block of fixed size, i.e., 4 points and explicitly inverting it within the iteration. Then the loss of sparseness which will inevitably occur when an explicit form of iterative method, i.e. the conjugate gradient method is used will be small and independent of the size of the given problem.

The main concern of this section is to construct new groupings of the mesh points of the network into small order groups or blocks of 4 points and to investigate their advantages when used explicitly in preconditioned iterative methods.

We consider one of the subdomains Ω_1 or Ω_2 (Fig.3.1) with Dirichlet boundary conditions.

For a large class of 2-dimensional linear elliptic differential equations in which a 5-point approximation scheme on a uniform mesh (Fig.3.2) is used, it is a simple procedure to approximate the partial derivatives of the PDE by suitable central finite difference expressions using Taylor series and quadrature techniques. Thus, the normalised finite difference equation at the point P has the form,

$$\alpha_4 u_{L,P} + \alpha_1 u_{B,P} + u_P + \alpha_2 u_{R,P} + \alpha_3 u_{T,P} = b_P \ , \tag{3.1}$$

for all points within a coordinate square region such as in Fig.3.1.

Here we have denoted the points on the network to the Left, Top, Right, and Bottom of the representative point P in Fig.3.2 by appropriate suffixes.

For such equations, the coefficients $\alpha_1, \alpha_2, \alpha_3$ and α_4 satisfy the relationship,

$$1 \geq \alpha_1 + \alpha_2 + \alpha_3 + \alpha_4 \ , \tag{3.2}$$

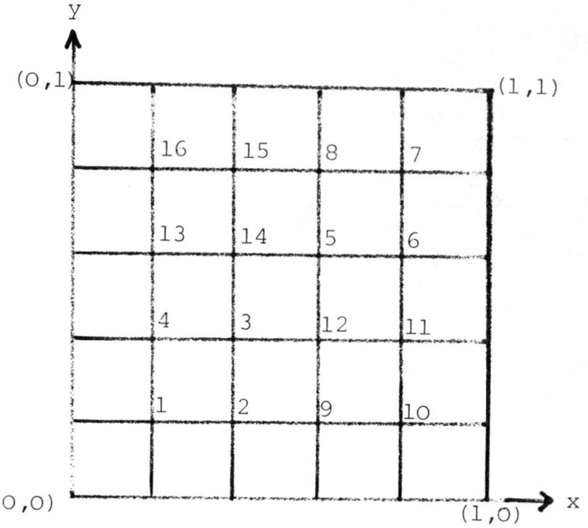

FIG. 3.1

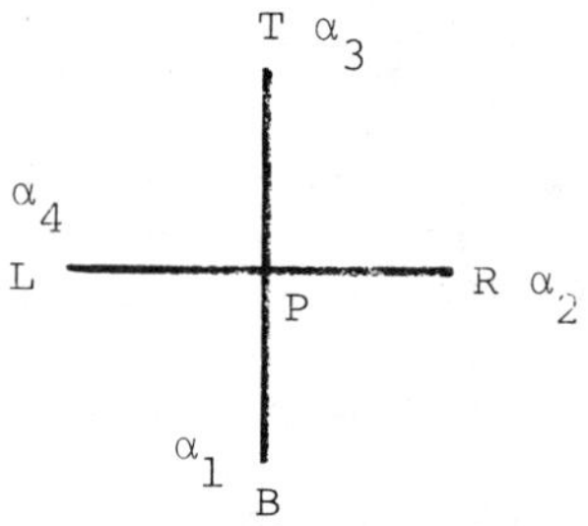

FIG. 3.2

and when grouped together by a row ordering of the points in matrix form will reduce the problem to one of solving a sparse system of linear difference equations $A\underline{u}=\underline{b}$ where the coefficient matrix A is of large order, $M=n^2$ non-singular, positive definite and symmetric of the well known form,

$$\begin{bmatrix} B_1 & \Lambda_3 & & & O \\ \Lambda_1 & B_2 & \ddots & & \\ & \ddots & \ddots & \ddots & \\ & & \ddots & \ddots & \Lambda_3 \\ O & & & \Lambda_1 & B_r \end{bmatrix}, \text{ with } B_i = \begin{bmatrix} 1 & \alpha_2 & & & O \\ \alpha_4 & \ddots & \ddots & & \\ & \ddots & \ddots & \ddots & \\ & & \ddots & \ddots & \alpha_2 \\ O & & & \alpha_4 & 1 \end{bmatrix}, i=1,2,\ldots,n, \tag{3.3}$$

and $\Lambda_i = \alpha_i I$, $i=1,3$. (Varga [1]).

However, for the simple case of the unit square and $\Delta x=\Delta y=1/5$ as illustrated in Fig.3.1, a red/black ordering of the 4-point groups will result in a coefficient matrix which has the block structure,

$$A = \left[\begin{array}{c|c} D_1 & C_1 \\ \hline C_2 & D_2 \end{array}\right], \tag{3.4}$$

where $D_1=D_2=R_0I_8$, with $I_8\equiv$the (8×8) identity matrix and,

$$C_1 = \left[\begin{array}{c|c} R_2 & R_3 \\ \hline R_1 & R_4 \end{array}\right], \qquad C_2 = \left[\begin{array}{c|c} R_4 & R_3 \\ \hline R_1 & R_2 \end{array}\right], \tag{3.5}$$

with,

$$R_0 = \begin{bmatrix} 1 & \alpha_2 & 0 & \alpha_3 \\ \alpha_4 & 1 & \alpha_3 & 0 \\ 0 & \alpha_1 & 1 & \alpha_4 \\ \alpha_1 & 0 & \alpha_2 & 1 \end{bmatrix}, \qquad R_1 = \begin{bmatrix} 0 & 0 & 0 & \alpha_1 \\ 0 & 0 & \alpha_1 & 0 \\ 0 & 0 & 0 & 0 \\ 0 & 0 & 0 & 0 \end{bmatrix}, \tag{3.6}$$

$$R_2 = \begin{bmatrix} 0 & 0 & 0 & 0 \\ \alpha_2 & 0 & 0 & 0 \\ 0 & 0 & 0 & \alpha_2 \\ 0 & 0 & 0 & 0 \end{bmatrix}, \qquad R_3 = \begin{bmatrix} 0 & 0 & 0 & 0 \\ 0 & 0 & 0 & 0 \\ 0 & \alpha_3 & 0 & 0 \\ \alpha_3 & 0 & 0 & 0 \end{bmatrix}$$

and

$$R_4 = \begin{bmatrix} 0 & \alpha_4 & 0 & 0 \\ 0 & 0 & 0 & 0 \\ 0 & 0 & 0 & 0 \\ 0 & 0 & \alpha_4 & 0 \end{bmatrix}.$$

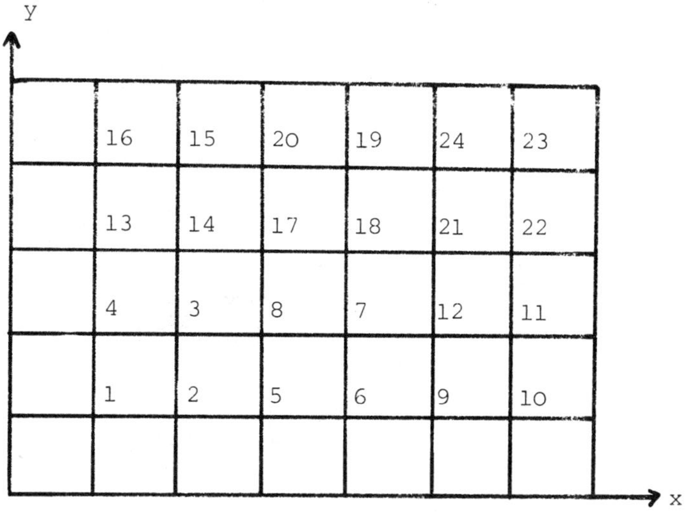

FIG. 3.3

Further for larger meshes, it is a simple matter to deduce that when the 4-point blocks are taken in natural row ordering (Fig.3.3) and because the block equations only refer to adjacent blocks then the coefficient matrix has the familiar structure, i.e.

$$A \equiv \begin{bmatrix} B_1 & C_1 & & & \bigcirc \\ A_1 & B_2 & C_2 & & \\ & \ddots & \ddots & \ddots & \\ & & & & C_{k-1} \\ \bigcirc & & & A_k & B_k \end{bmatrix}, \qquad (3.7)$$

where $C_i \equiv R_3 I$, $A_i \equiv R_i I$ and,

$$B_i \begin{bmatrix} R_0 & R_2 & & & \bigcirc \\ R_4 & \ddots & \ddots & & \\ & \ddots & \ddots & \ddots & \\ & & & & R_2 \\ \bigcirc & & & R_4 & R_0 \end{bmatrix}, \quad i=1,2,\dots,k.$$

Now any explicit block iterative method can be considered as a point preconditioned iterative method applied to a transformed matrix $A^E=[\text{diag}\{R_0\}]^{-1}A$ and vector $b^E=[\text{diag}\{R_0\}]^{-1}b$. Since R_0 is a small order diagonal submatrix which can be easily inverted. The matrix A^E and vector b^E (where the superfix E denotes the 'explicit' form) can be evaluated explicitly from a new computational molecule (Fig.3.4) and is a preconditioned form (diagonally block scaled) of the original matrix A. To determine the preconditioned linear system $A^E u=b^E$, with $A^E = I-L^E-U^E$, we proceed as follows:

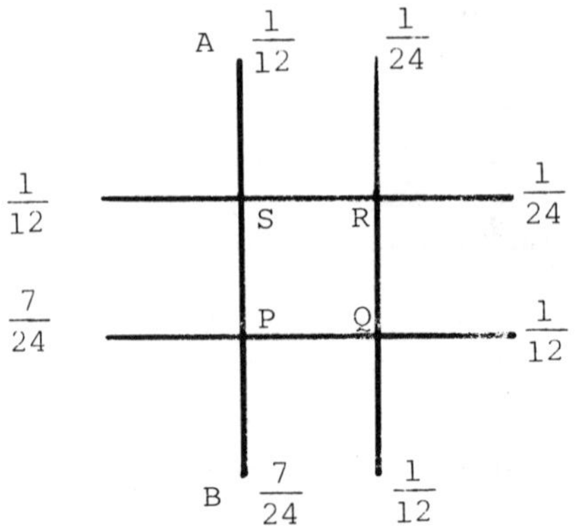

FIG. 3.4

The matrix $[\mathrm{diag}\{R_0\}^{-1}]$ is simply $\mathrm{diag}\{R_0^{-1}\}$ and the inverse of R_0 given by,

$$R_0^{-1} = (d^{-1}) \begin{bmatrix} \alpha_5 & -\alpha_2\alpha_6 & 2\alpha_2\alpha_3 & -\alpha_3\alpha_7 \\ -\alpha_4\alpha_6 & \alpha_5 & -\alpha_3\alpha_7 & 2\alpha_3\alpha_4 \\ 2\alpha_1\alpha_4 & -\alpha_1\alpha_7 & \alpha_5 & -\alpha_4\alpha_6 \\ -\alpha_1\alpha_7 & 2\alpha_1\alpha_2 & -\alpha_2\alpha_6 & \alpha_5 \end{bmatrix}, \qquad (3.8)$$

where $d=(\alpha_1\alpha_3-\alpha_2\alpha_4)^2-2(\alpha_1\alpha_3+\alpha_2\alpha_4)+1$ and $\alpha_5=1-\alpha_1\alpha_3-\alpha_2\alpha_4$, $\alpha_6=1+\alpha_1\alpha_3-\alpha_2\alpha_4$, $\alpha_7=1+\alpha_2\alpha_4-\alpha_1\alpha_3$.

The block structure of A^E is the same as that of A (i.e., eqn. (3.7)) with the submatrices R_0 replaced by identity matrices, and the submatrices R_i, replaced by $R_0^{-1}R_i$, i=1,2,3,4. In addition, where R_i, i=1,2,3,4 has a column or row of zeros so does $R_0^{-1}R_i$, and where an element α_i occurs as the (p,q)th element of R_i, the qth column of $R_0^{-1}R_i$ is the pth column of R_0^{-1}, multiplied by α_i. So, for example,

$$R_0^{-1}R_1 = (d^{-1}) \begin{bmatrix} 0 & 0 & -\alpha_1\alpha_2\alpha_6 & \alpha_1\alpha_5 \\ 0 & 0 & \alpha_1\alpha_5 & -\alpha_1\alpha_4\alpha_6 \\ 0 & 0 & -\alpha_1^2\alpha_7 & 2\alpha_1^2\alpha_4 \\ 0 & 0 & 2\alpha_1^2\alpha_2 & -\alpha_1^2\alpha_7 \end{bmatrix}. \qquad (3.9)$$

Similarly,

$$R_0^{-1}R_2 = (d^{-1}) \begin{bmatrix} -\alpha_2^2\alpha_6 & 0 & 0 & 2\alpha_2^2\alpha_3 \\ \alpha_2\alpha_5 & 0 & 0 & -\alpha_2\alpha_3\alpha_7 \\ -\alpha_1\alpha_2\alpha_7 & 0 & 0 & \alpha_2\alpha_5 \\ 2\alpha_1\alpha_2^2 & 0 & 0 & -\alpha_2^2\alpha_6 \end{bmatrix},$$

$$R_0^{-1}R_3 = (d^{-1}) \begin{bmatrix} -\alpha_3^2\alpha_7 & 2\alpha_2\alpha_3^2 & 0 & 0 \\ 2\alpha_3^2\alpha_4 & -\alpha_3^2\alpha_7 & 0 & 0 \\ -\alpha_3\alpha_4\alpha_6 & \alpha_3\alpha_5 & 0 & 0 \\ \alpha_3\alpha_5 & -\alpha_2\alpha_3\alpha_6 & 0 & 0 \end{bmatrix},$$

and

$$R_0^{-1}R_4 = (d^{-1})\begin{bmatrix} 0 & \alpha_4\alpha_5 & -\alpha_4\alpha_3\alpha_7 & 0 \\ 0 & -\alpha_4^2\alpha_6 & 2\alpha_3\alpha_4^2 & 0 \\ 0 & 2\alpha_1\alpha_4^2 & -\alpha_4^2\alpha_6 & 0 \\ 0 & -\alpha_1\alpha_4\alpha_7 & \alpha_4\alpha_5 & 0 \end{bmatrix}$$

For the model Laplacian problem and a square grid $\alpha_1=\alpha_2=\alpha_3=\alpha_4=-1/4$ so that,

$$R_0^{-1} = \frac{1}{6}\begin{bmatrix} 7 & 2 & 1 & 2 \\ 2 & 7 & 2 & 1 \\ 1 & 2 & 7 & 2 \\ 2 & 1 & 2 & 7 \end{bmatrix} \text{ and } R_0^{-1}R_1 = -\frac{1}{24}\begin{bmatrix} 0 & 0 & 2 & 7 \\ 0 & 0 & 7 & 2 \\ 0 & 0 & 2 & 1 \\ 0 & 0 & 1 & 2 \end{bmatrix},$$

from which we can establish the computational stencil at the point P to be as in Fig.3.4.

Further analyses will yield analogous relationships for the remaining points Q,R and S of the 4-point block.

The use of this computational stencil to derive the solution of self-adjoint P.D.E.'s is given in Evans [2].

Finally, the transformed matrix A^E has the block form, given by eqn.(3.7), where,

$$B_i = \begin{bmatrix} I & R_0^{-1}R_2 & & & 0 \\ R_0^{-1}R_4 & I & R_0^{-1}R_2 & & \\ & \ddots & \ddots & \ddots & \\ & & & & R_0^{-1}R_2 \\ 0 & & & R_0^{-1}R_4 & I \end{bmatrix}, \qquad (3.10)$$

$C_i = R_0^{-1}R_3$, $A_i = R_0^{-1}R_i$, $i=1,2,\ldots,k$.

Thus, in this section we have shown that when the system of difference equations,

$$A^E\underline{u} = \underline{b}^E, \qquad (3.11)$$

is derived from the new computational molecule given by Fig.3.4 then the matrix A^E has already been block diagonally scaled by the 4×4 block matrix R_0. Thus, this can be regarded as a form of *diagonal scaling* or *preconditioning*.

4. <u>Conjugate Gradient Acceleration</u>. Acceleration strategies to obtain the solutions of (3.11) have been shown to yield superior convergence rates in block JOR and SOR schemes, (Evans, [2]).

However, in this section we are concerned with the application of

Conjugate Gradient (CG) method to the Domain Decomposition technique.

Since the matrix A given in (3.1) is symmetric and positive definite then we can apply the C.G. algorithm in the form,

$$\left.\begin{aligned} r^{(n)} &= b-Au^{(n)} \\ d^{(n)} &= -r^{(n)} \end{aligned}\right\} n=1$$

$$\begin{aligned}
\sigma_n &= r^{(n)}.d^{(n)}/d^{(n)}.Ad^{(n)} \\
u^{(n+1)} &= u^{(n)}+\sigma_n d^{(n)} \\
r^{(n+1)} &= r^{(n)}-\sigma_n Ad^{(n)} \\
\tau_{n+1} &= -r^{(n+1)}Ad^{(n)}/d^{(n)}.Ad^{(n)} \\
d^{(n+1)} &= r^{(n+1)}+\tau_{n+1}d^{(n)} \\
&\text{until } |u^{(n+1)}-u^{(n)}|<5\times10^{-6}, \text{ otherwise } n=n+1
\end{aligned} \tag{4.1}$$

Further we can immediately see that since the system $A^E u=b^E$ has been *explicitly* obtained from Fig.3.4, then in this case, the Conjugate Gradient algorithm can also be applied directly to A^E to form a preconditioned C.G. method to yield an immediate improvement. Thus we have for the *first* preconditioning stage,

$$\left.\begin{aligned} r^{(n)} &= b^E-A^E u^{(n)} \\ d^{(n)} &= -r^{(n)} \end{aligned}\right\}$$

$$\begin{aligned}
\sigma_n &= r^{(n)}.d^{(n)}/d^{(n)}.A^E d^{(n)} \quad \longleftarrow n=n+1 \\
u^{(n+1)} &= u^{(n)}+\sigma_n d^{(n)} \\
r^{(n+1)} &= r^{(n)}-\sigma_n A^E d^{(n)} \\
\tau_{n+1} &= -r^{(n+1)}A^E d^{(n)}/d^{(n)}.A^E d^{(n)} \\
d^{(n+1)} &= r^{(n+1)}+\tau_{n+1}d^{(n+1)} \\
&\text{until } |u^{(n+1)}-u^{(n)}|<5\times10^{-6} \text{ otherwise}
\end{aligned} \tag{4.2}$$

Next a *second* preconditioning stage can be applied *implicitly* when we premultiply eqn.(3.11) by a non-singular matrix M^{-1} where M^{-1} is an approximate inverse of A. Thus, (3.11) is transformed into the preconditioned form,

$$M^{-1}A^E u = M^{-1}b \,, \tag{4.3}$$

where M is a conditioning matrix such that its inverse is easily computed.

Let us consider the possible forms for the matrix M:

a) Splitting of A^E

If we assume that A^E can be written (without loss of generality) as

$A^E=I-L^E-U^E$ where L^E and U^E are strictly lower and upper triangular matrices which can be derived from Fig.3.4 then a suitable sparse preconditioning form for M given by Evans [3],[4] is,

$$M = (I-\omega L^E)(I-\omega U^E) \ . \tag{4.4}$$

b) Approximate factorisation of A^E

Alternatively, we can consider the factorisation of A^E into easily invertible lower and upper triangular matrix factors for which some standard methods are well known, i.e. LU, LL^T or LDU. However in the factorisation procedure of a sparse matrix, large numbers of 'fill-ins' occur (zero entries which are replaced often by small insignificant real numbers inherently incorporating a round-off error). Such numbers also greatly increase the computational work and storage requirements of the matrix. (Evans [5]).

Thus, in order to reduce the fill-in terms of the triangular factors L,U to a minimum, many researchers have considered approximate factorisation techniques of the form,

$$M = L_r U_r \approx A^E \quad , \tag{4.5}$$

where L_r, U_r denote the corresponding sparse triangular factors in which r non-zero off-diagonal 'fill-in' vectors have been retained.

Finally, the (Preconditioned)^2 Conjugate Gradient method (P^2CG) is specifically formulated to efficiently solve the algebraic system $A^Eu=b^E$ resulting from the above discretization. The P^2CG method is an iterative algorithm in which the following steps are computed at each iteration. For $k\geq 1$,

$$\left.\begin{aligned} r^{(n)} &= M^{-1}b-M^{-1}A^Eu^{(n)} \\ d^{(n)} &= -r^{(n)} \end{aligned}\right\} n=1$$

$$\begin{aligned} \sigma_n &= r^{(n=}.d^{(n)}/d^{(n)}.M^{-1}A^Ed^n \\ u^{(n+1)} &= u^{(n)}+\sigma_n d^{(n)} \\ r^{(n+1)} &= r^{(n)}-\sigma_n M^{-1}A^Ed^{(n)} \qquad\qquad n=n+1 \\ \tau^{n+1} &= -r^{(n+1)}M^{-1}A^Ed^{(n)}/d^{(n)}M^{-1}A^Ed^{(n)} \qquad (4.6) \\ d^{(n+1)} &= r^{(n+1)}+\tau_{n+1}d^{(n)} \end{aligned}$$

to be continued until $|u^{(n+1)}-u^{(n)}|<5\times 10^{-6}$, otherwise

The convergence of the P^2CG method is determined both by the clustering of the eigenvalues and the condition number of $M^{-1}A^E$, and thus critically depends on the selection of M in (4.4) and (4.5). Good preconditioners M are symmetric and positive definite and significantly reduce the condition number of the system, are less expensive in solving $Mr^{(n)}=(b-A^Eu^{(n)})$, rather than Au=b, and do not significantly increase the amount of storage relative to the storage needed to solve Au=b. By appropriately preconditioning the system we greatly reduce the amount of work expended in the computation of the solution u.

5. Convergence of the Preconditioned Conjugate Gradient Method. For the Conjugate Gradient method we have that,

$$||u-u^{(n)}||_A \leq ||p_n(u-u^{(0)})||_A \leq \max_j |p_n(\lambda_j)| \cdot ||u-u_0||_A \quad ,$$

$$\forall\ p_n \in \hat{P}_n \quad , \tag{5.1}$$

where $\hat{P}_n$ is the set of polynomials $p_n(z) = \sum_{j=0}^{n} \beta_j z^j$, $\beta_j \in R$ of degree at most n with $\beta_0=1$.

To estimate the reduction of the initial error $||u-u^{(0)}||$ after n steps it is sufficient to construct a polynomial p_n of degree at most n such that $p_n(0)=1$ and p_n is as small as possible on the interval $[\lambda_1,\lambda_M]$ containing the eigenvalues of A, i.e. so that the quantity, i.e. the convergence rate

$$\gamma_n = \max |p_n(z)| \quad , \quad z \in [\lambda_1,\lambda_M] \quad , \tag{5.2}$$

is as small as possible.

The best polynomial is well known in approximation theory and is the Chebyshev polynomial with the corresponding value of γ_n

$$\gamma_n = 2\left|\frac{\sqrt{\kappa(A)}-1}{\sqrt{\kappa(A)}+1}\right|^n \quad , \quad n=0,1,2. \tag{5.3}$$

Thus, for a given $\varepsilon>0$ to satisfy,

$$||u-u^{(n)}||_A \leq \varepsilon ||u-u^{(0)}||_A \quad ,$$

it is sufficient to choose n such that $\gamma_n \leq \varepsilon$ or

$$n > \tfrac{1}{2}\sqrt{\kappa(A)}\ \log 2/\varepsilon \quad , \tag{5.4}$$

where $\kappa(A) = \frac{\lambda_{max}}{\lambda_{min}}$ with $\lambda_{max} = \max_j(\lambda_j)$ and $\lambda_{min} = \min_j(\lambda_j)$ is the condition number.

We thus conclude that the required number of iterations for the conjugate gradient method is proportional to $\sqrt{\kappa(A)}$.

Since in a typical FD/FE application, we have $\kappa(A)=O(h^{-2})$ then the required number of iterations for the C.G. method would be $O(h^{-1})$.

Now, if $\rho(J)$ and $\rho(L)$ and $\rho(2B)$ denote the dominant eigenvalues of the point l line and 2×2 block Jacobi schemes respectively then,

$$\rho(J) = 1-\pi^2h^2/2, \quad \rho(L) = 1-\pi^2h^2, \quad \rho(2B) = 1-\pi^2h^2 \quad ,$$

confirming that the matrix A^E substituted for A in the C.G. schemes will bring about an improvement of $\sqrt{2}$ (Evans [2]). Numerical experiments in progress confirm these expectations.

6. Schwarz Alternating Procedure for Overlapping Domains. We now consider the Schwarz Alternating Procedure (SAP) to a domain Ω which is decomposed into 2 overlapping sub-domains Ω_1 and Ω_2 in which the values along the internal boundaries Γ_i, i=1,2 are extrapolated (over-relaxed) by an acceleration factor ω.

Consider then the 2-dimensional Dirichlet problem,

$$-\Delta u = f \text{ in } \Omega=\{(x,y) \mid 0<x<1,\ 0<y<1\}, \qquad (6.1)$$
$$u|_\Gamma = \phi$$

where f and ϕ are known. The domain Ω (Fig.6.1) is decomposed into two overlapping subdomains $\Omega_1=\{(x,y) \mid 0<x<x_k, 0<y<1\}$ and $\Omega_2=\{(x,y) \mid x_\ell<x<1, 0<y<1\}$, where $x_\ell<x_k$ and $x_k=x_\ell+d$.

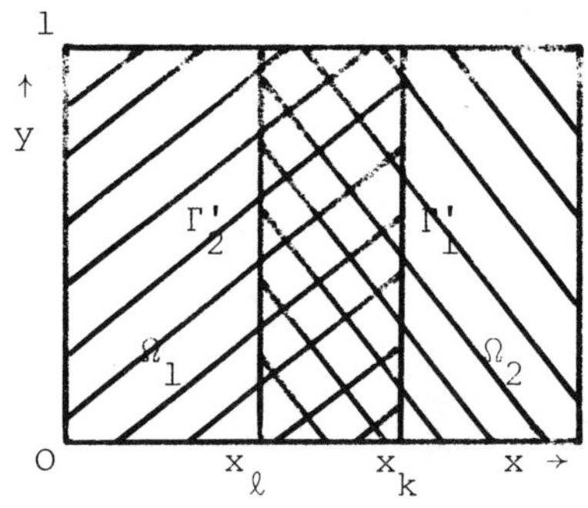

FIG.6.1

If we denote the boundaries of Ω_i by $\Gamma_i(\Gamma_i \subset \Gamma)$ and $\Gamma_i'(\Gamma_i' \subset \Omega_j)$, $(i\neq j)$, i=1,2, Γ_1' and Γ_2' are called the pseudo-boundaries along which we extrapolate.

Applying the Schwarz Alternating Procedure (SAP) with pseudo-boundary relaxation factor ω for solving problem (6.1), we obtain the following subproblems,

$$\left.\begin{aligned} -\Delta u^{(i+1)} &= f, \text{ in } \Omega_1 \\ u^{(i+1)} &= \phi, \text{ on } \Gamma_1 \\ u^{(i+1)} &= v^{(i-1)}+\omega(v^{(i)}-v^{(i-1)}) \text{ on } \Gamma_1' \end{aligned}\right\} \qquad (6.2)$$

$$\left.\begin{aligned} -\Delta v^{(i+1)} &= f, \text{ in } \Omega_2 , \\ v^{(i+1)} &= \phi, \text{ on } \Gamma_2 , \\ v^{(i+1)} &= u^{(i)}+\omega(u^{(i+1)}-u^{(i)}), \text{ on } \Gamma_2', \ i=0,1,2,\ldots , \end{aligned}\right\} \tag{6.3}$$

where, $v^{(-1)}(x_k,y)$, $v^{(0)}(x_k,y)$ and $u^{(0)}(x_\ell,y)$ are the initial guesses.

The errors $e^{(i)}(x,y)=u^*(x,y)-u^{(i)}(x,y)$ and $E^{(i)}(x,y)=u^*(x,y)-v^{(i)}(x,y)$ satisfy the following subproblems,

$$\left.\begin{aligned} -\Delta e^{(i+1)} &= 0 , \text{ in } \Omega_1 \\ e^{(i+1)} &= 0, \text{ on } \Gamma_1 , \\ e^{(i+1)} &= E^{(i-1)}+\omega(E^{(i)}-E^{(i-1)}), \text{ on } \Gamma_1' \end{aligned}\right\} \tag{6.4}$$

and,

$$\left.\begin{aligned} -\Delta E^{(i+1)} &= 0 , \text{ in } \Omega_2 , \\ E^{(i+1)} &= 0, \text{ on } \Gamma_2 , \\ E^{(i+1)} &= e^{(i)}+\omega(e^{(i+1)}-e^{(i)}), \text{ on } \Gamma_2', \ i=0,1,2,\ldots, \end{aligned}\right\} \tag{6.5}$$

where $E^{(-1)}(x_k,y)$, $E^{(0)}(x_k,y)$ and $e^{(0)}(x_\ell,y)$ are the initial errors.

If we assume,

$$E^{(i)}(x_k,y) = \sum_{m=1}^{\infty} b_m^{(i)} \sin m\pi y, \ i=-1,0. \tag{6.6}$$

and,

$$e^{(0)}(x_\ell,y) = \sum_{m=1}^{\infty} a_m^{(0)} \sin m\pi y , \tag{6.7}$$

then the solutions of problems (6.4) and (6.5) are given by,

$$e^{(i+1)}(x,y) = \sum_{m=1}^{\infty} a_m^{(i+1)} \gamma_m(x,x_k) \sin m\pi y , \tag{6.8}$$

and,

$$E^{(i+1)}(x,y) = \sum_{m=1}^{\infty} b_m^{(i+1)} \gamma_m(1-x,1-x_\ell) \sin m\pi y , \tag{6.9}$$

where,

$$\gamma_m(x,x_k) = \frac{\text{sh } m\pi x}{\text{sh } m\pi x_k} .$$

According to (6.8) and (6.9) and the pseudo-boundary conditions of problems (6.4) and (6.5) we have,

$$\begin{aligned} e^{(i+1)}(x_k,y) &= \sum_{m=1}^{\infty} a_m^{(i+1)} \sin m\pi y = (1-\omega)E^{(i-1)}(x_k,y)+\omega E^{(i)}(x_k,y) \\ &= \sum_{m=1}^{\infty} \gamma_m(1-x_k,1-x_\ell)[(1-\omega)b_m^{(i-1)}+\omega b_m^{(i)}] \sin m\pi y , \end{aligned}$$

i.e.,

$$a_m^{(i+1)} = \gamma_m(1-x_k,1-x_\ell)[(1-\omega)b_m^{(i-1)}+\omega b_m^{(i)}],\ i=0,1,2,\ldots, \tag{6.10}$$

and in the same way we get,

$$b_m^{(i+1)} = \gamma_m(x_\ell,x_k)[(1-\omega)a_m^{(i)}+\omega a_m^{(i+1)}],\ i=0,1,2,\ldots . \tag{6.11}$$

From (6.10) and (6.11) we have,

$$a_m^{(i+1)} = \rho_m(x_\ell,x_k)[\omega^2 a_m^{(i)}-2\omega(\omega-1)a_m^{(i-1)}+(\omega-1)^2 a_m^{(i-2)}],\quad i=2,3,4,\ldots, \tag{6.12}$$

and,

$$b_m^{(i+1)} = \rho_m(x_\ell,x_k)[\omega^2 b_m^{(i)}-2\omega(\omega-1)b_m^{(i-1)}+(\omega-1)^2 b_m^{(i-2)}],\quad i=1,2,3,\ldots, \tag{6.13}$$

where, for m=1,2,3,...

$$\rho_m(x_\ell,x_k) = \gamma_m(x_\ell,x_k).\gamma_m(1-x_k,1-x_\ell), \tag{6.14}$$

and, $a_m^{(0)}, b_m^{(-1)}$ and $b_m^{(0)}$ are known,

$$\begin{aligned} a_m^{(1)} &= \gamma_m(x_\ell,x_k)[(1-\omega)b_m^{(-1)}+\omega b_m^{(0)}], \\ b_m^{(1)} &= \gamma_m(1-x_k,1-x_\ell)[(1-\omega)a_m^{(0)}+\omega a_m^{(1)}], \\ a_m^{(2)} &= \gamma_m(x_\ell,x_k)[(1-\omega)b_m^{(0)}+\omega b_m^{(1)}]. \end{aligned} \tag{6.15}$$

Thus, from (6.8),(6.9),(6.12) and (6.13) we have,

$$\begin{aligned} ||e^{(i+1)}(x,y)||_{L_2}^2 &\leq ||e^{(i+1)}(x_k,y)||_{L_2}^2 \\ &= \int_0^1 [e^{(i+1)}(x_k,y)]^2 dy \\ &= \tfrac{1}{2}\sum_{m=1}^{\infty}[a_m^{(i+1)}]^2, \end{aligned} \tag{6.16}$$

$$\begin{aligned} &= \tfrac{1}{2}\sum_{m=1}^{\infty}\{\rho_m(x_\ell,x_k)[\omega^2 a_m^{(i)}-2\omega(\omega-1)a_m^{(i-1)} \\ &\qquad +(\omega-1)^2 a_m^{(i-2)}]\}^2 \\ &\leq \tfrac{1}{2}\sum_{m=1}^{\infty}\{\rho_1\omega^2|a_m^{(i)}-2a.a_m^{(i-1)}+a^2.a_m^{(i-2)}|\}^2, \end{aligned} \tag{6.17}$$

and,

$$||E^{(i+1)}(x,y)||_{L_2}^2 \leq \tfrac{1}{2}\sum_{m=1}^{\infty}\{\rho_1\omega^2|b_m^{(i)}-2ab_m^{(i-1)}+a^2b_m^{(i-2)}|\}^2, \tag{6.18}$$

where $a=(\omega-1)/\omega$, and,

$$\rho_1 = \rho_1(x_\ell,x_k) = \left(\frac{\text{sh}\ \pi x_\ell}{\text{sh}\ \pi x_k}\right)\left(\frac{\text{sh}\ \pi(1-x_k)}{\text{sh}\ \pi(1-x_\ell)}\right). \tag{6.19}$$

If we let $a_m^{(i)}=r^i$ (or $b_m^{(i)}=r^i$), then the characteristic equation of (6.12) (or (6.13)) is obtained in the form,

$$r^3-\rho_m(x_\ell,x_k)\omega^2(r-a)^2 = 0 \; . \qquad (6.20)$$

From (6.16) and (6.17) we can see that the convergence-rate is determined by the low-frequency component (see Evans, [6],[7]). So we define the convergence factor of procedures (6.2) and (6.3) as,

$$r(\rho,\omega) = \max_{1\leq j\leq 3} |r_j| \; , \qquad (6.21)$$

where r_1,r_2,r_3 are the roots of the cubic equation,

$$r^3-\rho_1\omega^2(r-a)^2 = 0 \; , \; 0\leq\rho_1<1 \; . \qquad (6.22)$$

By making the substitution,

$$r = (R+\frac{\rho\omega^2}{3}) \; , \qquad (6.23)$$

the reduced cubic equation is obtained in the form,

$$R^3+R(2a - \tfrac{1}{3}\rho\omega^2)\rho\omega^2+(\tfrac{2}{3}\rho\omega^2 a- \tfrac{2}{27}\rho^2\omega^4-a^2)\rho\omega^2 = 0 \; , \qquad (6.24)$$

or

$$R^3+bR+c = 0 \; , \qquad (6.25)$$

with,

$$b = \rho\omega^2(2a - \tfrac{1}{3}\rho\omega^2) \; , \qquad (6.26)$$

$$c = \rho\omega^2(\tfrac{2}{3}\rho\omega^2 a- \tfrac{2}{27}\rho^2\omega^4-a^2) \; ,$$

where $a=(\omega-1)/\omega$, and ρ_1=convergence factor at $\omega=1$.

The maximum rate of convergence of (6.22) is given by the value of ω which allows the discriminant of (6.25) to be zero, i.e.,

$$c^2/4 + b^3/27 = 0 \; . \qquad (6.27)$$

By substituting the values of b and c of (6.26) into (6.27), we obtain the expression which when simplified yields the relation,

$$a^3(27a-4\rho\omega^2) = 0 \; . \qquad (6.28)$$

Thus, the optimum value of ω, i.e. ω_0, which yields the maximum rate of convergence is given by the nonlinear relation $\omega^2=27a/4$ or the cubic equation,

$$\omega^3 - \frac{27}{4}\omega+ \frac{27}{4\rho} = 0 \; , \qquad (6.29)$$

which is already in its reduced form.

Now for $\omega>1$ the discriminant of (6.29) is ≤ 0 so the solution of (6.29) is given by the trigonometic relations,

$$\omega_1 = 2\sqrt{-\beta/3}\,\cos(s/3) \; ,$$

$$\omega_2 = 2\sqrt{-\beta/3}\,\cos(s/3+2\pi/3) \; , \qquad (6.30)$$

and

$$\omega_3 = 2\sqrt{-\beta/3}\,\cos(s/3+4\pi/3) \; ,$$

with,

$$s = \arc\cos(\beta/2\sqrt{-\beta^3/27}) \;, \tag{6.31}$$

where $\beta=-27/4$.

The optimum value of ω which maximises the convergence rate ($1<\omega<2$) is then given by,

$$\omega_o = 2\sqrt{-\beta/3}\cos(s/3+4\pi/3) \;, \tag{6.32}$$

or

$$\omega_o = (3/\sqrt{\rho})\cos[(s+4\pi)/3] \;, \tag{6.33}$$

with $s=\arc\cos(-\sqrt{\rho})$.

The optimum convergence factor ρ_o at the optimum value ω_o is determined from the solution of (6.25) at which the discriminant $c^2/4+b^3/27=0$.

The roots of (6.25) are given by,

$$\begin{aligned} R_1 &= A+B \\ R_2 &= -(A+B)/2+0.5(A-B)\sqrt{3}i \;, \\ R_3 &= -(A+B)/2-0.5(A-B)\sqrt{3}i \;, \end{aligned} \tag{6.34}$$

where

$$A = 3\sqrt{-c/2+\sqrt{c^2/4+b^3/27}} \;,$$

$$B = 3\sqrt{-c/2-\sqrt{c^2/4+b^3/27}} \;.$$

Also, at the optimum value ω_o, relation (6.28) holds from which we can establish that,

$$A_o-B_o = -3a_o/4 \;, \tag{6.35}$$

where $a_o=(\omega_o-1)/\omega_o$.

Thus the optimum factor ρ_o is given by the solution of (6.22) which is related to the solution of (6.25) by the transformation (6.23) and is,

$$\rho_o = 3a_o/4 + \rho\omega_o^2/3 \;, \tag{6.36}$$

and by virtue of (6.28) we have,

$$\rho_o = 3a_o = 3(\omega_o-1)/\omega_o \;.$$

Finally, the range of convergence of the Schwarz alternating method with overrelaxation factor ω on the pseudo boundaries can be determined as follows.

From an examination of the sign of the discriminant $(c^2/4+b^3/27)$ of (6.25) it can be established that at the ends of the convergence range, there exists one real root and two complex conjugate roots of (6.23). Thus, if we denote the roots of (6.23) as $\alpha_1, \alpha_2 \pm i\gamma_2$ and using the theory of equations, the following relations can be established,

$$\begin{aligned} \alpha_1 + 2\alpha_2 &= \rho\omega^2 \\ 2\alpha_1\alpha_2+(\alpha_2^2+\gamma_2^2) &= 2a\rho\omega^2 \\ \alpha_1(\alpha_2^2+\gamma_2^2) &= a^2\rho\omega^2 \ . \end{aligned} \tag{6.37}$$

Now the range of convergence is determined by the roots $|\alpha_2 \pm i\gamma_2|=1$ and consequently the relations (6.37) become,

$$\begin{aligned} \alpha_1 + 2\alpha_2 &= \rho\omega^2 \\ 2\alpha_1\alpha_2+1 &= 2a\rho\omega^2 \\ \alpha_1 &= a^2\rho\omega^2 \ , \end{aligned} \tag{6.38}$$

to yield,

$$(2a\rho\omega^2-1)/a^2\rho\omega^2 = \rho\omega^2(1-a) \quad , \tag{6.39}$$

or

$$\rho\omega^2 = 1/[a(1+a)] \ .$$

By the substitution of $a=(\omega-1)/\omega$, this relation simplifies to the quadratic equation,

$$(\omega-1)(2\omega-1) = 1/\rho \quad , \tag{6.40}$$

for the determination of the values $\omega=\omega_f$ for which the Schwarz alternating method is convergent and is,

$$\omega_f = 2(1-1/\rho)/(3\pm\sqrt{9-8(1-1/\rho)}) \ . \tag{6.41}$$

Convergence factor ρ_1 at $\omega=1$	Optimal overrelaxation parameter ω	Optimal convergence factor $\rho_0=\gamma(\rho,\omega_0)$	Convergence range	
			Lower ω_f	Upper ω_f
0.6	1.127	0.339	-0.196	1.696
0.65	1.144	0.378	-0.162	1.662
0.7	1.163	0.421	-0.131	1.631
0.75	1.185	0.468	-0.104	1.604
0.8	1.210	0.520	-0.079	1.579
0.85	1.240	0.581	-0.057	1.557
0.9	1.279	0.654	-0.036	1.536
0.95	1.334	0.752	-0.017	1.517
0.98	1.389	0.841	-0.007	1.507
0.99	1.419	0.887	-0.003	1.503

TABLE 6.1

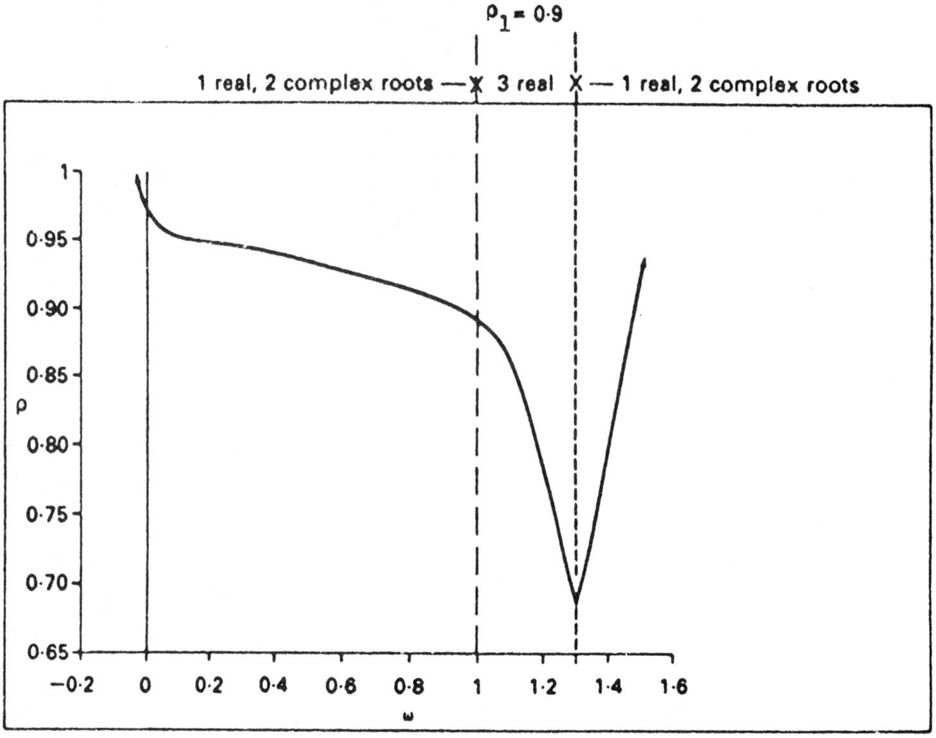

FIG. 6.2

Some typical values of the optimal overrelaxation factor ω_0, the maximum convergence factor ρ_0 and the convergence range of the Schwarz alternating method for various values of ρ are given in Table 6.1. Also, a plot of ρ against ω is given in Figure 6.2 for ρ_1=0.9 at ω=1.

REFERENCES

(1) R.S. VARGA, Matrix Iterative Analysis, Prentice Hall, Englewood Cliffs, N.J., 1963.

(2) D.J. EVANS, Implicit Block Explicit Overrelaxation Schemes, in Elliptic Problem Solvers II, edit. G. Birkhoff & A. Schoenstadt, Acad.Press, 1984, pp.285-300.

(3) D.J. EVANS, The Use of Preconditioning in Iterative Methods for Solving Linear Equations with Symmetric Positive Definite Matrices, J.I.M.A., 4, (1968), pp.295-314.

(4) D.J. EVANS, The Analysis and Application of Sparse Matrix Algorithms in the Finite Element Method, in MAFELAP 72, edit. J.R. Whiteman, Acad. Press, 1973, pp.427-447.

(5) D.J. EVANS, On Preconditioning Iterative Methods for Elliptic Partial Differential Equations, in Elliptic Problem Solvers I, edit. Martin H. Schultz, Acad.Press, 1981, pp.261-269.

(6) D.J. EVANS, JIANPING SHAO, LISHAN KANG and YUPING CHEN, The Convergence Rate of the Schwarz Alternating Procedure (II) - for 2 Dimensional Problems, Int.Jour.Comp.Math., 20 (1986), pp.325-339.

(7) D.J. EVANS, LISHAN KANG, YUPING CHEN and JIANPING SHAO, The Convergence Rate of the Schwarz Alternating Procedure (VI) - With Pseudo Boundary Relaxation Factor, Int.J.Comp.Math., 21 (1987), pp. 185-203.

CHAPTER 16

Domain Decomposition Techniques for Efficient Adaptive Local Grid Refinement

Richard E. Ewing*

ABSTRACT

In the multidimensional numerical simulation of certain multiphase fluid flow processes, many phenomena are sufficiently localized and transient that self-adaptive local grid refinement techniques are necessary to resolve the local physical behavior. For large-scale simulation problems, efficiency is the key to the choice of specific adaptive strategies. Purely local refinement techniques require complex data tree structures and associated specialized solution techniques. Although these tree structures of changing length are amenable to parallel computation, they are very difficult to use efficiently in a vector mode. Techniques which involve a relatively coarse macro-mesh with potential local refinement in each separate mesh will be discussed. The macro-mesh will be the basis for domain decomposition techniques and parallel solution algorithms. Uniform meshing in the subdomains will allow efficient vectorization as well as parallelization of the algorithms. Similarly, different solution processes can be applied to different sub-domains. A preconditioner, based upon the domain-decomposition techniques of Bramble, Pasciak, and Schatz is utilized to efficiently solve the combination domain-decomposition and local grid refinement problem. Techniques for applying this concept to resolve sharp, moving fluid interfaces in large-scale simulation problems will be discussed. Extensions to local time-stepping will also be presented.

1. INTRODUCTION

Domain decomposition and adaptive local grid refinement techniques possess enormous potential for local accuracy improvements in many large-scale problems. In order to illustrate this potential, we will discuss grid refinement

*Departments of Mathematics, Petroleum Engineering, and Chemical Engineering, University of Wyoming, Laramie, Wyoming 82071.

techniques in the context of large-scale simulation of petroleum recovery applications. We will also present ways to easily incorporate local grid refinement capabilities in large, existing codes.

The objective of reservoir simulation is to understand the complex chemical, physical and fluid flow processes occurring in a petroleum reservoir sufficiently well to be able to optimize the recovery of hydrocarbon. To do this, one must build mathematical and computational models capable of predicting the performance of the reservoir under various exploitation schemes. Many of the chemical and physical phenomena which govern enhanced recovery processes have extremely important local character. Therefore the models used to simulate these processes must be capable of resolving these critical local features.

Mathematical models of enhanced recovery processes involve coupled systems of nonlinear partial differential equations. In order to compare the results of these models with physical measurements to assess their validity and to make decisions based on these models, the partial differential equations must be discretized and solved on computers. Field scale hydrocarbon simulations normally involve reservoirs of such great size that uniform gridding on the length scale of the local phenomena would involve systems of discrete equations of such enormous size as to make solution on even the largest computers prohibitive. Therefore local grid refinement capabilities are becoming more important in reservoir simulation as the enhanced recovery procedures being used become more complex and involve more localized phenomena.

There are two distinct classes of local grid refinement techniques—fixed and dynamic. For problems with fixed wells, faults, pinchouts and large fractures, certain fixed local refinements have proven to be very effective. Dynamic and adaptive grid refinement to follow moving fluid interfaces is much more complex. Techniques which work well for fixed refinement can involve a data structure which is so complex that it can be very inefficient for dynamic applications. In this paper we present methods that can be applied to both fixed and dynamic refinement problems in an efficient and accurate manner. We present methods which are accurate discretizations on composite grids and are relatively easy to implement in existing simulators for various applications.

2. MODEL EQUATIONS FOR POROUS MEDIA FLOW

The miscible displacement of one incompressible fluid by another, completely miscible with the first, in a horizontal reservoir $\Omega \subset \mathbb{R}^2$ over a time period $J = [T_0, T_1]$, is given by

$$-\nabla \cdot \left(\frac{k}{\mu}\nabla p\right) \equiv \nabla \cdot \mathbf{u} = q, \quad x \in \Omega, \quad t \in J, \tag{2.1}$$

$$\phi\frac{\partial c}{\partial t} - \nabla \cdot (\mathbf{D}\nabla c - \mathbf{u}c) = q\tilde{c}, \quad x \in \Omega, \quad t \in J, \tag{2.2}$$

where p and $\mathbf{u}$ are the pressure and Darcy velocity of the fluid mixture, ϕ and k are the porosity and the permeability of the medium, μ is the concentration-dependent viscosity of the mixture, $\mathbf{D}$ is a diffusion/dispersion tensor, c is the concentration of the invading fluid, q is the external rate of flow, and $\tilde{c}$ is the inlet or outlet concentration. In addition to Equations (2.1) and (2.2), initial and flow boundary conditions are specified. The flow at injection and production wells is modeled in Equations (2.1) and (2.2) via point sources and sinks.

The equations describing two phase, immiscible, incompressible displacement in a horizontal porous medium are given by

$$\phi \frac{\partial S_w}{\partial t} - \nabla \cdot \left(k \frac{k_{rw}}{\mu_w} \nabla p_w \right) = q_w, \quad x \in \Omega, \quad t \in J, \tag{2.3}$$

$$\phi \frac{\partial S_o}{\partial t} - \nabla \cdot \left(k \frac{k_{ro}}{\mu_o} \nabla p_o \right) = q_o, \quad x \in \Omega, \quad t \in J, \tag{2.4}$$

where the subscripts w and o refer to water and oil respectively, S_i is the saturation, p_i is the pressure, k_{ri} is the relative permeability, μ_i is the viscosity, and q_i is the external flow rate, each with respect to the ith phase. The pressure between the two phases is described by the capillary pressure

$$p_c(S) = p_o - p_w. \tag{2.5}$$

Note that $\frac{dp_c}{dS} \leq 0$.

Although formally, the equations presented in (2.1) and (2.2) seem quite different from those in (2.3) and (2.4), the latter system may be rearranged in a form which very closely resembles the former system. In order to use the same basic simulator in our sample computations to treat both miscible and immiscible displacement, we define variables for total fluid pressure and Darcy velocity (see [7]). We combine Equations (2.3) and (2.4) to obtain

$$-\nabla \cdot (k\lambda(S)\nabla p) = q_w + q_o = q_t, \tag{2.6}$$

$$v_t = -k\lambda(S)\nabla p, \tag{2.7}$$

$$\phi \frac{\partial S}{\partial t} + \nabla \cdot \left(k\lambda(S)\overline{\lambda}_o \frac{dp_c}{dS} \nabla S \right) + \nabla \cdot (\overline{\lambda}_w v_t) = q_w, \tag{2.8}$$

where S is the water saturation. λ, λ_0 and λ_w, the mobilities of the total fluid and the oil and gas phases, respectively, are defined by ratios of the relative permeabilities and phase viscosities (see [13]).

The equations presented above describe both miscible and immiscible flow in porous media. They can be used to simulate various production strategies in an attempt to understand and hopefully optimize hydrocarbon recovery. However, in order to use these equations effectively, parameters that describe the rock and fluid properties for the particular reservoir application must be input. Since these rock and fluid properties cannot be measured directly in situ, they must be determined via history matching and reservoir characterization techniques (see [12]). The heterogeneities in the reservoir, which can be very localized often dominate the flow process and may require local grid refinement for adequate resolution [16].

A typical example of a fixed localized phenomenon which requires special treatment in simulation is fluid flow in the neighborhood of wells. If fluid flow rates are specified at injection or production wells, the use of Dirac delta functions as point sources and sinks in the mathematical equations has been shown to be a good model for well-flow behavior beyond some minimal distance away from the wells. In this case, the pressure (which determines the flow) grows like $\ln r$ where r is the distance to that well. A different well model, involving specification of a bottom hole pressure as a boundary condition, also gives rise to a logarithmic growth in pressure up to a finite specified pressure. Because of the rapidly changing behavior of the pressure in the vicinity of wells, accurate pressure approximations require local grid refinement.

We will consider a simple example problem to illustrate our local refinement techniques. From Equation (2.1) or Equation (2.6), the pressure p of a fluid in a horizontal reservoir $\Omega \subset \mathbb{R}^2$ satisfies

$$-\nabla \cdot \frac{k}{\mu} \nabla p = q \quad \text{in} \quad \Omega. \tag{2.9}$$

Assuming no flow boundary conditions, we have

$$\frac{k}{\mu} \frac{\partial p}{\partial \nu} = 0 \quad \text{on} \quad \partial\Omega, \tag{2.10}$$

where $\frac{\partial}{\partial \nu}$ is the outward normal on $\partial\Omega$. For the existence of p we assume that the mean value of q is zero and for uniqueness we impose that p have mean value zero. If fluid flow rates at injection and production wells are specified via Dirac delta functions at the N_w wells x_i with associated flow rates q_i, then

$$q = \sum_{i=1}^{N_w} \delta(x - x_i) q_i \ . \tag{2.11}$$

Several techniques which assume radial flow near the well have been used to obtain local properties of p. One such technique involves subtracting out the singular behavior of p around the wells [8,13,17,20,21]. A radial flow assumption is probably not bad around injection wells, but may be inadequate for production wells where different techniques, such as local grid refinement, are often needed. It has been shown [29] that appropriate local grid refinement around these singularities can greatly increase the accuracy throughout the reservoir.

Two different types of self-adaptive grid refinement have been applied in reservoir simulation. The first technique is a truly local grid refinement where an arbitrary level of refinement can be applied at any region in space. Certain SPE references [9,10,22,28–30] utilize this type of refinement. This technique requires a special data structure for effective matrix set-up as well as special algorithms for efficient solution. A data structure has been developed [9,10,14,15] that will support truly local refinement and dynamic "unrefinement" in both space and time. The special tree structure allows truly local grid refinement and is implemented via an efficient multi-linked list. The dynamic multi-linked list representation efficiently allows both placement and removal of local meshes. A local grid analysis triggers the dynamic changes in the trees for adaptivity.

The data structures have proven to be very effective for elliptic or time-independent partial differential equations, and for fixed refinement applications.

However, the complexity of the data structures and the associated solution processes make many of the truly local refinement procedures inefficient for large-scale, time-dependent problems. If different grids are used for each time-step in a large problem, the overhead associated with the data structures and the grid generation can easily dominate the overall computation times. For this reason, alternate techniques which do not require complex data structures or regeneration of the grid at each time-step are desirable.

A technique termed patch refinement [1,4,11] is an attractive alternative to truly local refinement. This method does not require as complex a data structure but does involve ideas of passing information from one grid to another. The idea of a local patch refinement method is to pick a patch that includes most of the critical behavior requiring better resolution, and use a special, possibly uniform, refinement within this patch. If a uniform fine grid is utilized in the patch, very fast solvers, perhaps utilizing vector-based algorithms, can be applied locally in this region using boundary data from the original coarse grid.

The local patch refinement techniques [4,19,24,26,27] have proven to be very effective for obtaining local resolution around fixed singular points such as wells in a reservoir. We will discuss the patch approximation technique first in the context of local refinement around a fixed point or region like a well and then extend the concept to dynamic problems.

We have developed fast solution methods for the approximation of problems requiring mesh refinement. These techniques are related to various domain decomposition methods [2–6,19,23–25]. High accuracy throughout the computational region is obtained by incorporating local refinements around wells. A composite grid is obtained by superimposing these refinements on a quasi uniform grid on the original domain. Previous techniques usually had no systematic way of dealing with such questions as interface interpolation, mass conservation, and degree of grid overlap. They also usually involve the solution of the coarse grid problems with the regions corresponding to the refinement removed. This destroys the banded structure and ease of vectorization of the coarse grid regions.

In the methods to be discussed below, the problem is formulated with a composite operator on the composite grid. The techniques are iterative procedures which drive the residual of this composite grid operator to zero. Composite grid operators for finite element discretization are common and relatively easy to describe and analyze. Examples of accurate finite difference based composite grid operators for variable coefficient problems are presented in [18]. Complete error analyses for these difference stars will appear elsewhere. A new domain decomposition variant is presented to efficiently solve the resulting matrix equations. This involves the development of a preconditioner. This preconditioner is novel in that the task of computing its inverse applied to a vector reduces to the solving of separate matrix systems for the local refinements and the matrix system for the quasi uniform grid on the original domain. Note that this quasi uniform grid overlaps the regions of local refinement and its corresponding matrix problem remains invariant when local refinements are dynamically added or removed. This local refinement technique can be incorporated in existing reservoir codes without extensive modification. Furthermore, if the nodes on the quasi uniform grid are chosen in a regular pattern, highly vectorizable algorithms for the solution of the corresponding matrix system can be developed.

In the next section, we describe our technique which allows ease of implementation in existing codes. First a composite grid, as well as a corresponding composite grid operator, is formed from a coarse uniform grid with superimposed refinements in subregions denoted collectively by Ω_2. The coarse grid remains in the region $\Omega_1 = \Omega/\Omega_2$. The discretization in this example is given via finite element techniques. However, by considering the matrix structure of this algorithm, we can see how the techniques can be extended to finite differences or other spatial discretizations.

3. LOCAL GRID REFINEMENT STRATEGY FOR FIXED POINTS

Multiplying (2.9) by an arbitrary (sufficiently regular) function ϕ, integrating by parts and using (2.10), we see that the solution p satisfies

$$A(p, \phi) = (f, \phi) \tag{3.1}$$

where

$$A(u, v) = \int_\Omega \frac{k}{\mu} \nabla u \cdot \nabla v dx$$

and

$$(u, v) = \int_\Omega uv \; dx \; .$$

The Galerkin approximation to (3.1) is to find a function P in a suitable finite dimensional subspace M_h of the Sobolev space $H^1(\Omega)$ such that

$$A(P, \phi) = (f, \phi), \quad \text{for all} \quad \phi \in M_h. \tag{3.2}$$

Since the bilinear form $A(\cdot, \cdot)$ corresponds to the composite operator, (3.2) is, in general, difficult to solve for P. Instead we will use a preconditioned iterative method to obtain P. We must then find a comparable form $B(\cdot, \cdot)$ such that, given g, the problem of finding $W \in M_h$ satisfying

$$B(W, \phi) = (g, \phi), \quad \text{for all} \quad \phi \in M_h. \tag{3.3}$$

is relatively easy.

As was described in [4], the problem of calculating the action of the inverse of the preconditioner essentially reduces to the solution of discrete mixed problems on the refined subgrids, and discrete Neumann problems on the original grid. Due to the regularity of the mesh geometry, such problems are generally easier to solve than the system resulting from the composite grid discretization.

We first split the bilinear form into parts $A(u, v) = A_1(u, v) + A_2(u, v)$, where

$$A_i(u, v) = \int_{\Omega_i} \frac{k}{\mu} \nabla u \cdot \nabla v \; dx \; . \tag{3.4}$$

Then we decompose any $V \in M_h(\Omega)$ as follows: $V = V_p + V_r$ where V_p equals V on Ω_1, $V_p \in M_h(\Omega_2)$ on Ω_2 and $V_r \in M_h(\Omega)$ satisfies

$$A(V_r, \phi) = 0 \ \text{ for all } \ \phi \in M_h(\Omega_2). \tag{3.5}$$

Then, as in [4], we see that for $V \in M_h(\Omega)$,

$$A(V,V) = A_1(V,V) + A_2(V_p,V_p) + A_2(V_r,V_r) \ . \tag{3.6}$$

The action of the inverse of (3.6) is not easy to obtain. However, by replacing $A(V,V)$ by

$$B(V,V) = A_1(V,V) + A_2(V_p,V_p) + A_2(V_c,V_c) \ , \tag{3.7}$$

where V_c is determined by the original coarse grid $M_h^c(\Omega)$ and satisfies $V_c = V$ in Ω, and

$$A_2(V_c, \phi) = 0 \ \text{ for all } \ \phi \in M_h^c(\Omega_2), \tag{3.8}$$

then the action of the inverse of (3.7) is relatively easy to obtain and the form $B(\cdot,\cdot)$ is comparable to the form $A(\cdot,\cdot)$ with comparability constants independent of the grid size h [5–6]. As described in [4], the following algorithm suffices for solving

$$B(W, \phi) = (g, \phi) \ \text{ for all } \ \phi \in M_h \ , \tag{3.9}$$

given g.

Algorithm For Computing W [4]:

1. Find U_p by solving mixed problems on the regions Ω_2.
2. Pass the local information to the right hand side of the original problem and compute any solution U_c of the coarse grid problem

$$A(U_c, \phi) = (g, \overline{\phi}) - A_2(U_p, \overline{\phi}) \ \text{ for all } \ \phi \in M_h^c \tag{3.10}$$

   where $\overline{\phi}$ is any function in M_h which equals ϕ on Ω_1.

3. Find U_r on Ω_2 by computing the discrete harmonic extension with respect to the refinement subspaces.
4. Compute $\overline{U}$, the mean value of $U = U_p + U_r$. Set $W = U - \overline{U}$.

Matrix Form of the Algorithm:

First, we consider the matrix A^c, generated by the finite element approximation of the equations (2.9)–(2.11) using a coarse quasi-uniform mesh. Let the solution P of the original coarse grid problem be decomposed in the form $P = (P_1, P_b, P_2)^T$, where P_1, P_2, and P_b are the parts of the coarse grid solution in Ω_1, Ω_2, and the intersection of the boundary of Ω_1 and Ω_2, respectively. The corresponding decomposition of the matrix A^c can be described in

$$A^c \begin{pmatrix} P_1 \\ P_b \\ P_2 \end{pmatrix} = \begin{pmatrix} A_{11}^c & A_{1b}^c & 0 \\ A_{b1}^c & A_{bb}^c & A_{b2}^c \\ 0 & A_{2b}^c & A_{22}^c \end{pmatrix} \begin{pmatrix} P_1 \\ P_b \\ P_2 \end{pmatrix} . \tag{3.11}$$

We assume that a code exists or can be easily written to solve (3.11) for a quasi-uniform grid which can take advantage of the banded structure of the matrix $\overline{A}^c$ which is equivalent to A^c except utilizing a standard lexicographical ordering of the unknowns.

Next assume that due to some localized process, grid refinement is desired in Ω_2. Let P_r be the new approximation on the refined grid in Ω_2 and A_{rr} be the local matrix on Ω_2. Let A_{br} and A_{rb} be the new connection matrices between the interface between Ω_1 and Ω_2 and the refined grid on Ω_2. Then, in order to maintain the sparsity of the composite grid matrix and a simple data structure obtained by concatonating P_r to P, we can write the composite matrix problem in the form

$$\tilde{A}\tilde{P} = \begin{pmatrix} A_{11}^c & A_{1b}^c & 0 & 0 \\ A_{b1}^c & A_{bb}^c & 0 & A_{br} \\ 0 & 0 & I & 0 \\ 0 & A_{rb} & 0 & A_{rr} \end{pmatrix} \begin{pmatrix} P_1 \\ P_b \\ P_2 \\ P_r \end{pmatrix} = \begin{pmatrix} f_1 \\ f_2 \\ 0 \\ f_3 \end{pmatrix}. \tag{3.12}$$

We note that the I on the diagonal of (3.12) and the zeroes in the corresponding row, column and right hand side enforce the removal of P_2 from the system without destroying the relationship of

$$\begin{pmatrix} A_{11}^c & A_{1b}^c \\ A_{b1}^c & A_{bb}^c \end{pmatrix} \tag{3.13}$$

to A^c and hence $\overline{A}^c$.

As an initial guess for P_r, denoted P_r^0, we solve the local problem on Ω_2 with zero Dirichlet conditions on the interface between Ω_1 and Ω_2 (equivalent to setting $P_b^0 = 0$):

$$P_r^0 = A_{rr}^{-1} f_3. \tag{3.14}$$

This problem can be solved exactly or approximately by some iterative technique. This step could be considered as the first part of a block Gauss-Seidel iterative procedure for the solution of (3.11). The next step would be to use the approximation for P_r^0 and then invert the block (3.13) to obtain an approximation for P_1 and P_b. Since this block involves a complex region and may not be well-conditioned, we use an alternate solution method which involves a preconditioner, denoted by B, for the composite matrix $\tilde{A}$.

Using B, we define, for each iterate n, and an iteration parameter τ

$$\tilde{P}^{n+1} = \tilde{P}^n + \tau B^{-1}(\tilde{f} - \tilde{A}\tilde{P}^n). \tag{3.15}$$

Let Q be the residual vector given by

$$\tilde{f} - \tilde{A}\tilde{P}^n = \begin{pmatrix} f_1 - A_{11}^c P_1^n - A_{1b}^c P_b^n \\ f_2 - A_{b1}^c P_1^n - A_{bb}^c P_b^n - A_{br} P_r^n \\ 0 \\ f_3 - A_{rb} P_b^n - A_{rr} P_r^n \end{pmatrix} \equiv \begin{pmatrix} Q_1^n \\ Q_2^n \\ 0 \\ Q_4^n \end{pmatrix}. \tag{3.16}$$

Next we solve the original coarse grid problem

$$A^c \begin{pmatrix} W_1^{n+1} \\ W_b^{n+1} \\ W_2^{n+1} \end{pmatrix} = \begin{pmatrix} Q_1^n \\ Q_2^n - A_{br}A_{rr}^{-1}Q_4 \\ 0 \end{pmatrix} \tag{3.17}$$

(or its rearranged equivalent problem using $\overline{A}^c$ to take advantage of banding of $\overline{A}^c$) for W_1^{n+1} and W_b^{n+1}. We have in essence inverted (3.13) in an efficient and vectorizable manner. Then using W_b^{n+1}, we complete the block Gauss-Seidel analogy on (3.12) and obtain W_r^{n+1} by solving

$$A_{rr}W_r^{n+1} = Q_4 - A_{rb}W_b^{n+1}. \tag{3.18}$$

Finally, from (3.15), we set

$$\tilde{P}^{n+1} = \begin{pmatrix} P_1^n \\ P_b^n \\ 0 \\ P_r^n \end{pmatrix} + \tau \begin{pmatrix} W_1^n \\ W_b^n \\ 0 \\ W_r^n \end{pmatrix}.$$

Since this algorithm only requires two separate solutions of mixed problems on the subregions (each subregion problem possibly being solved via a different parallel processor) and one solution on the original, uniform coarse grid, it is relatively easy to perform. Similarly, no complex data structure is required and the algorithm can be implemented in existing large-scale codes without severely disrupting the solution process. Promising numerical results for the algorithm appeared in [4]. These results will be extended to more general reservoir simulation problems in a forthcoming paper in the Tenth SPE Symposium on Reservoir Simulation by Ewing, Boyett, Babu, and Heinemann.

As stated, the algorithm in its most general form involves two separate solutions on the subregions at each step. This comes from the desire to have a symmetric preconditioner from the form $B(\cdot,\cdot)$. As is mentioned in [23], the FAC algorithm [19,23,24] involves only one subregion solution per iteration. See [23] for a comparison of FAC and this algorithm [4] and their theories.

By considering the domain decomposition techniques presented in [5] which led to this algorithm, we can see that if the subregion problems ((3.14) and its sequels with updated guesses for P_b) are solved exactly, then Q_4 in (3.16) and (3.17) is identically zero and the action of the preconditioner is symmetric with only one subregion solution per iteration (from (3.18)). Preliminary computations indicate that if the subregion problem is solved iteratively with its own preconditioner, the full algorithm with two subregions solved will converge faster for some problems. Iterative solution of the unrefined region causes no difficulty with either version of the algorithm. This is an important consideration for the reservoir simulation applications when iterative solution of the unrefined problem is essential due to their size, while direct solution of the refined region problems is usually possible.

4. FINITE DIFFERENCE METHODS

In practical reservoir simulation applications, the finite difference system of equations for the unknown values of the fluid pressure at the grid points will be nonlinear and will be coupled to other nonlinear partial differential equations like (2.2) and (2.8). Using a Newton method, we need to solve, for each iteration, a system with the associated Jacobian or linearization matrix. If the problem is linear, then the Jacobian matrix is the matrix of the finite difference scheme itself. Here we would like to explain how the local refinement technique can be naturally incorporated with the philosophy of the already existing codes for reservoir simulation and easily implemented without destroying their data structure.

We first consider the case for fixed local spatial refinement and a single partial differential equation. We write the approximation to the given equation and boundary conditions for each time step on the whole grid (coarse and fine) as a system of equations for the unknown values of the pressure Φ at the grid points at that time step with $\Phi^n(x) = \Phi(x, t^n)$. Consider the system

$$L\Phi^n = f, \tag{4.1}$$

where L is a positive definite Jacobian matrix, possibly different for each time step, resulting from the finite difference approximation described above, and f represents nonhomogeneous, initial, and boundary terms. The non-zero structure of the matrix L for most of the local grid refinement structures appearing in the SPE literature has been presented explicitly [9,22,29]. For these orderings of the unknowns, the band structure and corresponding efficiency of solution is lost. The solution algorithms presented in these papers have, in general, lost the potential for vectorization which can be so beneficial for the enormous problems encountered. We shall present solution ideas which maintain vectorization benefits and add important parallel capabilities which will be important for the emerging parallel computer architectures.

The local refinement approximations in all cases considered here and in [18] actually replace a coarse grid cell by a group of refined cells. Then the construction of the finite difference scheme may be considered to be created in the following manner: (a) construct within a global strategy of the existing code the finite difference approximation on the coarse grid; (b) treat all cells where the local refinement is introduced as dead cells by enforcing zero pressure values in these cells—the resulting matrix will be L_{CC}; (c) add to the system new equations approximating the problem on the fine grid. Let the matrix L_{FF} describe the regular difference stencil on the interior of the refined regions and the matrix L_{FC} describe the non-standard connections between coarse and refined grid cells described in [18].

This procedure can then be described in a matrix form as

$$\begin{pmatrix} L_{CC} & L_{CF} \\ L_{FC} & L_{FF} \end{pmatrix} \begin{pmatrix} \Phi_L \\ \Phi_F \end{pmatrix} = \begin{pmatrix} f_L \\ f_F \end{pmatrix} \tag{4.2}$$

where Φ_L are the unknowns in the coarse grid cells and Φ_F in the refined cells.

Of course, this ordering of the nodes will destroy the banded structure of L but will maintain the banded structure of L_{CC} and L_{FF}. Thus, any direct method for (4.2) will lead to a large amount of fill-in during the elimination process and will greatly reduce the efficiency of the model. However, we can define an iterative procedure which takes advantage of the banded structure of L_{CC} and L_{FF} to allow full vectorization of the inversion of each in the algorithms.

We note that Equation (4.2) is not to be stored and used directly, but is intended to motivate our solution technique. The residual produced by (4.2) in an iterative procedure will be the same as the residual produced by the composite grid matrices presented in other papers [9,22,29] depending, of course, on the composite grid stencils. It is this residual that we want to drive to zero. However, there are efficient ways [4] to evaluate this residual without forming the matrix form (4.2). We next define an efficient and vectorizable preconditioner for use in an iterative solution for (4.2). The matrix solution for this preconditioned iterative procedure is very similar to the matrix solution for the finite element based algorithm presented in Section 3.

This procedure is the following: (a) for a given inital guess, solve the problem $L_{FF}\Phi_F = f_F$; (b) with the computed Φ_F solve $L_{CC}\Phi_C = f_C - L_{CF}\Phi_F$; (c) update the guess for the coarse grid and repeat by solving the problem $L_{FF}\Phi_F = f_F - L_{FC}\Phi_L$ using the new iterate for Φ_L. This block Gauss-Seidel iterative technique is clearly very similar to the local refinement algorithm by Bramble, Ewing, Pasciak, and Schatz [4].

Of course, as in the algorithm described in Section 3, the solutions of L_{cc} may not be very well-conditioned. Therefore, instead of a straightforward block Gauss-Seidel iterative procedure for the solution of (4.2), we suggest following the algorithm given in Section 2 by replacing finite element matrices by corresponding finite difference matrices. The extension of the finite element theory to the corresponding finite difference theory will be presented elsewhere.

In determining the data structure for this algorithm, the additional unknowns representing the local grid refinement are simply concatonated on the end of the existing vector of coarse grid unknowns with a pointer set to the beginning location and a length given for the vector of refined cell unknowns. If the coarse grid and refined grid are ordered with a lexicographical ordering, we maintain the banded structure of L_{CC} and L_{FF} which allows efficient vectorization. It is essential to point out that both systems involving L_{CC} and L_{FF} can be solved within the strategy of the existing codes with their data structure and the effective methods that have been developed.

In many current simulations, when wells are brought on line or shut in, strong local transients are generated which adversely affect the convergence of the Newton iterations for the nonlinear systems. The common remedy for this problem, to cut the time-step size across the entire reservoir, is wasteful since the transients are very local in both time and space. This motivates the use of local time-stepping techniques around the wells in combination with the local spatial refinement. In order to get more accurate initial guesses for the Newton method around the wells to enable convergence with the original large time-steps, local time-stepping problems can be defined using the same preconditioning techniques as for local spatial grid refinement. A full discussion with error analysis of this local time-stepping technique will be presented by Ewing, Lazarov, Pasciak, and Jacobs in another paper.

5. ADAPTIVE REFINEMENT FOR DYNAMIC PHENOMENA

Since the local reservoir processes are often dynamic, efficient numerical simulation requires the ability to perform dynamic self-adaptive local grid refinement. The need for adaptive techniques has provided the impetus for the development of local grid refinement software tools, some of which are used in day to day applications for small to mid size problems. Software and engineering tools capable of dynamic local-grid refinement need to be developed for large-scale, fluid flow applications. The adaptive grid refinement algorithms must also be closely matched with the architectural features of the new advanced computers to take advantage of possible vector and parallel capabilities.

For time-dependent problems, often there is much information which can be used from preceding time-steps to help drive our adaptivity process. In parabolic problems, where the solution changes smoothly in time, the "optimal" grid used at the previous time-step should be a very good approximation to the desired grid at the next time step. Thus beginning with a new coarse grid at each time step and using the elliptic techniques of error estimators to drive the local refinement would be wasteful. For small parabolic problems, when the grid is changing very slowly in time, a much better technique would be to take the grid from the last time-step, apply a grid analysis to determine where new grid is needed and where grid is no longer needed, and then change only the grid that indicates need for change. For large time-dependent problems, iterative solution processes are much more efficient than direct solution techniques. For problems with fairly smoothly changing solutions, the same preconditioner can generally be used for several time steps, because the matrices change smoothly, greatly saving in computational effort. If the size of the grid and hence the number of unknowns is constantly changing, clearly the preconditioner must be changed. Similarly, as mentioned earlier, changing the number of unknowns greatly hinders vectorization techniques. Therefore, a considerably more efficient alternative to constantly changing the grid is to use a larger refined area within which the action is maintained for several time steps and to move the patch less frequently, after several steps.

For hyperbolic or transport-dominated parabolic partial differential equations arising in fluid flow problems, sharp fluid interfaces move along characteristic or near-characteristic directions. The computed fluid velocities determine both the local speed and direction of the regions where local refinement will be needed at the upcoming time steps. This information can be utilized in the adaptive method to move the local refinement with the front. We are currently experimenting with using the computed fluid velocities to move the patch grids in quantum jumps. The analysis description and analyses for these methods are given in [11]. The techniques described in the last section are applied at the macro-cell level and the relationship with more general domain decomposition techniques are most apparent. In these techniques, great care must be taken to preserve mass balance when grid is removed and the flow properties must be averaged and described on the new coarser grid.

ACKNOWLEDGEMENTS

This research was supported in part by U.S. Army Research Office Contract No. DAAG29–84–K–0002, by U.S. Air Force Office of Scientific Research Contract No. AFOSR–85–0117, and by the National Science Foundation Grant Nos. DMS–8504360 and DMS–8712021.

REFERENCES

1. M.J. Berger and J. Oliger, *Adaptive mesh refinement for hyperbolic partial differential equations*, Man. NA-83-02, Computer Science Department, Stanford University, 1983.

2. P.E. Bjorstad and O.B. Widlund, *Iterative methods for the solution of elliptic problems on regions partitioned into substructures*, preprint.

3. P.E. Bjorstad and O.B. Widlund, *Solving elliptic problems on regions partitioned into substructures*, in *Elliptic Problem Solvers II*, G. Birkhoff and A. Schoenstadt, eds., Academic Press, 1984, pp. 245-256.

4. J.H. Bramble, R.E. Ewing, J.E. Pasciak, and A.H. Schatz, *A Preconditioning technique for the Efficient Solution of Problems with Local Grid Refinement*, Computer Methods in Applied Mechanics and Engineering, (to appear).

5. J.H. Bramble, J.E. Pasciak, and A.H. Schatz, *An Iterative Method for Elliptic Problems on Regions Partitioned into Substructures*, Math. Comput., 46 (1986), pp. 361–370.

6. J.H. Bramble, J.E. Pasciak, and A.H. Schatz, *The construction of preconditioners for elliptic problems by substructuring, I*, Math. Comput., 47 (1986), pp. 103–134.

7. G. Chavent, *A new formulation of diphasic incompressible flows in porous media*, in *Lecture Notes in Mathematics*, No. 503, Springer-Verlag, 1976.

8. B.L. Darlow, R.E. Ewing, and M.F. Wheeler, *Mixed finite element methods for miscible displacement in porous media*, Proc. Sixth SPE Symposium on Reservoir Simulation, New Orleans, 1982, pp. 137-146; and Soc. Pet. Eng. J., 4 (1984), pp. 391-398.

9. J.C. Diaz, R.E. Ewing, R.W. Jones, A.E. McDonald, L.M. Uhler, and D.U. von Rosenberg, *Self-Adaptive Local Grid Refinement for Time-Dependent, Two-Dimensional Simulation*, in Finite Elements in Fluids, Vol. VI, Wiley, New York (1985), pp. 279–290.

10. J.C. Diaz and R.E. Ewing, *Potential of HEP-like MIMD Architectures in Self-Adaptive Local Grid Refinement for Accurate Simulation of Physical Processes*, Proc. Workshop on Parallel Processing Using the HEP, Norman, Oklahoma, 1985, pp. 209–226.

11. M. Espedal and R.E. Ewing, *Characteristic Petrov-Galerkin subdomain methods for two-phase immiscible flow*, Computer Methods in Applied Mechanics and Engineering, 64 (1987), pp. 112–115.

12. R.E. Ewing, *Determination of coefficients in reservoir simulation*, in *Numerical Treatment of Inverse Problems for Differential and Integral Equations*, P. Deuflhardt and E. Hairer, eds., Birkhauser, Berlin, 1982, pp. 206-226.

13. R.E. Ewing, *Problems arising in the modeling of processes for hydrocarbon recovery*, in *The Mathematics of Reservoir Simulation*, R.E. Ewing, ed., Frontiers in Applied Mathematics, Vol. 1, SIAM, Philadelphia, 1983, pp. 3-34.

14. R.E. Ewing, *Adaptive Mesh Refinement in Large-Scale Fluid Flow Simulation*, Accuracy Estimates and Adaptivity for Finite Elements, Ch. 16, I. Babuska, O. Zien- kiewicz, and E. Oliveira, eds., John Wiley and Sons, New York, 1986, pp. 299-314.

15. R.E. Ewing, *Efficient Adaptive Procedures for Fluid Flow Applications*, Comp. Meth. Appl. Mech. Eng., 55 (1986), pp. 89–103.

16. R.E. Ewing, *Adaptive grid-refinement techniques for treating singularities, heterogeneities, and dispersion, IMA Volume 11, Numerical Simulation in Oil Recovery*, M.F. Wheeler, ed., Springer-Verlag, New York, 1988, pp. 133–148.

17. R.E. Ewing, J.V. Koebbe, R. Gonzalez, and M.F. Wheeler, *Mixed finite element methods for accurate fluid velocities*, Finite Elements in Fluids, Vol. 4, John Wiley, (1985) pp. 233-249.

18. R.E. Ewing and R.D. Lazarov, *Adaptive locl grid refinement*, Proceedings SPE Rocky Mountain Regional Meeting, SPE No. 17806, Casper Wyoming, May 11–13, 1988.

19. R.E. Ewing, S. McCormick, and J. Thomas, *The fast composite grid method for solving differential boundary-value problems*, Proc. Fifth ASCE Specialty Conference, Laramie, Wyoming, 1984, pp. 1453-1456.

20. R.E. Ewing, T.F. Russell, and M.F. Wheeler, *Simulation of miscible displacement using mixed methods and a modified method of characteristics*, Proc. Seventh SPE Symposium on Reservoir Simulation, SPE No. 12241, San Francisco, November 15-18, 1983, pp. 71-82.

21. R.E. Ewing, T.F. Russell, and M.F. Wheeler, *Convergence analysis of an approximation of miscible displacement in porous media by mixed finite elements and a modified method of characteristics*, Computer Meth. Appl. Mech. Eng., R.E. Ewing, ed., 47 (1984), pp. 73-92.

22. Z.E. Heinemann, G. Gerken, and G. von Hantelmann, G., *Using Local Grid Refinement in a Multiple Application Reservoir Simulator*, paper SPE 12255 presented at the Seventh SPE Symposium on Reservoir Simulation, San Francisco, November, 1983.

23. J. Mandel and S. McCormick, *Iterative solution of elliptic equations with refinement: the two-level case*, Proceedings of Second International Symposium on Domain Decomposition Methods, Los Angeles, California, January, 1988, (to appear).

24. S. McCormick and J. Thomas, *The fast adaptive composite grid method for elliptic boundary value problems*, Math. Comp., 46 (1986), pp. 439–456.

25. J. Pasciak, *Domain decomposition preconditioners for elliptic problems in two and three dimensions: First approach*, First International Symposium on Domain Decomposition Methods for Partial Differential Equations, R. Glowinski, G. Golub, G. Meurant, and J. Periaux, eds., SIAM, Philadelphia, 1988.

26. O.A. Pedrosa, Jr., *Use of Hybrid Grid in Reservoir Simulation*, Ph.D. Thesis, Stanford University, California, December, 1984.

27. O.A. Pedrosa and K. Aziz, *Use of Hybrid Grid in Reservoir Simulation*, paper SPE 13507 presented at the SPE 1985 Middle East Technical Conference, Bahrain, March, 1985.

28. P. Quandalle and P. Besset, *Reduction of Grid Effects Due to Local Sub-Gridding in Simulations Using a Composite Grid*, paper SPE 13527 presented at the SPE 1985 Reservoir Simulation Symposium, Dallas, February, 1985.

29. D.U. Von Rosenberg, *Local Grid Refinement for Finite Difference Methods*, paper SPE 10974 presented at the 57th Annual Fall Technical Conference, New Orleans, September, 1982.

30. M.L. Wasserman, *Local Grid Refinement for Three-Dimensional Simulators*, paper SPE 16013 presented at the Ninth SPE Symposium on Reservoir Simulation, San Antonio, February, 1987.

Domain Decomposition Method and Parallel Algorithms

Kang Li-shan*

Abstract: This paper introduces the results on DDM and PA which were obtained recently by the Parallel Computation Research Group at Wuhan University.

1 Introduction

In 1980, using the Domain Decomposition Method (DDM) (see [1]) we began to design a class of asynchronous parallel algorithms, S-CR (Schwarz-chaotic relaxation), for solving mathematical physics problems while the multiprocessor system WuPP-80 was designed at Wuhan University.

In 1982, the WuPP-80, an MIMD machine with 4 processors, was put into operation at Wuhan University. By using the S-CR, a new class of asynchronous parallel DDM, many mathematical physics problems were solved on the machine and successful computing results were obtained (see [2]).

During the period from 1982 to 1985, a systematic theory on the DDM as the foundation of the asynchronous parallel algorithms for solving P.D.E.'s was developed (see [3],[4]).

In the Spring of 1986 D.J. Evans visited Wuhan University and worked with members of our Group on DDM and a series of extremely deep results of the convergence of the Schwarz alternating procedure (SAP) for the model problems were obtained (see Evans and Kang, et al. [5]—[10]). In the autumn of 1986, G. Rodrigue visited Wuhan University and suggested the use of mixed boundary conditions on the pseudo-boundaries of the subdomains. In this direction, many interesting results on the convergence of the DDM were obtained (see [13]—[17]).

In 1987, we began to study the DDM without overlapping. In this case, the symmetric DDMs are used for solving the symmetric problems in a symmetric

* Wuhan University, Wuhan, Hubei, China

domain. In this way, special kinds of parallel algorithms can be developed by which we can get the solution in two steps. These algorithms are constructed on the basis of the symmetric principle of errors. We regard this as a major breakthrough in the theory of DDM and it should influence physics, mathematics and mechanics, as well as parallel computing (see [18]—[21]).

In addition, we have also studied the algebraic DDM and DDM with other techniques (see [22]—[25]).

2 Convergence Rate of the SAP

(a) Consider the two point–boundary value problem

$$(2.1)\qquad \begin{cases} Lu \equiv -\dfrac{d^2u}{dx^2} + q^2u = f(x) & \Omega = \{x \mid 0 < x < 1\} \\ u(0) = a, \quad u(1) = b \end{cases}$$

The Ω is decomposed into two subdomains

$$\Omega_1 = \{x \mid 0 < x < x_k\} \text{ and } \Omega_2 = \{x \mid x_m < x < 1\}$$

where

$$x_k > x_m,\; x_k = x_m + d.$$

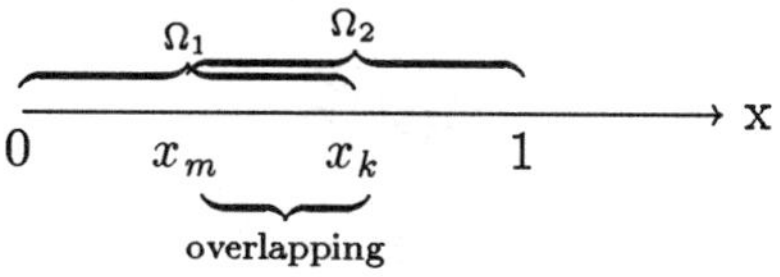

SAP:

$$\begin{cases} Ly^{(i+1)} = f(x) & \text{in } \Omega_1 \\ y^{(i_1)}(0) = a, \quad y^{(i+1)}(x_k) = z^{(i)}(x_k) \end{cases}$$

$$\begin{cases} Lz^{(i+1)}(x_k) = f(x) & \text{in } \Omega_2 \\ z^{(i+1)}(x_m) = y^{(i+1)}(x_m), \quad z^{(i+1)}(1) = b \end{cases}$$

where $z^{(0)}(x_k)$ is the initial guess.

Theorem 2.1 *(Evans, Kang, Shao and Chen (1986)) the convergence factor of SAP is*

$$\rho_q(x_m, x_k) = \frac{\operatorname{sh} q x_m}{\operatorname{sh} q x_k} \cdot \frac{\operatorname{sh} q(1 - x_k)}{\operatorname{sh} q(1 - x_m)} = \rho_q(x_m, x_m + d)$$

where $\operatorname{sh} x = (e^x - e^{-x})/2$. *Moreover,*

$$\rho(x_m, x_m + d) = \frac{\operatorname{sh} q x_m}{\operatorname{sh} q(x_m + d)} \cdot \frac{\operatorname{sh} q(1 - x_m - d)}{\operatorname{sh} q(1 - x_m)} < 1,$$

and d *is the size of overlapping.*

Theorem 2.1 gives the exact relationship between the convergence factor and the geometric character of the domain decomposition. For fixed d, we have

$$\max_{x_m \in \Omega} \mathrm{sh}_q(x_m, x_m + d) = \rho_q((1-d)/2) = \left\{ \frac{\mathrm{sh}\, q[(1-d)/2]}{\mathrm{sh}\, q[(1+d)/2]} \right\}$$

This means that for fixed d, the worst case of SAP is the symmetric decomposition: $\mathrm{mes}\, \Omega_1 = \mathrm{mes}\, \Omega_2$.

For the discrete form of (2.1), we have

Theorem 2.2 *(Evans, Kang, Shao and Chen, 1986)*
The convergence factor of numerical SAP is

$$\rho_q^*(m,k) = \rho_q^*(m, m+D) = \frac{s(m)}{s(m+d)} \cdot \frac{s(N-m-D)'}{s(N_D)}$$

where $s(x) = (r_1^x - r_2^x)/2$, *and*

$$r_1 = Q + \sqrt{Q^2 - 1}, \quad r_2 = Q - \sqrt{Q^2 - 1}$$

and

$$Q = (2 + q^2 h^2)/2.$$

It is easy to prove that $\lim_{h \to 0} \rho_q(x_m . x_k)$.

(b) For the two dimensional problem

$$(2.2) \qquad \left\{ -\Delta u + q^2 u = f \quad \text{in } \Omega = \{(x,y) \mid 0 < x < 1,\ 0 < y < 1 \right\}$$

we decompose Ω into two subdomains

$$\begin{aligned} \Omega_1 = &\ \{(x,y) \mid 0 < x < x_k,\ 0 < y < 1\} \\ \Omega_2 = &\ \{(x,y) \mid x_m < x < 1,\ 0 < y < 1\}. \end{aligned}$$

Theorem 2.3 *(Evans, Kang, Shao and Chen 1986) The convergence factor of the SAP is*

$$\rho_q(x_m, x_k) = \frac{\mathrm{sh}\sqrt{\pi^2 + q^2}x_m \ \mathrm{sh}\sqrt{\pi^2+q^2}(1-x_k)}{\mathrm{sh}\sqrt{\pi^2+q^2}x_k \ \mathrm{sh}\sqrt{\pi^2+q^2}(1-x_m)}.$$

Theorem 2.4 *The convergence factor of the numerical SAP is*

$$\bar\rho_q(m,k) = \bar\rho_q(m, m+S) = \frac{\bar S(m)}{\bar S(m+D)} \cdot \frac{\bar S(N-m-D)}{\bar S(N-m)},$$

where

$$\begin{aligned} \bar S(x) = &\ (\bar r_1^x - \bar r_2^x)/2, \\ \bar r_1 = &\ (\bar Q + \sqrt{\bar Q^2 - 1},\ \bar r_2 = \bar Q - \sqrt{\bar Q^2 - 1}, \\ \bar Q = &\ 1 + 2\sin^2 \tfrac{\pi h}{2} + \tfrac{h^2 q^2}{2}. \end{aligned}$$

For the n-dimensional problem, we have

Theorem 2.5 *(Chen and Kang, 1987)*

$$\rho_q(x_m, x_k) = \frac{\operatorname{sh}\sqrt{(n-1)\pi^2+q^2}x_m \operatorname{sh}\sqrt{(n-1)\pi^2+q^2}(1-x_k)}{\operatorname{sh}\sqrt{(n-1)\pi^2+q^2}x_k \operatorname{sh}\sqrt{(n-1)\pi^2+q^2}(1-x_m)}.$$

(2) Neumann problems
1-dimensional problem:

$$(2.3)\qquad \begin{cases} -\dfrac{d^2u}{dx^2}+q^2u &= f(x) \quad 0<x<1 \\ \frac{du(0)}{dx} = a, & \frac{du(1)}{dx} = b \end{cases}$$

SAP-1: If we use the Dirichlet conditions on the pseudo-boundaries, we have

Theorem 2.6 *(Kang and Evans, 1986): The convergence factor of SAP-1 is*

$$\rho_q(x_m, x_k) = \frac{\operatorname{ch} qx_m}{\operatorname{ch} qx_k}\cdot\frac{\operatorname{ch} q(1-x_k)}{\operatorname{ch} q(1-x_m)}.$$

where $\operatorname{ch} x = (e^x + e^{-x})/2$.

If $x_m \to 0$, then $\rho_q(x_m, x_m+d) \to \frac{\operatorname{ch} q(i-1)}{\operatorname{ch} qd \operatorname{ch} q} \neq 0$.
So we must change the pseudo-boundary conditions.
SAP-2: We use the Neumann conditions on the pseudo-boundaries, we have

Theorem 2.7 *(1986), The convergence factor of SAP-2 is*

$$\rho_q(x_m, x_k) = \frac{\operatorname{sh} qx_m}{\operatorname{sh} qx_k}\cdot\frac{\operatorname{sh} q(1-x_k)}{\operatorname{sh} q(1-x_m)}.$$

Similar results hold for the Dirichlet problem.
For the two-dimensional problem

$$(2.4)\qquad \begin{cases} -\Delta u + q^2 u &= f \quad \text{in } \Omega = \{(x,y) \mid 0<x<1,\ 0<y<1\} \\ \dfrac{\partial u}{\partial n} &= y \quad \text{on } \Gamma \end{cases}$$

A we have

Theorem 2.8 *The convergence factor of SAP-1 is*

$$\rho_q(x_m, x_k) = \frac{\operatorname{ch}\sqrt{\pi^2+q^2}x_m \operatorname{ch}\sqrt{\pi^2+q^2}(1-x_k)}{\operatorname{ch}\sqrt{\pi^2+q^2}x_k \operatorname{ch}\sqrt{\pi^2+q^2}(1-x_m)}.$$

Theorem 2.9 *The convergence factor of SAP-2 is*

$$\rho_p(x_m, x_k) = \frac{\operatorname{sh}\sqrt{\pi^2+q^2}x_m \operatorname{sh}\sqrt{\pi^2+q^2}(1-x_k)}{\operatorname{sh}\sqrt{\pi^2+q^2}x_k \operatorname{sh}\sqrt{\pi^2+q^2}(1-x_m)}.$$

The above results have been extended to the cases of more than two subdomains and to moving pseudo-boundaries for the purpose of balancing the load of multiprocessors.

3 Acceleration

For accelerating the convergence of SAP there are several ways.

(a) SAP with Pseudo-boundary Relaxation Factor

Consider the problem (2.1). We introduce the factor ω in the pseudo-boundary conditions as follows:

$$\begin{aligned} u^{(i+1)} &= v^{(i+1)} + \omega(v^{(i)} - v^{(i-1)}) \quad \text{on } \Gamma_1' \subset \Omega_2 \\ v^{(i+1)} &= u^{(i)} + \omega(u^{(i+1)} - u^{(i)}) \quad \text{on } \Gamma_2' \subset \Omega_1 \\ i &= 1, 2, 3, \ldots \end{aligned}$$

where $v^{(0)}$ and $v^{(1)}$ on Γ_1' are given.

Denote the convergence factor of SAP ($\omega = 1$) by ρ_q, then we have

Theorem 3.1 *(Evans, Kang, Chen and Shao, 1986)*
The optimal overrelaxation factor is given by

$$\omega_{\text{opt}} = (3/\sqrt{\rho_q})\cos((s + 4\pi)/3),$$

where

$$s = \arccos(-\sqrt{\rho_q})$$

and the corresponding convergence factor is

$$\rho_{\text{opt}} = 3(\omega_{\text{opt}} - 1)/\omega_{\text{opt}}.$$

(b) SAP with Mixed Pseudo-boundary Conditions

For one-dimensional problem (2.1) with $q = 0$, we use the mixed pseudo- boundary conditions

$$c_3 z^{(i+1)} + c_4 \frac{dz^{(i+1)}}{dx} = c_3 y^{(i+1)} + c_4 \frac{dy^{(i+1)}}{dx} \qquad \text{at } x = x_m$$

where $z^{(0)}(x_k)$ and $\frac{dz^{(0)}(x_k)}{dx}$ are the initial guesses.

Theorem 3.2 *(Lin, Wu, Rodrigue and Kang,(1987))*
The convergence factor of the SAP with parameters is

$$\rho_0 = \left| \frac{(c_1(x_k - 1) + c_2)(c_3 x_m + c_4)}{(c_1 x_k + c_2)(c_3(x_m - 1) + c_4)} \right|$$

We can easily choose the parameters c_i $(i = 1, 2, 3, 4)$ such that

$$\rho_0 = 0.$$

In these cases, the SAP in two steps gives the exact solution. If $q \neq 0$ we have

Theorem 3.3 *The convergence factor of the SAP with parameters is*

$$\rho_q = \left| \frac{(c_1 \operatorname{sh} q(x_k - 1) + c_2 q \operatorname{ch} q(x_k - 1))(c_3 \operatorname{sh} q x_m + c_4 \operatorname{ch} q x_m)}{(c_3 \operatorname{sh} q(x_m - 1) + c_4 q \operatorname{ch} q(x_m - 1))(c_1 \operatorname{sh} q x_k + c_2 q \operatorname{ch} q x_k)} \right|.$$

If we choose the parameters c_i $(i=1,2,3,4)$ such that $\rho_q = 0$, then the SAP in two steps gives the exact solution.

For two dimensional problem (2), we use the mixed pseudo–boundary conditions. We have

Theorem 3.4 *(Lin, Wu, Kang, and Rodrigue)*
The convergence factor of SAP with parameters is

$$\rho = \max_{x \epsilon N+} \left| \frac{(c_1 \operatorname{sh} Q_i(x_k - 1) + c_2 Q_i(x_k - 1))(c_3 \operatorname{sh} Q_i x_m + c_4 \operatorname{ch} Q_i x_m)}{(c_3 \operatorname{sh} Q_i(x_m - 1) + c_4 Q_i(x_m - 1))(c_1 \operatorname{sh} Q_i x_k + c_2 \operatorname{ch} Q_i x_k)} \right|$$

where

$$Q_i = (i^2\pi^2 + q^2)^{\frac{1}{2}}.$$

For solving Neumann problem (4), we have

Theorem 3.5 *The convergence factor of SAP with parameters is*

$$\bar{\rho} = \max_{x \epsilon N+} \left| \frac{(c_1 \operatorname{sh} Q_i(x_k - 1) + c_2 Q_i(x_k - 1))(c_3 \operatorname{ch} Q_i x_m + c_4 \operatorname{sh} Q_i x_m)}{(c_3 \operatorname{ch} Q_i(x_m - 1) + c_4 Q_i(x_m - 1))(c_1 \operatorname{ch} Q_i x_k + c_2 \operatorname{sh} Q_i x_k)} \right|$$

For n–dimensional problems the conclusion is almost the same if we replace Q_i in the formulae by $\bar{Q}_i = ((n-1)i^2 + q^2)^{\frac{1}{2}}$ (see Chen and Kang(1987)).

For more than two subdomains(DDM)]

Consider the one–dimensional problem (1). The Ω is decomposed into m subdomains

$$\Omega_j = \{x \mid x_j^{(1)} < x < x_j^{(2)}\} \quad j = 1,2,\ldots,m$$

$$x_j^{(1)} < x_{j+1}^{(1)} < x_j^{(2)} < x_{j+1}^{(2)} \quad j = 1,2,\ldots,m; x_1^{(1)} = 0 \text{ and } x_m^{(2)} = 1.$$

Algorithm:

$$\begin{cases} -\dfrac{d^2 u_j^{(i)}}{dx^2} + q^2 u_j^{(i)} = f & \in \Omega_j \\[2ex] c_j^{(1)} u_j^{(i)} + d_j^{(1)} \dfrac{du_j^{(i)}}{dx} = c_j^{(1)} j_{j-1}^{(i)} + d_j^{(1)} \dfrac{du_{j-1}^{(i)}}{dx} & \text{at } x = x_j^{(1)} \\[2ex] c_j^{(2)} u_j^{(i)} + d_j^{(2)} \dfrac{du_j^{(i)}}{dx} = c_j^{(2)} u_{j+1}^{i-1} + d_j^{(2)} \dfrac{du_{j+1}^{(i-1)}}{dx} & \text{at } x = x_j^{(2)} \\[2ex] i = 1; \quad j = 1,2,\ldots,m, & \\[2ex] i = 2; \quad j = m-1, m-2, \ldots, 1, \quad u_m^{(2)} = u_m^{(1)}, & \end{cases}$$

where $c^{(k)}, d_j^{(k)}$ $(k = 1,2;\ j = 1,1,\ldots,m)$ are parameters, $u_j^{(0)}$ and $\frac{du^{(0)}}{dx}$ on $x_j^{(2)}$ $(j = 1,2,\ldots,m-1)$ are initial guesses.

Theorem 3.6 *(Wu, Lin, Rodrigue and Kang (1987))*
If the parameters are chosen such that

$$c_{jh}^{(1)} \operatorname{sh} qx_j^{(1)} + d_j^{(1)} q \operatorname{ch} qx_j^{(1)} = 0 \qquad j = 2, 3, \ldots, m$$

$$c_{jh}^{(1)} \operatorname{sh} qx_j^{(1)} + d_j^{(1)} q \operatorname{ch} qx_j^{(1)} = 0 \qquad j = 2, 3, \ldots, m$$

then the exact solution can be obtained in two steps.

For the discrete form (DDDM) we denote $x_j^{(1)} = J_1 h, x_j^{(2)} = J_2 h.$

Theorem 3.7 *If the parameters are chosen such that*

$$\begin{cases} hc_j^{(1)} - d_j^{(1)} + d_j^{(1)}(r_1^{J_1+1} - r_2^{J_1-1})/(r_1^{J_1} - r_2^{J_1}) & = 0 \\ hc_j^{(2)} - d_j^{(2)} + d_j^{(2)}(r_1^{J_2+1} - r_2^{J_2-1})/(r_1^{J_2} - r_2^{J_2}) & \neq 0, \end{cases}$$

where

$$r_1 = \ 1 + q^2h^2/2 + ((2qh)^2 + (qh)^4)^{\frac{1}{2}}/2$$

$$r_1 = \ 1 + q^2h^2/2 - ((2qh)^2 + (qh)^4)^{\frac{1}{2}},$$

then the exact solution can be obtained in two steps.

The similar technique can be used for accelerating the convergence of DDM used for solving the multi–dimensional problems.

(c) Extrapolation Techniques

Assume the function–sequence $\{u^{(i)}\}$ is obtained by SAP.

Denote the exact solution by u^*, and the errors by

$$e^{(i)} = u^* - u^{(i)}$$

Theorem 3.8 *(Lin, Liu and Kang (1987))*
If there exists a constant $P \neq 1$, such that

$$e^{(i+1)} \quad = Pe^{(i)},$$

then

$$u^* \quad = \tfrac{1}{1-P} u^{(i+1)} - \tfrac{P}{1-P} u^{(i)}.$$

For one–dimensional problems with constant coefficients, we usually can get the exact solution in two steps no matter whether the original sequence of the SAP converges.

For solving two–dimensional problem (2), if $x_k = x_m$, we use the SAP with parameters $c_1 = 1.c_2 = 0, c_3 = 0$ and $c_4 = 1$. We have

$$e^{(i+1)} = -e^{(i)}.$$

It means that the original sequence of functions does not converge. But we can get the exact solution as follows:

$$u^* = (u^{(1)} + u^{(2)})/2.$$

4 Symmetric DDM without Overlapping

(a) Symmetric Principle of Errors

Consider the linear problem

$$(*)\begin{cases} Lu = f & \text{in } \Omega \\ u = \phi & \text{on } \Gamma = \partial\Omega, \end{cases}$$

the domain Ω is symmetric w.r.t. Γ' and the operator L is symmetric w.r.t. the domain decomposition, means that $Lu(x) = L\tilde{u}(x'), \forall\ x\varepsilon\Omega_1$, where $x'\varepsilon\Omega_2$ is the symmetric point of x w.r.t. Γ'. We have the following theorem.

Theorem 4.1 *(Rao, 1987), If problem (*) has unique solution u^* then*

$$u^* = \begin{cases} (u^{(1)} + u^{(2)})/2 & \text{on } \bar{\Omega}_1 \\ \\ (v^{(1)} + v^{(2)})/2 & \text{on } \bar{\Omega}_2, \end{cases}$$

where $v^{(i)}$ and $u^{(i)}$ satisfy the following problems:

$$\begin{cases} Lu^{(1)} = f & \text{in } \Omega_1 \\ \\ u^{(1)} = \phi & \text{on } \Gamma \\ \\ u^{(1)} = q & \text{on } \Gamma' \end{cases} \qquad \begin{cases} Lv^{(1)} = f & \text{in } \Omega_2 \\ \\ v^{(1)} = \phi & \text{on } \Gamma \\ \\ v^{(1)} = g & \text{on } \Gamma' \end{cases}$$

and

$$\begin{cases} Lu^{(2)} = f & \text{in } \Omega_1 \\ \\ u^{(2)} = \phi & \text{on } \Gamma \\ \\ \frac{\partial u^{(2)}}{\partial n} = \frac{\partial v^{(1)}}{\partial n} & \text{on } \Gamma' \end{cases} \qquad \begin{cases} Lv^{(2)} = f & \text{in } \Omega_2 \\ \\ v^{(2)} = \phi & \text{on } \Gamma \\ \\ \frac{\partial v^{(2)}}{\partial n} = \frac{\partial u^{(1)}}{\partial n} & \text{on } \Gamma' \end{cases}$$

where n is the outer normal direction of Γ'.

In 1987, the symmetric principle of errors was first discovered by Rao Chuan-xia and widely extended to many applications, especially to the parallel computing (see [17] and [18]).

The discrete form of the Symmetric Principle of Errors has been established by Shao, Wu, etc. [19].

(b) Symmetrization Principle

Denote

$f^+(x) = (f(x) + f(x'))/2, \qquad f^-(x) = (f(x) - f(x'))/2.$

Theorem 4.2 *(Lu, 1987), The solution of $(*)$ is*

$$u^*(x) = u^+(x) - u^-(x' \ \text{ on } \bar{\Omega}_1, \qquad u^*(x) = u^+(x) + u^-(x) \ \text{ in } \Omega_2,$$

where u^+ and u^- are the solutions of problems

$$\begin{cases} Lu^{(+)} = f^{+} & \text{in } \Omega_1 \\ u^{(+)} = \phi^{+} & \text{on } \Gamma \\ \frac{\partial u^{(+)}}{\partial n} = 0 & \text{on } \Gamma' \end{cases}$$

$$\begin{cases} Lu^{(-)} = f^{-} & \text{in } \Omega_2 \\ u^{(-)} = \phi^{-} & \text{on } \Gamma \\ u^{(-)} = 0 & \text{on } \Gamma' \end{cases}$$

respectively.

For the discrete form of $(*)$

$$\begin{cases} L_h u_i = f_i, & \forall i \ \varepsilon \ \Omega_h \\ u_j = \phi_j, & \forall j \ \varepsilon \ \Gamma_h, \end{cases}$$

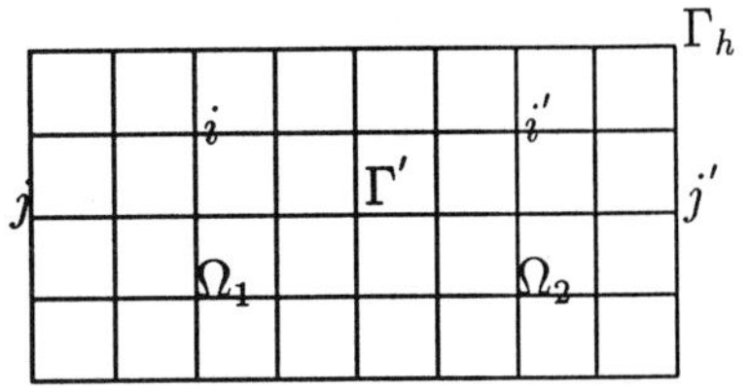

If $L_h u_{i'} = L_h u_i$, $\forall i \ \varepsilon \ \Omega_i$, i' is the symmetric point of i w.r.t. Γ', then we have

$$u_i = u_i^+ + u_i^- \quad \text{and} \quad u_{i'} = u_i^+ - u_i^-$$

where

$$\begin{cases} Lu_i^+ = f_I^+ & = (f_i + f_{i'})/2 \\ u_i^+ + \phi_j^+ & \forall j \ \varepsilon \ \Gamma_h \\ u_k^+ = u_{k'}^+ & \forall k \ \varepsilon \ \Omega_1 \end{cases} \qquad \begin{cases} Lu_i^- = f_i^- & = (f_i - f_{i'})/2 \ \forall i \ \varepsilon \ \Omega_1 \\ u_j^- = \psi_j^- & \forall j \ \varepsilon \ \Gamma_h \\ u_k^- = 0 & \forall k \ \varepsilon \ \Gamma'. \end{cases}$$

$\forall i \varepsilon \Omega_1 \bigcup \Gamma'$.

5 DDM with Other Techniques

(a) Schwarz-Projection Method for Non-linear Problems (see [25].
(b) Schwarz-Multigrid Method(see [22]).
(c) DDM-Operator Splitting Method (see [24])
(d) Numerical SAP (see [21] and [23]).

6 Numerical Experiments

The S-CR algorithms, a class of asynchronous parallel algorithms based upon the theory of DDM, were implemented and tested on the multicomputer system UwPP-80 installed at Wuhan University in 1982.

References

[1] Kang Li-shan, *The generalization of the Schwarz alternating procedure*, Wuhan University Journal, No. 4, pp. 11-23 (1979)

[2] Kang Li-shan, chen Yu-ping, et al., *Research of Distributed Parallel Processing System*, Wuhan University Press, (1984)

[3] Kang Li-shan, Sun Le-lin, Chen Yu-ping, *Asynchronous Parallel Algorithms for Solving Mathematical Physics Problems*, China Academic Publishers, Beijing, (1985)

[4] Kang Li-shan, Chen Yu-ping, Sun Le-lin and Quan Hui-yun, *The asynchronous parallel algorithms S-COR for solving P.D.E.'s on multiprocessors*, International Journal of Computer Mathematics, vol. 18, pp 163-172, (1985)

[5] Evans, D.J., Kang Li-shan, Shoa Jian-ping and Chen Yu-ping, *The convergence rate of the Schwarz alternating procefure (I)–For one-dimensional problems*, Ibid., vol 20, pp 157-170 (1986)

[6] Evans, D.J., Shoa Jian-ping, Kang Li-shan and Chen Yu-ping, *The convergence rate of the Schwarz alternating procedure (Ii)—For two-dimensional problems*, Ibid., vol. 20, pp 325-339, (1986)

[7] Kang Li-shan and Evans, D.J., *The convergence rate of the Schwarz alternating procedure (III)—For Neumann problem*, Ibid., vol. 21, pp 85-108, (1987)

[8] Evans, D.J., Kang Li-shan, Chen Yu-ping and Shao Jian-ping, *The convergence rate of the Schwarz alternating procedure (IV)— With pseudo-boundary relaxation factor*, Ibid., vol 21, pp 185-203, (1987)

[9] Kang Li-shan and Evans, D.J., *The convergence rate of the Schwarz alternating procedure (V)—For more than two subdomain*, Ibid., vol. 23, (1987)

[10] Shao Jian-ping, Kang Li-shan, Chen Yu-ping and Evans, D.J., *The convergence rate of Schwarz alternating procedure—For unsymmetric problems*, Ibid., vol. 25, (1988)

[11] Lin Guang-ming, Wu Zhi-jian, Rodrigue, G. and Kang Li-shan, *Domain decomposition method (DDM) with mixed pseudo-boundary conditions—For one dimensional problems*, Parallel Algorithms and Domain Decomposition, Ed. by Kang Li-shan, Wuhan University Press, pp 93-116 (1987)

[12] Wu Zhi-jian, Lin Guang-ming, Rodrigue, G. and Kang Li-shan, *Domain decomposition method with mixed pseudo-boundary conditions—For more than two subdomains*, Ibid., pp 117-125 (1987)

[13] Lin Guang-ming, Wu Zhi-jian, Kang Li-shan and Rodrigue, G., *Domain decomposition method (DDM) with pseudo-boundary conditions —For two - dimensional problems*, Ibid., pp 126-133, (1987)

[14] Liu Yu-hui, Lin Guang-ming, Kang Li-shan and Rodrigue, G., *Discrete domain decomposition method (DDDM) with mixed pseudo- boundary conditions—For discrete two–dimensional problems,*, Ibid., pp 134-140, (1987)

[15] Chen Lu-juan and Kang Li-shan, *The convergence factor of SAP for multi-dimensional problems*, Ibid., pp 141-148, (1987)

[16] Chen Lu-juan and Kang Li-shan, *The convergence factor of SAP for multi-dimensional problems*, Ibid., pp 141-148, (1987)

[17] Kang Li-shan, Sun Le-lin and Chen Yu-ping,*Asynchronous parallel algorithm for general linear problems*, Ibid., pp 84-89, (1987)

[18] Rao Chuan-zia, *Symmetric domain decomposition (SDD) and exact procedure for solving linear PDE's*, Ibid., pp 161-176, (1987)

[19] Shao Jian-ping and Kang Li-shan, *Symmetric domain decomposition for linear operator equations*, Ibid., pp 177-185, (1987)

[20] Shao Jian-ping, Wu Zhi-jian, Rao chuan-zia, Kang Li-shan and Chen Yu-ping, *Numerical symmetric domain decomposition for solving linear systems*, Ibid., pp 186-193, (1987)

[21] Lin Guang-ming, Liu Yu-hui and Kang Li-shan, *Acceleration of the domain decomposition method*, Ibid., pp 194-201, (1987)

[22] Shao Jian-ping, Kang Li-shan, Chen Yu-ping and Evans, D.J., *The convergence factor of numerical Schwarz algorithm for linear system*, Ibid., pp 205-222, (1987)

[23] Shao Jian-ping, *Schwarz alternating method and multigrid method*, Ibid., pp 223-236, (1987)

[24] Liu Yu-hui and Rodrigue, G., *Convergence rate of discrete Schwarz alternating algorithms for solving elliptic P.D.E.'s*, Ibid., pp 249-264, (1987)

[25] Chen Lu-juan, *A parallel alternating direction algorithm*, Ibid., pp 279–289, (1987)

[26] Sun Le-lin and Quan Hui-yun, *Schwarz–projection method for solving some nonlinear P.D.E.'s*, Ibid., pp 265-278, (1987)

[27] Kang Li-shan, Qiu Yu-lan, Chen Yu-ping and Pen De-chen, *The Schwarz algorithms for multiprocessors (I)—For solving linear wlliptic boundary value problems*, Research of Distributed Parallel Processing System, Wuhan University Press, pp 71-79, (1984)

[28] Kang Li-shan, Chen Yu-ping and Qiu Yu-lan, *The Schwarz algorithm for multiprocessors (Ii)—For solving mildly nonlinear elliptic boundary value problems*, Ibid., pp 80-91, (1984)

[29] Kang Li-shan and Chen Yu-ping, *The Schwarz algorithm for multiprocessors (III)—For solving 2-dimensional steady state Navier-Stokes equations*, Ibid., pp 92-101, (1984)

[30] Chen Yu-ping and Kang Li-shan, *The Schwarz algorithm for multiprocessors (IV)—For solving the elliptic problems with singularities*, Ibid., pp 102-114, (1984)

[31] Kang Li-shan, Chen Yu-ping and Sun Le-lin, *The Schwarz algorithm for multiprocessors (V)—For solving unsteady mathematical physics problems*, Ibid., pp 115-122, (1984)

[32] Kang Li-shan, *A class of new asynchronous parallel algorithms*, Ibid., pp 60-70, (1984)

[33] Shao Jian-ping and Kang Li-shan, *An asynchronous parallel mixed algorithm for linear and nonlinear equations*, Parallel Computing, 5, pp 313-321, (1987)

[34] Lu Tao, *Improved and generalized symmetric domain decomposition algorithms*, to appear in The Proceedings of the Symposium on Parallel Algorithms held in Beijing, Nov. 16-18, 1987

Incomplete Domain Decomposition Preconditioners for Nonsymmetric Problems

Gérard Meurant*

Abstract. The aim of this paper is to derive incomplete Domain Decomposition preconditioners that can be used on parallel computers with the Orthomin(1) method for solving non symmetric linear systems. We mainly show how to extend the techniques which have been presented in [13] and we present numerical results that demonstrate the usefulness of the preconditioners described in this paper.

1. Introduction. In the last years, there has been a great development of domain decomposition preconditioners for the conjugate gradient method. This new interest mainly comes from the fact that these methods can be easily and efficiently used on parallel computers with a large number of processors. Up to now, the research has been almost essentially directed towards finding good preconditioners for symmetric linear systems arising from finite difference or finite element discretizations of elliptic partial differential equations in two and three dimensional domains. Several papers adressing these issues have recently appeared : Bjorstad & Widlund [1], Bramble, Pasciak & Schatz [2], Golub & Mayers [9], Chan & Resasco [3], Meurant [13], [14], [15].

In this paper, we will show how to extend the techniques of [13] to the solution of non symmetric problems arising from the finite difference discretization of diffusion convection equations in two dimensional domains. As an acceleration of the basic linear iteration we will use the Orthomin(1) method (see for instance [7]). This method is not the best to solve this problem , but it will be just enough for our main purpose which is constructing preconditioners.

The outline of the paper is as follows. Section 2 introduces the model problem we are solving. In Section 3 we briefly present the tools we are using to construct the preconditioners, mainly how to approximate the inverse of a non symmetric tridiagonal matrix and the generalization of the basic block preconditioner INV (see [5]). Section 4 motivates our derivation exhibiting an exact DD solver and then, Section 5 shows how to derive an incomplete decomposition whose symmetric form was first described in [13]. We conclude in Section 6 with some numerical experiments for model problems with different ratios of the diffusion and convection coefficients.

2. The Model Problem. The problem we want to solve is a linear elliptic PDE,

$$-\frac{\partial}{\partial x}(a(x,y)\frac{\partial u}{\partial x}) - \frac{\partial}{\partial y}(b(x,y)\frac{\partial u}{\partial y}) + 2\alpha(x,y)\frac{\partial u}{\partial x} + 2\beta(x,y)\frac{\partial u}{\partial y} + cu = f,$$

* CEA, Centre d'Etudes de Limeil-Valenton, BP 27, 94190 Villeneuve St Georges, France

$$\text{in } \Omega \subset R^2, \quad u|_{\partial\Omega} = 0 \quad \text{or} \quad \frac{\partial u}{\partial n}|_{\partial\Omega} = 0,$$

Ω being a rectangle. With standard finite differences schemes (5 point) and row–wise ordering, this leads to a block tridiagonal linear system :

$$A\,x \;=\; b,$$

with

$$A = \begin{pmatrix} D_1 & B_1 & & & \\ A_2 & D_2 & B_2 & & \\ & \ddots & \ddots & \ddots & \\ & & A_{n-1} & D_{n-1} & B_{n-1} \\ & & & A_n & D_n \end{pmatrix}.$$

With a centered scheme, as we suppose $c \geq 0$, when $a \equiv 1$, $b \equiv 1$ and α, *beta* are constant, D_i is point tridiagonal strictly diagonally dominant if $\alpha h < 2 + \frac{ch^2}{2}$, A is diagonally dominant if $\alpha h < 1 + \frac{ch^2}{2}$ and $\beta h < 1 + \frac{ch^2}{2}$. A_i is a diagonal matrix. Under these conditions A is a non symmetric M-matrix.

The Orthomin(1) algorithm that we are using for solving our problem is the following,

$$x^0, r^0 = b - \overline{A}x^0, p^0 = r^0,$$

$$\alpha_k = \frac{(r^k, \overline{A}p^k)}{(\overline{A}p^k, \overline{A}p^k)},$$

$$x^{k+1} = x^k + \alpha_k p^k,$$

$$r^{k+1} = r^k - \alpha_k \overline{A}p^k,$$

$$\beta_k = -\frac{(\overline{A}r^{k+1}, \overline{A}p^k)}{(\overline{A}p^k, \overline{A}p^k)},$$

$$p^{k+1} = r^{k+1} + \beta_k p^k,$$

$$\overline{A}p^{k+1} = \overline{A}r^{k+1} + \beta_k \overline{A}p^k.$$

We use this method with either $\overline{A} = M^{-1}A$ or $\overline{A} = AM^{-1}$ where M is the preconditioner.

3.Tools. The first tool we use is how to solve tridiagonal linear systems with sparse right hand sides. This has been described in [13]for symmetric matrices. As there is no fundamental difference when the matrix is non symmetric, we refer the reader to this paper.

The second technique, which was developped for the symmetric case in Concus, Golub & Meurant [5] concerns approximating the inverse of a tridiagonal matrix. When T is a symmetric tridiagonal matrix, T^{-1} is approximated by a tridiagonal matrix $trid(T^{-1})$ whose non zero elements are the same as the corresponding ones of T^{-1}. The justification of this technique is given in [6]. Now let T be tridiagonal non symmetric,

$$T = \begin{pmatrix} a_1 & b_1 & & & \\ c_2 & a_1 & b_2 & & \\ & \ddots & \ddots & \ddots & \\ & & c_{n-1} & a_{n-1} & b_{n-1} \\ & & & c_n & a_n \end{pmatrix},$$

it is well known that we can symmetrize T by left multiplying with a diagonal matrix D,

$$D = \begin{pmatrix} 1 & & & \\ & \frac{b_1}{c_2} & & \\ & & \ddots & \\ & & & \frac{b_1 \cdots b_{n-1}}{c_2 \cdots c_n} \end{pmatrix}.$$

Then, $\overline{T} = DT$ is symmetric and we can apply the same techniques as before to $\overline{T}$.
If T is line diagonally dominant, so is $\overline{T}$, then the elements of T^{-1} decrease away from the diagonal on each column. If T is column diagonally dominant then the elements of T^{-1} decrease away from the diagonal on each line. We can approximate T^{-1} with $trid(\overline{T}^{-1})D$. For this approximation, we only need to store D and 2 diagonals of $\overline{T}^{-1}$. Another possibility will be to right symmetrize.
The third tool is the block (incomplete) factorization INV of Concus, Golub & Meurant [5] which can be extended to non symmetric matrices as follows . Suppose A is a block tridiagonal matrix, then the block Cholesky factorization of A can be written as

$$A = (\Delta + L)\,\Delta^{-1}\,(\Delta + U)$$

$$\Delta = \begin{pmatrix} \Delta_1 & & & \\ & \ddots & & \\ & & \ddots & \\ & & & \Delta_n \end{pmatrix},$$

$$L = \begin{pmatrix} 0 & & & \\ A_2 & 0 & & \\ & \ddots & \ddots & \\ & & A_n & 0 \end{pmatrix}, U = \begin{pmatrix} 0 & B_1 & & \\ & 0 & B_2 & \\ & & \ddots & \ddots \\ & & & 0 \end{pmatrix},$$

with

$$\begin{cases} \Delta_1 = D_1, \\ \Delta_i = D_i - A_i\,\Delta_{i-1}^{-1}\,B_{i-1}. \end{cases}$$

To construct an INV incomplete decomposition, we simply replace the inverse with a tridiagonal approximation,

$$\Delta_i = D_i - A_i\; trid(\Delta_{i-1}^{-1})\; A_{i-1}.$$

It follows that all the Δ_i are tridiagonal matrices. The stability of this incomplete factorization will be adressed in another paper [16].

4. An exact DD solver. To develop a DD method, as for the symmetric case, we partition the domain Ω into strips $\Omega_i, i = 1, \ldots, k$ and we renumber the unknowns in such a way that the components of x related to the subdomains appear first and then the ones for the interfaces. With this (block) ordering, the system can be written as

$$\begin{pmatrix} B_1 & & & & & C_1 & & & \\ & B_2 & & & & E_2 & C_2 & & \\ & & B_3 & & & & E_3 & \ddots & \\ & & & \ddots & & & & \ddots & C_{k-1} \\ & & & & B_k & & & & E_k \\ & & & & & & & & \\ Q_1 & R_2 & & & & B_{1,2} & & & \\ & Q_2 & R_3 & & & & B_{2,3} & & \\ & & \ddots & \ddots & & & & \ddots & \\ & & & Q_{k-1} & R_k & & & & B_{k-1,k} \end{pmatrix} \begin{pmatrix} x_1 \\ x_2 \\ \vdots \\ \vdots \\ x_k \\ \\ x_{1,2} \\ x_{2,3} \\ \vdots \\ x_{k-1,k} \end{pmatrix} = \begin{pmatrix} b_1 \\ b_2 \\ \vdots \\ \vdots \\ b_k \\ \\ b_{1,2} \\ b_{2,3} \\ \vdots \\ b_{k-1,k} \end{pmatrix},$$

where each B_i is related to a subdomain Ω_i,

$$B_i = \begin{pmatrix} D_i^1 & B_i^1 & & & \\ A_i^2 & D_i^2 & B_i^2 & & \\ & \ddots & \ddots & \ddots & \\ & & A_i^{m_i-1} & D_i^{m_i-1} & B_i^{m_i-1} \\ & & & A_i^{m_i} & D_i^{m_i} \end{pmatrix}, \quad i = 1, \ldots, k.$$

D_i^j and $B_{i,j}$ are point tridiagonal, m_i is the number of mesh lines in Ω_i.
The matrices C_i, E_i, Q_i and R_i have a very special structure,

$$C_i = \begin{pmatrix} 0 \\ \vdots \\ 0 \\ C_i^{m_i} \end{pmatrix}, \quad E_i = \begin{pmatrix} E_i^1 \\ 0 \\ \vdots \\ 0 \end{pmatrix}, \quad Q_i = (0 \ \cdots \ 0 \ Q_i^{m_i}), \quad R_i = (R_i^1 \ 0 \ \cdots \ 0).$$

$C_i^{m_i}$, E_i^1, $Q_i^{m_i}$ and R_i^1 are diagonal matrices.
To derive an exact DD solver, we eliminate $x_1, \ldots, x_k$ to get a reduced system involving only the unknowns for the interfaces.

$$\begin{pmatrix} B'_{1,2} & G_1 & & & \\ F_2 & B'_{2,3} & G_2 & & \\ & \ddots & \ddots & \ddots & \\ & & F_{k-2} & B'_{k-2,k-1} & G_{k-2} \\ & & & F_{k-1} & B'_{k-1,k} \end{pmatrix} \begin{pmatrix} x_{1,2} \\ x_{2,3} \\ \vdots \\ x_{k-2,k-1} \\ x_{k-1,k} \end{pmatrix} = \begin{pmatrix} b'_{1,2} \\ b'_{2,3} \\ \vdots \\ b'_{k-2,k-1} \\ b'_{k-1,k} \end{pmatrix}$$

It is easy to see that we have the following formulas for $i = 1, \ldots, k-1$,

$$B'_{i,i+1} = B_{i,i+1} - Q_i \ B_i^{-1} \ C_i - R_{i+1} \ B_{i+1}^{-1} \ E_{i+1},$$

$$F_i = -Q_i \ B_i^{-1} \ E_i,$$

$$G_{i-1} = -R_i \ B_i^{-1} \ C_i,$$

$$b'_{i,i+1} = b_{i,i+1} - Q_i \ B_i^{-1} \ b_i - R_{i+1} \ B_{i+1}^{-1} \ b_{i+1}.$$

As for the symmetric case, to simplify these expressions, we use 2 factorizations for the matrices B_i coresponding to a subdomain, a block LU one (top–down) which can be written as

$$B_i = (\Delta_i + L_i) \ \Delta_i^{-1} \ (\Delta_i + U_i),$$

$$\begin{cases} \Delta_i^1 = D_i^1, \\ \Delta_i^j = D_i^j - A_i^j \ (\Delta_i^{j-1})^{-1} \ B_i^{j-1}. \end{cases}$$

and a block UL one (bottom–up),

$$B_i = (\Sigma_i + U_i) \ \Sigma_i^{-1} \ (\Sigma_i + L_i),$$

$$\begin{cases} \Sigma_i^{m_i} = D_i^{m_i}, \\ \Sigma_i^j = D_i^j - B_i^j \ (\Sigma_i^{j+1})^{-1} \ A_i^{j+1}. \end{cases}$$

With these notations, we have the following result,

Theorem

For $i = 1, \ldots, k-1$,

$$B'_{i,i+1} = B_{i,i+1} - Q_i^{m_i} \ (\Delta_i^{m_i})^{-1} \ C_i^{m_i} - R_{i+1}^1 \ (\Sigma_{i+1}^1)^{-1} \ E_{i+1}^1.$$

If

$$S_i^1 = (\Sigma_i^1)^{-1} \ E_i^1,$$

$$S_i^l = -(\Sigma_i^l)^{-1} \ A_i^l \ S_i^{l-1}, \ l = 2, \ldots, m_i,$$

then,

$$F_i = -C_i^{m_i} \ S_i^{m_i},$$

Similarly

$$T_i^{m_i} = (\Delta_i^{m_i})^{-1}\, C_i^{m_i},$$

$$T_i^l = -(\Delta_i^l)^{-1}\, B_i^l\, T_i^{l+1},\ l = m_i - 1, \dots, 1,$$

and,

$$G_{i-1} = -R_i^1\, T_i^1,$$

5. Domain Decomposition Preconditioners. From this exact factorization we can derive an approximation that will give a Domain Decomposition preconditioner. To do this, we approximate the inverses of tridiagonal matrices in the same way we were doing in INV.
The motivations have been explained in [15] on a two strips example, so here we directly derived the preconditioner for the k strips case, $k > 2$. We choose M as

$$M = L \begin{pmatrix} M_1^{-1} & & & & & & \\ & M_2^{-1} & & & & & \\ & & \ddots & & & & \\ & & & M_k^{-1} & & & \\ & & & & M_{1,2}^{-1} & & \\ & & & & & \ddots & \\ & & & & & & M_{k-1,k}^{-1} \end{pmatrix} U,$$

$$L = \begin{pmatrix} M_1 & & & & & & & & \\ & M_2 & & & & & & & \\ & & \ddots & & & & & & \\ & & & \ddots & & & & & \\ & & & & M_k & & & & \\ Q_1 & R_2 & & & & M_{1,2} & & & \\ & Q_2 & R_3 & & & H_2 & M_{2,3} & & \\ & & \ddots & & & & \ddots & \ddots & \\ & & & Q_{k-1} & R_k & & & H_{k-1} & M_{k-1,k} \end{pmatrix},$$

$$U = \begin{pmatrix} M_1 & & & & C_1 & & & \\ & M_2 & & & E_2 & C_2 & & \\ & & \ddots & & & \ddots & \ddots & \\ & & & \ddots & & & E_{k-1} & C_{k-1} \\ & & & M_k & & & & E_k \\ & & & & M_{1,2} & J_1 & & \\ & & & & & M_{2,3} & J_2 & \\ & & & & & & \ddots & \ddots \\ & & & & & & & M_{k-1,k} \end{pmatrix}.$$

In the lower right corners of L and U are the factors of an incomplete block Cholesky decomposition of the reduced system. As in [13],[15], we choose the approximations

$$M_{1,2} = B_{1,2} - Q_1^{m_1}\, trid[(\Delta_1^{m_1})^{-1}]\, C_1^{m_1} - R_2^1\, trid[(\Sigma_2^1)^{-1}]\, E_2^1.$$

$$M_{i,i+1} = B_{i,i+1} - Q_i^{m_i}\, trid[(\Delta_i^{m_i})^{-1}]\, C_i^{m_i} - R_{i+1}^1\, trid[(\Sigma_{i+1}^1)^{-1}]\, E_{i+1}^1 - H_i\, trid(M_{i-1,i}^{-1})\, J_{i-1},$$

$$S_i^1 = diag[(\Sigma_i^1)^{-1} E_i^1],$$
$$S_i^l = -diag[(\Sigma_i^l)^{-1} A_i^l S_i^{l-1}], \quad l = 2, \ldots, m_i,$$
$$H_i = -Q_i^{m_i} S_i^{m_i},$$

$$T_i^{m_i} = diag[(\Delta_i^{m_i})^{-1} C_i^{m_i}],$$
$$T_i^l = -diag[(\Delta_i^l)^{-1} B_i^l T_i^{l+1}], \quad l = m_i - 1, \ldots, 1,$$
$$J_{i-1} = -R_i^1 T_i^1.$$

In these formulas *diag* defines a diagonal approximation. Then, H_i and J_i are diagonal matrices. M_i is chosen as an INV block LU or UL approximation of B_i. Whatever is the approximation, we can solve independantly for the M_i's i.e. for each subdomain, but we have a block recursion for the reduced system i.e. the interfaces. We call this method INVDD.
As for the symmetric case there are many other possibilities giving more parallelism :
1) take $H_i = 0$, $\forall i$; then everything is parallel as there is no more recursion within the interfaces (INVDDH).
2) take only "some" $H_i = 0$, as needed by the number of available processors (INVDDS).
3) use an incomplete twisted factorization for the approximate reduced system [15].
Notice that these approaches are purely algebraic and are feasible for any diagonally dominant block tridiagonal M-matrices regardless of their origins.
Moreover, one can use different approximations (in place of INV) for the subdomains, like FFT-based preconditioners or point preconditioners. Modified (i.e. zero row sums) preconditioners are also possible.

6. Numerical results. We solve the following problem in the domain $\Omega =]0,1[\text{ x }]0,1[$,

$$-\Delta u + 2\alpha \frac{\partial u}{\partial x} + 2\beta \frac{\partial u}{\partial y} = f \quad \text{in } \Omega, \quad u \mid_{\partial\Omega} = 0.$$

We discretize with the 5 point scheme and central differences for the first order term. The right hand side is the same as in [15], the starting guess is a random vector, the stopping criterion is $\|r^k\|_2 \leq 10^{-6} \|r^0\|_2$ and the value of h is $\frac{1}{101}$.
We are interested in looking, for a fixed mesh size, at the number of iterations of the Orthomin(1) conjugate gradient-like method as a function of the number of subdomains k, for different values of α and β. We give only results for INVDD, the results for more parallel methods (including the number of operations) will be given in another paper [16].

k	$\alpha = 0, \beta = 0$	$\alpha = 1, \beta = 1$	$\alpha = 25, \beta = 50$	$\alpha = 100, \beta = -1$	$\alpha = 300; \beta = -1$
2	24	32	14	18	13
4	25	34	15	19	13
8	25	32	18	19	13
16	27	36	31	20	14
24	32	40	40	24	16
32	33	41	40	27	17
40	36	48	42	31	19
50	34	39	38	28	20

We can see that even if the increase in the number of iterations can be sometimes a little bit larger than for the symmetric case, it is still very slight so the use of these domain decomposition can be useful provided the more parallel techniques developed in [15] are used.

7. Conclusions. The Domain Decomposition methods presented in this paper offer a great deal of parallelism when used with iterative methods. Because of the algebraic nature of these preconditioners they can also be used with discontinuous coefficient problems as they are based on robust approximations.

References

[1] P. BJORSTAD & O.B. WIDLUND, *Iterative methods for the solution of elliptic problems on regions partioned into substructures.* SIAM J. on Numer. Anal. v 23, n 6, (1986) pp 1097–1120.

[2] J.H. BRAMBLE, J.E. PASCIAK & A.H. SCHATZ, *The construction of preconditioners for elliptic problems by substructuring. I.* Math. of Comp. v 47, n 175, (1986) pp 103–104.

[3] T. CHAN & D. RESASCO, *A domain decomposition fast Poisson solver on a rectangle.* Yale Univ. report YALEU/DCS/RR 409 (1985).

[4] T. CHAN, *Proceedings of the second international symposium on domain decomposition methods for partial differential equations.* SIAM (1988).

[5] P. CONCUS, G.H. GOLUB & G. MEURANT, *Block preconditioning for the conjugate gradient method.* SIAM J. Sci. Stat. Comp., v 6, (1985) pp 220–252.

[6] P. CONCUS & G. MEURANT, *On computing INV block preconditionings for the conjugate gradient method.* BIT v 26 (1986) pp 493–504.

[7] H. ELMAN, *Iterative methods for large sparse non symmetric systems of linear equations.* Ph.D. thesis, Technical Report 229, Yale University (1982).

[8] R. GLOWINSKI, G.H. GOLUB, G. MEURANT & J. PERIAUX, *Proceedings of the first international symposium on domain decomposition methods for partial differential equations.* SIAM (1988).

[9] G.H. GOLUB & D. MAYERS, *The use of preconditioning over irregular regions.* In "Computing methods in applied science and engineering VI", R. Glowinski & J.L. Lions Eds, North–Holland (1984).

[10] G. MEURANT, *The block preconditioned conjugate gradient method on vector computers.* BIT v 24 (1984) pp 623–633.

[11] G. MEURANT, *The conjugate gradient method on supercomputers.* Supercomputer v 13 (1986) pp 9–17.

[12] G. MEURANT, *Multitasking the conjugate gradient method on the CRAY X-MP/48.* Parallel Computing 5 (1987) pp 267–280.

[13] G. MEURANT, *Domain decomposition vs block preconditioning.* In [8].

[14] G. MEURANT, *Conjugate gradient preconditioners for parallel computers* In "Proceedings of the third SIAM conference on parallel processing for scientific computing", Los Angeles 1987, SIAM (1988).

[15] G. MEURANT, *Domain decomposition preconditioners for the conjugate gradient method.* submitted to Calcolo (1988)

[16] G. MEURANT, *Numerical experiments with domain decomposition preconditioners for non symmetric problems.* to appear

CHAPTER 19

A Domain Decomposition Method for Boundary Layer Problems*

Garry Rodrigue†
Edna Reiter††

Abstract: In this paper we analyze the behavior of a specific domain decomposition technique for solving a boundary value problem of the type

$$L_\varepsilon[u] = \frac{\partial}{\partial x}u + \varepsilon\Delta u = 0, \quad (x, y) \in \Omega$$

We are concerned primarily in problems where ε is sufficiently small and where the boundary conditions yield ordinary and parabolic boundary layers in the solution. The global domain is decomposed into subdomains according to the particular layers and the global solution is obtained by piecing together the different subdomain solutions. An algorithm for locating the layers and consequently the subdomains will be constructed. Numerical results will be presented.

1 Introduction

In this paper we study the application of a domain decomposition technique to the convection-diffusion equation

$$\frac{\partial u}{\partial x} - \varepsilon\left[\frac{\partial^2 u}{\partial x^2} + \frac{\partial^2 u}{\partial y^2}\right] = 0 \tag{1.1}$$

where $0 < \varepsilon << 1$ and $(x, y) \in \Omega = [0,1] \times [0,1]$. Given certain boundary conditions on (1.1), the solution is known to possess boundary layers where, for example, there can be regions of Ω such that

* This work was performed under the auspices of the U.S. Department of Energy at the Lawrence Livermore Laboratory under Contract W-7405-Eng-48.

† University of California, Davis, and Lawrence Livermore National Laboratory, Livermore, California 94550

‡ Department of Mathematics, California State University, Hayward, California 94542

$$\begin{aligned} u &\simeq \textstyle\sum_{n=0}^{m} \varepsilon^n \phi_n \\ \frac{\partial \phi_0}{\partial x} &= 0 \\ \frac{\partial \phi^n}{\partial x} &= \varepsilon \left[\frac{\partial^2 \phi_{n-1}}{\partial x^2} + \frac{\partial^2 \phi_{n-1}}{\partial y^2} \right], n > 0 \end{aligned}$$

(called ordinary boundary layers) or there can be regions such that

$$\begin{aligned} u &\simeq \textstyle\sum_{n=0}^{\ell} \varepsilon^n \Psi_n \\ \frac{\partial \Psi_0}{\partial x} - \frac{\partial^2 \Psi_0}{\partial \xi^2} &= 0 \\ \frac{\partial \Psi_n}{\partial x} - \frac{\partial^2 \Psi_n}{\partial \xi^2} &= \varepsilon \frac{\partial^2 \Psi_{n-1}}{\partial x^2}, \quad n > 0 \end{aligned}$$

(called parabolic boundary layers) and ξ is a local "stretched" coordinate, [1] .

In the following sections of the paper, we will define a domain decomposition on Ω based on the boundary layer behavior of the solution and then apply a variant of the Schwarz Alternating Procedure to obtain a numerical approximation of the solution of (1.1)

2 Parabolic Layers

In this paper, the boundary conditions we impose on (1.1) are

$$(2.1) \qquad \begin{aligned} &(a) \quad u(x,1) = 1, \qquad 0 \le x \le 1, \\ &(b) \quad u(x,0) = 0, \qquad 0 \le x \le 1, \\ &(c) \quad u(0,y) = 1, \qquad 0 < y < 1, \\ &(d) \quad \frac{\partial u}{\partial x}(1,y) = 0, \quad 0 < y < 1. \end{aligned}$$

In [2], it was established that the sequence of functions $\{u^{(n)}\}$ defined by the iteration

$$(2.2) \qquad \begin{cases} \dfrac{\partial u^{(0)}}{\partial x} - \varepsilon \dfrac{\partial^2 u^{(0)}}{\partial^2 y} = 0 \\[2ex] \dfrac{\partial u^{(n)}}{\partial x} - \varepsilon \dfrac{\partial^2 u^{(n)}}{\partial y^2} = \varepsilon \dfrac{\partial^2 u^{(n-1)}}{\partial x^2}, \quad n > 0 \end{cases}$$

$$\begin{aligned}
(2.2a)\quad u^{(n)}(x,1) &= 1, \quad 0 \le x \le 1 \\
(2.2b)\quad u^{(n)}(x,0) &= 0, \quad 0 \le x \le 1 \\
(2.2c)\quad u^{(n)}(\alpha,y) &= g(y), \quad 0 < y < 1, \ \alpha = \text{fixed}
\end{aligned}$$

satisfies

$$\|u - u^{(n)}\|_\infty = O(\varepsilon^{n+1})$$

when

$$\frac{d^k g}{\partial y^k}(0) = \frac{d^k g}{\partial y^k}(1) = 0, \quad k = 0,1,\ldots,n. \tag{2.3}$$

Since the boundary conditions (2.1a-c) do not satisfy (2.3), a domain decomposition strategy would be to split Ω into the subregions $\Omega = \Omega_1 \bigcup \Omega_2$ where

$$\begin{aligned}
\Omega_1 = &\ [0,\ell_1]\times[0,1], \quad \ell_1 > 0, \\
\Omega_2 = &\ [\ell_2,1]\times[0,1], \\
\ell_2 < &\ \ell_1,
\end{aligned}$$

(see Figure 1) and numerically solve (1.1)—(2.1a–d) on Ω_1. This is followed by numerically carrying out the iteration (2.2) with conditions (2.2a-c) and using

$$g(y) = u(\ell_2, y) \tag{2.4}$$

where u is the computed solution in Ω_1.

To test the feasibility of this approach, we take several values of $\ell_1 > 0$ and for each of these values we solve (1.1) with $\varepsilon = .0005$, the boundary conditions (2.1a-c) and

$$\frac{\partial u}{\partial x}(\ell_1, y) = 0, \quad 0 < y < 1. \tag{2.5}$$

We then carry out the iteration (2.2) on $\Omega_2 = [\ell_2,1]\times[0,1]$, $\ell_2 = \ell_1 - \Delta x$, and

$$\begin{aligned}
u^{(n)}(x,1) &= 1, & \ell_2 \le x \le 1 \\
u^{(n)}(x,0) &= 0, & \ell_2 \le x \le 1 \\
u^{(n)}(\ell_2,y) &= u(\ell_2,y), & 0 \le y \le 1.
\end{aligned}$$

On Ω_1, the equation (1.1) is approximated by 2nd-order finite differences on a grid with $\Delta x = .0015$ and $\Delta y = .001$ and the resultant linear system is solved with a direct matrix solver. (2.2), on the other hand, requires the solution of inhomogeneous heat equations with the x-variable interpreted as the time variable. In this situation, the Crank-Nicholson method is used on a grid with $\Delta x = .01, \Delta y = .001$. Table 1 lists the results. Each entry of the table is $\|U^{(n)} - U^{(n-1)}\|_\infty$ where

$U^{(n)}$ is the computed approximation to $u^{(n)}$. Note that convergence in all cases is quite rapid.

In [2], it is established that for fixed Δx and Δy, there exists $\varepsilon > 0$ so that divergence occurs. In order to determine such values,we use the same grid structure as before, take $\ell_2 = 0.041$, vary the ε and list the errors in Table 2. As can be seen, larger values of ε will result in divergence.

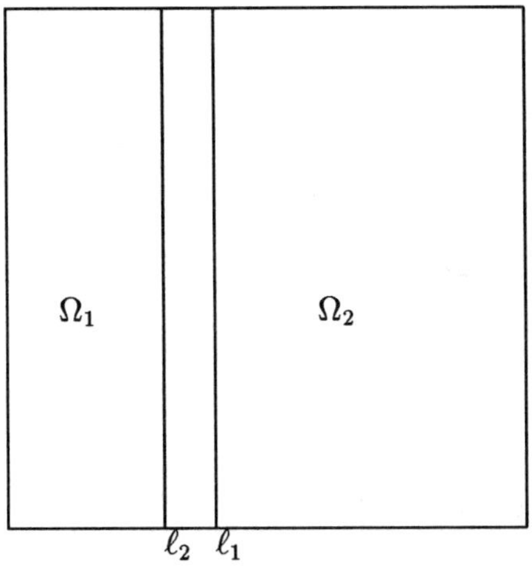

Figure 1

Table 1:

$n \backslash \ell_1$	.003	.018	.033	.048	.063
1	.90401	.77714	.71635	.67007	.63575
2	.01379	.00287	.00143	.00100	7.8×10^{-4}
3	.00601	1.15×10^{-4}	4.3×10^{-5}	2×10^{-5}	1.3×10^{-5}
4	3.9×10^{-5}	6.1×10^{-6}	2.08×10^{-6}	8.9×10^{-7}	5×10^{-7}
5	3.1×10^{-6}	3.8×10^{-7}	1.2×10^{-7}	5×10^{-8}	2×10^{-8}
6	2.8×10^{-7}	3×10^{-8}	1×10^{-8}		
7	2×10^{-8}				

Table 2:

$n \backslash \varepsilon$	5×10^{-5}	5×10^{-4}	5×10^{-3}	5×10^{-2}
1	.73911	.63575	.277827	.037807
2	.000178	.00078	.010518	0.8499
3	4.1×10^{-7}	1.3×10^{-5}	.00195	2.6304
4	3×10^{-9}	5×10^{-7}	.00085	21.944
5		2×10^{-8}	4.6×10^{-4}	184.6
6			2.9×10^{-4}	∞

3 Ordinary Layers

We consider the problem (1.1) with boundary conditions (2.1a-d). In [3] it was established that the sequence of functions $u^{(n)}$ defined by the iteration

$$(3.1)\qquad \begin{aligned} &(a)\quad u^{(0)}(x,y) &&= 1 \\ &(b)\quad \frac{\partial u^{(n)}}{\partial x} &&= \varepsilon\left[\frac{\partial^2 u^{(n-1)}}{\partial x^2} + \frac{\partial^2 u^{(n-1)}}{\partial y^2}\right],\ n > 0 \\ &(c)\quad u^{(n)}(x,0) &&= f(x), 0 \le x \le 1 \end{aligned}$$

$$(3.2)\qquad \frac{d^k f}{dx^k}(0) = 1,\ k = 0, 1, \ldots, n,$$

satisfies

$$\|u - u^{(n)}\|_\infty = O(\varepsilon^{n-1})$$

(u is the solution to (1.1)).

As before, since the boundary conditions (2.1c) does not satisfy (3.2), a domain decomposition strategy would be to split Ω into the subregions $\Omega = \Omega_1 \bigcup \Omega_2$ where

$$(3.3)\qquad \begin{aligned} \Omega_1 &= [0,1] \times [0, \ell_1],\ \ \ell_1 > 0 \\ \Omega_2 &= [0,1] \times [\ell_2, 1],\ \ \ell_2 < \ell_1, \end{aligned}$$

(see Figure 2) and numerically solve (1.1) on Ω_1, followed by numerically carrying out the iteration (3.1a-c) with

$$(3.4)\qquad f(x) = u(x, \ell_2)$$

where u is computed solution in Ω_1

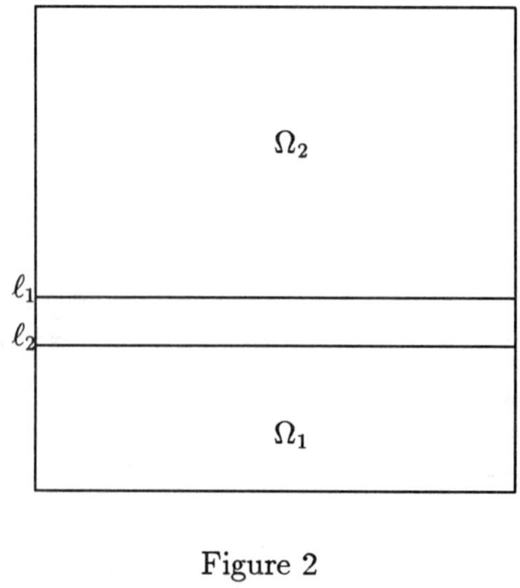

Figure 2

To test the feasibility of this approach, we take several values of $\ell_2 > 0$ and solve (3.1a-c)—(3.4). As a test case, we take $\varepsilon = 2\times10^{-4}$, use the exact solution of (1.1) on Ω_1 (cf. [2]) and then carry out (3.1a–c) -(3.4) on Ω_2 using a Backward Euler method on a grid with $\Delta x = \Delta y = 10^{-2}$. Convergence to an error of $\|U^{(n)} - U^{(n-1)}\|_\infty < 10^{-5}$ occurred at 30 iterations for each value of $\ell_2 = 10^{-2}, 10^{-1}, 4\times10^{-1}$. Also, for $\ell_2 = 10^{-2}$, convergence occurred for $\varepsilon = 10^{-4}$ whereas divergence occurred for $\varepsilon = 5\times10^{-4}$.

4 Schwarz Method

In this section we develop a Schwarz Alternating Procedure for solving (1.1) based on the results of the previous two sections. That is, we split $\Omega = \Omega_1 \bigcup \Omega_2 \bigcup \Omega_3$ where

$$\begin{aligned} \Omega_1 = &\ [0,1]\times[b,1] \\ \Omega_2 = &\ [0,r]\times[0,t] \quad ,b<t, \\ \Omega_3 = &\ [\ell,1]\times[0,t] \quad ,\ell<r, \end{aligned}$$

(see Figure 3). Let $u_1^{(0)} = u_2^{(0)} = u_3^{(0)} = 1$, on $\Omega_1, \Omega_2, \Omega_3$ respectively. We then define the sequences $\{u_1^{(i)}\}, \{u_2^{(i)}\}, \{u_3^{(i)}\}$ as follows: for $i = 1, 2, \ldots,$

1) $u_2^{(i)}$ solves (1.1) on Ω_2 with

$$\begin{aligned} u_2^{(i)} &= u_1^{(i)} && \text{on} \quad [0,r]\times\{t\} \\ u_2^{(i)} &= 1 && \text{on} \quad \{0\}\times[0,t] \\ u_2^{(i)} &= 0 && \text{on} \quad [0,r]\times\{0\} \\ \frac{\partial}{\partial x}u_2^{(i)} &= 0 && \text{on} \quad \{r\}\times[0,t] \end{aligned}$$

2) $u_3^{(i)}$ solves (2.2) on Ω_3 with

$$\begin{aligned} u_3^{(i)} &= u_2^{(i)} && \text{on} \quad \{\ell\}\times[0,t] \\ u_3^{(i)} &= u_1^{(i-1)} && \text{on} \quad [\ell,1]\times\{t\} \\ u_3^{(i)} &= 0 && \text{on} \quad [\ell,1]\times\{0\} \end{aligned}$$

3) $u_1^{(i)}$ solves (3.1b) on Ω_1 with

$$\begin{aligned} u_1^{(i)} &= u_2^{(i)} \text{ on } && [0,r]\times\{b\} \\ u_1^{(i)} &= u_3^{(i)} \text{ on } && [r,1]\times\{b\} \end{aligned}$$

We carried out the above iterations under different scenarios to examine its convergence behavior to the solution of (1.1). In all cases, the following mesh sizes were used:

$$\Omega_1 \; : \; \Delta x = \Delta y = 10^{-2}$$

$$\Omega_2 \; : \; \Delta x = \Delta y = 10^{-3}$$

$$\Omega_3 \; : \; \Delta x = 10^{-2}, \; \Delta y = 10^{-3}$$

The numerical method of solution was the same as that in sections 2 and 3.

In the first experiment, the lower boundary of Ω_1 is fixed and the boundary between Ω_2 and Ω_3 is varied. In this case

$$\begin{aligned} \varepsilon &= 10^{-4} \\ b &= .03 \\ t &= .031 \\ r &= 0.41 \\ \ell &= r - k \times 10^{-3} \end{aligned}$$

Table 3 records the results.

In the second experiment, the boundary between Ω_2 and Ω_3 is fixed and the lower boundary of Ω_1 is varied. In this case,

$$\begin{aligned} \varepsilon &= 2 \times 10^{-4} \\ t &= 0.31 \\ b &= t - k \times 10^{-2} \\ r &= 0.41 \\ \ell &= .04 \end{aligned}$$

Table 4 records the results.

In the final experiment, the boundaries of $\Omega_1, \Omega_2, \Omega_3$ are held fixed and the value of ε is varied. In this case,

$$\begin{aligned} b &= 0.3 \\ t &= 0.31 \\ r &= 0.41 \\ \ell &= 0.30 \end{aligned}$$

Table 5 records the results.

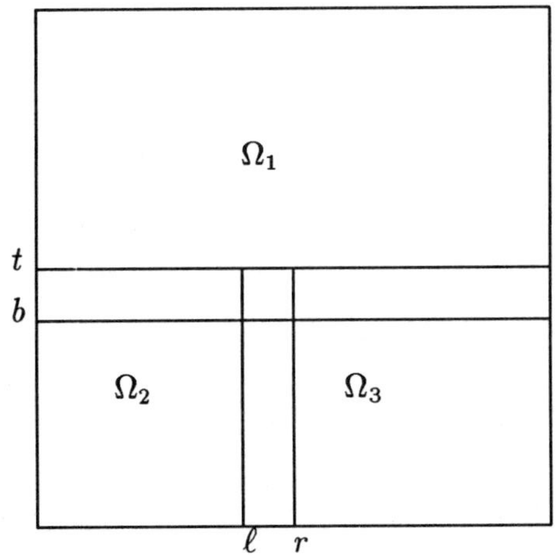

Figure 3

Table 3:

$n\backslash k$	1	.20	.40
1	.64325	.72092	.891205
2	.0022	.00213	.002195
3	5.26×10^{-4}	5.05×10^{-4}	4.53×10^{-4}
4	1.68×10^{-4}	1.59×10^{-4}	1.39×10^{-4}
5	5.5×10^{-5}	5.2×10^{-5}	4.3×10^{-5}
6	1.8×10^{-5}	1.6×10^{-5}	1.3×10^{-5}
7	5×10^{-6}	5×10^{-6}	4×10^{-6}
8	2×10^{-6}	1×10^{-6}	1×10^{-6}

Table 4:

$n\backslash k$	1	10	20
1	.2580	.3363	.5799
3	.6844	.2859	.0682
5	.9212	.4923	.0846
7	.7625	.4753	.0623
9	.4705	.3134	.0325
11	.2133	.1572	.0132
13	.0734		
15	.02		

Table 5:

$n\backslash\varepsilon$	5×10^{-5}	2×10^{-4}	2.5×10^{-4}	$3. \times 10^{-4}$	3.5×10^{-4}
1	.6684	.6718	.6659	.6561	.6432
2	10^{-4}	.0184	.0304	.0502	.0802
3	7×10^{-5}	.0146	.0428	.0978	.1907
4	8×10^{-7}	.0148	.0571	.1634	.3838
5	10^{-7}	.0129	.0658	.2344	.6615
6	10^{-8}	.0101	.0676	.2986	1.0091
7		.0072	.0629	.3437	∞
8		.0047	.0536	.3621	
9		.003	.0438	.3647	
10		.0018	.0337	.3459	
11		.001	.0241	.3044	

References

[1] Eckhaus, W., *Asymptotic Analysis of Singular Perturbations*, North-Holland, 1979

[2] Rodrigue, G., Hedstrom, G., Chin, C., and Reiter,E., *Domain Decomposition for Parabolic Boundary Layer Problems*; to appear

[3] Eckhaus, W. and de Jaeger, E.M., *Asymptotic Solutions of Singular Perturbation Problems for Linear Differential Equations of Elliptic type*, Arch. National Mech. Anal., Vol.23, p.p. 26-86

Wavefront Elimination and Renormalization*

Wei Pai Tang†

Abstract. Recently, a new class of optimal fast solver for the model problem –wavefront elimination ($\mathcal{WE}$)– has been developed using template operators. The complexity of this type of new algorithms is only one fraction of the cost of the multigrid method applied to the same problem. These algorithms have potential for an efficient parallel implementation and also as preconditioning operators for general elliptic problems.

A more interesting fact is that this algorithm shows some conceptual connection with two new theories – renormalization in physics and fractals in mathematics. In this paper, we will demonstrate these rather interesting relations between the wavefront elimination, the renormalization theory and fractals.

Key Words. Wavefront elimination ($\mathcal{WE}$), renormalization, fractal, template operator ($\mathcal{TO}$), fast solver.

1. Introduction. The study of fast solvers is always an important part of research in domain decomposition [5]. Using template operator ($\mathcal{TO}$) - a new structure of the linear operator in finite dimensional space [6] - some optimal fast solvers are presented here. This result has answered an open question, namely, can a direct approach for solving the model problem achieve an optimal complexity? In particular, the complexity of our new algorithm, called wavefront elimination ($\mathcal{WE}$), is even better than the complexity of the multi-grid method for solving the same problem. Different from a traditional approach, there is more than one discrete $\mathcal{TO}$ which is used on the same grid point. The combination of these different $\mathcal{TO}$'s makes the sparsity of the consequent $\mathcal{TO}$'s during the $\mathcal{WE}$ process on the same grid point possible. A more interesting fact is that this new algorithm exhibits some amusing conceptual relationships with two new important theories - renormalization in physics and fractals in mathematics. Renormalization theory was discovered by K.G. Wilson at Cornell[7]. The basic idea of this theory can be understood as a successive thinning out of the degree of freedom in the partition function. The N-particle problem is transformed into an N'-particle

* This research was supported by the Natural Sciences and Engineering Research Council of Canada.

† Department of Computer Science, University of Waterloo, Waterloo, Ontario, Canada N2L 3G1.

problem with $N' < N$, whereby the temperature T and the magnetic field H may also have to be renormalized. The wavefront elimination process can be viewed as a renormalization process from different scales. During the elimination process, the template operators display a sequence of self-similar structure on the different scales. In the backward solution process of the $\mathcal{WE}$, we see a procedure surprisingly similar to the recursively detailing a fractal. These interesting facts contribute another convincing example of a fundamental principle which organizes a whole universe. As we know this principle has been fascinatingly demonstrated by fractal geometry and renormalization theory. As we will see in this paper this principle has also provided us with guidance for designing a whole new class of algorithm, for the numerical solution of P.D.E's.

2. Template Operators. In [6] the template operator was first introduced for identifying the problems for which the Schwarz Splittings are most suitable. In the same paper it was also successfully applied to obtain a "good" splitting when the Schwarz approach is used. The unique features of this structure are as follows. First, in a template operator the artificial sequential constraints in the matrix structure are removed. The original topological frame of the continuous problem from which the discrete operator is derived is well preserved. Second, in particular, the locality of the operator and the proximity of the variables are also maintained in this new structure. Therefore, many physical phenomena which are related to the topology of the solution region can be easily presented in this form. That is the key to the successful application of $\mathcal{TO}$ in the study of Schwarz Splittings. A more important feature is that the $\mathcal{TO}$ provides us a "graphical structure" to think in pictures.

> *In graphical representation, natural processes can be comprehended in their full complexity by intuition. New ideas and associations are stimulated, and the creative potential of all those who think in pictures is awakened.*
>
> —The Beauty of Fractals

In this paper, we will show how the $\mathcal{TO}$ can provide us a "graphical" structure for designing a new class of fast solver.

In [6] the template operator is presented as a structure which is mathematically equivalent to the matrix form, in other words, we can find a one-to-one mapping between a template operator and a matrix. Here a similar idea is used, but the formality is different though we can present the following algorithm in terms of the form we used in [6] or even in terms of matrices. In order to display the interesting connection between our fast algorithm and its physical interpretations, we will adapt the definition of the template operator in this paper for this new context[1].

Consider the Dirichlet problem

$$\begin{cases} \Delta u(x,y) = f(x,y) \\ u(x,y)|_{\Gamma_\Omega} = g(x,y) \end{cases}$$

where Ω is a unit square in the (x,y) plane. Let us lay an equally spaced mesh on this square and let the mesh size be $h = \frac{1}{n+1}$. Here we always assume $n = 2^L - 1$. The following figure shows a grid with $n = 2^3 - 1$.

[1] Since we are mainly discussing the Dirichlet problem of the 2-dimensional Poisson equation on a unit square region in this paper, the pictures of templates are mostly demonstrated on a square region with equally spaced mesh. But the same idea can be directly applied to any irregular region and irregular mesh such as finite element triangulations (see [6]).

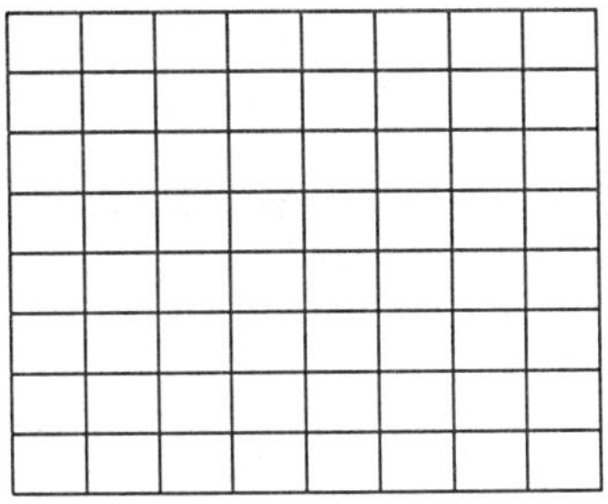

The traditional numerical technique for this problem is first to discretize the differential operator on every interior node, then to form a matrix equation and finally to solve this matrix equation by some efficient algorithm. Here a different approach is used. There is no explicit system of linear equations involved. Each node can have as many operators as needed. These operators all represent some local physical principles under some conditions. In order to preserve the locality of these operators and the topological relations of the variables, a new structure of the discrete operators is introduced as follows: we define a template $\mathcal{T}$ as:

$$
\mathcal{T} = \begin{bmatrix}
b_{0,0} & b_{0,1} & b_{0,2} & \cdots & b_{0,n} & b_{0,n+1} \\
b_{1,0} & O_{1,1} & O_{1,2} & \cdots & O_{1,n} & b_{1,n+1} \\
b_{2,0} & O_{2,1} & O_{2,2} & \cdots & O_{2,n} & b_{2,n+1} \\
\vdots & \vdots & \vdots & & \vdots & \vdots \\
b_{n,0} & O_{n,1} & O_{n,2} & \cdots & O_{n,n} & b_{n,n+1} \\
b_{n+1,0} & b_{n+1,1} & b_{n+1,2} & \cdots & b_{n+1,n} & b_{n+1,n+1}
\end{bmatrix}
$$

where b_{ij} represent the boundary nodes while O_{ij} represent the interior nodes. A template vector

$$
U = \begin{bmatrix}
g_{0,0} & g_{0,1} & g_{0,2} & \cdots & g_{0,n} & g_{0,n+1} \\
g_{1,0} & x_{1,1} & x_{1,2} & \cdots & x_{1,n} & g_{1,n+1} \\
\vdots & \vdots & \vdots & & \vdots & \vdots \\
g_{n,0} & x_{n,1} & x_{n,2} & \cdots & x_{n,n} & g_{n,n+1} \\
g_{n+1,0} & g_{n+1,1} & g_{n+1,2} & \cdots & g_{n+1,n} & g_{n+1,n+1}
\end{bmatrix}
$$

can be defined on this template, where g_{ij} are the boundary values on the boundary nodes b_{ij}, which are known, and x_{ij} are the values of the unknown $u(x,y)$ of the continuous problem at node O_{ij} [2].

Given a finite difference approximation, for example a five-point stencil, a template operator Δ_{ij} for node O_{ij} can be defined as:

$$
\begin{array}{c}
j \\
\Delta_{ij} = \dfrac{1}{h^2} \begin{bmatrix}
0 & \cdots & 0 & 0 & 0 & \cdots & 0 \\
\vdots & \vdots & \vdots & \vdots & \vdots & & \\
0 & \cdots & 0 & -1 & 0 & \cdots & 0 \\
0 & \cdots & -1 & 4 & -1 & \cdots & 0 \\
0 & \cdots & 0 & -1 & 0 & \cdots & 0 \\
\vdots & \vdots & \vdots & \vdots & \vdots & & \\
0 & \cdots & 0 & 0 & 0 & \cdots & 0
\end{bmatrix}
\end{array}
$$

---

[2] For simplicity, we only define one function value on each node. There is no difficulty in generalizing this definition to include the cases where both the function value and its derivatives are needed on each node. In [6], for a generalized version of $\mathcal{T}$ even a state vector can be defined on each node.

We will also only discuss the finite difference approximation in this paper for the same reason.

where $i = 1, \cdots n$; $j = 1, \cdots n$ and 4 is located in position (i, j). Note that the values of x_{ij} in the template vector U are sampled from the true solution of the continuous Poisson equation.

The operation of Δ_{ij} on U is to multiply the elements of Δ_{ij} with the corresponding elements of U, and the summation of these products is the result of this operation. Thus we have

$$\Delta_{ij} \circ U = \frac{1}{h^2}[-x_{i-1,j} - x_{i+1,j} - x_{i,j-1} - x_{i,j+1} + 4x_{ij}].$$

The Poisson operator at O_{ij} can be written in terms of the TO

(1) $$\Delta_{ij} \circ U = f_{ij} + \tau_{ij},$$

where $f_{ij} = f(ih, jh)$ and τ_{ij} is the truncation term for the five-point stencil. It is known that $\tau_{ij} \approx O(h^2)$.

Since most of the elements in Δ_{ij} are zeros, a compact notation for Δ_{ij} is convenient. Let

$$\Delta(i, j, 1, \frac{1}{h^2}) = \frac{1}{h^2} \begin{bmatrix} 0 & -1 & 0 \\ -1 & 4 & -1 \\ 0 & -1 & 0 \end{bmatrix}_{i,j,1}$$

denote Δ_{ij}, where i, j is the center position of the template operator; 1 denotes the increment of the indices and $\frac{1}{h^2}$ is the constant factor of the operator. Then (1) can be rewritten as:

$$\begin{aligned} \Delta(i, j, 1, \tfrac{1}{h^2}) \circ U &= \frac{1}{h^2} \begin{bmatrix} 0 & -1 & 0 \\ -1 & 4 & -1 \\ 0 & -1 & 0 \end{bmatrix}_{i,j,1} \circ U \\ &= \frac{1}{h^2}(-x_{i-1,j} - x_{i+1,j} - x_{i,j-1} - x_{i,j+1} + 4x_{ij}) \\ &= f_{ij} + \tau_{ij} \end{aligned}$$

The compact form for the template operator derived from a skewed five point stencil at node O_{ij} is:

$$\tilde{\Delta}_{ij} = \tilde{\Delta}(i, j, 1, \frac{1}{2h^2}) = \frac{1}{2h^2} \begin{bmatrix} -1 & 0 & -1 \\ 0 & 4 & 0 \\ -1 & 0 & -1 \end{bmatrix}_{i,j,1}.$$

Then the same Poisson operator at O_{ij} can also be expressed by $\tilde{\Delta}_{ij}$ as:

$$\begin{aligned} \tilde{\Delta}_{ij} \circ U &= \frac{1}{2h^2}(-x_{i-1,j-1} - x_{i+1,j-1} - x_{i-1,j+1} - x_{i+1,j+1} + 4x_{ij}) \\ &= f_{ij} + \tilde{\tau}_{ij} \end{aligned}$$

where $\tilde{\tau}_{ij}$ is the truncation term for the skewed five point stencil at node O_{ij}. Similarly, we have the template operators derived from a nine-point and a skewed nine-point stencil at node O_{ij}:

$$\Delta'_{ij} = \Delta'(i, j, 1, \frac{1}{6h^2}) = \frac{1}{6h^2} \begin{bmatrix} -1 & -4 & -1 \\ -4 & 20 & -4 \\ -1 & -4 & -1 \end{bmatrix}_{i,j,1}$$

$$\Delta''_{ij} = \Delta''(i,j,1,\frac{1}{12h^2}) = \frac{1}{12h^2}\begin{bmatrix} 0 & 0 & -1 & 0 & 0 \\ 0 & -4 & 0 & -4 & 0 \\ -1 & 0 & 20 & 0 & -1 \\ 0 & -4 & 0 & -4 & 0 \\ 0 & 0 & -1 & 0 & 0 \end{bmatrix}_{i,j,1}$$

The last two operators have an accuracy of $O(h^4)$. Note, if the template operator derived from the five-point stencil at node O_{ij} uses mesh size $2h$, it can be written as

$$\Delta(i,j,2,\frac{1}{4h^2})\circ U = \frac{1}{4h^2}\begin{bmatrix} 0 & -1 & 0 \\ -1 & 4 & -1 \\ 0 & -1 & 0 \end{bmatrix}_{i,j,2}$$

$$\begin{aligned} \Delta(i,j,2,\tfrac{1}{4h^2})\circ U &= \frac{1}{4h^2}(-x_{i-2,j} - x_{i+2,j} - x_{i,j-2} - x_{i,j+2} + 4x_{ij}) \\ &= f_{ij} + O[(2h)^2] \end{aligned} \tag{2}$$

As we mentioned above, in a traditional approach, only one template operator at each node is used. But in our wavefront elimination process, more than one, even several different template operators at each node are used simultaneously. This is one key difference between the $\mathcal{WE}$ and conventional methods.

3. Wavefront Eliminations. The influencing and influenced wavefronts for each node in a template operator are introduced in [6]. These concepts characterize the propagation of the influences between nodes. Here the concept of wavefronts is used in a different context, namely to characterize the change of the non-zero pattern during the elimination process.

The first wavefront of any node O_{ij} in a template operator is defined here as the set of nodes where the elements of the template operator Δ_{ij} are non-zero except the node O_{ij} itself. Since we will use two or even more different template operators at each node O_{ij}, the wavefront will certainly be referred to the corresponding template operator at O_{ij}. For example, the first wavefront W_{ij} for Δ_{ij} is

$$W_{ij} := \{O_{i,j-1}, O_{i,j+1}, O_{i+1,j}, O_{i-1,j}\},$$

while the first wavefront $\widetilde{W}_{ij}$ for $\widetilde{\Delta}_{ij}$ is

$$\widetilde{W}_{ij} := \{O_{i-1,j-1}, O_{i-1,j+1}, O_{i+1,j-1}, O_{i+1,j+1}\}.$$

Similarly denote W'_{ij} as the wavefront for Δ'_{ij} and W''_{ij} for Δ''_{ij}, where

$$W'_{ij} := W_{ij} \cup \widetilde{W}_{ij},$$

and

$$W''_{ij} := \widetilde{W}_{ij} \cup \{O_{i-2,j}, O_{i+2,j}, O_{i,j+2}, O_{i,j-2}\}.$$

There are many ways of performing wavefront eliminations. It is actually a whole class of direct approaches for the numerical solution of elliptic P.D.E's. To illustrate the basic idea of the wavefront eliminations let us introduce the simplest version of wavefront eliminations below. It is easy to see from the following description of the new algorithms: if we change the combination of the template operators and chose the different wavefronts, we may obtain many different algorithms. They all have an optimal complexity though the constants for each approach will be slightly different. The simplest case of a wavefront elimination only uses two kinds of template operators at each node O_{ij}, namely Δ_{ij} and $\widetilde{\Delta}_{ij}$.

Let us define the addition $\Delta_1+\Delta_2$ of two template operators Δ_1 and Δ_2 to be a new template operator such that the elements of $\Delta_1+\Delta_2$ are the sum of the corresponding elements of Δ_1 and Δ_2. Likewise, define the scalar product $\alpha\Delta$ be a new template operator such that the elements of $\alpha\Delta$ are the products of α with the corresponding elements of Δ.

One step of the wavefront elimination is to eliminate the non-zero elements of a template operator, say Δ_{ij}, by the template operators at the nodes of its first wavefront. Let

$$\Delta_{ij} \circ U = f_{ij} + \tau_{ij}, \tag{3}$$

$$\Delta_{i-1,j} \circ U = f_{i-1,j} + \tau_{i-1,j}, \tag{4}$$

$$\Delta_{i+1,j} \circ U = f_{i+1,j} + \tau_{i+1,j}, \tag{5}$$

$$\Delta_{i,j-1} \circ U = f_{i,j-1} + \tau_{i,j-1}, \tag{6}$$

$$\Delta_{i,j+1} \circ U = f_{i,j+1} + \tau_{i,j+1}, \tag{7}$$

where $\tau_{ij}, \tau_{i+1,j}, \cdots$ are the truncation terms.

It is known that $\tau_{ij} \approx O(h^2), i, j = 1, \cdots, n$. Then one step of the wavefront elimination for Δ_{ij} can be written as

$$\begin{aligned} \Delta_{ij} &+ \frac{1}{4}(\Delta_{i-1,j} + \Delta_{i+1,j} + \Delta_{i,j-1} + \Delta_{i,j+1}) \\ &= \frac{1}{4h^2}\begin{bmatrix} 0 & 0 & -1 & 0 & 0 \\ 0 & -2 & 0 & -2 & 0 \\ -1 & 0 & 12 & 0 & -1 \\ 0 & -2 & 0 & -2 & 0 \\ 0 & 0 & -1 & 0 & 0 \end{bmatrix}_{i,j,1} \end{aligned}$$

To simplify the notation, let

$$\begin{aligned} \mathcal{W}_1(y_{ij}) &= y_{ij} + \frac{1}{4}(y_{i-1,j} + y_{i+1,j} + y_{i,j-1} + y_{i,j+1}), \\ \mathcal{W}_2(y_{ij}) &= y_{ij} + \frac{1}{4}(y_{i-1,j-1} + y_{i+1,j-1} + y_{i+1,j+1} + y_{i-1,j+1}) \end{aligned}$$

where y_{ij} is defined on node O_{ij}. Note that y can be a template operator or a grid function. Denote

$$\Delta^{(0.5)}(i,j,1,\frac{1}{4h^2}) = \frac{1}{4h^2}\begin{bmatrix} 0 & 0 & -1 & 0 & 0 \\ 0 & -2 & 0 & -2 & 0 \\ -1 & 0 & 12 & 0 & -1 \\ 0 & -2 & 0 & -2 & 0 \\ 0 & 0 & -1 & 0 & 0 \end{bmatrix}_{i,j,1}.$$

Then we have

$$\mathcal{W}_1(\Delta_{ij}) = \Delta^{(0.5)}(i,j,1,\frac{1}{4h^2}).$$

If the wavefront elimination is carried out on the equations (3) – (7) we will have

$$\mathcal{W}_1(\Delta_{ij}) \circ U = \mathcal{W}_1(f_{ij}) + \mathcal{W}_1(\tau_{ij}).$$

Comparing the non-zero pattern of $\mathcal{W}_1(\Delta_{ij})$ with $\widetilde{\Delta}_{ij}$, it is easy to see that in one more step of elimination the non-zero elements of $\mathcal{W}_1(\Delta_{ij})$ at position $(i+1, j+1)$, $(i+1, j-1)$, $(i-1, j+1)$, $(i-1, j-1)$ can be annihilated as follows:

$$\Delta^{(0.5)}(i,j,1,\tfrac{1}{4h^2}) - \widetilde{\Delta}_{ij} = \frac{1}{4h^2}\begin{bmatrix} 0 & 0 & -1 & 0 & 0 \\ 0 & 0 & 0 & 0 & 0 \\ -1 & 0 & 4 & 0 & -1 \\ 0 & 0 & 0 & 0 & 0 \\ 0 & 0 & -1 & 0 & 0 \end{bmatrix}_{i,j,1} \tag{8}$$

$$= \frac{1}{4h^2}\begin{bmatrix} 0 & -1 & 0 \\ -1 & 4 & -1 \\ 0 & -1 & 0 \end{bmatrix}_{i,j,2} \tag{9}$$

$$= \Delta(i,j,2,\frac{1}{(2h)^2}) \tag{10}$$

To summarize the above steps, we have

$$\mathcal{W}_1(\Delta_{ij}) - \widetilde{\Delta}_{ij} = \Delta(i,j,2,\frac{1}{(2h)^2}).$$

Let

$$\begin{aligned} \Delta_{ij}^{(2)} &= \Delta(i,j,2,\frac{1}{(2h)^2}), \\ f_{ij}^{(1)} &= f_{ij}, \\ \widetilde{f}_{ij}^{(1)} &= \widetilde{f}_{ij}, \\ \tau_{ij}^{(1)} &= \tau_{ij}, \\ \widetilde{\tau}_{ij}^{(1)} &= \widetilde{\tau}_{ij}, \\ f_{ij}^{(2)} &= \mathcal{W}_1(f_{ij}) - \widetilde{f}_{ij}^{(1)}, \\ \tau_{ij}^{(2)} &= \mathcal{W}_1(\tau_{ij}) - \widetilde{\tau}_{ij}^{(1)}. \end{aligned}$$

We have

$$\begin{aligned} \Delta_{ij}^{(2)} \circ U &= \frac{1}{(2h)^2}\begin{bmatrix} 0 & -1 & 0 \\ -1 & 4 & -1 \\ 0 & -1 & 0 \end{bmatrix}_{i,j,2} \circ U \\ &= \frac{1}{4h^2}(-x_{i-2,j} - x_{i+2,j} - x_{i,j-2} - x_{i,j+2} + 4x_{ij}) \\ &= f_{ij}^{(2)} + \tau_{ij}^{(2)}. \end{aligned}$$

Applying a similar procedure to $\widetilde{\Delta}_{ij}$, we will have

$$\begin{aligned} \widetilde{\Delta}_{ij}^{(2)} &= \mathcal{W}_2(\widetilde{\Delta}_{ij}) - \Delta_{ij}^{(2)} \\ &= \widetilde{\Delta}(i,j,2,\frac{1}{8h^2}) \\ &= \frac{1}{8h^2}\begin{bmatrix} -1 & 0 & -1 \\ 0 & 4 & 0 \\ -1 & 0 & -1 \end{bmatrix}_{i,j,2}, \end{aligned}$$

and

$$\tilde{\Delta}_{ij}^{(2)} \circ U = \tilde{f}_{ij}^{(2)} + \tilde{\tau}_{ij}^{(2)},$$

where

$$\begin{aligned} \tilde{f}_{ij}^{(2)} &= \mathcal{W}_2(f_{ij}^{(1)}) - f_{ij}^{(2)}, \\ \tilde{\tau}_{ij}^{(2)} &= \mathcal{W}_2(\tau_{ij}^{(1)}) - \tau_{ij}^{(2)}. \end{aligned}$$

This is one complete step of a wavefront elimination. Here a new set of template operators at node O_{ij} is obtained:

$$\begin{aligned} \Delta_{ij}^{(2)} \circ U &= \frac{1}{(2h)^2} \begin{bmatrix} 0 & -1 & 0 \\ -1 & 4 & -1 \\ 0 & -1 & 0 \end{bmatrix}_{i,j,2} \circ U = f_{ij}^{(2)} + \tau_{ij}^{(2)}, \\ \tilde{\Delta}_{ij}^{(2)} \circ U &= \frac{1}{8h^2} \begin{bmatrix} -1 & 0 & -1 \\ 0 & 4 & 0 \\ -1 & 0 & -1 \end{bmatrix}_{i,j,2} \circ U = \tilde{f}_{ij}^{(2)} + \tilde{\tau}_{ij}^{(2)}. \end{aligned}$$

The increment of the index in these operators is 2 now. If we recursively proceed with this process we will obtain

$$\begin{aligned} \Delta_{ij}^{(k)} \circ U &= \frac{1}{(2^{k-1}h)^2} \begin{bmatrix} 0 & -1 & 0 \\ -1 & 4 & -1 \\ 0 & -1 & 0 \end{bmatrix}_{i,j,2^{k-1}} \circ U = f_{ij}^{(k)} + \tau_{ij}^{(k)}, \\ \tilde{\Delta}_{ij}^{(k)} \circ U &= \frac{1}{2(2^{k-1}h)^2} \begin{bmatrix} -1 & 0 & -1 \\ 0 & 4 & 0 \\ -1 & 0 & -1 \end{bmatrix}_{i,j,2^{k-1}} \circ U = \tilde{f}_{ij}^{(k)} + \tilde{\tau}_{ij}^{(k)}. \end{aligned}$$

Here a sequence of self-similar template operators is generated for the same node at different stages of the elimination. If we call them *W*–template operators [3], they look exactly the same as an ordinary template operator which is derived from a large mesh size. The key difference here is the right hand side of the equation derived from wavefront elimination – it is derived from a *renormalization* process which we will discuss in the next section. The right hand side of the traditional template operator is simply the function value of the source term at the given node. Notice that we have always kept the truncation term with the template operator during the above discussion. The motivation is to let both template operators $\Delta_{ij}^{(k)}$ and $\tilde{\Delta}_{ij}^{(k)}$ operate on the same U. Furthermore, the analysis of the error propagation during the elimination process can be easily shown by the growth of τ_{ij}^{k} and $\tilde{\tau}_{ij}^{(k)}$. So far the rigorous error analysis has not been completed but the following informal discussion provides us with a very rough picture. From the definition of the wavefront elimination process, we have the recurrence relations

$$\begin{aligned} \tau_{ij}^{(k)} &= \mathcal{W}_1(\tau_{ij}^{(k-1)}) - \tilde{\tau}_{ij}^{(k-1)}, \\ \tilde{\tau}_{ij}^{(k)} &= \mathcal{W}_2(\tilde{\tau}_{ij}^{(k-1)}) - \tau_{ij}^{(k)}. \end{aligned}$$

[3] *W* - stands for wavefront

Because the numbers of the elements in $\tau_{ij}^{(k)}$ and $\tilde{\tau}_{ij}^{(k)}$ are reduced by a factor of 4 at each step, we can not apply some traditional techniques for the recurrence relations to this problem. But from this relation we may observe a rough estimate of the growth factor of $\tilde{\tau}_{ij}^{(k)}$ and $\tilde{\tau}_{ij}^{(k)}$. As we know that the eigenvalues of the operator $\mathcal{W}_1$ are:

$$\lambda_{ij} = 1 + \frac{1}{2}\left(\cos\frac{i\pi}{2^{L-k}} + \cos\frac{j\pi}{2^{L-k}}\right), \tag{11}$$

$$i = 1, 2, \cdots, (2^{L-k} - 1), \tag{12}$$

$$j = 1, 2, \cdots, (2^{L-k} - 1). \tag{13}$$

Thus the maximum growth factor for $\mathcal{W}_1(\tau_{ij})$ is less than 2. After subtracting $\tilde{\tau}_{ij}^{(k-1)}$ from $\mathcal{W}_i(\tau_{ij}^{(k-1)})$, the rough estimate of the growth factor of $\tau_{ij}^{(k)}$ is less than 1 or close to 1. We may obtain a similar rough estimate for $\tilde{\tau}_{ij}^{(k)}$. Preliminary numerical tests agree with this estimate but the result is by no means concrete.

Since the increment of the index is doubled at each step of the elimination, the number of the nodes involved in the elimination process is reduced by a factor of 4. A compact description of the above discussion is as follows:

1. Let $k = 1$

$$\begin{aligned} f_{ij}^{(1)} &= f((i-1)h, (j-1)h) \\ \tilde{f}_{ij}^{(1)} &= f((i-1)h, (j-1)h) \\ & \qquad i = 1, \cdots, 2^L - 1; \\ & \qquad j = 1, \cdots, 2^L - 1; \end{aligned}$$

2. For nodes $p = 2^k i$, $q = 2^k j$ where $i = 1, \cdots, 2^{L-k} - 1$, $j = 1, \cdots, 2^{L-k} - 1$.

$$\begin{aligned} f_{pq}^{(k+1)} &= \mathcal{W}_1(f_{pq}^{(k)}) - \tilde{f}_{pq}^{(k)}, \\ \tilde{f}_{pq}^{(k+1)} &= \mathcal{W}_2(\tilde{f}_{pq}^{(k)}) - f_{pq}{}^{(k+1)}. \end{aligned}$$

3. k=k+1, if $k < L$ go to step 2.

This is a forward elimination process. After $L - 1$ steps of wavefront eliminations, the template operator $\Delta_{ii}^{(L-1)}$ for the center node O_{ii} where $i = 2^{L-1} - 1$ has reached the boundary of the solution region, namely, the increment of the index in $\Delta_{ii}^{(L-1)}$ is 2^{L-1}:

$$\Delta_{ii}^{(L-1)} \circ U = \frac{1}{(2^{L-1}h)^2} \begin{bmatrix} 0 & -1 & 0 \\ -1 & 4 & -1 \\ 0 & -1 & 0 \end{bmatrix}_{i,j,2^{L-1}} \circ U = f_{ii}^{(L-1)}.$$

Since the boundary values are known, we can obtain the result of x_{ii} directly from this operator. Then a backwards solution process can be started. The complexity of the forward process is $\frac{10}{3}N$ where $N = n^2 = (2^L - 1)^2$.

The backward process has the same complexity as the forward process. We will need another $3N$ operations to obtain the value of x_{ij} at the finest level. Therefore the complexity of the whole process is approximately $10N$. As we mentioned above, for each node O_{ij} there are many different template operators with different accuracies.

Due to the symmetry of these operators, many combinations of these operators can be chosen to perform the wavefront elimination. A more detailed discussion will be presented in a future report. The other algorithms also have an optimal complexity but the constants will be slightly bigger. The idea of using this algorithm as a preconditioner for a general elliptic equation is also considered. We are also studying the possibility of directly generalizing these algorithms to more general problems. Another interesting application of this algorithm is in combination with other efficient algorithms. For example, three or four steps of wavefront elimination are first executed then an other efficient algorithm is applied to solve the reduced system. This strategy can be used in case the truncation term grows during the elimination. If three steps of wavefront elimination are applied, the size of the reduced system is only $\frac{1}{64}$th of the original one. A great saving in computational cost can be achieved.

Some other generalizations can also be derived from this algorithm. First, the application of the $\mathcal{WE}$ to problems on an irregular solution region is considered. Suppose a solution of the Dirichlet problem on an irregular region Ω is wanted:

$$\begin{cases} \Delta U(x,y) = f(x,y) \\ U(x,y)|_{\Gamma_\Omega} = g(x,y) \end{cases}$$

We can enclose Ω by a larger rectangular region Ω' and let

$$f'(x,y) = \begin{cases} f(x,y) & \text{if } (x,y) \in \Omega \\ 0 & \text{if } (x,y) \in \Omega' - \Omega \end{cases}$$

The wavefront elimination can be applied to the problem

$$\begin{cases} \Delta \tilde{U}(x,y) = f'(x,y) \\ \tilde{U}(x,y)|_{\Gamma_\Omega} = 0. \end{cases}$$

Then apply an optimal algorithm due to Greengard and Rakhlin [1], [4] to the Dirichlet problem of the Laplace equation

$$\begin{cases} \Delta \tilde{U}(x,y) = 0 \\ \tilde{U}(x,y)|_{\Gamma_\Omega} = g(x,y) - \tilde{U}(x,y). \end{cases}$$

It is easy to verify that $\tilde{U} - U'$ is the solution of the original problem and the complexity of this procedure is also optimal. The generalization of this idea to a three dimensional problem is also considered. Unfortunately, there is no optimal algorithm in analogy with the two-dimensional version. The best result we can obtain so far has a complexity of $O(N \log n)$, where $N = n^3$. How to find a renormalization mapping for three dimensional problems is a challenging open problem.

From the above discussion, we can see that the study of the new algorithms does create a challenging new direction of research for fast solvers. It is more important than the optimum of the complexity per se.

4. Renormalization Theory and Wavefront Eliminations. In the last section a new class of algorithm $\mathcal{WE}$ was introduced. During the elimination process a sequence of self-similar W-template operators is derived on the same node. Let's see how the process can be interpreted in terms of physics. Imagine the solution U to be temperature, therefore the right hand side of the Poisson equation ought to be the heat source. It is not difficult to see that Δ_{ij} or $\tilde{\Delta}_{ij}$ can be interpreted as a conservation relation between heat source and heat flux around the small cell where node O_{ij} is located. After one step of wavefront elimination, a new W-template operator

$$\Delta_{ij}^{(2)} \circ U = \begin{bmatrix} 0 & -1 & 0 \\ -1 & 4 & -1 \\ 0 & -1 & 0 \end{bmatrix}_{i,j,2} \circ U = f_{ij}^{(2)}$$

where

$$f_{ij}^{(2)} = \mathcal{W}_1(f_{ij}^{(1)}) - \tilde{f}_{ij}^{(1)}, \tag{14}$$

is obtained.

Imagine this process as the result of zooming our observation position a little further away from the solution plane. This new equation can also be viewed as another conservation relation for node O_{ij} except the scale is different. In order to make the conservation valid without changing the solution then the heat source has to be renormalized. The mapping (14) actually can be viewed as the renormalization transformation. If we compare this process with K.G. Wilson's renormalization theory for the magnet, there are surprising similarities between the two completely different subjects. In his theory, the same magnet of given temperature, when viewed on different scales, looks as if it were at different temperature [4]. Consider a magnet of N atoms with inter-atomic distance a and temperature T. On a coarse scale where the elementary block is taken to have sidelength $a' = b \cdot a$ and comprises b^3 atoms, the magnet looks like one with $N' = N/b^3$ atoms but with another *renormalized* temperature T'. The relation $T' = R_b(T)$ is called renormalization transformation. In a recent development of this theory, it can also be derived that the pattern of fluctuation at the critical value of temperature is self-similar. This basic idea eventually led to quantitative results and explained the physics of phase transitions in a satisfying way. L. P. Kadanoff first discovered the scaling law in 1966 [2], but it was K. G. Wilson who finally surmounted the difficulties and developed the method of renormalization into a technical instrument that has proven its worth in innumerable applications. It is not surprising that renormalization theory has recently led to fractal phase boundary. The book "The Beauty of Fractals" [3] cites a dictum from V. F. Weisskopf,

> *There's a fog of events and suddenly you see a connection. It expresses a complex of human concern that goes deeply to you that connects things that were always in you that were never put together before.*

From the sequence of hierarchical self-similar W-template operators [5] we can also sense the principle which Mandelbrot discovered that organizes a whole universe of self-similar structure. If the forward process of the wave-front elimination is equivalent to a renormalization of the heat source term, then the backward solution process is surprisingly similar to recursively detailing the solution when our "camera" is zooming into the fine scales. This interesting connection between wavefront elimination, renormalization theory and fractals has inspired us to generalize the wavefront elimination to

---

[4] In wavefront elimination, the same temperature were produced from a different heat source if the temperature field is viewed from different scales

[5] It is also worth-while to mention that template operators provide us a right structure to exhibit their self-similarity and renormalization relations?

It would be very difficult to observe this fact from the structure of a matrix other kinds of elliptic equations. Some preliminary study shows that this is a promising new direction for research into fast algorithms.

Before we conclude this section, it is also very interesting to compare wavefront elimination with the multi-grid method. It is not difficult to see that one basic idea behind the two very different approaches is the same, namely, both are using the idea of hierarchical computation in order to achieve the optimal complexity. But to achieve the same goal, the approaches they use are two extreme examples. In the wavefront elimination a renormalization of the source term, in other words the right hand side, is used to march from the fine grid level to coarse while in multi-grid the error of the solution is projected from a fine grid to a coarser one. In the wavefront elimination case we are able to complete the solution process in one scan while the latter need a few scans.

5. Conclusion. A whole new class of fast solvers is briefly discussed in this paper. A great many interesting open problems remain to be studied in this area. In particular, the discussion of the error analysis in section 3 is informal. Even though there appears to be no instability problem in our preliminary numerical tests, more studies are needed for a concrete result regarding the error analysis of the $\mathcal{WE}$.

Acknowledgment. The author wishes to thank G. Rodrigue and T. Chan, for helpful discussion. Thanks are also due to Mrs. Z. Kaszas for her help in latexing my paper.

REFERENCES

[1] L. GREENGARD, *The rapid evaluation of potetial fields in particle systems*, PhD thesis, Yale Univ., Computer Science Dept., Yale Univ., 1987.

[2] L. KADANOFF, *Scaling laws for Ising models near t_c*, Physics, 2 (1966), pp. 263–272.

[3] H. O. PEITGEN AND P. RICHTER, *The beauty of fractals*, Springer-Verlag, New York, 1986.

[4] V. ROKHLIN, *Rapid solution of integral equations of classical potential theory*, Journal of Computational Physics, 60 (1983), pp. 187–207.

[5] P. N. SWARZTRAUBER, *A direct method for the discrete solution of separable elliptic equations*, SIAM J. Numer. Anal., 11 (1974), pp. 1136–1150.

[6] W. TANG, *Schwarz splitting and template operators*, PhD thesis, Stanford University, Computer Science Dept., Stanford, CA94305, 1987.

[7] K. G. WILSON, *Renormalization group and critical phenomena (i), (ii)*, Phys. Rev. B4, (1971), pp. 3174–3183, 3184–3205.

PART III

Parallel Implementation

Efficiency of Multicolor Domain Decomposition on Distributed Memory Systems

Luigi Brochard*

Abstract. This paper presents a model for the performance prediction of some domain decomposition methods on MIMD distributed memory systems. This performance analysis is based on a deterministic model of the computation, communication and control complexities of the algorithm on a given architecture. The classes of domain decomposition we are studying are based on multicolor techniques where the convergence properties of the iterative solver are independent of the number of subdomains. This allows us to have a good understanding of the overheads of a parallel computation without dealing with the numerical behavior of the parallel method. Using this analysis, the efficiency of multicolor SOR and multicolor multigrid methods is studied, depending on the domain decomposition (slice, square, cube) and on different parameters. Special attention will be given to the multigrid methods and to some variants.

1.Introduction.

Analyzing and predicting the performance of a multiprocessor system is very complex, since many factors jointly determine the behavior of the system. In this paper, we present a deterministic model which can provide a good understanding of the interactions between the algorithm and the architecture. As a case study, we have chosen domain decomposition methods which seem well adapted to parallel computing, and particularly a class of domain decomposition methods which provides a fully parallel algorithm and a convergence rate independent of the number of subdomains.
The basic idea of domain decomposition methods is very simple: instead of solving a partial differential equation on a given domain, we decompose the initial domain in smaller ones and obtain the overall solution by solving smaller problems on these subdomains. Though this idea is rather old, and related to the Schwarz Alternating Method [27], it has triggered in the past several years a burst of new studies in this area, mainly due to the good fitting between the domain decomposition idea and parallel computing. A lot of these new studies are related to a class of domain decomposition techniques which can be called Preconditoned Gradient Method where the original operator on the whole domain is reduced to an operator on the interfaces of the subdomains which is solved using the iterative preconditioned conjugate gradient method. Several preconditioners have been proposed in the past years, see [19] for a review of some of them, and some new ones have been presented at this conference. A good advantage of this class of methods is that, using good preconditioners, the number of iterations can be kept small. Unfortunately, the condition number

* IBM Scientific Center, 1530 Page Mill Road, Palo Alto, CA 94304
Permanent address:
Ecole Nationale des Ponts et Chaussees, La Courtine B.P 105, 93194 Noisy le Grand Cedex, France

depends on the domain decomposition (slice, square, cube), and on the number of subdomains [9].
Therefore, as our main goal is the study of the parallel efficiency of the method, and of the different overheads related to its parallel implementation, we will move our attention to a particular class of domain decomposition methods where the number of iterations is independent of the number of the subdomains.
The outline of this paper is the following. In section 2, we present the domain decomposition technique we will be studying, the multicolor SOR method and its accelerated version using a multigrid adaptation. In section 3, we present the distributed memory systems we will be dealing with, and the performance analysis model we will be using. This deterministic model is based on the algorithm computation, communication and control complexities on the specific network of the MIMD distributed memory machine. On this basis, the general speed-up behavior of such parallel numerical algorithms will be discussed. In section 4, the performance analysis will be applied to the two parallel solvers we have presented and their efficiency will be compared depending on the domain decomposition and on the number of subdomains. A modified version of the standard multigrid algorithm will be also presented, and compared to the original one. It should be noted that throughout this paper we assume the number of processors and its memory size are large enough to have each subdomain associated with a specific processor.

2. Multicolor domain decomposition.

The basic idea of domain decomposition method is that instead of solving the initial problem, $f(u) = g$ on a domain Ω, the problem is split into P subproblems $f_{|_i}(u) = g_{|_i} = i = 1...P$, where $f_{|_i}$ and $g_{|_i}$ are the restrictions of f and g to the subdomains Ω_i , with $\Omega = \bigcup \Omega_i$, and a coupling condition at the subdomain interfaces.

2.1 Multicolor SOR method. In the following, we consider methods where every subdomain is only coupled to its neighbors through the exchange of boundary values between adjacent regions at every iteration. It can be applied as in the original method introduced by Schwarz [27] , as an alternate method, where every subdomain will be successively solved, or using multicolor relaxation schemes. With such coloring techniques, it is well known that for a large class of partial differential equations, there exists c color ordering schemes which decompose the naturally ordered one step Gauss Seidel method into an equivalent c step Jacobi method, where each step is parallel contrary to the sequential original Gauss Seidel method [1,21,22] . The best example is the red black ordering applied to the five point symmetric stencil, which leads to a two step Jacobi equivalent to the natural ordered Gauss Seidel method [31]. Using such an ordering, all the same color points can be processed simultaneously. With the Zebra method, as shown on Fig. 1, it is easy to see that depending on the processor allocation, you can get a scattered or gathered allocation.

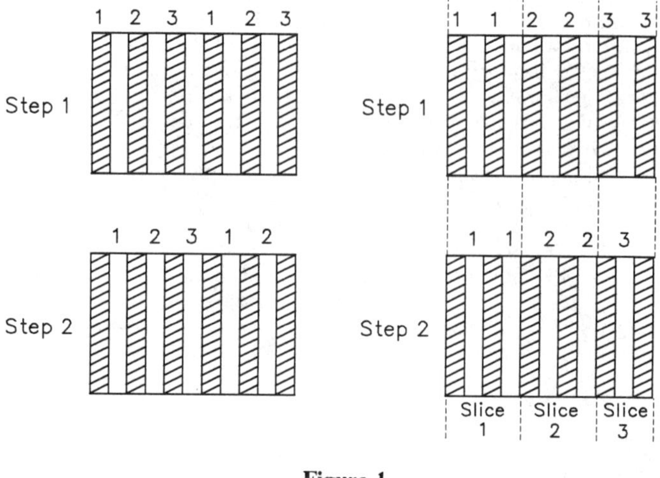

Figure 1
Scattered and gathered processor allocation of a 2 color Zebra method.

The scattered allocation leads to a heavy communication scheme since at every step each processor has to fetch one column from a neighbor processor, while with the gathered allocation the only interface column is be needed. This can be generalized to c color orderings and different geometries, though the coloring and load balancing may become more difficult [2,22]. This gathered processor allocation applied to the multicolor relaxation scheme will be called multicolor domain decomposition.

2.2 Multicolor multigrid method. In this paper, we only consider the standard multigrid method where the $\ell+1$ grid levels of the multigrid method are processed from finer to coarser and numbered 0 to ℓ. On each grid level a smoothing operator, typically several SOR relaxation steps, is applied and reduction and interpolation operators are used to project the residual of the equation on the the finer grid level to the coarser level, and vice versa. Fig. 2 presents different cycles that can be used between the different grid levels. With an appropriate choice of operators and cycles, it has been proved the error of this iterative method at every step is reduced by a constant factor independent of the mesh size, leading to an optimal sequential method [29].

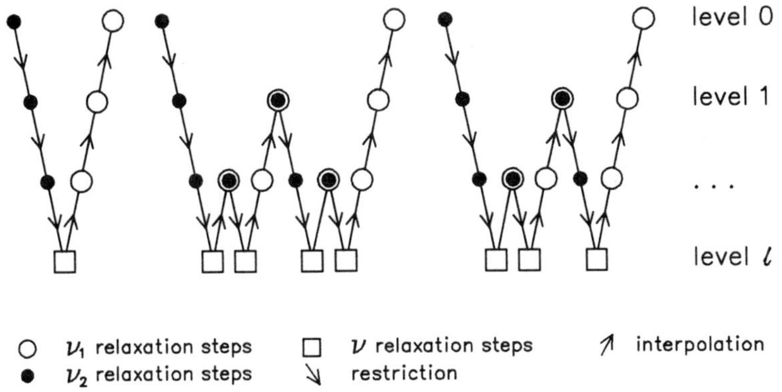

Figure 2
V, W and F cycles of a 4 level multigrid method.

We define the multicolor multigrid method as a parallel multigrid method where each grid level is performed sequentially as in the original sequential algorithm, but where the grid points within each level are performed simultaneously using the multicolor domain decomposition presented above. Embedding of the different subdomains is assumed on all grid levels, cf Fig. 3.

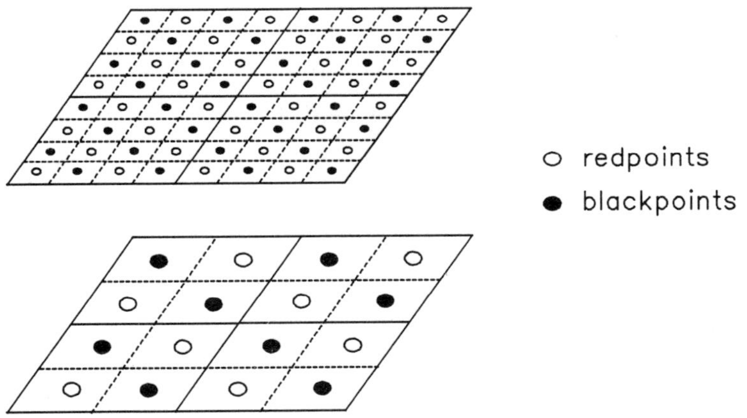

Figure 3
Subgrid projection of a 2 level and 2 color multigrid method.

This standard multigrid method [3], can be seen as horizontal parallel method, in opposition to some recent adaptation where a vertical parallelism can be applied to the method leading to a data driven algorithm where all grid levels are performed simultaneously [8,12,16] .
In the following we will deal with this only standard multigrid method, where each processor handles a subset of the finest grid level and its projection on the coarser levels, and try to understand the impact of communication on this method and on a variant where the last level is processed by a single processor trying to avoid the coarsest grid bad effect.

3. Performance analysis model

3.1 Speed-up evaluation. Throughout the literature, two different definitions of speed-up can be found: an absolute and a relative one. In this paper, as we are mainly interested in the communication and synchronization overheads of a parallel computation, we will use the relative speed-up definition of a P-processor computation of N data:

$$Sp(N,P) = \frac{T(N,1)}{T(N,P)}$$

where $T(N,1)$ denotes the execution time of the sequential algorithm and $T(N,P)$ the execution time of the studied parallel algorithm using P processors. Clearly with such a definition, there are deterministic algorithms which can give super-linear speed-up [10,23] , i.e:

$$Sp(N,P) > P$$

but we will restrict our study to algorithms leading to speed-up bounded by P.
In the following, we will suppose the execution time using P processors is given by the equality:

$$T(N,P) = T_{cpu}(N,P) + T_{io}(N,P), + T_{ctr}(N,P)$$

where T_{cpu} , T_{io} and T_{ctr} denote the execution times of the computation, communication and control parts of the parallel process. In this definition, the computation time is related to the arithmetic complexity of the algorithm, the communication time to its data movement complexity on a given architecture and the control time to tasks spawning, synchronization and termination detection of a distributed process. Clearly this equation states a non-overlapping condition between the various components of a parallel synchronous process. Following our definition we also state:

$$T(N,1) = T_{cpu}(N,1)$$

meaning there is neither communication nor control overhead in a sequential process.
Concerning the arithmetic complexity of a parallel algorithm, three different types can be found. In the first one, we have:

$$T_{cpu}(N,P) = \frac{T_{cpu}(N,1)}{P} \tag{1}$$

which means the algorithm is fully parallel with a perfect load balance. Matrix matrix multiplication [11] , multicolor relaxation methods [4], Gauss-Jordan matrix elimination [22] can be of this type.
In the second one, we find algorithms which are not fully parallel or where load is unbalanced. We have then:

$$T_{cpu}(N,P) > \frac{T_{cpu}(N,1)}{P}$$

For example, Gauss matrix eliminations [24] and reduction operations [27] belong to this large category. Those algorithms present a non constant degree of parallelism which causes a degraded speed-up independent of the parallel overheads.
At last, we could be interested finding parallel algorithms such that:

$$T_{cpu}(N,P) < \frac{T_{cpu}(N,1)}{P}$$

leading to a potential super-linear speed-up. Typically, this could arise using non-linear complexity algorithms which have been parallelized by partitioning techniques such as aggregation-disaggregation, sub-structuring or domain decomposition methods. For example if we suppose an $O(N^2)$ algorithm, its computation time will be divided by 4 on a two $N/2$ subsets partition and can lead to a super-linear relative speed-up if the aggregation part of the algorithm does not waste the gain. Matrix rank one tearing and updating belongs to this category [10], which does not imply a better absolute speed-up [20]. As we are mainly interested in the communication and control overheads of a parallel computation, we will suppose a fully parallel algorithm and a perfect load balance (1). Under those hypotheses the relative speed-up is bounded by P and can be expressed:

$$Sp(N,P) = \frac{P}{1 + a(N,P) + b(N,P)} \tag{2}$$

where $a(N,P)$ and $b(N,P)$ are positive functions of N and P measuring the parallel overhead. More precisely, let be a P-processor, with P given by:

$$P = Q^k$$

where k is the dimension of the processors network and Q the number of processors per direction , see paragraph 3.2. Then if we suppose the communication and control algorithms complexities can be expressed as rational functions of Q such as:

$$T_{io}(N,P) = \sum_{i=-k}^{m} \alpha_i(N).Q^i$$

and

$$T_{ctr}(N,P) = \sum_{i=-k}^{m} \beta_i(N).Q^i$$

where k, m, α and β would have to be determined, the relative speed-up $Sp(N,P)$ can be written:

$$Sp(N,P) = \frac{P}{1 + (a_0 + a_1.Q + a_2.Q^2 + ... + a_k.Q^k) + (b_1.Q^{k+1} + ... + b_m.Q^{k+m})} \tag{3}$$

where a and b are respectively polynomial functions in Q of degree $\leq k$ and $> k$ with coefficients a_i and b_i given by:

$$a_i = \frac{\alpha_{i-k} + \beta_{i-k}}{T_{cpu}(N,P)}, \; i = 0...k$$

and

$$b_i = \frac{\alpha_i + \beta_i}{T_{cpu}(N,P)}, \; i = 1...m$$

This formulation is very general since different behaviors can be seen depending on the function $a(N,P)$ and $b(N,P)$. When $b(N,P) \equiv 0$, speed-up is a monotone increasing function of P . Then depending on the property of $a(N,P)$ we can see that if $a_k \neq 0$, the speed-up is asymptotic to $1/a_k$, while if $a_k = 0$ there is no asymptote, as we have:

$$Sp(P) \sim \frac{Q^{k-j}}{a_j}$$

where j is the greatest index of the non-null coefficients a_i. In the following, multicolor domain decomposition methods will present an application of this property. On the contrary when $b(N,P) \not\equiv 0$, formulation (3) leads to non-monotone speed-up with a critical number of processors P_c beyond which it decreases. Several examples related to global communication effects show this property: global convergence detection of a distributed computation is one [5], the modified multigrid method is another we will study below.
In the following, the execution time of the computational part of the algorithms will be given by the formulation:

$$T_{cpu}(N,P) = arith(N,P) \, . \, t_{cpu}$$

where t_{cpu} denotes the elemental time to perform an arithmetic operation and $arith(N,P)$ the arithmetic complexity of the parallel algorithm. For simplicity, no arithmetic start-up time has been included to the formulation.

3.2 Nearest neighbor network and communication complexity. In this study, we only consider nearest neighbor networks without dealing with the other various classes [17] which are much less used in MIMD distributed memory systems.

Following [15], we introduce the $NN(k,P)$ processor which is a P-processor with $P = Q^k$, $P = 2^p$, $Q = 2^q$, and where each processor is connected to its nearest neighbors with wraparound connection. With such a definition, a $NN(1,P)$ is a ring of P-processors where each one is connected to its two neighbors, a $NN(2,P)$ is a $\sqrt{P} \times \sqrt{P}$ mesh of processors connected to its four neighbors with wraparound connection, and a $NN(p,P)$ is a p-binary cube connected processor, also called hypercube of dimension p, where every processor is connected to its $\log_2 P$ neighbors. An important feature is whether it is possible in practice to use simultaneously the $2.k$ or p channels of a given node on a $NN(k,P)$, leading to communication transfer routed through different paths. In the following , we will suppose a $NN(k,P)$ processor can receive (or send) simultaneously $2.k$ messages to (or from) its $2.k$ neighbors if $k \neq p$, and send (or receive) p messages to (or from) its p neighbors on a hypercube.

A first way to give an evaluation of a static network is the maximum distance between two nodes, related to the diameter of the corresponding graph, which represents the maximum number of steps for a single message to reach one processor. It can be seen easily [15] this diameter d is given by:

$$d(k,P) = k.Q/2 \tag{8}$$

if the communication links are multidirectional, while given by:

$$d(k,P) = k.(Q - 1) \tag{9}$$

if the communication links are mono-directional. For example on a $NN(1,P)$ we have $d = P - 1$ with a one-way round algorithm, while $d = P/2$ with a two-way round algorithm which sends simultaneously data in the two opposite directions of the ring, reducing the maximum distance by a factor of two. On a $NN(2,P)$, d will be either $2(\sqrt{P} - 1)$ or $\sqrt{P}$, while it will be $\log_2 P$ on a $NN(p,P)$, since the same link cannot be used in two opposite directions at the same time.

A different way to evaluate those networks is to give the complexity of some specific communication function. We just recall here a few results that will be needed in the following, the shift and gather-scatter transfers, and refer to [6,14,18,25] for a detailed presentation. Those evaluations will be given using the two parameters t_{start} and t_{com}, where t_{start} represents the start-up time to send a message and t_{com} represents the elemental transfer time, i.e the inverse of the communication throughput of the link. With such a definition, the time to shift N data from one processor to any of its nearest neighbor is given by:

$$T_{shift} = t_{start} + N.t_{com}$$

The values t_{start} and t_{com} depend specifically on the machine design. For example, on the Intel iPSC1 hypercubes where the packet length is large (1K byte), or on the LCAP system [7], we have $t_{start} >> t_{com}$ and the start-up term is dominant, while on the FPS-T and Caltech [30] hypercubes, where the packet length is small (1 and 8 bytes respectively), we have $t_{start} \sim t_{com}$ leading to a dominant throughput term.

This shift function is used when only local to local communication is needed, which is the case with the multicolor domain decomposition methods. But if some global communication is needed, broadcast, multibroadcast, gather and scatter transfers can be usefull. In the following we present the gather-scatter algorithm, which is the only one we need in the modified multigrid method.

Gathering data consists of moving pieces of data of the same length from all processors to a particular one. This kind of transfer is used for example in the inner product computation when every processor computing partial inner products of its vector chunk , one must sum these partial results to get the global inner-product. Scattering data is the reverse operation in which a particular processor sends a different piece of information to every processor, and has therefore the same complexity. On a P-ring using a one-way round algorithm, the scatter algorithm is performed in $P - 1$ steps where at each step a different packet is sent in an assembly line manner beginning by the most distant one, giving complexity:

$$T_{scatter}(N) = (P - 1)(t_{start} + \frac{N}{P} t_{com}) = d.t_{start} + \frac{P-1}{P} .N.t_{com}$$

Consistently on a NN(k,P), the same method can be applied where at each step $i, i = 0...k-1$, scatter algorithms are applied simultaneously to N/Q^{k-i} data on each Q-ring of the network in the i *th* direction, leading to the complexity:

$$T_{scatter}(N) = \sum_{i=0}^{k-1}(Q-1)\left(t_{start} + \frac{N}{Q^{k-i}}.t_{com}\right) = d.t_{start} + \frac{P-1}{P}.N.t_{com}$$

where d is the maximum distance k.(Q-1). Note that contrary to multibroadcast transfer, scatter and gather algorithms can be performed in a two-way round fashion, leading to the complexity:

$$T_{scatter}(N) = \sum_{i=0}^{k-1}(Q/2)\left(t_{start} + \frac{N}{Q^{k-i}}.t_{com}\right) = d.t_{start} + \frac{Q}{2(Q-1)}.\frac{P-1}{P}.N.t_{com}$$

where d is the maximum distance k. Q/2. For simplicity, in the following we assume the gather and scatter complexities are:

$$T_{scatter}(N) = d.t_{start} + N.t_{com} \tag{14}$$

4. Efficiency of multicolor domain decomposition

We introduce a 1-dimension decomposition as a slice or strip decomposition, a 2-dimension decomposition as a square or box decomposition and a 3-dimension decomposition as a cube decomposition of the s dimension problem , $s = 1...3$. In the following, as we are not interested in the mapping problem, we use the natural mapping and map the r-dimension decomposition into a NN(r,P), leading to the possible re-mapping of the initial NN(k,P) network into the NN(r,P).

4.1 Multicolor SOR methods. Let s be the physical domain dimension and r be the domain decomposition dimension, with $r \leq s$. Assuming the exchange of data between adjacent regions is reduced to the interface grid points, and a perfect load balance between processors, we define $V(N,P)$ as number of grid points to compute and $S(N,P)$ number of grid points to exchange. We have then:

$$V(N,P) = \frac{n^s}{Q^r} = \frac{N}{P} \tag{15}$$

and

$$S(N,P) = 2.r.\frac{n^{s-1}}{Q^{r-1}} \tag{16}$$

Note that our evaluation assumes a two-color red-black relaxation and that if extended to a c-color scheme $S(N,P)$ would become $c.r.\frac{n^{s-1}}{Q^{r-1}}$.

Following our definitions, at every step we have:

$$T_{cpu}(N,P) = arith(V(N,P)).t_{cpu}$$

and:

$$T_{io}(N,P) = 2.r.t_{start} + S(N,P).t_{com}$$

As the work per iteration is constant throughout the process and as hypothesis (1) holds, the speed-up is then given by:

$$Sp(N,P) = \frac{P}{1 + \frac{2.r.}{n}.\frac{t_{com}}{t_{cpu}}.Q + \frac{2.r}{N}.\frac{t_{start}}{t_{cpu}}.P} \tag{17}$$

This algorithm is very efficient since the function $b(N,P)$ is identical to zero due to the only local effect of communication, leading to a monotone increasing speed-up. The function $a(N,P)$ has two components. The first related to the start-up time is linear in P and decreases as N^{-1} with the problem size The second, related to the communication throughput, is linear in Q and decreases as n^{-1}, which is due to the perimeter effect coming from the ratio of data to compute and data to exchange. It is obvious that for a start-up bound problem, the best efficiency is reached for the smallest domain decomposition dimension, while with a throughput bound problem, the best efficiency is reached with the highest dimension domain decomposition. Although this speed-up function has theoretically an asymptote we note that, if the start-up term can be neglected, a square decomposition decomposition leads for example to a non-asymptotic function increasing as $\sqrt{P}$. Having a slice decomposition, the function $a(N,P)$ is a linear function of P and is equal to:

$$a(N,P) = (\frac{2}{n}.\frac{t_{com}}{t_{cpu}} + \frac{2}{N}.\frac{t_{start}}{t_{cpu}}).P$$

In this paper, we do not study the effect of convergence checking and refer to [6,26] for a more detail study. A comparison between a linear control algorithm and experimental results is presented in [5] on the LCAP system [7] with start-up bound problems.

4.2 The standard parallel multigrid method. Let ν_1 and ν_2 be., respectively, the number of relaxation steps per grid i, $i = 0 \ldots \ell - 1$, in transition from the finer to the coarser grids and vice versa, and ν the number of relaxations on the coarsest grid level ℓ, meaning the coarsest grid level is not solved directly, but using the same relaxation scheme as the smoothing part. In the following, we suppose the restriction and interpolation operators complexities are negligible compared to the smoothing parts and each coarse grid level has 2^s less grid points than the previous one. Then, assuming the previous hypotheses of the multicolor relaxation method, and defining $V(N,P,i)$ and $S(N,P,i)$ as the counterparts of $V(N,P)$ and $S(N,P)$ on each grid level, we have:

$$V(N,P,i) = \left(\frac{n}{2^i}\right)^s.\frac{1}{P} \tag{18}$$

and

$$S(N,P,i) = 2.r.\left(\frac{n}{2^i}\right)^{s-1}.\frac{1}{Q^{r-1}} \tag{19}$$

Restricting our study to V-cycle, and summing on each grid level, the computing and exchange times at each iteration step are then:

$$T_{cpu}(N,P) = \sum_{i=0}^{\ell-1}(\nu_1 + \nu_2)T_{cpu}(V(N,P,i)) + \nu.T_{cpu}(V(N,P,\ell))$$

and

$$T_{io}(N,P) = \sum_{i=0}^{\ell-1}(\nu_1 + \nu_2).T_{io}(S(N,P,i)) + \nu.T_{io}(S(N,P,\ell))$$

where $T_{cpu}(V(N,i,j))$ and $T_{io}(S(N,i,j))$ denote the computing and exchange times for each sub-domain of level j using i processors. Then, as hypothesis (1) holds, the speed-up of a V-cycle and ℓ-level multigrid method can be written:

$$Sp(N,P) = \frac{P}{1 + \left(\frac{2.r}{n}.\frac{A(s-1,\ell)}{A(s,\ell)}.\frac{t_{com}}{t_{cpu}}\right).Q + \left(\frac{2.r}{N}.\frac{(\nu_1+\nu_2).\ell + \nu}{A(s,\ell)}.\frac{t_{start}}{t_{cpu}}\right).P} \tag{20}$$

with

$$A(s,\ell) = (\nu_1 + \nu_2).\frac{1 - 2^{-s\ell}}{1 - 2^{-s}} + \nu.\frac{1}{2^{s\ell}}$$

We note first this speed-up formulation is related, as the multicolor domain decomposition (17), to the case where the function $b(N,P)$ of (2) is $\equiv 0$, and therefore leads to a monotone increasing speed-up function coming from the only local communication effect. We will now consider two cases, depending on the number of grid level ℓ.

With complete multigrid methods, ℓ is of the order of $\log_2 n$, and as n is usually large, we have $2^{-s\ell} \sim N^{-1} \sim 0$, leading to the approximation:

$$A(s,\ell) \sim \nu_1 + \nu_2.\frac{1}{1-2^{-s}}$$

and to the corresponding speed-up:

$$Sp(N,P) \sim \frac{P}{1 + \left(\frac{2.r}{n}.\frac{1-2^{-s}}{1-2^{-s+1}}.\frac{t_{com}}{t_{cpu}}\right).Q + \left(\frac{2.r}{N}\left(1-\frac{1}{2^s}\right)\left(\ell + \frac{\nu}{\nu_1+\nu_2}\right).\frac{t_{start}}{t_{cpu}}\right).P} \tag{21}$$

It shows clearly the communication overhead of a ℓ-level multigrid method compared to the corresponding relaxation method, multiplies the start-up effect by a factor $\ell + \nu/(\nu_1+\nu_2)$ while the throughput effect remains roughly unchanged, leading to a more start-up bound problem and to preferable small dimension domain decomposition.
With incomplete multigrid methods, where the number of level remains small, as in the original two-grid method, the number of relaxation on the coarsest grid ν has to be large to solve the equation exactly, we have then $(\nu_1+\nu_2).\ell << \nu$, and the approximation:

$$A(s,\ell) \sim \nu.\frac{1}{2^{s\ell}}$$

leading to the corresponding speed-up:

$$Sp(N,P) \sim \frac{P}{1 + \left(2^{\ell}.\frac{2.r}{n}.\frac{t_{com}}{t_{cpu}}\right).Q + \left(2^{s\ell}.\frac{2.r}{N}.\frac{t_{start}}{t_{cpu}}\right).P} \tag{22}$$

We note this evaluation is exactly the speed-up of the relaxation method on the coarsest grid level. Thinking to more general cycles, as shown on Fig. 2, it is now obvious that the main drawback of multigrid methods being the relaxation steps on the coarsest grid, V-cycles should be preferred to W and F-cycles, at least speaking about relative speed-up and parallel efficiency. Identically, the complete multigrid method should be prefered to the incomplete one.

4.3 Complexity of modified multigrid method. A good strategy to minimize the effect of the coarsest grid level could be to perform this last step using one single processor [13] . Therefore the V- cycle method remains unchanged all the way down the ℓ first levels, then a gather transfer is done on the coarsest grid level prior to its computation by a single processor, and followed by a scatter transfer before the computation goes as previously all the way up. Its complexity is then:

$$T_{cpu}(N,P) = \sum_{i=0}^{\ell-1}(\nu_1+\nu_2).T_{cpu}(V(N,P,i)) + \nu.T_{cpu}(V(N,1,\ell))$$

and

$$T_{io}(N,P) = \sum_{i=0}^{\ell-1}(\nu_1+\nu_2).T_{io}(S(N,P,i)). + 2.T_{scatter}(V(N,P,\ell))$$

where $2.T_{scatter}(V(N,P,\ell))$ is the complexity of the sub-domains gather and scatter transfer before and after the resolution on the coarsest grid level. Obviously, this strategy will be more efficient under the condition:

$$\nu.T_{cpu}(V(N,1,\ell)) + 2.T_{scatter}(V(N,P,\ell)) \le \nu.T_{cpu}(V(N,P,\ell)) + \nu.T_{io}(S(N,P,\ell))$$

Assuming $\ell = \log_2 n$ and $t_{start} >> t_{com}$ we can prove the following result:
If $d \ge \nu.r$, the modified multigrid is not worth compared to the standard one.
If $d < \nu.r$, the modified multigrid method is faster than the standard one, under the condition:

$$\frac{t_{start}}{t_{cpu}} \ge \frac{1}{2.\left(r - \frac{d}{\nu}\right)}$$

If $t_{start} = t_{com}$ the result is still valid with d + 1 instead of d.
It should be noted the result is not valid when $t_{start} < t_{com}$, which seems unlikely in reality.
In any cases, the speed-up of the modified complete multigrid method is given: by:

$$Sp(N,P) = \frac{P}{1 + c.Q + \left(\frac{2}{N}\left(1 - \frac{1}{2^s}\right)\left(r.\ell + \frac{d}{v_1 + v_2}\right)\frac{t_{start}}{t_{cpu}} + \frac{1}{N}.\frac{v}{v_1 + v_2} + \frac{2}{N}.\frac{t_{com}}{t_{cpu}}\right).P} \tag{23}$$

where c is the coefficient of Q in the standard multigrid speed-up formulation (21). Note that the maximum distance d is a function of P given by (8 & 9) , so that the speed-up is not only a function of Q and P, but also of $k.Q^{k+1}$ leading to a non-monotone increasing function which comes from the global effect of gather-scatter transfers. With this kind of global communication problem, we remark the hypercube topology can be useful to reduce this maximum distance d, leading to a possible efficient algorithm for large value of P, which is impossible using a ring of processors.

Conclusion

The present study is an attempt to achieve a better understanding of the relationship between parallel algorithms and architectures. The above speed-up cannot be taken as certified quantitative results, since several hypotheses have been taken which restrict or ignore certain phenomena, but more as a qualitative approach. The speed-up formulation we have proposed, based on the functions $a(N,P)$ and $b(N,P)$, permits a formal classification of different algorithm behavior, but also provides a good way to understand the different overhead of a parallel computation. Function $a(N,P)$, which is a polynomial function in Q of degree less or equal to the network dimension, is related only to the local communication effects and implies a monotone increasing speed-up. Its asymptotic or non-asymptotic behavior has been also shown. Function $b(N,P)$, which is a polynomial function in Q of degree greater to the network dimension, is related to the global communication effects and implies a non-monotone speed-up with an optimal number of processors and a maximum speed-up beyond which it decreases. We have seen also that, on a distributed memory system using message passing, the communication overhead has two different components related to the through-put of the communication link and to the time to initialize a message. More precisely, the through-put term of the communication overhead is proportional to Rc the ratio of the elemental transfer time over the elemental arithmetic time, t_{com}/t_{cpu} , times the ratio of the communication and arithmetic complexities, $com(N,P)/arith(N,P)$. The start-up term is proportional to the ratio Rs of the time to initialize a message over the elemental arithmetic time, t_{start}/t_{cpu} , times the ratio of the distance that characterizes the communication transfers over the arithmetic complexity, $d/arith(N,P)$. This distance can be an horizontal distance measuring the diameter of the graph, as with global communication problem, and therefore related to $k.Q$ leading to a non monotone increasing speed-up. It can be also a vertical distance between the different grid levels, as with the complete multigrid method, leading to a still monotone increasing speed-up but with a stronger start-up term and a lower asymptote.
These shows clearly the importance of both architecture and algorithms parameters, leading to a "condition number", $Rs \times d/arith(N,P) + Rc \times com(N,P)/arith(N,P)$, which has to be as small as possible. We can therefore speak of a local or global problem, depending on the value of d, of a computation or communication bound problem, depending on the value of $com(N,P)/arith(N,P)$, and of a start-up or through-put bound problem, depending on the value of $Rs \times d/arith(N,P)$ compared to $Rc \times com(N,P)/arith(N,P)$.

References

[1] Adams L., Ortega J. A multicolor SOR method for parallel computation. IEEE Proceedings of International Conference on Parallel Processing, 1982, pp. 53-56.
[2] Berger M., Bokhari S. A partitioning strategy for non uniform problems on multiprocessors. ICASE Report No. 85-55, ICASE NASA Langsley Research Center, Hampton Va., 1985.
[3] Brandt A. Multigrid solvers on parallel computers. Elliptic problems solvers. Schultz M. Ed., Academic Press, New York, 1981, pp 39-84.
[4] Brochard L. Domain decomposition and relation methods. Parallel Algorithms and Architectures, Cosnard M. Ed., Elsevier Sc. Publ., North-Holland, 1986, pp. 61-72.

[5] Brochard L. Communication and control costs of domain decomposition on loosely coupled multiprocessors. Proceedings of First International Conference on Supercomputing '87, Polychronopoulos C. Ed., Lecture Notes in Computer Science, Springer-Verlag, New York, to appear.
[6] Brochard L. Efficiency of some numerical algorithms on distributed systems. Research Report CERIA. Ecole Nationale des Ponts et Chaussees, Paris, 1987.
[7] Clementi E., Detrich J. Large scale paralll computing on LCAP systems. Experimental parallel computing architectures. Dongarra J. Ed., Elsevier Sc. Publ., Amsterdam, 1986, pp. 141-176.
[8] Chan T., Schreiber R. Parallel networks for multi-grid algorithms: architectue and complexity. Siam J. Sci. Comput., Vol 6, No 3, 1985, pp 698-713.
[9] First International Symposium on Domain Decomposition Methods for Partial Differential Equations. Paris 1987. Glowinski, Golub, Meurant, Periaux Ed., SIAM 1988.
[10] Dongarra J.J., Sorensen D.C. A fully parallel algorithm for the symmetric eigenvalue problem. SIAM J. Sci. Stat. Comp. Vol 8, No 2 , 1987, s 139-154.
[11] Fox G.C., Otto S.W., Hey A.J. Matrix algortm hypercube 1: Matrix multiplication. Parallel Computing. Vol.4, No. 2, 1987, pp. 17-32.
[12] Gannon D., Van Rosendale J. On the structure of parallelism in a highly concurrent PDE solver. Journal of Parallel and Distributed Computing, 3, pp 106-135, 1986.
[13] Hempel R. Parallel implementations of the multigrid method. The SUPRENUM project, ICCIAM, Paris, 1987.
[14] Ho C.T., Johnsson S.L. Distributed routing algorithms for broadcasting and personalized communication in the hypercube. Technical Report, YALEU/DCS/483, 1986.
[15] Hockney R.W., Jesshope C.R. Parallel computers, Adam Hilger, Bristol, 1981.
[16] Hoppe H.C., Muhlenbein H. Parallel adaptative full-multigrid methods on message based multiprocessors. Parallel Computing Vol 3, 1986, pp 269-287.
[17] Hwang K., Briggs F. Computer architecture andparallel processing. Mc. Graw Hill, 1984.
[18] Ipsen I.C., Saad Y., Schultz M.H. Complexity of linear system solution on a multiprocessor ring. Research report YALEU/DCS/349, 1985.
[19] Keyes D., Gropp W. A comparison of decomposition technique for elliptic partial differential equations and their parallel implementation. SIAM J. Sci. Stat. Comp. Vol 8, No 2, 1987, s 166-202.
[20] Lo S.S., Philippe B. The symmetric eigenvalue problem on a multiprocessor. Algorithms and Architectures, Cosnard M. Ed. Elsevier Sci. Publ., North-Hollond, 1986, pp. 31-43.
[21] O'Leary D.P. Ordering schemes for parallel of processing on certain mesh problems. SIAM J. Sci. Stat. Vol 5, No 3, 1984, pp 620-632.
[22] Ortega J.M., Voigt R.G. Solutions of diferential equations on vector and parallel computers. SIAM Review, Vol. 27, No. 2, 1985, pp. 149-240.
[23] Parkinson D. Parallel efficiency can be greater than unity. Parallel Computing, Vol. 3, No. 3, 1986, pp 261-262.
[24] Saad Y. Gaussian eliminations on hypercubes. Algorithms and Architectures, Cosnard M. Ed., Elsevier Sci. Publ., North-Holland, 1986, pp. 5-19.
[25] Saad Y, Schultz H. Data communication in hypercubes. Research report YALEU/DCS/428, 1985.
[26] Saltz J., Naik V., Nicol D. Reduction of the effect of the communication delays in scientific algorithms on message passing MIMD architecture. ICASE Report No. 86-4, ICASE NASA Langsley Research Center, Hampton Va., 1986.
[27] Schwarz H.A. Uber einige Abbildungsaufgaben. Ges. Math. Abh. Vol 11, 1869, pp 65-83.
[28] Schwartz J.T. Ultracomputers. ACM Transaction on programming languages and systems, Vol. 2, No. 4, 1980, pp. 484-521.
[29] Stuben K., Trottenberg U. Multigrid Fundamental algorithm, model problem analysis and application. Proceedings of the Multigrid Methods Conference, Koln 1981, Lectures Notes in Mathematics No. 960, Springer-Verlag, Berlin, 1982, pp. 1-176.
[30] Thole C.A Experiments with multigrid methods on the Caltech hypercube. GMD Studieren Nr.103, St. Augustin, 1985.
[31] Young D. Iterative solution of large linear system. Academic Press, New York, 1971.

CHAPTER 22

Domain Decomposition on Parallel Computers

William D. Gropp*
David E. Keyes†

Abstract. We consider the application of domain decomposition techniques for the solution of sparse linear systems on parallel computers. We consider two representative types of MIMD parallel computer; a message passing and a shared memory architecture. For each we develop complexity estimates and compare these against actual computations. Various tradeoffs in parallel computation costs are discussed. Our complexity estimates are tested for a variety of methods, decompositions, and problem sizes. Results from both an Intel iPSC Hypercube and an Encore Multimax are presented.

1. Introduction

Domain decomposition techniques appear to be a natural way to distribute the solution of large sparse linear systems across many parallel processors. In this paper we develop complexity estimates for several types of "real" parallel computer architectures, and validate those estimates on representative machines, with particular emphasis on the case of large numbers of processors and large problems. We also look at the tradeoffs between various forms of preconditioning, categorized by the efficiency of their parallel implementation.

Parallel computers may be divided into two broad classes: distributed memory and shared memory. In a distributed memory parallel processor, each processor has its own memory and no direct access to memory on any other processor. Such machines are usually termed "message passing" computers since interprocessor communication is accomplished through the sending and receiving of messages. In a shared memory parallel processor, each processor has direct, random access to the same memory space as every other processor. Interprocessor communication is conducted directly in the shared memory. In practice, of course, most shared memory machines have local memory, called the cache, and communication is through messages, called cache faults. However, each type of

*Department of Computer Science, Yale University, New Haven, CT 06520. The work of this author was supported in part by the Office of Naval Research under contract N00014-86-K-0310 and the National Science Foundation under contract number DCR 8521451.

† Department of Mechanical Engineering, Yale University, New Haven, CT 06520. The work of this author was supported in part by the National Science Foundation under contract number EET-8707109.

parallel processor is optimized for a different interprocessor communication pattern, and we consider the effects of these optimizations on domain decomposition.

We consider only partitioned matrix methods applied to preconditioned conjugate gradient techniques such as those reviewed in [5]. Briefly, the matrix is partitioned into subdomains connected by small (lower dimension) interface regions. For a single domain decomposed into two subdomains (1 and 2) connected by an interface (3), the partitioned matrix would look like

$$A = \begin{pmatrix} A_{11} & 0 & A_{13} \\ 0 & A_{22} & A_{23} \\ A_{13}^T & A_{23}^T & A_{33} \end{pmatrix},$$

where A_{11} and A_{22} come from the interior of the subdomains, A_{33} from *along* the interface, and A_{13} and A_{23} from the interactions bewteen the subdomains and the interfaces. We are interested in various preconditioners for A, based on their efficacy and on their parallel limitations.

The choice of preconditioner is critical in domain decomposition, as with any iterative method. In the context of parallel computing, the main distinction is between preconditioners which are purely local, those which involve neighbor communication, and those which involve global communication. (We use the word "communication" here in a general sense; in a shared memory machine, this refers to shared access to memory.) As example preconditioners we consider a block diagonal matrix for the purely local case and a preconditioner based on FFT solves along the interfaces for the neighbor communication case. As an example involving global communication, we consider the Bramble *et al* preconditioner [1], which requires the solution of a linear system for the cross points. This method involves only low bandwidth global communication (that is, the size of the messages scales with the number of processors); we will not consider any method which uses high bandwidth communication (on the order of the size of the problem).

We develop a complexity model for both types of parallel computer which is based on two major contributions: floating point work and "shared memory access". This latter term measures the cost of communicating information between processors. In a distributed memory system, this is represented by a communication time. In a shared memory system, there are several contributions, including cache size and bandwidth, and the number of simultaneous memory requests which may be served.

2. Comments on Parallelism Costs

In any parallel algorithm, there are a number of different costs to consider. The most obvious of these are intrinsically serial computations. For example, the dot products in the conjugate gradient method involves the reduction of values to a single sum; this takes at least $\log p$ time. More subtle are costs from the implementation: both software and hardware. An example software cost is the need to guarantee safe access to shared data; this is often handled with barriers or more general critical regions. Example hardware costs include bandwidth limits in shared resources such as memory buses and startup and transfer speeds in communication links. Perhaps the most subtle cost lies in algorithmic changes to "improve" parallelism; by choosing a poor algorithm with better parallelism than another, less parallel algorithm, artificially good results can be found. An example is computing the forces in the n-body problem; the naive algorithm is almost perfectly parallel but substantially slower than the linear in n algorithm (which contains some reductions and hence some intrinsically serial computations) [3].

We can identify these costs with the domain decomposition algorithms that we are considering:

- dot (inner) products. These involve a reduction, and hence at least $\log p$ time; in addition, there may be some critical sections (depending on the implementation).

- Matrix-vector products. These involve shared data, and hence may introduce some constraints on shared hardware resources.
- Preconditioner solves. These depend on the preconditioner chosen, and hence gives us the most freedom in trading off greater parallelism against superior algorithmic performance.

Note that the sharing of data is not random; most of the data sharing occurs between neighbors.

2.1. Message passing and Shared memory models

Two methods for achieving parallelism in computer hardware for MIMD machines include message passing and shared memory. In both of these, the software and hardware costs discussed above show up in the cost to access shared data. Each of these methods is optimized for a different domain, and these optimizations are reflected in the actual costs. In the following, to simplify the notation will will express all times in terms of the time to do a floating point operation. Further, we will drop constants (like 2) from our estimates.

In a message passing machine, each processor has some local memory and a set of communication links to some (usually not all) other processors (called nodes). Each processor only has access to its own local memory. Communication of shared data is handled (usually by the programmer) by explicitly delivering data over the communication links. This takes time $s + rn$ for n words, where s is a start up time (latency) and r is the time to transfer a single word. This is good for local or nearest neighbor communication. For more global communication (such as a dot product), times depend on the interconnection network. For a hypercube, the global time is $(s + rn)\log p$; for a mesh, it is $(s + rn)\sqrt{p}$.

In a shared memory machine, each processor may directly access a shared global memory. Communication (access to) shared data is handled by simply reading the data. However, the actual implementation of this introduces a number of limits. For example, if the memory is on a common bus, then there is a limit to the number of processors that can simultaneously read from the shared memory. One way to model this cost is as $1 + p/\min(p, P)$ [4]. Here p is the number of processors and P is the maximum number of processors that may use the resource at one time.

In addition, the access to the shared data must be controlled; this can add additional costs in terms of barriers or critical sections. These can add additional terms which are proportional to $\log p$.

3. Complexity Estimates

We can estimate the computational complexity for these two models for several forms of domain decomposition. We note that these are rather rough estimates, good (because of their generality) for identifying trends.

3.1. Message Passing

In this case we can easily separate out the computation terms and the communication terms. For each part of the algorithm, we will place a computation term above the related communication term. In the formulas below, the constants in front of each term have been dropped for clarity.

For strips, we have

$$\begin{array}{lllllll} Ax \text{ multiply} & + & \text{dot products} & + & \text{subdomain solves} & + & \text{interface solves} \\ \frac{n^2}{p} & + & \frac{n^2}{p} + \log p & + & \frac{n^2}{p}\log\frac{n}{p} & + & n\log n \quad + \\ s + rn & + & (s+r)\log p & + & s + rn & + & s + rn \end{array}$$

and for boxes, we have

Ax multiply	+	dot products	+	subdomain solves	+	interface solves	
$\frac{n^2}{p}$	+	$\frac{n^2}{p} + \log p$	+	$\frac{n^2}{p} \log \frac{n}{\sqrt{p}}$	+	$\frac{n}{\sqrt{p}} \log \frac{n}{\sqrt{p}}$	+
$s + r\frac{n}{\sqrt{p}}$	+	$(s+r)\log p$	+	$s + r\frac{n}{\sqrt{p}}$	+	$s + r\frac{n}{\sqrt{p}}$	

These costs are all per iteration. It is assumed that all neighbor-neighbor interactions can occur simultaneously.

3.2. Shared Memory

In this case, a detailed formula depends on the specific design tradeoffs made in the hardware. The formula here applies to bus-oriented shared memory machines; a different formula would be needed for machines like the BBN Butterfly. These formulas are dominated by bandwidth limitations (the $\min(p, P)$ terms) and barrier costs (the $\log p$ terms).

For strips, we have

$$\frac{n^2}{p}\left(a + \frac{bp}{\min(p, P_1)}\right) + 2\left(\frac{n^2}{p} + \log p\right) + 2\left(\frac{n^2}{p}\log\frac{n}{p}\right)\left(c + \frac{dp}{\min(p, P_2)}\right) + n\log n + 3\log p.$$

A similar formula holds for boxes. Here, a, b, c, d, P_1, and P_2 are all constants that depend on the particular hardware and implementation. P_i give a limit on the number of processors that can effectively share a hardware resource. The ratios a/b and c/d reflect the ratio of local work to use of the shared resource (such as memory banks or memory bus).

3.3. Implications

For domain decomposition, we can trade iteration count against work and parallel overhead. We will consider three representative tradeoffs:

1. No communication. This amounts to diagonal preconditioning. Call the number of iterations $I_{\text{decoupled}}$.
2. Local communication only. The "$K^{1/2}$" preconditionings such as those in [1], where we can expect the iteration count to be proportional to p for strips and $\sqrt{p}$ for boxes. Call the number of iterations I_{local}.

   The cost of the "$K^{1/2}$" preconditioning includes an extra subdomain solve to symmetrize the preconditioner and is roughly

   $$n\log n + \frac{n^2}{p}\log\frac{n}{p} + 2(s + rn)$$

   for a strip decomposition and a message passing system. The preconditioning is dominated by the extra solve for the "harmonic" component; the communication costs are the same as for the matrix-vector product. Thus the local communication preconditioning is effective if

   $$2I_{\text{local}} \leq I_{\text{decoupled}}.$$

3. Global communication. In the case of box-wise decompositions, cross points occur at the intersections of the interfaces. The cross-points form a global linear system that is discussed in [1]. The iteration count is roughly constant; call it I_{global}.

In this case, the additional cost is that of a cross-point solve and the communication of the entries in the cross-point problem and its solution. For the case of a message passing system, we can do this in $p^2 + (s+r)\log p$ time if each processor solves the cross-point system and in time $p^{3/2} + p(s + r\sqrt{p})$ time if a parallel algorithm is used for the solve, assuming a straightforward approach based on gaussian elimination on a ring. More sophisticated approaches for hypercubes which are $p + (s+r)\log p$ are known [2].

Comparing this to the cost of the local computation of $n^2/p \log n/\sqrt{p} + (s+n)\log p$, the floating point work is negligible unless $n \approx p$ (when the problem is essentially all cross-points). Assuming then that we use the local method, then the additional cost of the global communication makes the full cross-point method better whenever

$$I_{\text{global}}(2-e) \le I_{\text{local}},$$

where e is the parallel efficiency, equal to 1 minus the ratio of communication time to computation time.

For the shared memory case, if barriers and the dot product reduction are the dominant parallelism costs, the result is similar.

4. Experiments

The standard test problem considered was

$$\nabla^2 u = g$$

where $g = 32(x(1-x) + y(1-y))$ on the unit square. The first set of experiments was conducted on an Encore Multimax 320 shared memory parallel computer with 18 processors; we used only 16, allowing the remaining 2 processors to handle various system functions. The experiments on the Encore Multimax were done in double precision. The results are shown in Tables 1–2. We note that the Encore is a time sharing machine, so these times are accurate to only about 10%. The tables show the iteration count I, an estimate of the condition number κ, the time in seconds T, and the relative speedup s. The relative speedup is defined as the ratio of the time from the previous *column* with the time from the current column. The times do not include initial setup, including the cost of the initial matrix vector multiply and preconditioner solve. While this slightly distorts the total time, it does allow the time per iteration to be determined by dividing the time by the iteration count.

The actual computations were performed with no other users on the machine; however, various system programs (mailers, network daemons) used some resources. In addition, the programmer can not force each process to run on a different processor.

The next set of experiments was run on a 64 node Intel Hypercube, with 4.5 Megabytes of memory on each node. All runs were in single precision to allow a large problem to fit in this memory space. The results are shown in Tables 3–6.

The programs in both of these cases were nearly the same. Only the code dealing with shared data was changed to use either messages or shared memory (in particular, the Encore implementation is *not* a message passing implementation using the shared memory to simulate message; it is a "natural" shared memory code. The Intel code is a "natural" message passing code.)

There are some differences between the results for the Encore and the Intel implementations. These differences seem to be due to the difference in floating point arithmetic on the two machines. Double precision was required on the Encore to get the fast Poisson

$h^{-1}\backslash p$		1	2	4	8	16
16	I	1	3			
	κ	1.00	1.26			
	T	0.2	0.22			
	s		0.91			
32	I	1	2	5		
	κ	1.00	1.09	3.29		
	T	0.97	0.80	0.78		
	s		1.21	1.03		
64	I	1	2	4	8	
	κ	1.00	1.09	3.29	13.0	
	T	4.95	4.13	3.33	2.63	
	s		1.2	1.24	1.27	
128	I	1	2	4	8	16
	κ	1.00	1.09	3.29	13.0	51.9
	T	24.3	20.0	16.9	14.2	12.1
	s		1.22	1.18	1.19	1.17
256	I	1	2	4	8	16
	κ	1.00	1.09	3.29	13.0	51.9
	T	108	105	84.6	71.9	63.7
	s		1.03	1.24	1.18	1.13
512	I	1	2	4	8	16
	κ	1.00	1.09	3.29	13.0	51.9
	T	513	481	420	353	327
	s		1.07	1.15	1.19	1.08

Table 1: Results for strips on the Encore Multimax 320.

$h^{-1}\backslash p$		1	4	16
64	I	1	6	7
	κ	1.00	10.7	10.2
	T	4.95	6.17	1.55
	s		0.76	3.98
128	I	1	6	7
	κ	1.00	11.0	14.1
	T	24.3	31.2	7.7
	s		0.78	4.05
256	I	1	6	8
	κ	1.00	13.6	18.6
	T	108	143	43.4
	s		0.76	3.29
512	I	1	7	8
	κ	1.00	16.7	23.7
	T	513	864	205
	s		0.59	4.21

Table 2: Results for boxes on the Encore Multimax 320, using full vertex coupling.

$h^{-1}\backslash p$	1	2	4	8	16	32
32 I	1	2	5			
κ	1.00	1.09	3.29			
T	6.08	5.14	5.22			
s		1.18	0.98			
64 I	1	2	4	8		
κ	1.00	1.09	3.29	13.0		
T	29.9	25.9	21.2	17.3		
s		1.15	1.22	1.23		
128 I	1	2	4	8	18	
κ	1.00	1.09	3.29	13.0	51.9	
T	142	125	105	85.9	78.9	
s		1.13	1.19	1.22	1.09	
256 I	1	2	4	8	18	41
κ	1.00	1.09	3.29	13.0	51.9	207.5
T	652	588	508	425	390	361
s		1.11	1.16	1.20	1.09	1.08

Table 3: Results for strips on the Intel Hypercube.

$h^{-1}\backslash p$	1	4	16	64
32 I	1	5	7	
κ	1.00	7.88	7.01	
T	6.08	6.93	3.29	
s		0.88	2.11	
64 I	1	6	7	6
κ	1.00	10.7	10.2	7.16
T	29.9	40.4	10.9	4.5
s		0.74	3.71	2.42
128 I	1	6	7	7
κ	1.00	11.04	14.1	10.5
T	142	195	48.4	12.6
s		0.73	4.03	3.84
256 I	1	6	8	8
κ	1.00	13.57	18.6	14.5
T	652	925	262	57.7
s		0.70	3.53	4.54

Table 4: Results for boxes on the Intel Hypercube, using full vertex coupling.

solver we used to work for $h = 1/512$. Single precision was required on the Intel to fit the problems in memory.

5. Comments

While the overall results for strips may seem poor, they actually represent very good speedup on a *per iteration* basis.

A number of the full vertex coupling (global) results show superlinear relative speedup. This is a real effect, which derives from the superlinear growth in the cost of solving a single domain as a function of the number of points in the domain. This speedup is of course available to a single processor algorithm. In fact, the slight relative speedups seen for the strip decompositions are due almost entirely to this effect alone, since the number of iterations is proportional to the number of processors.

$h^{-1}\backslash p$		1	4	16	64
32	I	1	6	11	
	κ	1.00	12.1	25.3	
	T	6.08	8.15	4.34	
	s		0.75	1.88	
64	I	1	6	12	17
	κ	1.00	15.9	32.2	94.4
	T	29.9	40.2	17.4	8.44
	s		0.74	2.31	2.06
128	I	1	6	13	19
	κ	1.00	20.1	40.5	119.7
	T	142	195	88.9	30.1
	s		0.73	2.19	2.95
256	I	1	7	13	33
	κ	1.00	24.8	49.6	794.8
	T	652	1080	425	231
	s		0.60	2.54	1.84

Table 5: Results for boxes on the Intel Hypercube, without vertex coupling but with interfaces.

5.1. Comparison with the theory

In the case of the message passing results (Intel Hypercube), it is possible to fit the theoretical complexity estimate to the measured times. Taking only the highest order terms in the latency s and the computation suggests a fit for the strip decomposition of

$$a_1 \frac{n^2}{p} + a_2 \frac{n^2}{p} \log \frac{n}{p} + a_3 + a_4 \log p.$$

We have ignored the r terms because $s \gg rn$ for given n on the Intel hypercube. To avoid any problems with small n/p (such as not being in the asymptotic regime for the FFT and fast Poisson solvers), we eliminated the data with $1/(hp) = 8$. A least squares fit to the data yields $a_1 = 0.00048$, $a_2 = 0.0012$, $a_3 = 0.027$, and $a_4 = 0.030$. The relative residual is 0.038. With these values (or directly from the data), the efficiency *per iteration* can be shown to be around 90% for the larger problems.

The data for the box decompositions is harder to fit (in part because our model does not include some implementation effects on the Intel and because the communication terms are small). However, the formula

$$0.00021 \frac{n^2}{p} + 0.0013 \frac{n^2}{p} \log \frac{n}{\sqrt{p}} + 0.0023 \frac{n}{\sqrt{p}} + 0.072 \log p$$

is a reasonable fit with relative residual 0.092. The efficiency *per iteration* is lower, because of the increase communication overhead, but is still above 70% for the larger problems. This makes the strip decomposition superior for moderate numbers of processors (where the iteration counts are similar).

This leads to the results in Tables 5 and 6, where a simpler communication strategy has been traded against larger iteration counts. The results are summarized in Table 7, which shows the optimal choice of preconditioners for our implementation on the Hypercube. For significant parallelism, the globally coupled preconditioner prevails in spite of its higher communication overhead.

For the shared memory machine, the complexity estimates are harder to demonstrate. In part, this is due to the design of the shared memory machines; the number of processors

$h^{-1}\backslash p$		1	4	16	64
32	I	1	12	18	
	κ	1.00	29.1	49.2	
	T	6.08	8.62	3.95	
	s		0.71	2.18	
64	I	1	17	25	32
	κ	1.00	61.0	103.6	187.9
	T	29.9	60.6	20.1	10.1
	s		0.49	3.01	1.99
128	I	1	23	35	44
	κ	1.00	124.9	212.6	397.6
	T	142	395	128	39.7
	s		0.36	3.09	3.22
256	I	1	27	48	210
	κ	1.00	252.9	430.8	23480.
	T	652	2170	829	789
	s		0.30	2.62	1.05

Table 6: Results for boxes on the Intel Hypercube, using diagonal blocks only (no coupling.)

$p\backslash h^{-1}$	16	32	64	128	256
4	Decoupled	Global	Local	Local	Global
16		Global	Global	Global	Global
64			Global	Global	Global

Table 7: Optimal choice of algorithm for the given problem and implementation on the Intel Hypercube, from the choices of decoupled block diagonal, locally coupled interfaces, and full vertex coupling preconditioners for the box decomposition.

is deliberately limited to roughly what the hardware (i.e., memory bus) can support. The dominant effect is usually load balancing or the intrinsically serial parts of the computation (synchronization points and dot products).

References

[1] J. H. Bramble, J. E. Pasciak, and A. H. Schatz, *The Construction of Preconditioners for Elliptic Problems by Substructuring, I,* Mathematics of Computation, 47 July (1986), pp. 103–134.

[2] T. F. Chan, Y. Saad, and M. H. Schultz, *Solving Elliptic Partial Differential Equations on the Hypercube Multiprocessor,* Technical Report YALE/DCS/RR-373, Yale University, Department of Computer Science, March 1985.

[3] L. Greengard and W. D. Gropp, *A Parallel Version of the Fast Multipole Method,* To appear in the proceedings of the Third SIAM Conference on Parallel Processing for Scientific Computing, December 1–4, 1987.

[4] H. F. Jordan, *Interpreting Parallel Processor Performance Measurements,* SIAM Journal on Scientific and Statistical Computing, 8/2 (1987), pp. s220–s226.

[5] D. E. Keyes and W. D. Gropp, *A Comparison of Domain Decomposition Techniques for Elliptic Partial Differential Equations and their Parallel Implementation,* SIAM Journal on Scientific and Statistical Computing, 8/2 (1987), pp. s166–s202.

Parallel Efficiency of a Domain Decomposition Method

M. Haghoo*
Wlodzimierz Proskurowski*

Abstract. Domain decomposition techniques for elliptic PDEs in a rectangular region are considered. By dividing the region into square boxes, the original problems are replaced by a set of subproblems. These subproblems are either independent or can be easily decoupled. More than 80% of execution time is spent to solve these independent and decoupled subsystems. Therefore this technique is well suited for parallel processing.
We study the speed-up and efficiency factors as functions of mesh size and the number of boxes. We also study the dependence of the overall execution time on the number of boxes and processors.

1. Introduction.

While many algorithms have to be radically modified for parallel implementation, or the parallelism has to be realized on very low level, as in dot products of two vectors, the domain decomposition techniques are well suited for parallel processing (see also **[3]**).

Substructuring of the domain transforms the large original system into a set of subproblems of smaller and equal dimensions. Application of a proper preconditioner that decouples these connected subproblems increases substantially the rate of convergence. Then these subsystems lend themselves for a balanced, high level parallel processing.

We consider elliptic PDEs with Dirichlet boundary conditions in a region that is divided into subregions, subsequently called boxes. In the Neumann-Dirichlet preconditioner we partition the region in a chessboard like manner in which white and black squares correspond to Neumann and Dirichlet boxes. Half of the boxes (the Dirichlet ones) form a set of disjoint subproblems and remaining (Neumann) boxes can be easily decoupled.

*Department of Mathematics, University of Southern California, Los Angeles, CA 90089-1113

The dominant portion of the execution time is spent on solving subproblems on these boxes. Therefore, these expensive computations can be reduced substantially in a parallel implementation.
We study how well the problem is suited for parallel processing. We use Intel's hypercube of 16 nodes and large memory. We perform extensive experiments to find such parameters (number of mesh points, boxes, and processors) for which the maximum speed-up and efficiency and the minimum execution time can be achieved.

2. Problem statement. We seek the numerical solution $u(x,y)$ on $\Omega = (0,1)\times(0,1)$ so that:

$$-\Delta u + cu = F \quad \text{on } \Omega,$$
$$u = G \quad \text{on } \partial\Omega.$$

By using the finite difference method and applying the 5-point stencil approximation to the Laplacian, we obtain the linear system

$$Au=f.$$

The goal is to solve this system effectively. Three factors are considered:
1. Substructuring the domain.
2. Using an effective preconditioner in the Conjugate Gradient method.
3. Implementing the problem on a parallel processor.

Figure 1 shows schematically the relation between these factors.

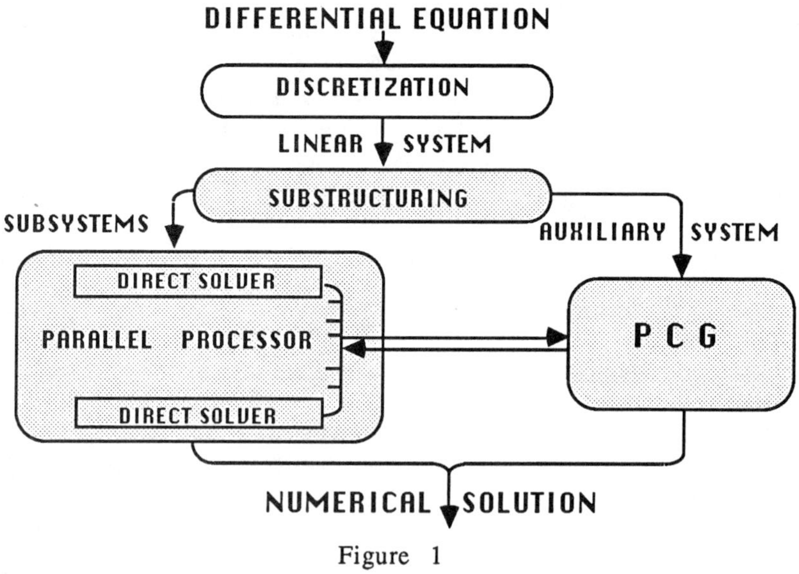

Figure 1

3. Substructuring. The domain Ω in which the problem is defined, is partitioned by mesh lines parallel to coordinate system into boxes. We consider four types of unknowns (see Figure 2):
1. Points inside white boxes,
2. Points inside shaded boxes,
3. Points on separator lines,
4. Cross points.

We call *Dirichlet boxes* all points in *1*, while we call *interconnected Neumann boxes* the totality of points belonging to *2*, *3*, and *4*. We further divide the second group into the the cross points and decopuled Neumann boxes which we call just *Neumann boxes*. This notation is used for the Neumann-Dirichlet preconditioner.

4. Preconditioned Conjugate Gradient Method. The Neumann-Dirichlet preconditioner, represented by matrix B, is a related problem to A in which the Neumann conditions are imposed on the sides of every other box, and Dirichlet conditions on the remaining boxes. Here, as shown in Figure 2, the region is partitioned so that the white squares correspond to the Dirichlet boxes, and the shaded squares together with the separator lines correspond to the Neumann boxes. The sides of Neumann boxes that coincide with the boundary of the region keep their original Dirichlet conditions.

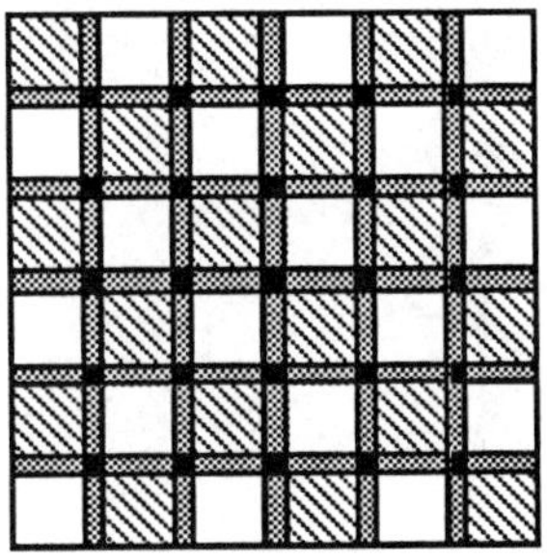

Figure 2

The original system $Au=f$ can be replaced by the following three steps:

$$Bv = h, \qquad (1)$$
$$Cw = g, \qquad (2)$$
$$Bu = h+Sw, \qquad (3)$$

where $g = S^T(f-AB^{-1}h) = S^T(f-Av)$, $C = S^TAB^{-1}S$, and $S = [0, I]^T$ is a n^2 by q matrix, n is the number of mesh points on each side of the region, and q is the number of total points in the inter-connected Neumann boxes. Here h is the same as f except on separator lines where components of h can be chosen arbitrarily.

Steps (1) and (3) are solved directly, and the capacitance system (2) is solved by the preconditioned conjugate gradient method in which B is used as a preconditioner. Forming C is extremely expensive. Fortunately, C need not to be generated. Instead, for a given vector w_i, one needs to form Cw_i in each step of the conjugate gradient iterations. Considering the fact that $Cw_i = (S^TA)B^{-1}Sw_i$, one obtains Cw_i by solving the system $Bz = Sw_i$ and then by applying S^TA to z. Thus solving (1)-(3) involves only systems with B.

It should be noted that the conjugate gradient method iterations are carried out on vectors defined only on separator points (the residuals and other vectors are zero in the interior of the boxes). Therefore, the corresponding computations are reduced to points belonging to separator lines only.

5. Sequential Implementation. As we discussed the main task is to solve a system with B repeatedly. In this section we detail solving a B-system sequentially and in section 6 we describe how to solve this system in parallel. Let us denote the matrices A and B by a 4x4 block matrices as following:

$$A = \begin{pmatrix} A_{11} & 0 & A_{13} & 0 \\ 0 & A_{22} & A_{23} & 0 \\ A_{13}^T & A_{23}^T & A_{33} & A_{34} \\ 0 & 0 & A_{34}^T & A_{44} \end{pmatrix} \qquad B = \begin{pmatrix} A_{11} & 0 & A_{13} & 0 \\ 0 & A_{22} & A_{23} & 0 \\ 0 & 2A_{23}^T & A_{33} & A_{34} \\ 0 & 0 & A_{34}^T & A_{44} \end{pmatrix}$$

-- Dirichlet pts

-- Neumann inner pts

-- Separator pts

-- Cross pts

Note that B is different from A only in the darkened boxes.
Let A_N denote the right bottom submatrix of B, and define B_N as preconditioner for A_N :

$$A_N = \begin{pmatrix} A_{22} & A_{23} & 0 \\ 2A_{23}^T & A_{33} & A_{34} \\ 0 & A_{34}^T & A_{44} \end{pmatrix} \qquad B_N = \begin{pmatrix} A_{22} & A_{23} & 0 \\ 2A_{23}^T & A_{33} & A_{34} \\ 0 & 0 & I \end{pmatrix}$$

Then computing v in $Bv=h$ consists of solving:

$$A_N v_N = h_N \qquad (4)$$

$$A_{11} v_1 = h_1 - A_{13} v_3 \qquad (5)$$

where v_N and h_N correspond to the Neumann boxes and the cross points. While (5) is a system of disjoint subsystems, (4) is a coupled large system and needs to be decoupled first. We, therefore, use the capacitance system once more, where the capacitance matrix C_N is defined on cross points only.
By using this preconditioner B_N, we replace (4) by three steps similar to those described in Section 4.

$$B_N z_N = k_N \qquad (6)$$

$$C_N w_N = g_N \qquad (7)$$

$$B_N v_N = k_N + S_N w \qquad (8)$$

where g_N , C_N , and S_N are defined similarly to g, C, and S in Section 4.
It is important to notice that in equations (6)-(8) the subsystems on Neumann boxes are decoupled into a set of subproblems on individual Neumann boxes, and thus all of these equations can be solved directly.
By solving (6), (7), and (8) one obtains the solution to (4). Using this solution, one solves (5), which completes the process of solving a system with B .
We now briefly describe how to solve a subproblem on a Neumann box. The Neumann boxes are of three types (Figure 3): boxes with four corners removed, boxes with two corners removed, boxes with one corner removed. Figure 4 shows the corresponding matrices of the subproblems of inner dimensions $m=3$. These matrices are banded positive definite, where half of

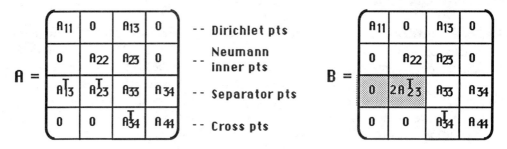

the bandwidth is equal to the width of corresponding box. Shifts of the off diagonals of these matrices are caused by the missing corners.

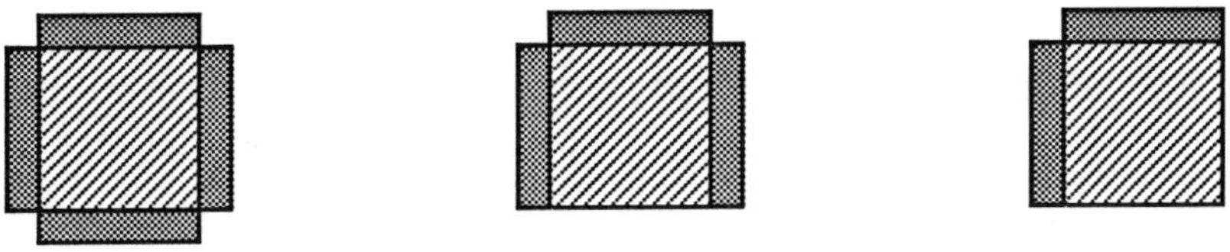

Figure 3

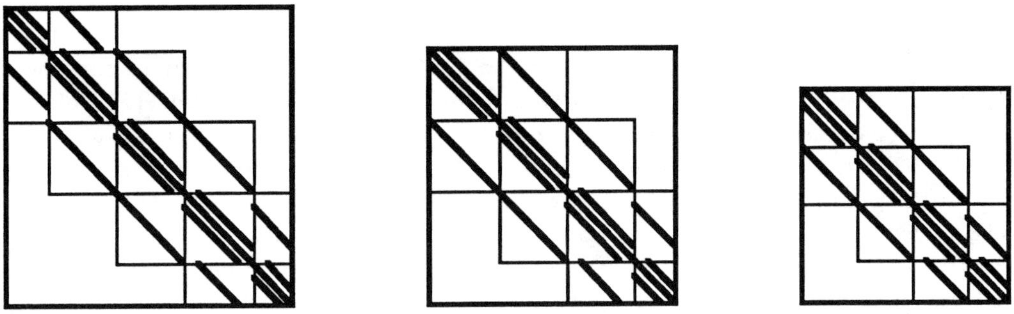

Figure 4

To solve subproblems on Neumann boxes, (as well as those on Dirichlet boxes) one can use a direct solver or a modified fast solver.
With a general purpose package like LINPACK, one can solve subproblems on individual boxes directly without further modifications. On the other hand, FISHPACK routines can be applied only to separable problems on a rectangular domain. Although our model problem (9) is separable, but the domain (in case of a Neumann box) is not a rectangle. One needs to modify the problem by extending the domain to a complete rectangle, and using the capacitance system at the third level. For more detail see [1]. Asymptotically, the computational complexity of FISHPACK routines is much smaller than those of LINPACK.

6. Parallel Implementation. The problem is well suited for parallel processing. From previous section we can conclude that the main computation task is to solve a linear system with B, repeatedly. As discussed previously, this system consists of a set of disjoint Dirichlet subsystems and a set of Neumann subsystems inter-connected through cross points. While the first subproblems are immediately applicable for parallel processing, the latter ones can easily be decoupled. Thus, all the subproblems readily lend themselves for a balanced high level parallel processing.
The parallel algorithm is based on the sequential one. The following computations occur repeatedly in the algorithm:
(a) Solving a system with B.
(b) Small steps such as multiplying $S^T A$ by a vector, computing inner products.
We denote the number of mesh points in each coordinate direction by n, the number of boxes by n_0, the total number of separator points by s, and the size

of a box by m. Thus, the total number of points is n^2, the number of boxes is n_0^2, and the number of points in a box is m^2. This notation is used throughout the paper.

Each preconditioned conjugate gradient iteration contains one instance of operations (a) and a few instances of (b). Denote the number of operations for (b) by s. The number of separator points is bounded from above by n_0^2 x $2m$, since each box contributes at most $2m$ points. Thus:

$$s < n_0^2 \times 2m = (n/m)^2 \times 2m = 2n^2/m,$$

and the operation counts for (b) is proportional to n^2/m.

In contrast to (b) the cost of solving a system with B is significant and it dominates all the others combined. The operation count for a banded symmetric positive definite subsystem on a box is of order m^4 for the decomposition, and of order m^3 for the substitution. Since our model problem (9) has constant coefficients, only three different types of boxes arise, and thus three different subsystems needed to be formed, factored, and saved at the preprocessing stage with negligible cost and small amount of storage. Then, using the factored form, one needs only substitution to solve a subsystem on a box. Thus, this operation count during the iterations is reduced to m^3. Therefore, the number of operations for solving a B system is proportional to:

$$n_0^2 \times m^3 = (n/m)^2 \times m^3 = n^2 m.$$

Consequently, the ratio of the operation count for a B-solver over the other computations in each iteration is proportional to $n^2 m/(n^2/m) = m^2$, which is substantial for all values of m ($4 <= m <= 17$) used.

The experiments confirm that B-solvers are the dominant parts of the computations and show that well over 80% of CPU time is spent for B-systems (see further in Table 3). We, therefore, solve these systems in parallel.

A system with B is solved by performing (6), (7), (8), and (5), in that order, by direct methods. In this process there are three instances of data exchanges, those corresponding to (6), (8), and (5).

The parallel algorithm proceeds as follows. First, after solving the Neumann subsystems of (6), processors have to communicate to form g_N, the right hand side of (7), that is solved directly and sequentially in the host (C_N has already been formed in preprocessing steps). Then, the processors solve (8) in parallel and distribute the data for Dirichlet subsystems. Finally, after solving (5) in parallel, the processors exchange data to form the inner products for conjugate gradient iterations.

We assign one or more of the subsystems to each node of the parallel processor. The communication for solving one subsystem consists of sending the right hand side of the subsystem to a node and receiving the solution from the node.

There are some sequential computation in the preprocessing stage and in conjugate gradient iterations. The volume of these computations is relatively small.

We, therefore, can divide the computations into three categories: computations for solving systems with B, other repeated computations, such as computing inner products, adding two vectors, etc, and one-time occurring computations, such as forming and factoring C_N.

We apply parallelism only to the first category. The volume of computations of second and third ones are small. We justify this further in the next section (see also Table 3).

7. Numerical Results. The programs were developed in FORTRAN-77. The sequential version was run on a DEC-KL10, and the parallel version on a 16-node Intel's hypercube, iPSC (Intel's Personal Super Computer).
As a model problem we chose

$$-\Delta u + u = F \quad (9)$$

in (0,1)x(0,1) with Dirichlet boundary condition. The right hand side F is chosen so that the exact solution is $u = 20\sin(\pi y)x(1-x)$.
Let as before n and n_0 be the number of mesh points and boxes in each coordinate direction, respectively. We ran the sequential program for $n = 16, 24, 32, 48, 64, 96, 128$, and $n_0 = 2, 4, 6, 8, 12, 16$. All possible combinations were executed. Table 1 shows the number of iterations and estimate of the condition number of the capacitance matrix C for different values of n and n_0. The table contains the main results obtained using the sequential version.

CONVERGENCE RATE					
	Number of				
Mesh Points	Boxes	Separator Points	Cross Points	Number of Iterations	Estimate of Condition #
32^2	4^2	168	9	7	7.08
	8^2	336	49	8	4.91
48^2	4^2	264	9	8	8.79
	6^2	420	25	9	7.54
	8^2	560	49	8	6.39
	12^2	792	121	7	4.92
	16^2	960	225	7	4.01
64^2	4^2	360	9	8	10.11
	8^2	784	49	9	7.68
	16^2	1440	225	7	4.94
96^2	4^2	552	9	8	12.11
	6^2	900	25	10	10.81
	8^2	1232	49	9	9.56
	12^2	1848	121	9	7.63
	16^2	2400	225	8	6.45
128^2	4^2	744	9	8	13.61
	8^2	1680	49	9	11.02
	16^2	3360	225	9	7.65

Table 1

For most of our experiments we chose $u = 20\sin(\pi y)x(1-x)$. To study the effect of smoothness of the solution on the rate of convergence, we executed the same problem as in Table 1, this time with a solution randomly generated. The number of iterations have increased by an average of 24% (with maximum deviation of 11%) . The results are in Table 2. This increase is by no means significantly large, and therefore all the rest of the experiments were carried out for the smooth solution.

ITERATION NUMBER OF SMOOTH AND RANDOM SOLUTION

Number of		Iteration Number of		Increment (in %)
Mesh Points	Boxes	Smooth Solution	Random Solution	
32^2	4^2	7	9	29
	8^2	8	9	13
48^2	4^2	8	9	13
	6^2	9	12	33
	8^2	8	10	25
	12^2	7	9	29
	16^2	7	8	14
64^2	4^2	8	9	13
	82	9	11	22
	16^2	7	9	29
96^2	4^2	8	10	25
	6^2	10	12	20
	8^2	9	12	33
	12^2	9	11	22
	16^2	8	10	25
128^2	4^2	8	10	25
	8^2	9	12	33
	16^2	9	11	22

Table 2

We group the computations into three categories: computation for solving systems with **B**, computations for other repeated steps, and computation for those steps that occur only once. We show the execution time for each category (in %) in Table 3; columns 3, 4 and 5, respectively. As Table 3 shows, up to 87% (and an average of 80%) of CPU time is spent for solving the systems with **B**. We can conclude that the computation for these systems play a dominant role, and the rest is relatively small. We, therefore, apply parallelism only to solve the systems with **B**.

Let **p** be the number of nodes of the hypercube that are used in a parallel execution. And let, as before, **n** and n_0 be the number of mesh points and boxes along each side of the rectangular domain, respectively. We let these three factors vary. The only restrictions are that **n** be divisible by n_0 and $n_0^2/2$ be divisible by **p**. The first restriction is obvious. The second restriction allows us to assign an equal number of boxes to each processor (n_0^2 is the total number of boxes, half of them Dirichlet, the other half Neumann boxes). We executed the parallel version for all combinations of the sequential version, for **p = 2, 3, 4, 6, 8, 9, 12, 16.** We measured the execution times for all different runs and compared them against the execution time of the sequential version. The results are in Table 4. The elements in this table are the ratios of the sequential time over the parallel execution time.

PARALLELIZABILITY				
Number of		CPU time (in %) for solving		
Mesh Points	Boxes	B-systems	Small Steps	One-time steps
32^2	4^2	80	10	10
	8^2	74	20	6
48^2	4^2	82	8	10
	6^2	82	10	8
	8^2	79	15	6
	12^2	73	20	7
	16^2	64	20	16
64^2	4^2	83	5	11
	8^2	82	5	12
	16^2	71	12	16
96^2	6^2	86	7	7
	8^2	85	5	9
	12^2	82	8	10
	16^2	79	12	9
128^2	8^2	87	6	6
	16^2	82	10	10

Table 3

From Table 4, one can observe that the speed-up factors are small in the upper right portion of the table. The speed-up improves as we move to the bottom parts of the table. The standard definition of parallel efficiency, E, is given as the ratio of the speed-up over the number of nodes, $E = sp/p$, where sp represents speed-up. Table 5 shows these efficiency factors (in %) based on the data from Table 4.

There are three factors that affect the size of the entries in Tables 4 and 5: the number of mesh points, the number of boxes, and the number of processors, denoted by n, n_0, and p, respectively. Below we discuss the effects n and n_0 on speed-up and efficiency. We study the influence of each one while the other two variables are kept fixed. The dependence of speed-up and efficiency as functions of p is discussed in [2] (see also [4]).

First, we study the efficiency as a function of n while keeping n_0 and p constant. We find that the efficiency is an increasing function on n. This can be explained as following. In the complexity count the highest order term, m^3, is connected with the cost of B-solvers. Other parts of the algorithm (an overhead) are of lower order. Thus, for fixed n_0, as n (and consequently m) increases, the share of the B-solver is increased. This is confirmed by the experimental data in Table 3. The portions of the algorithm connected with B-solvers are implemented in parallel, hence higher speed-up and efficiency are achieved as n grows. Moreover, let us define r, as the ratio of communication-time over execution-time for the portion of the algorithm which is implemented in parallel. We analyze the influence of n on r. The cost of solving a subsystem on a box (using SPBSL of LINPACK) is proportional to m^3, and the communication exchange is proportional to the size of the right hand side, namely m^2. Therefore, r is proportional to $1/m = n_0/n$ (as $m = n/n_0$). For a fixed n_0, as n increases, r decreases and, therefore, the speed-up and efficiency improve. Figure 5 shows the speed-up as a function of n for fixed n_0 and p.

SPEED-UP FACTORS					
Number of		Number of processors			
Mesh Points	Boxes	2	4	8	16
32^2	4^2	1.48	2.07	2.07	—
	8^2	1.35	1.72	1.72	1.29
48^2	4^2	1.59	2.22	2.68	—
	8^2	1.51	2.11	2.58	2.16
	16^2	1.29	1.58	1.78	1.61
64^2	4^2	1.65	2.36	3.00	—
	8^2	1.59	2.32	2.93	3.04
	16^2	1.38	1.82	2.14	2.14
96^2	8^2	1.66	2.38	3.08	3.90
	16^2	1.52	2.14	2.71	2.95
128^2	8^2	—	2.56	3.54	4.45
	16^2	—	2.35	3.10	3.57

TABLE 4

Next, we fix n and p and discuss the affect of n_0 on efficiency. Here we see that as n_0 increases, the efficiency becomes smaller. Again, examining Table 3, we see that as n_0 increases, the percentage of CPU time required to solve B systems decreases, therefore the time percentage of parallel parts becomes smaller and thus efficiency decreases. Additionally, r, the communication overhead, is proportional to n_0/n. Thus, r increases together with n_0 and, as a consequence, efficiency deteriorates.

Thus, we conclude that speed-up and efficiency are increasing functions of n and decreasing functions of n_0. Here, a few comments are in place. One may get a wrong impression that having a large number of boxes is a disadvantage. This is not the case. In fact, from the point of view of the overall CPU time the opposite is true. The total execution time decreases as the number of boxes increases. This is true for all sizes of our problem in the sequential version and also in the parallel version (except when n is small, and n_0 and p are simultaneously approaching their extreme values). In all, increasing n_0 reduces the execution time and this reduction is substantial for large n and small p. Moreover, a large n_0, and consequently small m, reduces the storage requirements needed to solve subsystems on the boxes. And finally, for large values of p, one must have at least as many boxes as processors. Figure 6 shows speed-up as a function of n_0 for fixed n and p.

PARALLEL EFFICIENCY, in %

Number of Mesh Points	Number of Boxes	Number of processors 2	4	8	16
32^2	4^2	74	52	26	—
	8^2	68	43	22	8
48^2	4^2	80	56	34	—
	8^2	76	53	32	14
	16^2	65	40	22	10
64^2	4^2	83	59	38	—
	8^2	80	58	37	19
	16^2	69	46	27	13
96^2	8^2	83	60	39	24
	16^2	76	54	34	18
128^2	8^2	—	64	44	28
	16^2	—	59	39	22

Table 5

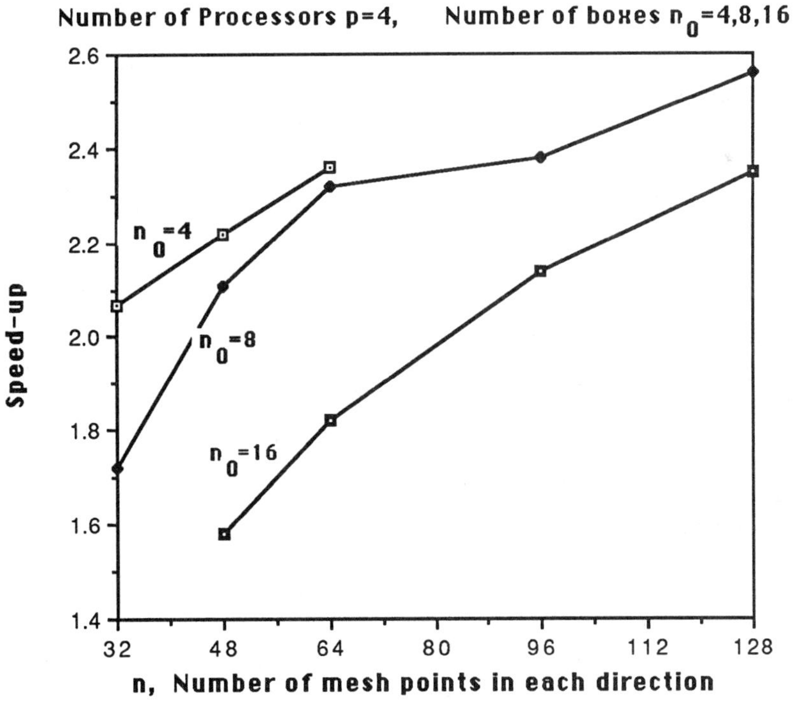

Figure 5

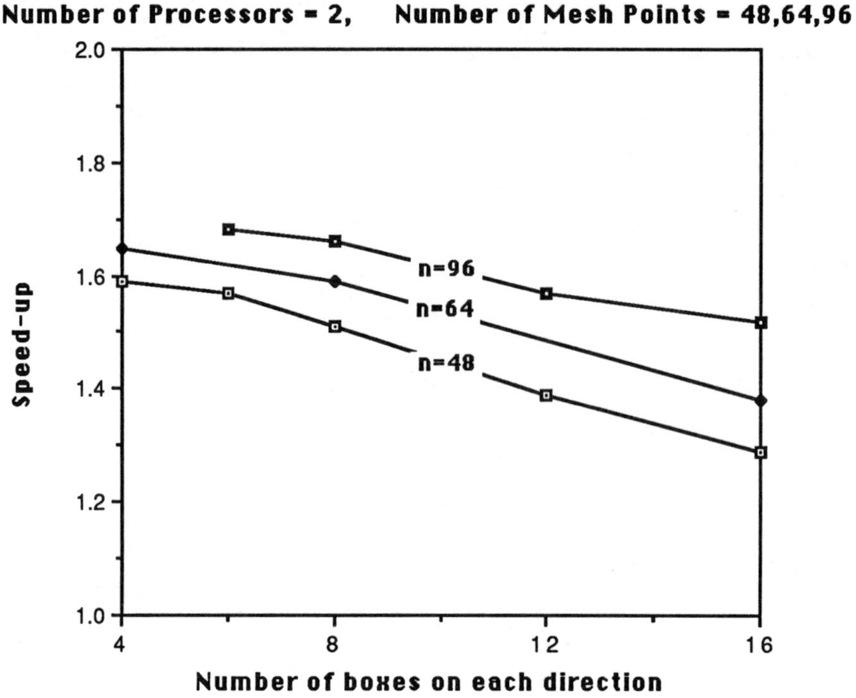

Figure 6

Efficiency substantially decreases with **p**. The decrease is even sharper for small problems (see **[2]**).

To summarize our discussion: The efficiency improves as **n** increases, and deteriorates as n_0 and **p** become larger. On the other hand, the total execution time decreases as these three factors increase, except when n_0 and **p** approach simultaneously their extreme values while **n** is relatively small. We should note that the given problem defines the value of **n**, and one needs only to choose the values of n_0 and **p** that minimize the execution time.

7. Conclusions. The results obtained using the sequential version (Table 1) are consistent with the similar results in **[1]**.

The main conclusions obtained are:

1. The problem is well suited for parallel processing (Table 3). The speed-up factor for two processors is as high as 1.65, and the speed-up is larger than $\sqrt{p}$, in most cases, where **p** is the number of processors (Table 4) .
2. Efficiency is an increasing function on **n** and a decreasing function of n_0 and **p** (Table 5).
3. For a given problem of size **n**, the total execution time reduces as n_0 and **p** increase, unless **n** is relatively small and n_0 together with **p** have their extreme values.

References

[1] M. Dryja, W. Proskurowski, and O. Widlund, *Numerical experiments and implementation of domain decomposition method with cross points for the model problem*, in "Advances in computer methods for PDEs-VI", IMACS, 1987, pp. 23-27.

[2] M. Haghoo and W. Proskurowski, *Parallel implementation of domain decomposition techniques on Intel's hypercube*, Proceedings of the third conference on hypercube, ACM Publications, 1988.

[3] D. E. Keyes and W. D. Gropp, *A comparison of domain decomposition techniques for elliptic partial differential equations and their parallel implementations*, SIAM J. Sci. Stat. Comput., v. 8, No. 2, March 1987, pp. s166-s202.

[4] O. A. McBryan and E. Van de Velde, *Hypercube algorithms and implementations*, SIAM J. Sci. Stat. Comput., v. 8, No. 2, March 1987, pp. s227-s286.

PART IV

Applications

Multi-Block Euler Method Using Patched Grids*

Torsten Berglind†

Abstract. A general purpose Multi-block Euler Solver for Compressible flows **MESC1** is developed. The Euler solver is of finite-volume type and uses explicit multi-stage Runge-Kutta time stepping. The basic concept is to express all boundary conditions in terms of image values of the flow variables. An arbitrary topology can then be specified by assigning boundary conditions on all outer surfaces of the computational blocks. A windtunnel model consisting of a 65 deg round leading edge delta wing with a sting is used as a test case. A patched C^1-continuous multi-block grid around the wing-sting configuration is constructed by transfinite interpolation. The wing is enveloped by an O-O type of grid with parabolic singularities at the wing tip, and the sting is enveloped by an O-O type of grid with polar singularities at the nose and the tail , with a slit for the wing grid. This grid is divided into eight computational blocks, so that each side of a computational block contains precisely one type of boundary condition. The flow at 10 deg angle of attack and at a free stream Mach number of 0.85, serves as a test case. The computation of the flow around a wing-fuselage shows substantially better agreement with measured data than the corresponding computation around a wing alone. This comparison serves as a first validation of the overall computational procedure.

1. Introduction. In recent years, an increasing interest has been shown in flow computations on complex flow regions such as a complete aircraft [9-12]. These flow regions are too complex to be mapped onto a single-block grid, with reasonable skewness and grid smoothness restrictions. Numerical methods using structured data, based on single-block mapping, have to be extended to multi-block transformations. The advantage is not only that grid smoothness properties are improved, but also that grid points can be concentrated in regions of interest more efficiently. The flow region is therefore broken into a set of blocks where each block is a hexahedron in the computational space.The decomposition of the

---

*This work was sponsored by the Defense Materiel Administration of Sweden.

†FFA, The Aeronautical Research Institute of Sweden, S-161 11 BROMMA

flow region implies that instead of one structured grid, a number of coupled sub-grids are applied. The coupling introduces interface boundaries between contiguous blocks. Grid lines of two adjoining regions may or may not align [8]. A more general type of interface is overlapping grids. The disadvantage with overlapping techniques is that it is difficult to preserve conservation. This work is restricted to patched C^1-continuous grids [12], which means that continuity conditions are imposed on grid points, grid spacing and grid line orientation at grid interfaces between adjoining blocks as well as any other grid point. This implies that boundary grid cells at an interface boundary can be computed as interior grid cells. Bookkeeping of the addresses of adjoining grid cells is needed at the interface boundaries.

2. Grid Generation. A windtunnel model consisting of a 65 deg round leading edge delta wing with a sting, proposed for the International Vortex Flow Symposium in Stockholm on October 1 to 3, 1986 [14], serves as a test case for the multi-block flow solver. The sting is spool-shaped and is rounded off at the tail in order to enable an O-O topology around the both the sting and the wing. The O-O topology is shown to be very efficient for Euler computations around delta wings [2,5] . The sting geometry is described in terms of analytical expressions, whereas the wing geometry is described by spanwise cuts.

The purpose of grid generation is to distribute grid points in a way such that flow gradients are resolved properly. Grid defects like sudden changes in grid line spacing or grid line orientation as well as very skewed grid cells contribute to additional truncation errors. A combination of high flow gradients and metric discontinuities may generate false vorticity which can destroy the solution. The surface grid at the wing is generated by MESHP44, a standard code [3] for grid generation around delta wings. Spanwise grid lines on the sting surface must conform to both grid points, grid size and grid line orientation at the wing-sting cut. Preliminary grid points close to the sting are generated by transfinite interpolation based on the boundary grid points. The preliminary grid points are then projected onto the sting surface [4] . The surface grid for the wing-sting configuration is depicted in Fig. 1 below.

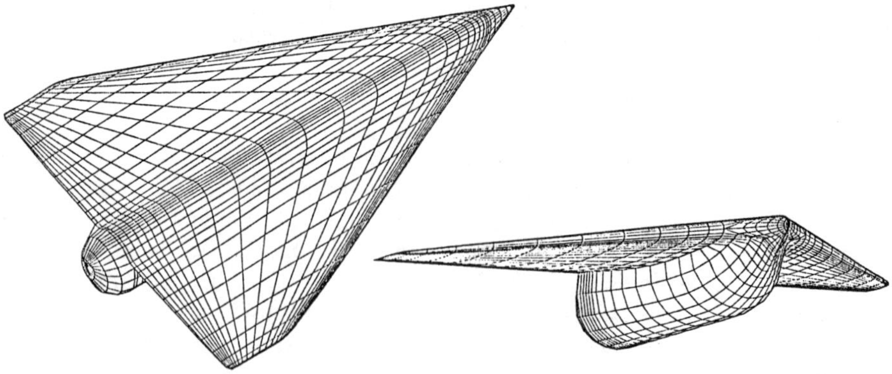

Figure 1. *Surface grid for the wing-sting configuration*

Besides the parabolic singularities at the wing tips and the polar singularities at the nose and the tail of the sting an additional type of mapping singularity appears at the wing-sting cut at both the leading and the trailing edge. The farfield boundary is chosen as a half sphere with a radius of five times the root chord of the wing. The corresponding surface grid is generated in an algebraic fashion, Fig 2.

The interior grid is generated by transfinite interpolation using the far-field and configuration surface grids and the configuration out-of-surface derivative as interpolation conditions [1,2,4]. The sudden change in grid line orientation at the intersection between the wing and the sting on the lower side propagates into the interior domain. The slope discontinuity is smoothed out with a post processor.

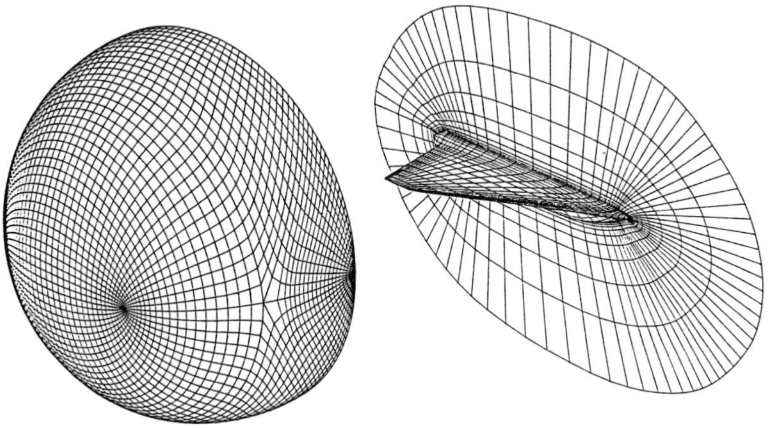

Figure 2. *Oblique view of the 3D grid.*

The topology described above is especially suitable for wing-fuselage configurations for which the fuselage extends far from the wing, because you can add spanwise gridlines on the nose and tail without increasing the size of the wing grid. The polar singularities implies good resolution of the nose and tail where high flow gradients occur. Generally an O-O grid around a delta wing has a slope discontinuity where the spanwise coordinate lines meet the symmetry plane. With a multi-block grid this can be avoided, i.e. the spanwise coordinate lines are all essentially perpendicular to the symmetry plane. Finally, the resolution at the nose and tail is very good due to the polar singularities. This is advantageous since very high flow gradients appear in these regions.

3. Euler Flow Model. The flow is restricted to that of a perfect gas with globally constant stagnation enthalpy h_0, so that the static pressure can be derived from the Bernoulli equation,

$$p = \frac{1}{2}\kappa\rho\,(2h_0 - u^2 - v^2 - w^2) \tag{1}$$

where u,v and w are the cartesian velocity components, ρ is the density and $\kappa = \frac{(\gamma-1)}{\gamma}$, $(\gamma = \frac{C_p}{C_v})$. The energy equation is therefore not needed, unless heat transfer is taken into account. The Euler equations for compressible 3D flow can then be written in the integral form as follows:

$$\int_{\Omega} \frac{\partial \vec{Q}}{\partial t}\, dV + \oint_{\partial\Omega} (n_x\, \vec{FX} + n_y\, \vec{FY} + n_z\, \vec{FZ})\, dS = 0 \tag{2}$$

with,

$$\vec{Q} \equiv \begin{pmatrix} \rho \\ \rho u \\ \rho v \\ \rho w \end{pmatrix}$$

$$\vec{FX} \equiv \begin{pmatrix} \rho u \\ \rho u^2 + p \\ \rho uv \\ \rho uw \end{pmatrix}, \quad \vec{FY} \equiv \begin{pmatrix} \rho v \\ \rho uv \\ \rho v^2 + p \\ \rho vw \end{pmatrix}, \quad \vec{FZ} \equiv \begin{pmatrix} \rho w \\ \rho uw \\ \rho vw \\ \rho w^2 + p \end{pmatrix}$$

where Ω is an arbitrary finite region, and $d\vec{S}$ is the outward pointing normal surface vector. The governing system of Eqs. (2) is hyperbolic.

4. Numerical Method. The numerical method used to integrate the Euler equations is a centered finite volume method for the spatial discretization together with an explicit one-step four-stage time integration scheme [5-7]. Since only steady solutions are of interest the local time step is used to accelerate the convergence to steady state.

4.1 Spatial Discretization. A centered finite volume discretization is obtained by applying the integral formulation of the governing Eqn. (3) to each grid cell and assuming that $\frac{\partial \vec{Q}}{\partial t}$ is constant within each grid cell and that the fluxes are constant across each cell surface,

$$VOL_{i,j,k}\, \frac{\partial \vec{Q}}{\partial t} + \vec{FI}_{i+\frac{1}{2},j,k} - \vec{FI}_{i-\frac{1}{2},j,k} + \vec{FJ}_{i,j+\frac{1}{2},k} -$$

$$-\vec{FJ}_{i,j-\frac{1}{2},k} + \vec{FK}_{i,j,k+\frac{1}{2}} - \vec{FK}_{i,j,k-\frac{1}{2}} = 0 \tag{3}$$

where,

$$\vec{FI} \equiv SIX\ \vec{FX} + SIY\ \vec{FY} + SIZ\ \vec{FZ}$$

$$\vec{FJ} \equiv SJX\ \vec{FX} + SJY\ \vec{FY} + SJZ\ \vec{FZ}$$

$$\vec{FK} \equiv SKX\ \vec{FX} + SKY\ \vec{FY} + SKZ\ \vec{FZ}$$

are the integrated fluxes and, SIX, SIY, SIZ, SJX, SJY, SJZ, SKX, SKY and SKZ are the metric coefficients, defined as the cartesian components of the cell surface vectors. The non-integer indices refer to cell surfaces whereas integer indices refer to cell centers. Non-integer index values of the fluxes $\vec{FX}, \vec{FY}$ and $\vec{FZ}$ are taken as the arithmetic means of neighboring cell centered values. Thus, the only quantities needed for the coordinate transformation are the x, y and z components of the grid points.

4.2 Artificial Viscosity. The Euler equations need artificial viscosity to avoid nonphysical shocks and sawtooth waves (with a wavelength of two grid lengths) in the solution. The model for artificial viscosity uses a combination of a variable-coefficient second order difference operator and a constant-coefficient fourth order difference operator [6]. The artificial viscosity terms are added to the semi-discrete scheme so that the resulting system becomes

$$\frac{\partial \vec{Q}}{\partial t} = N(\vec{Q}) - \chi\ (\delta_i[SWI(\vec{Q})\delta_i] + \delta_j[SWJ(\vec{Q})\delta_j] + \delta_k[SWK(\vec{Q})\delta_k])\vec{Q} -$$

$$-\beta\ CFL\ (\delta_i^4 + \delta_j^4 + \delta_k^4)\ \vec{Q} \tag{4}$$

where N is a nonlinear vector function, χ and β are filter constants, SWI, SWJ and SWK are coefficients proportional to the second derivative of the pressure and δ^2 and δ^4 are second and fourth order centered difference operators.

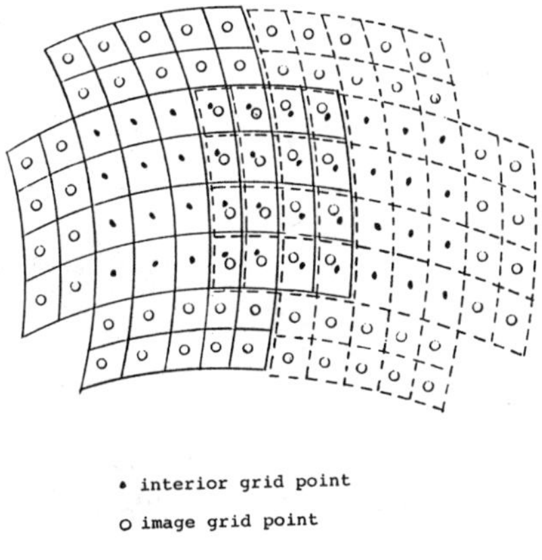

Figure 3. *Merging of two blocks.*

5.1 Multi-Block Implementation. To tailor new codes for each new topology is both time consuming and increases the possibilities of errors that are difficult to find. A general purpose multi-block flow solver greatly facilitates the application of the flow solver for different topologies [11,12]. The basic concept is to add to each block extra layers of grid cells outside the block, so that boundary grid points can be treated as interior grid points.

Since the second order filter operates on the pressure, we need image values of not only the flow variables but also the pressure.

Both the fourth and the second order filter requires two extra layers of image values outside the flow region. Each image grid cell will share the same storage cell as an interior grid cell of the contiguous block, Fig. 3. All anomalies of the scheme can then be located to subroutines dealing with boundary conditions. Subroutines updating the time step, flow variables and fourth and second order differences are completely symmetric in all curvilinear coordinate directions i, -i, j, -j, k and -k. The decomposition of the flow field into computational blocks is done so that each side of a computational block has precisely one type of boundary condition [10,12]. This means that we introduce more computational blocks than necessary but the handling of the boundary conditions is simplified. Subroutines dealing with boundary conditions are written to handle any coordinate surface. This means that the curvilinear coordinates can be chosen completely arbitrarily by specifying input parameters.

5.2 Boundary Conditions. As mentioned above, all boundary conditions are to be expressed in terms of image values of the flow variables outside the computational block. In our case four different types of boundary conditions are applied: solid wall, symmetry plane, inflow-outflow boundary and interface boundary.

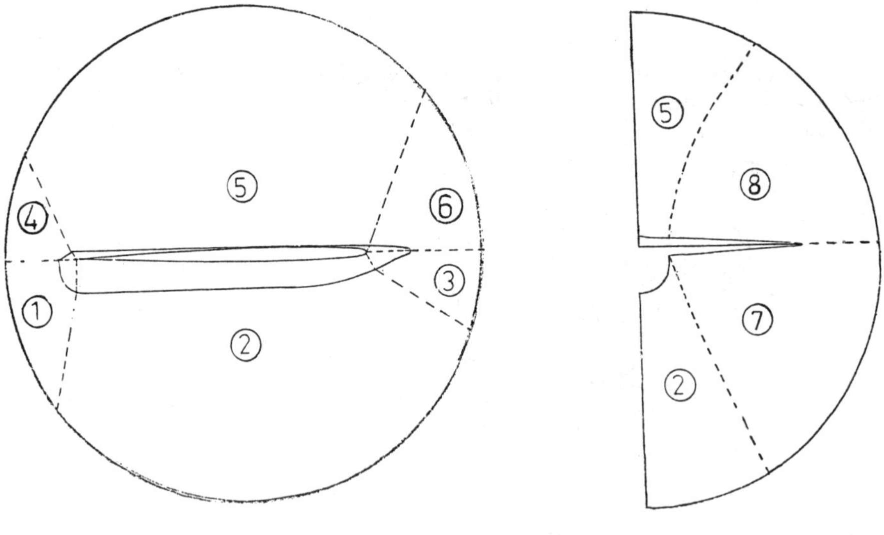

Figure 4. *Decomposition of the physical domain.*

At a solid wall the mass flux is zero while the pressure forces balances the momentum flux. As far as the fluxes are concerned, a symmetry plane boundary condition is the same as a solid wall boundary condition. At the inflow-outflow boundary absorbing boundary conditions [5] in their simplest form are applied. The governing equations are linearized locally and the characteristic variables are computed in the normal direction. The characteristic variables advected into the domain are fixed to the free-stream values whereas those which are advected out of the domain are linearly extrapolated from the interior to the boundary. The resulting set of characteristic variables is then transformed back to the primitive variables. The numerical boundary conditions for the second and fourth order filters is described in ref. [6] .

Finally, we have interface boundaries between computational blocks contiguous in the physical space. Since the same continuity restrictions of the computational grid are imposed at block interfaces as for the interior grid, these boundary grid cells can be treated as any other grid cell. The flux boundary conditions and the numerical boundary conditions for the fourth and second order difference operators automatically fall out by using image values from adjacent blocks. This can be obtained by either using indirect addressing and simply assigning the same address for corresponding grid cells or by updating the image values each stage of the time step. As shown in Fig. 4 above the flow region around the wing-sting configuration is decomposed into eight blocks so that each block surface has precisely one type of boundary condition. A schematic representation of the corresponding multiply-connected computational blocks is shown in Fig. 5. An interface boundary is specified by a pairwise enumeration of block surfaces contiguous in the physical space. However two coordinate surfaces can be merged together in essentially four different ways. It is consequently necessary for each coupling to also assign the type of merging.

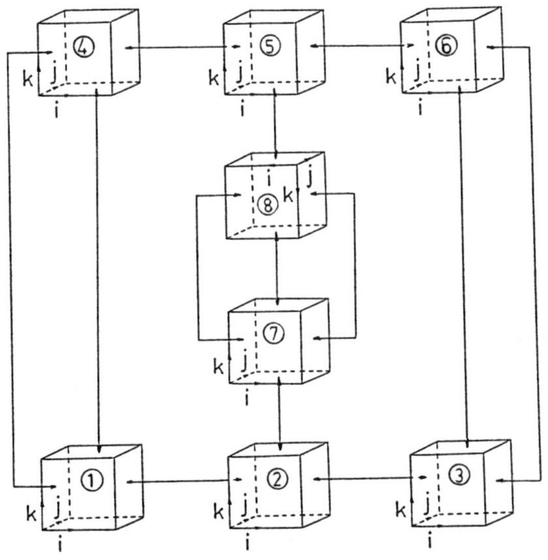

Figure 5. *Flow chart of couplings between computational blocks*

6. Computational Results. The flow around the wing-sting configuration was computed at 10 deg angle of attack and a free stream Mach number of 0.85. The grid contains 111,000 interior grid points and an oblique view of the grid was shown in Fig 2. The flow calculations were carried out with $\chi = 0.2$, $\beta = 0.01$ and $CFL = 1.0$. The initial flow field was extrapolated from computations on a coarser grid. Velocity vectors in three vertical planes are shown in Fig. 6. The flow properies have been interpolated onto intersection points between the grid and the plane. The velocity components have further been interpolated to an equispaced grid in this plane. The velocity vector plots show clearly the development of the vortex on the upper side of the wing. Figure 5 reveals that the vortex starts somewhere downstream of the 0.3 section.

Measurements of surface pressure at a Reynolds number of 1.3×10^7 [14], have been made along the previously mentioned chord sections. Computations around the wing-sting configuration are also compared with a similar computation around a wing alone with 286,000 grid points [13]. The boundary layer in the experiment separates under the primary vortex and rolls into a secondary vortex which reduces the strength of the primary vortex. The suction peaks will therefore tend to be higher in the Euler calculations than in the experiments.

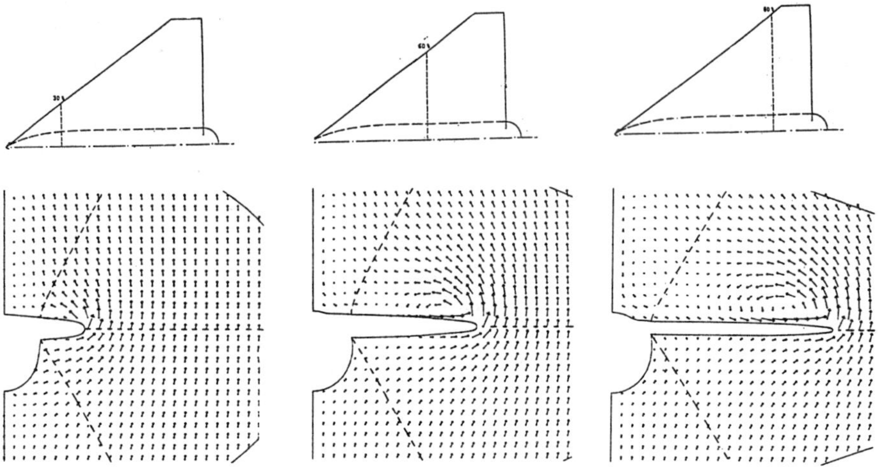

Figure 6. *Velocity vectors at* $\frac{x}{c_{root}} = 0.3, 0.6,$ *and* 0.8 .

Figure 7 shows measured and computed C_p- distributions at the sections $\frac{x}{c_{root}} = 0.3, 0.6$ and 0.8. As expected, it can be seen that for 0.3 section, the C_p- curve on the lower side shows better agreement with experiment than the corresponding curve for the wing alone computation. On the upper side of the wing the difference is very small between the wing and the wing-sting computations. Also at the sections $\frac{x}{c_{root}} = 0.6$ and 0.8, the agreement to the experiments is improved compared with corresponding computation around a wing alone, especially on the lower side of the configuration where the sting extends most. Another thing that is striking is that the suction peak for the wing-sting computation is much lower than for the alone wing computation. This can be explained by the fact that the strength of the vortex grows with a better resolution i.e., with the number of grid points, however the position of the vortex does not change. Also the absence of sharp bends on the the upper C_p- curve for the wing-sting computation probably has to do with the coarser grid. Finally a C_p- distribution on the configuration surface can be seen in Fig. 8. Traces of the vortex can be seen on the wing.

7. Conclusions. A general purpose Euler code which offers possibilities to compute generic multi-block grid topologies by specifying just a few input parameters, is developed. The multi-block grid approach is an important tool in generating efficient computational grids around complex configurations, such as a realistic aircraft. The computational results for a wing-sting configuration show that the grid topology used here efficiently resolves high flow gradients especially at the nose and tail and captures the vortex on the upper side of the wing. The grid topology is especially appropriate simple wing-fuselage configurations for which the fuselage extends far from the wing, such as space shuttles. Comparisons between wing-fuselage and wing alone computations show, as expected, that the former gives a better agreement with experimental values. Euler computations around more complex configurations remains for the future. Even though the effort in describing the grid topology to the Euler code is very small, the amount of work to generate a C^1- continuous multiblock grid is substantial. To get rid of the continuity restriction at grid interfaces would not only facilitate the grid generation a lot, but also make it possible to refine the grid in each block independently. A desirable extension for the future of the multi-block Euler code is therefore implementation of a more general interfacing.

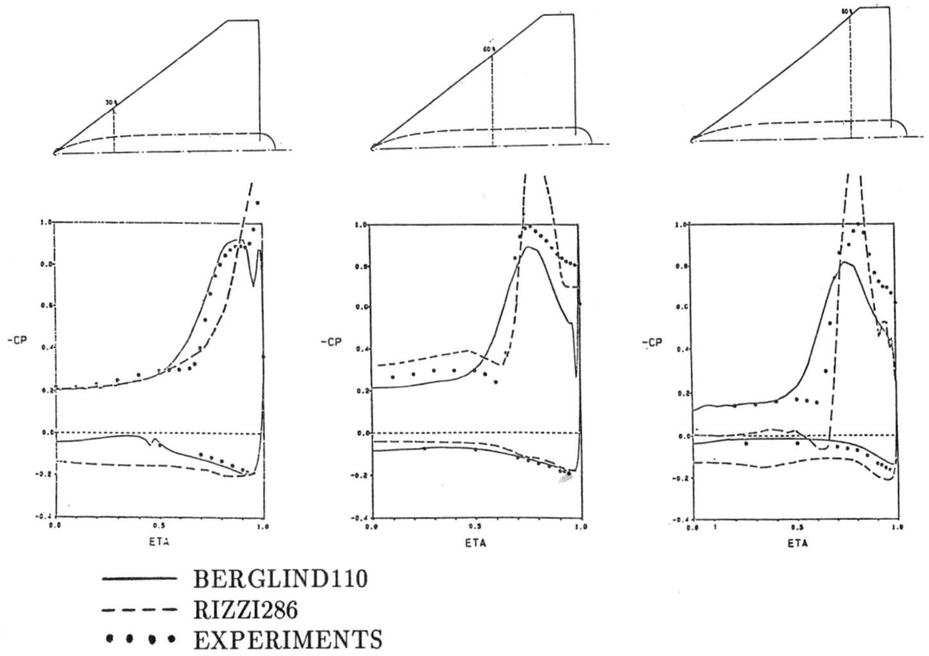

Figure 7. C_p- *distributions at the sections* $\frac{x}{c_{root}}$= 0.3, 0.6, *and* 0.8.

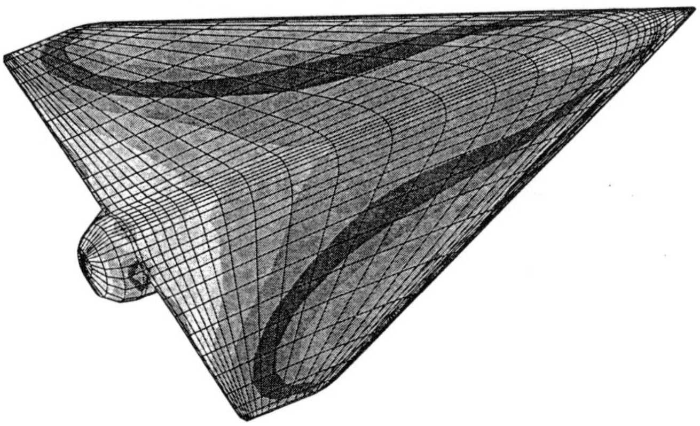

Figure 8. C_p- *distribution on the upper side of the wing-sting configuration.*

Acknowledgements. The author would like to thank Dr. Arthur Rizzi for valuable advices in the code development and Peter Eliasson and Tomas Strid for use of their flow visualization codes and for their help in extending them to multi-block topologies.

REFERENCES

(1) Eriksson, L.E. - Three-Dimensional Spline-generated Coordinate Transformations for Grids around Wing-Body Configurations, Numerical Grid Generation Techniques, NASA CP 2166, 1980.

(2) Eriksson, L.E. - Generation of Boundary-Conforming Grids around Wing-Body Configurations Using Transfinite Interpolation, AIAA Journal, Vol. 20, 1982, pp. 1313-1320.

(3) Eriksson, L.E. - FFA 3D Mesh Generation Application - Users Manual, FFA - 830307.

(4) Berglind T. - A Comparison of Single-Block and Multi-Block Grids around Wing-Fuselage Configurations, FFA TN 1986-42, Stockholm 1986.

(5) Rizzi, A. and Eriksson, L-E. - Computation of Flow around Wings Based on the Euler Equations, J. Fluid. Mech., Vol. 148, pp 45-71, 1985.

(6) Eriksson L.E. - Boundary Conditions for Artificial Dissipation Operators, FFA TN 1984-53, Stockholm 1984.

(7) Eriksson L.E. and Rizzi A. - Computer-Aided Analysis of the Convergence to Steady State of Discrete Aproximation Problems, J. Comp. Phys. 2, 12-26, 1967.

(8) Rai M. - A Conservative Treatment of Zonal Boundaries for Euler Equation Calculations, AIAA Paper No. 84-0164, AIAA 22nd Aerospace Sciences Meeting, Reno, NV, 1984.

(9) Fritz W. and Leicher S. -Numerical Solution of 3-D Inviscid Flow Fields around Complete Aircraft Configurations, Proceedings of the 15th ICAS Congress, London, UK, 7-12th Sept. 1986.

(10) Coleman, R. M. and Brabanski, M. L. - Numerical Grid Generation For Three-Dimensional Geometries Using Segmented Computational Regions, Proceedings of the International Conferens on Numerical Grid Generation in Computational Fluid Dynamics, Landshut, W. Germany, 14-17th July, 1986.

(11) Eberle A. - 3D Euler Calculations Using Characteristic Flux Extrapolation, AIAA-85-0119, 23rd Aerospace Science Meeting, Reno, NV, 1985.

(12) Berglind T. - Compressible Euler Solution on a Multi-Block Grid around a Wing-Fuselage Configuration, FFA TN 1987-46, Stockholm 1987.

(13) Rizzi A., Drougge G., Purcell C.J., Computations with the Euler Code WINGA2 for Vortex Flow around the 65 Degree Delta, Proceedings of the International Vortex Flow Experiment on Euler Code Validation, Stockholm, Sweden, 1-3 October 1986.

(14) Boersen S.J., Elsenaar A., Tests on the 65 Degree Delta Wing at NLR: A Study of Vortex Flow Development between Mach=.4 and 4., Proceedings of the International Vortex Flow Experiment on Euler Code Validation, Stockholm, Sweden, 1-3 October 1986.

Domain Decomposition and Mixed Finite Elements for the Neutron Diffusion Equation

Françoise Coulomb*†

Abstract. Among the classical methods used for solving the neutron diffusion equation, an iterative power method combined with a finite element method allows an efficient numerical treatment. A domain decomposition method seems well suited to the structure of a parallel computer. As the domains and data are often almost symmetrical, the mixed elements method yields well uncoupled systems. Some decompositions along the axes of symmetry are considered and numerically treated on two examples of reactors.

1. Introduction. The steady state formulation of the multigroup diffusion equation is the following (4,5):

$$- \mathrm{Div}(D_g(r)\nabla\Phi_g(r)) + \Sigma_g^t(r)\ \Phi_g(r) = \sum_{g'=1}^{G} \left[\frac{1}{\lambda}\chi_g \nu\Sigma_{g'}^f(r) + \Sigma_{g'g}^t(r)\right]\Phi_{g'(r)} \tag{1}$$

for g = 1 ... G

with: Φ_g = neutron flux in group g
D_g = diffusion coefficient in group g
Σ_g^t = total removal cross section in group g
Σ_g^f = macroscopic fission cross section for group g
$\Sigma_{gg'}^t$ = macroscopic scattering cross section from group g to group g'

* Service d'Etudes de Réacteurs et de Mathématiques Appliquées, Centre d'Etudes Nucléaires de Saclay, 91191 Gif-sur-Yvette Cedex, France.
† Laboratoire d'Analyse Numérique, Université Pierre et Marie Curie, 75005 Paris, France.

χ_g = fission spectrum for prompt neutrons
ν = average number of neutrons produced per fission
λ = effective multiplication factor
G = total number of energy groups
r = spacial dependence.

The boundary conditions are of Neumann-Dirichlet type.

This is generally solved by an iterative power method (4,5) combined with a finite element method ; it is reminded that the power method for solving an equation of the form: $M\Phi = \frac{\nu}{\lambda} F\Phi$, can be written:

$$\begin{cases} \psi^{(n+1)} = \nu\, M^{-1}\, F\Phi^{(n)} \\ \lambda^{(n+1)} = \dfrac{(1,\ F\,\psi^{(n+1)})}{(1,\ F\Phi^{(n)})} \\ \Phi^{(n+1)} = \dfrac{1}{\lambda^{(n+1)}}\,\psi^{(n+1)} \end{cases} \qquad \text{(outer)iteration } n+1$$

A parallel computer with four processors on-line has already been built for demonstration [1,2] ; it runs now with a Lagrange finite elements method ; the inner iterations technique is presently a parallel version of the block S.O.R. method (more precisely, an odd-even block S.O.R. method) ; the domain is partitioned into lines in 2D (planes in 3D), and each processor treats a set of contiguous lines (or planes).

As the matrices obtained through mixed elements methods are not positive definite, few results ensure the convergence of block Jacobi and block S.O.R. methods. In the case of complete symmetry, with an appropriate choice of the initial vector, the convergence of the block Jacobi and block S.O.R. methods holds in 1 iteration ; that is why, when the data are almost symmetrical, a domain decomposition along the axes of symmetry is used to get a parallel preconditionner ; by doing so and by choosing an appropriate initial vector for the iterations, a fast convergence of the previous iterative methods is expected. In practice, there are four axes (planes) of symmetry, so the domain is naturally splitted into eight parts ; in a new parallel computer project there will be eight processors ; we intend to assign a subdomain to each processor, this explains why we concentrate on the block Jacobi method.

The contents of the paper are as follows: in section 2, the mixed-dual variational formulation of the problem and its approximations are reminded ; in section 3, some decompositions are described and justified ; in section 4, they are tested on two examples of reactors and the numerical results are interpreted.

2. The mixed-dual method (6,7).

2.1. The variational formulation. Each iteration of the power method yields problems of the following form:

$$- \mathrm{Div}(D\nabla u) + \Sigma u = S \qquad \text{in } \Omega \tag{2}$$

with Dirichlet-Neumann boundary conditions: $\begin{cases} u = 0 & \text{on } \Gamma_0 \\ D\dfrac{\partial u}{\partial n} = 0 & \text{on } \Gamma_1 \end{cases}$

Ω being a bounded domain of $\mathbb{R}^n$ (n = 2,3) representing the core and $\partial\Omega = \Gamma_0 \cup \Gamma_1$, $\Gamma_0 \cap \Gamma_1 = \Phi$.

In the mixed-dual formulation, the flux u and the current "$\vec{p} = D\vec{\nabla}u$" appear as independant variables ; the variational formulation is the following:

$$\left|\begin{array}{l} \text{Find } (\vec{p},u)\in H_{0,\Gamma_1}(\text{div},\Omega) \times L^2(\Omega) \text{ so that:} \\ \displaystyle\int_\Omega \frac{1}{D}\vec{p}.\vec{q} + \text{div}\,\vec{q}\,u = 0 \qquad \forall\vec{q}\in H_{0,\Gamma_1}(\text{div},\Omega) \\ \displaystyle\int_\Omega -\text{div}\,\vec{p}\,v + \Sigma uv = \int_\Omega Sv \qquad \forall v\in L^2(\Omega) \end{array}\right. \tag{3}$$

where

$$\left|\begin{array}{l} H_{0\Gamma_1}(\text{div},\Omega) = \{\vec{q}\in H(\text{div},\Omega) \;/\; \langle\gamma_1(\vec{q}),v\rangle = 0 \quad \forall v\in H^1_{0,\Gamma_0}(\Omega)\} \\ \gamma_1(\vec{q}) = \text{normal derivative of } \vec{q} \\ H^1_{0,\Gamma_0}(\Omega) = \{v\in H^1(\Omega) \;/\; \gamma_0(v) = 0 \quad \text{on } \Gamma_0\} \\ \gamma_0(v) = \text{trace of } v \end{array}\right.$$

It is well known that under the assumptions:

$$\left|\begin{array}{l} D,\Sigma\in L^\infty(\Omega) \\ 0 < \nu \leqslant D(x) \text{ and } 0 \leqslant \Sigma(x) \quad \text{a.e. in } \Omega \\ f\in L^2(\Omega) \\ \partial\Omega \text{ "regular", meas}(\Gamma_0) > 0 \end{array}\right. \tag{4}$$

problem (3) has a unique solution ($\vec{p} = D\vec{\nabla}u,u$), where u is the solution of the classical primal problem:

$$\left|\begin{array}{l} \text{Find } u\in H^1_{0,\Gamma_0}(\Omega) \text{ so that:} \\ \displaystyle\int_\Omega D\,\vec{\nabla}u.\vec{\nabla}v + \Sigma uv = \int_\Omega Sv \qquad \forall v\in H^1_{0,\Gamma_0}(\Omega) \end{array}\right. \tag{5}$$

2.2. Approximation of the solution. Equation (3) is approximated by a finite element technique ; the approximation spaces are:

$$\left|\begin{array}{l} Q_h = \left\{\vec{q}_h\in H_{0,\Gamma_1}(\text{div},\Omega)/\vec{q}_h\in Q_K \quad \forall K\in T_h\right\} \\ V_h = \left\{v_h\in L^2(\Omega) \qquad /v_h\in P_K \quad \forall K\in T_h\right\} \end{array}\right.$$

T_h being a triangulation, P_K and Q_K polynomial spaces.

Let us suppose that Ω is a union of rectangles or parallelepiped rectangles, and T_h a regular family of triangulations constituted of (parallelepiped) rectangles. Let T_h be one of the triangulations. For all given integers k, ℓ, m, let us denote $P_{k,\ell}$ and $P_{k,\ell,m}$ the following spaces:

$$\begin{cases} P_{k,\ell} = \{P \in \mathbb{R}[X,Y] \ / \ \deg_x P \leqslant k \quad \deg_y P \leqslant \ell\} \\ P_{k,\ell,m} = \{P \in \mathbb{R}[X,Y,Z] \ / \ \deg_x P \leqslant k \quad \deg_y P \leqslant \ell \ \deg_z P \leqslant m\} \end{cases}$$

In the method of Raviart and Thomas, for each K:

$$\begin{cases} Q_K = P_{k+1,k} \times P_{k,k+1} & P_K = P_{k,k} & \text{in 2D} \\ Q_K = P_{k+1,k,k} \times P_{k,k+1,k} \times P_{k,k,k+1} & P_K = P_{k,k,k} & \text{in 3D} \end{cases}$$

k being a given integer ; the approximations of u and $\vec{p}$ are of order k+1. A modified method gives: $\begin{cases} Q_K = P_{\ell+1,0} \times P_{0,\ell+1} \\ P_K = P_{\ell,0} + P_{0,\ell} \end{cases}$; this method is of order 1 whatever integer ℓ may be.

2.3. Description of a mixed-dual element of order 1 in 2D: MXOL5 (3). The finite element basis can be represented in terms of unknowns as shown in fig. 1.

$P_K = P_{0,0}$ $\qquad\qquad$ $Q_K = P_{1,0} \times P_{0,1}$

FIG. 1: *Mixed element of order 1 MXOL5. There is one internal node for the flux, and the current unknowns are the constant values of* $\vec{p}_h \cdot \vec{n}$ *on each edge.*

This element MXOL5 is so-called because on each rectangle the approximation of u is constant, the approximation of $\vec{p}$ is linear, and there are 5 unknowns.

3. Domain decomposition. For the sake of simplicity, we restrict ourselves to the 2D case ; in 3D, there are analogous results.

In practice, we are often in the presence of a domain of the form indicated in fig. 2, which has four axes of symmetry (in 3D the domains have four planes of symmetry) and the data are almost symmetrically distributed ; here, only the horizontal and vertical axes are considered.

Decompositions along the axes of symmetry are the purpose of this study.

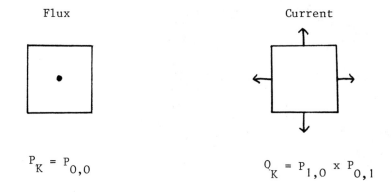

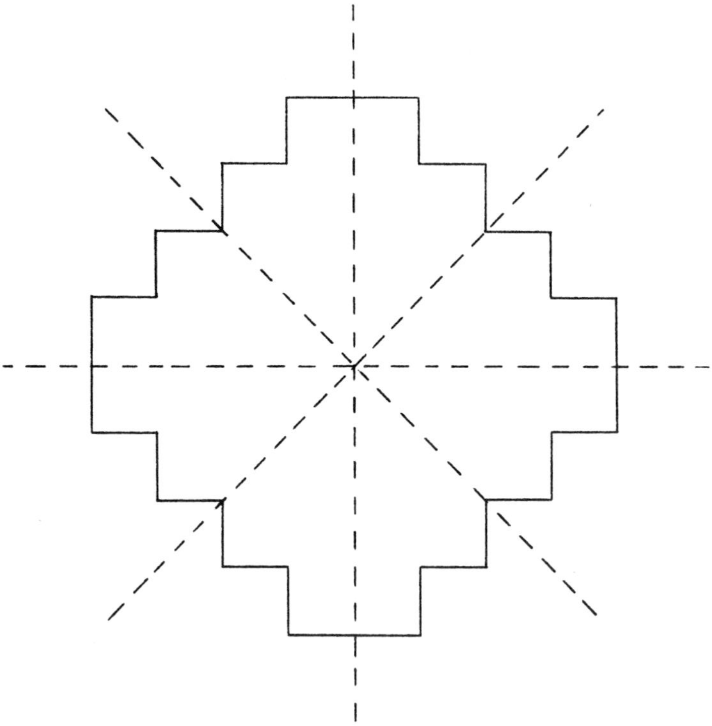

FIG. 2: *General form of the domain. The axes of symmetry are indicated in dotted-lines.*

In the sequel, it is supposed that:

(h1) Ω has an axis of symmetry Δ, and, Γ_0 and Γ_1 are symmetric with respect to Δ.

For each point $y \in \mathbb{R}^2$, $\tilde{y}$ denotes the mirror image of y with respect to Δ, and for each vector $\vec{z}$, $\hat{\vec{z}}$ denotes the mirror image of $\vec{z}$ with respect to the direction of Δ : $\vec{\Delta}$.

3.1. **Symmetry of the solution.** Let us suppose that the data D, Σ and S are symmetric with respect to Δ (condition (h2)). Then the solution satisfies:

$$\begin{cases} \vec{p}(\tilde{x}) = \hat{\vec{p}}(x) \\ u(\tilde{x}) = u(x) \end{cases}$$

The same result holds for the approximate solution, provided that:

(h3) $w_h \in V_h \iff W_h \in V_h$

$\vec{q}_h \in Q_h \iff \vec{Q}_h \in Q_h$

where: $W_h(x) = w_h(\tilde{x})$ and $\vec{Q}_h(x) = \vec{q}_h(\tilde{x})$.
Under the following assumptions:

(h4) the edges of the rectangles composing Ω are parallel to the axes of coordinates.
(h5) Δ is the x-axis or the y-axis.
(h6) T_h is symmetric with respect to Δ (see fig. 3).
(h7) the finite element of reference is of the R. and T. type.

property (h3) holds.

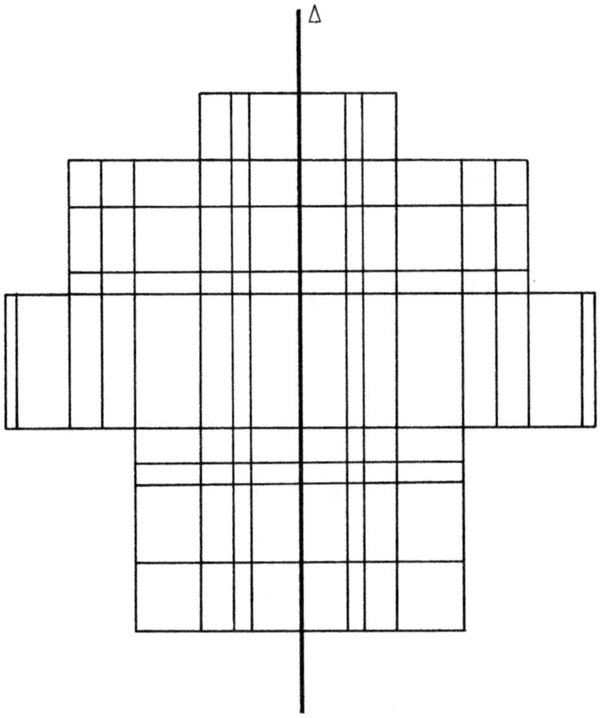

FIG. 3: *Symmetry of the triangulation* T_h

3.2. The block Jacobi method. The domain will be decomposed along the axes of symmetry.

Assume that (h4) and (h5') are verified, (h5') being the condition: Δ is the y-axis. Ω is splitted into two parts:

$$\Omega_1 = \left\{ z = \begin{pmatrix} x \\ y \end{pmatrix} \in \Omega \;/\; x < 0 \right\}$$
$$\Omega_2 = \left\{ z = \begin{pmatrix} x \\ y \end{pmatrix} \in \Omega \;/\; x > 0 \right\}$$

The interface $\left\{ z = \begin{pmatrix} x \\ y \end{pmatrix} \in \bar{\Omega} / x = 0 \right\}$ is denoted Γ_3 $(\Gamma_3 \subset \Delta)$. We have the choice of considering Γ_3 separately or as part of one of the subdomains.

The nodes are numbered in the following order: first those of $\bar{\Omega}_1 \backslash \Gamma_3$ then those of $\bar{\Omega}_2 \backslash \Gamma_3$, and to end those of Γ_3.

Γ_3 is supposed to be composed of edges of rectangles $K\in T_h$ (condition (h8)) (see fig. 4).

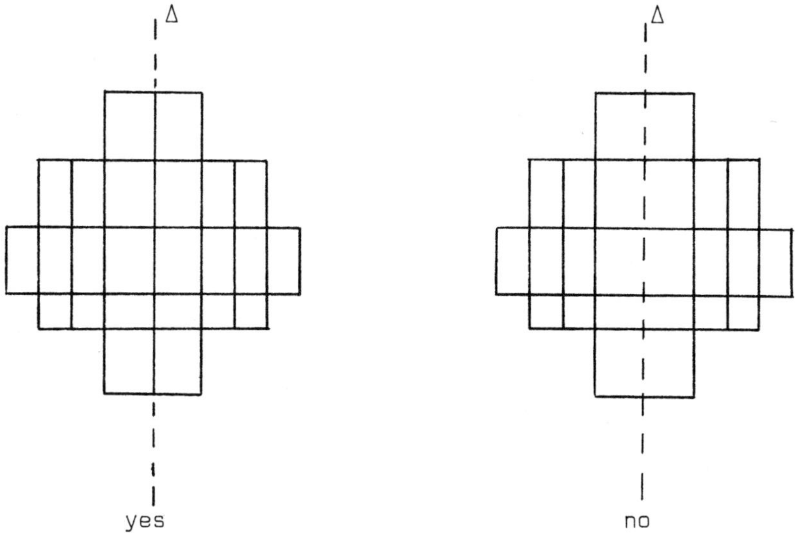

FIG. 4: *The domain is splitted into two parts with an interface composed of edges of rectangles* $K\in T_h$ *on the left figure ; the figure on the right shows a bad splitting.*

The matrix of the linear system has the following form:

$$A = \begin{pmatrix} A_{11} & 0 & A_{13} \\ 0 & A_{22} & A_{23} \\ {}^tA_{13} & {}^tA_{23} & A_{33} \end{pmatrix} \tag{6}$$

and we have to solve $A\begin{pmatrix} x_1 \\ x_2 \\ x_3 \end{pmatrix} = \begin{pmatrix} B_1 \\ B_2 \\ B_3 \end{pmatrix} = B$, with $B_3 = 0$, for all the nodes on Γ_3 are current nodes.

A_{11} is the matrix of the approximate problem corresponding to the equation:
$$\begin{cases} -\operatorname{div}(D\nabla u) + \Sigma u = S & \text{in } \Omega_1 \ ; \\ u = 0 & \text{on } \Gamma_0\cap\partial\Omega_1 \\ D\dfrac{\partial u}{\partial n} = 0 & \text{on } (\Gamma_1\cap\partial\Omega_1)\cup\Gamma_3 \end{cases}$$

Meas $(\Gamma_0) > 0$ and Γ_0 is symmetrix with respect to Δ (hypothesis h1), so meas $(\Gamma_0\cap\partial\Omega_1) > 0$; then A_{11} is invertible. For the same reason A_{22} is invertible. As A_{33} is the matrix of the positive definite bilinear form: $(\vec{p},\vec{q}) \to \int_\Omega \frac{1}{D}\vec{p}.\vec{q}$ on a subspace of Q_h, A_{33} is invertible. So, the block Jacobi method can be considered.

Let (h9) be the condition:

> For the element of reference $\hat{K}$, the unknowns are symmetrically distributed with respect to its vertical axis of symmetry.

Proposition: Under the assumptions (h2), (h6), (H7) and (h9), and if the initial vector $X^{(0)}$ satisfies:

(h10)
$$\begin{cases} X^{(0)}(i) = -X^{0}(j) & \text{if i and j are gradient nodes corresponding to the x-component of } p_h \text{, and are symmetric with respect to } \Delta. \\ X^{(0)}(i) = X^{0}(j) & \text{in the other case if i and j are symmetric with respect to } \Delta. \end{cases}$$

then the block Jacobi method converges in 1 iteration.

proof: it is not restrictive to suppose that the numbering of the nodes is done as follows:

. in domain 1, the first nodes to be numbered are those corresponding to the flux ; for the current nodes, we first number the nodes corresponding tothe x-component of p_h, then those corresponding to the y-component,

. the nodes of domain 2 are numbered symmetrically with respect to the axis of symmetry.

Then A has the following form:

$$A = \begin{pmatrix} A^F_{11} & A^{FX}_{11} & A^{FY}_{11} & 0 & 0 & 0 & A^F_{13} \\ {}^tA^{FX}_{11} & A^X_{11} & 0 & 0 & 0 & 0 & A^X_{13} \\ {}^tA^{FY}_{11} & 0 & A^Y_{11} & 0 & 0 & 0 & 0 \\ 0 & 0 & 0 & A^F_{11} & -A^{FX}_{11} & A^{FY}_{11} & -A^F_{13} \\ 0 & 0 & 0 & -{}^tA^{FX}_{11} & A^X_{11} & 0 & A^X_{13} \\ 0 & 0 & 0 & {}^tA^{FY}_{11} & 0 & A^Y_{11} & 0 \\ {}^tA^F_{13} & {}^tA^X_{13} & 0 & -{}^tA^F_{13} & {}^tA^X_{13} & 0 & A_{33} \end{pmatrix} = \begin{pmatrix} A_{11} & 0 & A_{13} \\ 0 & A_{22} & A_{23} \\ {}^tA_{13} & {}^tA_{23} & A_{33} \end{pmatrix}$$

We have also:

$$B = \begin{pmatrix} B^F_1 \\ B^X_1 \\ B^Y_1 \\ B^F_2 \\ B^X_2 \\ B^Y_2 \\ B_3 \end{pmatrix} = \begin{pmatrix} B^F_1 \\ 0 \\ 0 \\ B^F_1 \\ 0 \\ 0 \\ 0 \end{pmatrix} \qquad X^{(0)} = \begin{pmatrix} X^{(0)}_{1F} \\ X^{(0)}_{1X} \\ X^{(0)}_{1Y} \\ X^{(0)}_{1F} \\ -X^{(0)}_{1X} \\ X^{(0)}_{1Y} \\ 0 \end{pmatrix} = \begin{pmatrix} X^{(0)}_1 \\ X^{(0)}_2 \\ 0 \end{pmatrix}$$

At the first iteration, we have:

$$\begin{cases} A_{11}\, X^{(1)}_1 = B_1 - A_{13}\, X^{(0)}_3 = B_1 \\ A_{22}\, X^{(1)}_2 = B_2 - A_{23}\, X^{(0)}_3 = B_2 \end{cases}$$

but, $X_3 = 0$ and so $\begin{cases} A_{11}\, X_1 = B_1 \\ A_{22}\, X_2 = B_2 \end{cases}$; then $\begin{cases} X^{(1)}_1 = X_1 \\ X^{(1)}_2 = X_2 \end{cases}$.

$$A_{33}\, X_3^{(1)} = 0 - {}^tA_{13}\, X_1^{(0)} - {}^tA_{23}\, X_2^{(0)}$$

$$A_{33}\, X_3^{(1)} = 0 - \begin{pmatrix} {}^tA_{13}^F & {}^tA_{13}^X & 0 \end{pmatrix} \begin{pmatrix} X_{1F}^{(0)} \\ X_{1X}^{(0)} \\ X_{1Y}^{(0)} \end{pmatrix} + \begin{pmatrix} - A_{13}^F & {}^tA_{13}^X & 0 \end{pmatrix} \begin{pmatrix} X_{1F}^{(0)} \\ -X_{1X}^{(0)} \\ X_{1Y}^{(0)} \end{pmatrix}$$

$$A_{33}X_3^{(1)} = - {}^tA_{13}^F\, X_{1F}^{(0)} - {}^tA_{13}^X\, X_{1X}^{(0)} + A_{13}^F\, X_{1F}^{(0)} + {}^tA_{13}^X\, X_{1X}^{(0)} = 0$$

So $X_3^{(1)} = 0 = X_3$, and $X^{(1)} = X$.

We notice that the Gauss-Seidel method converges in one iteration with the only assumption that $X_3^{(0)} = 0$.

If Γ_3 is considered as part of one of the subdomains, we have no more this property of convergence in one iteration ; the hypothesis should be ${}^tA_{23}\left(X_2^{(0)} - X_2\right) = 0$ for example and that cannot be satisfied if some of the components of X_2 are not known.

There is an analogous result if there are two axes of symmetry and if the interface is considered as a fith domain.

The previous result explains why, if Ω has axes of symmetry, we decompose it along these axes ; when the data are almost symmetrical, we hope to have a fast convergence with the decomposition of dissection type (the interface is considered as a separate domain).

On a parallel computer, we shall assign a domain to each processor, and (eventually) the interface to the host processor.

4. Numerical tests. We have restricted ourselves to the 2D case and have carried out tests on two examples of reactor, choosing the mixed element MXOL5 ; some decompositions of the core have been considered. The numerical results led us to a study of the block Jacobi method on a simple example.

4.1. Description of the two reactors

4.1.1. The 2D IAEA Benchmark. It corresponds to a median plane of a 3D problem representing an idealized model of a Pressurized Water Reactor. This reactor contains four homogeneous regions (see fig. 5):

(1) a region of high enrichment
(2) a region of slight enrichment
(3) a region where the control absorbants have been mixed with the fuel
(4) a light water reflector.

4.1.2. The 2D Tihange. This problem represents a reactor at the begining of the second cycle. The heterogeneity of the core induces a checkerboard effect on the power distribution. It is a good representation of the different real reactor types ; its geometry is represented in fig. 6.

4.2. Description of the tests. The tests have been carried out with the mixed element MXOL5, on a CRAY-XMP.

In the two examples, Ω is symmetric with respect to the coordinates axes ; in the Benchmark case, the data are symmetric, but they are not in the Tihange case.

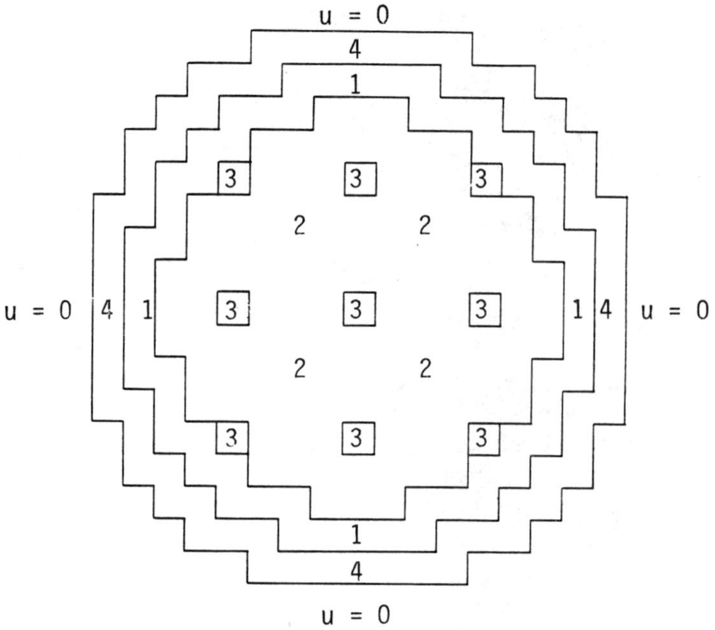

Size of an assembly : 20 cm

FIG. 5: *The 2D IAEA Benchmark*

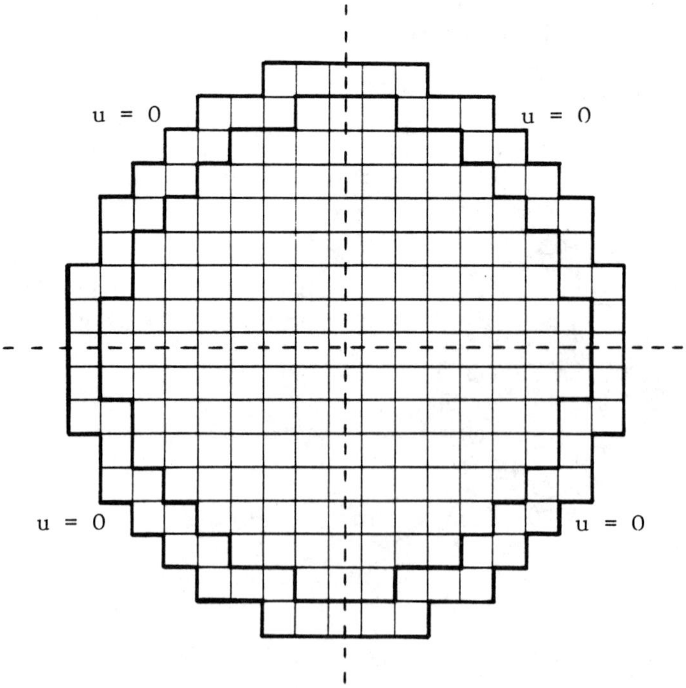

Size of an assembly : 21,5313 cm

FIG.6: *The 2D Tihange. Geometry: representation of the complete core.*

The meshes that are used are the following:

. 1 block per assembly, except for the assemblies on the axes of symmetry which are divided by these ones ;

. 4 blocks per assembly ; we note: "2×2" (see fig. 7).

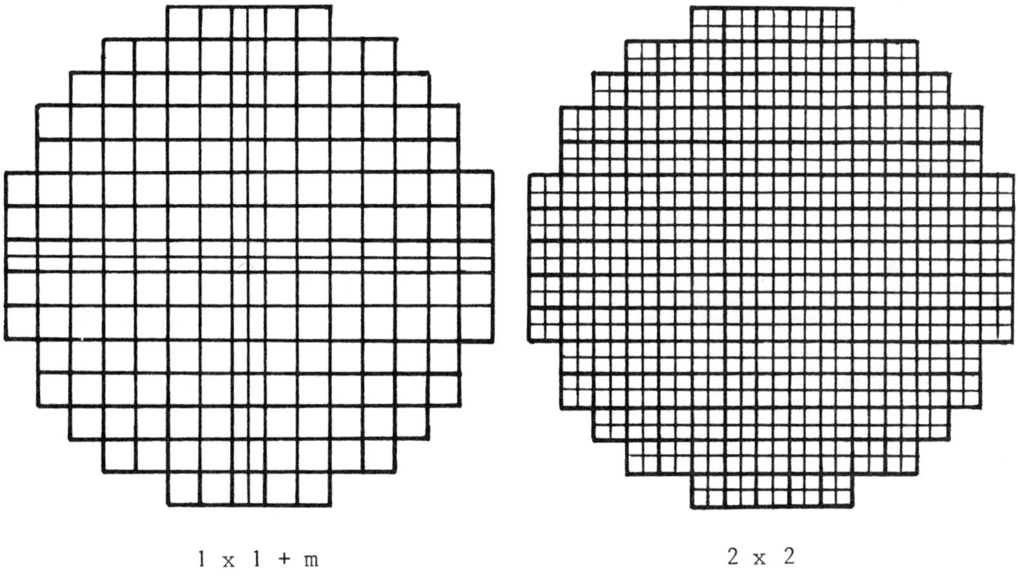

FIG. 7: *The two meshes: "1×1+m" and "2×2" for the reactor Tihange.*

Ω is split up either into two or into four subdomains ; we have the choice of considering the interface separately or as part of the subdomains ; that gives four decompositions named "2B", "2B+I", "4B" and "4B+I" (see fig. 8).

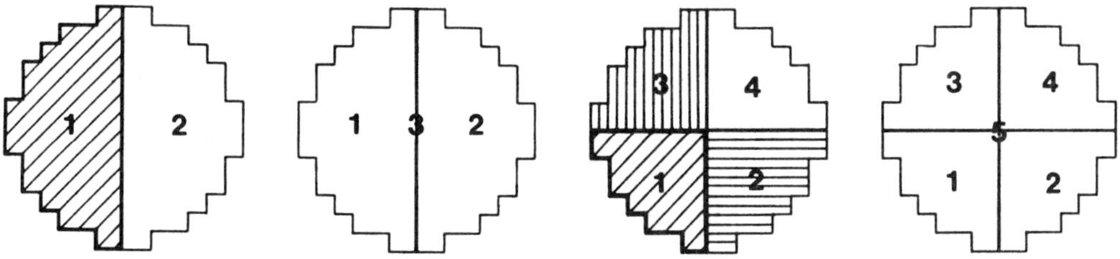

FIG.8: *The four decompositions of the domain.*

The stopping criteria are:

$$\left\| \frac{u_n}{\|u_n\|} - \frac{u_{n-1}}{\|u_{n-1}\|} \right\| \leqslant \text{epsf} \qquad \text{for the flux}$$

$$\frac{|\lambda_n - \lambda_{n-1}|}{|\lambda_n|} \leqslant \text{epsv} \qquad \text{for the current}$$

For the first test (Benchmark), we know that on the coordinates axes we have: $\vec{p}_h.\vec{n} = 0$ theoretically ; it was decided that $\vec{p}_h.\vec{n}$ should be numerically considered equal to zero if $|\vec{p}.\vec{n}| < 10^{-5}$.

4.3. Numerical results. Some numerical results are shown in Tables I.1, I.2, I.3 for the Benchmark, in Tables II.1 and II.2 for the Tihange.

TYPI indicates the solving method for the inner iterations: J stands for Jacobi, G.S. for Gauss-Seidel (S.O.R.) and C for a direct solving.

2 inner iterations are performed for each outer iteration, and the process is stopped after 150 iterations if the convergence is not yet obtained.

Table I.1

epsf = 5.10^{-4} epsv = 10^{-4}

Mesh	TYPI	Eigenvalue	Memory place (words)	Iter. time (c)	nb. iter.	Decomp.	Remarks
1 × 1 + m	C	1,03320059	67444	35	20	/	┼ 0
2 × 2	C	1,03336754	244200	89	15	/	┼ 0
1 × 1 + m	J	1,03318790	39266	146	51	2B	├→ 10^{-4} ⊥ 0
1 × 1 + m	G.S.	1,03320060	39266	58	20	2B	┼ 0
2 × 2	J	1,03332931	173374	479	49	2B	├→ 4.10^{-5} ⊥ 0
1 × 1 + m	J	1,03320060	39154	57	20	2B + I	├→ $1,2.10^{-5}$ ⊥ 0
1 × 1 + m	G.S.	1,03320060	39154	61	20	2B + I	┼ 0
2 × 2	J	1,03336754	172718	143	15	2B + I	┼ 0
1 × 1 + m	J	1,03319254	38500	210	73	4B	├→ $7,6.10^{-4}$ ⊥ $7,6.10^{-4}$
1 × 1 + m	G.S.	1,03317949	38500	73	24	4B	├→ $2,5.10^{-4}$ ⊥ $7,6.10^{-4}$
2 × 2	J	1,03333493	170288	274	69	4B	├→ $4,5.10^{-5}$ ⊥ $7,6.10^{-4}$
1 × 1 + m	J	0,53648906	38260	442	150	4B + I	maxi. preci.: (iter.18) flux: 8.10^{-4} λ: $1,6.10^{-5}$

In the last column ('Remarks'), we indicate especially the order of magnitude of $|\vec{p}_h.\vec{n}|$ on the interface:

$\mapsto$ r indicates that $|\vec{p}_h.\vec{n}|$ is at a maximum of r on the vertical interface.

It is to note that, as we are only testing the method, the storage has not been optimized: the matrices have been stockpiled in a "profil" way ; we intend to stock the diagonal blocks of the matrices in this way, and the off-diagonal matrices by a 'Morse' storage.

For the initialization, the current was taken equal to 0 and the flux equal to 1.

Table I.2

epsf = 10^{-6} epsv = 10^{-6}

Mesh	TYPI	Eigenvalue	Memory place (words)	Iter time (c)	nb. iter	Decomp.	Remarks
1 × 1 + m	J	0,79205594	39154	437	150	2B + I	$\underline{\uparrow}$ 0 $\mapsto$ 1 maxi. preci.: (around iter.19) flux: $3{,}5.10^{-4}$ eigenvalue: 2.10^{-6}
1 × 1 + m	J	1,03325707	39266	420	150	2B	$\underline{\uparrow}$ 0 $\mapsto$ 2.10^{-5} after iter. 145, the prec. is of: 10^{-8} for λ 10^{-6} for the flux ($> 10^{-6}$)

Table I.3

epsf = 5.10^{-3} epsv = 10^{-4}

Mesh	TYPI	Eigenvalue	Memory place (words)	Iter time (c)	nb. iter	Decomp.	Remarks
1 × 1 + m	J	1,03202216	39154	35	12	2B + I	$+$ 0
1 × 1 + m	J	1,03198796	39266	62	22	2B	$\mapsto$ 9.10^{-4} $\underline{\uparrow}$ 0
1 × 1 + m	J	1,03202216	38260	35	12	4B + I	$+$ 0
1 × 1 + m	J	1,03226591	38500	86	30	4B	$\mapsto$ 5.10^{-3} $\underline{\uparrow}$ 0

Table II.1

epsf = 10^{-5} epsv = 10^{-5}

Mesh	TYPI	Eigenvalue	Memory place (words)	Iter time (c)	nb. iter	Decomp.	Remarks
2 × 2	C	1,00331257	220469	210	36	/	
2 × 2	J	1,00331263	159578	631	66	2B	if epsf=10^{-4}, conv. is reached at iter 36
2 × 2	J	0,92635364	158864	1432	150	2B + I	maxi. preci.: flux: 8.10^{-5} λ: 5.10^{-7} between iter 47 and 60, $\lambda \approx 1{,}003315$
2 × 2	J	1,00331274	156422	1424	150	4B	if epsf=10^{-4}, conv. is reached at iter 51
2 × 2	J	0,53890438	155334	1445	150	4B + I	max. prec.: flux: $1{,}4.10^{-3}$ λ: 4.10^{-6} between iter 15 and 18 $1{,}003314 \leqslant \lambda \leqslant 1{,}003320$

Table II.2

epsf = 2.10^{-3} epsv = 10^{-5}

Mesh	TYPI	Eigenvalue	Memory place (words)	Iter time (c)	nb. iter	Decomp.	Remarks
2 × 2	C	1,00332450	220469	58	10	/	
2 × 2	J	1,00330566	159578	124	13	2B	$\frac{124}{2} = 62 \approx 58$
2 × 2	J	1,00332981	158864	95	10	2B + I	$\frac{95}{2} = 47{,}5 < 58$
2 × 2	J	1,00331549	156422	247	19	4B	$\frac{247}{4} = 61{,}75 \approx 58$
2 × 2	J	1,00332985	155334	116	12	4B + I	$\frac{116}{4} = 29 \ll 58$

4.4. Commentaries and interpretations

4.4.1. Benchmark test. The basic mesh for the interpretation is the mesh "1×1+m".

We first stated: epsf = 5.10^{-4} and epsv = 10^{-4}, as it is generally done in our code for the Benchmark. Firstly, we see that as stated mathematically, in the two cases of direct solving, there are properties of symmetry for the flux and the current, and $\vec{p}.\vec{n}$ is nil on the axes of symmetry. We notice that for the decomposition 4B + I, there are some problems ; that is why, we wanted to see what happened in the case 2B + I for a greater requested precision, and also for all the cases for a lower precision.

For a requested precision of 5.10^{-3} for the flux and 10^{-4} for the eigenvalue, the decompositions for which the interface is separated (2B + I, 4B + I) are by far the fastest ones, and they give a best approximation of $\vec{p}.\vec{n}$ on the interface. That still holds for a precision of 5.10^{-4} and 10^{-4} for the case 2B + I.

For the decompositions 2B + I and 4B + I, we note that for each outer iteration (at least for the first ones), the result is the same for the first and second iterations of the block Jacobi method ; this confirms the mathematical result that tells that the Jacobi method converges in one iteration with an appropriate initialization. In these two cases, it is noticed that after having reached a certain precision during several iterations:

$$\begin{cases} 5.10^{-4} \quad \text{and} \quad 2.10^{-6} & \text{for } 2B + I \\ 5.10^{-3} \quad \text{and} \quad 5.10^{-5} & \text{for } 4B + I \end{cases}$$

the results deteriorate, and become quite different from the solution. If we look more precisely, we notice that the property of converging in one iteration for the Jacobi method holds exactly only during $\begin{cases} \text{19 iterations for 2B + I} \\ \text{14 iterations for 4B + I} \end{cases}$, and the error at the second iteration increases, the process being more accentuated for the case 4B + I ; so, if at that point the requested precision is not reached, problems may appear. That can be partly explained by the fact that numerical results are not exact results ; here, after a certain number of iterations, we may loose slightly properties of symmetry and so, at the beginning of the Jacobi iterations, the starting vector has no more the property for the convergence in one iteration.

Increasing the number of inner iterations doesn't provide better results for the cases 2B and 4B, and may occur difficulties for the two others, because of the process discussed about above.

In the case of a requested precision of 5.10^{-4} for the flux and 10^{-4} for the eigenvalue, we note that the Gauss-Seidel method is faster than the Jacobi method and gives a better approximation of $\vec{p}_h.\vec{n}$ on the axes of symmetry.

4.4.2. Tihange test. The previous phenomena are amplified. With the mesh "2×2" the decompositions with a separated interface can only reach a precision of:

$$\begin{cases} 10^{-4} \text{ for the flux,} \quad 10^{-6} \text{ for the eigenvalue} & \text{in the case } 2B + I \\ 2.10^{-3} \text{ for the flux,} \quad 4.10^{-6} \text{ for the eigenvalue} & \text{in the case } 4B + I \end{cases}$$

4.5. Third test. The previous results led to a study of the block Jacobi method for a very simple case:

. Ω is a square, $\Gamma_0 = \partial\Omega$
. the triangulation is obtained by a splitting into four parts
. the functions D, Σ and S are set constant
. the mixed element is still MXOL5.

The unknowns are numbered as indicated on fig. 9.

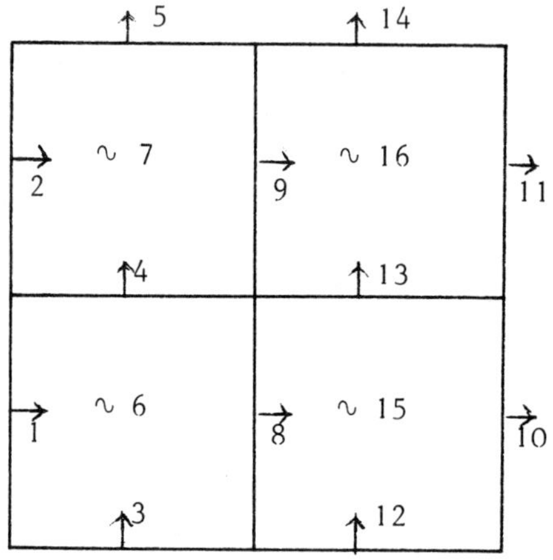

FIG.9: *Numbering of the nodes.*

The decomposition that are considered are the same as before:

- 2B : domain 1 $\leftrightarrow$ nodes: 1,2,3,4,5,6,7,8,9
  domain 2 $\leftrightarrow$ nodes: 10,11,12,13,14,15,16
- 2B + I: domain 1 $\leftrightarrow$ nodes: 1,2,3,4,5,6,7
  domain 2 $\leftrightarrow$ nodes: 10,11,12,13,14,15,16
  domain 3 $\leftrightarrow$ nodes: 8,9
- 4B : domain 1 $\leftrightarrow$ nodes: 1,8,3,4,6
  domain 2 $\leftrightarrow$ nodes: 10,12,13,15
  domain 3 $\leftrightarrow$ nodes: 2,9,5,7
  domain 4 $\leftrightarrow$ nodes: 11,14,16
- 4B + I: domain 1 $\leftrightarrow$ nodes: 1,3,6
  domain 2 $\leftrightarrow$ nodes: 10,12,15
  domain 3 $\leftrightarrow$ nodes: 2,5,7
  domain 4 $\leftrightarrow$ nodes: 11,14,16
  domain 5 $\leftrightarrow$ nodes: 8,9,4,13

The matrix of the linear system is the following, with $\alpha = \frac{1}{6D}$:

$$A = \begin{bmatrix}
2\alpha & & & & & -1 & & \alpha & & & & & & & & \\
 & 2\alpha & & & & & -1 & & \alpha & & & & & & & \\
 & & 2\alpha & \alpha & & -1 & & & & & & & & & & \\
 & & \alpha & 4\alpha & \alpha & 1 & -1 & & & & & & & & & \\
 & & & \alpha & 2\alpha & & 1 & & & & & & & & & \\
-1 & & -1 & 1 & & -\Sigma & & 1 & & & & & & & & \\
 & -1 & & -1 & 1 & & -\Sigma & & 1 & & & & & & & \\
\alpha & & & & & 1 & & 4\alpha & & \alpha & & & & & -1 & \\
 & \alpha & & & & & 1 & & 4\alpha & & \alpha & & & & & -1 \\
 & & & & & & & \alpha & & 2\alpha & & & & & 1 & \\
 & & & & & & & & \alpha & & 2\alpha & & & & & 1 \\
 & & & & & & & & & & & 2\alpha & \alpha & & -1 & \\
 & & & & & & & & & & & \alpha & 4\alpha & \alpha & 1 & -1 \\
 & & & & & & & & & & & & \alpha & 2\alpha & & 1 \\
 & & & & & & & -1 & & 1 & & -1 & 1 & & -\Sigma & \\
 & & & & & & & & -1 & & 1 & & -1 & 1 & & -1
\end{bmatrix}$$

The interest concerned particularly the decomposition 4B + I, because in the previous examples, it was with this decomposition that some problems arised.

We took: $\begin{cases} D = \frac{1}{600}, \frac{1}{6}, \frac{100}{6} \\ \Sigma = 1 \\ h = 1 \end{cases}$; the problem to solve is: AX = B.

The tests were run on an IBM.

4.5.1. Decomposition 4B + I. As the nodes are numbered domain by domain, the new ordering is: 1,3,6/10,12,15/2,5,7/11,14,16/ 8,9,4,13.

Suppose $D = \frac{100}{6}$; the solution of AX = B, with:

$B = {}^t(0, 0,-202, 0, 0,-202, 0, 0,-202, 0, 0,-202,0,0,0,0)$

$X = {}^t(100,100, 2,100,100, 2,100,100, 2,100,100, 2,0,0,0,0)$.

We take $X^{(0)} = X$ and perform 50 iterations. We should have $X^{(n)} \approx X$ (numerically) at each iteration ; but after a certain number of iterations, $X^{(n)}$ becomes different from X, and the error increases (see fig. 10).

This phenomenon doesn't appear in the two other cases.

In the case $D = \frac{1}{6}$, we wanted to see the influence of a slight modification of the initial vector. We took:

$B = {}^t(0,0,-4,0,0,-4,0,0,-4,0,0,-4,0,0,0,0)$ so that:

$X = {}^t(1,1, 2,-1,1,2,1,-1,2,-1,-1,2,0,0,0,0)$; we choose:

$X^{(0)} = {}^t(0,0,8,0,0,8,0,0,8,0,0,8,0,0,0,0)$ to have a convergence in one iteration, and $X_\varepsilon^{(0)} = (0,0,8,0,0,8,0,0,8,0,0,8,\varepsilon,\varepsilon,\varepsilon,\varepsilon)$ with $\varepsilon = 0{,}1$. During the first iterations, the results are different ; the error reduces, but at least during the ten first iterations, $|X^{(n)}(i) - X_\varepsilon^{(n)}| > 10^{-4} \quad \forall i$.

4.5.2. Decomposition 4B. The phenomenon described above doesn't appear. We note that the performance deteriorate as D increases. In the case $D = \frac{100}{6}$, the best approximation of 0 on the interface

ITERATION 1

I=	X(I)=	error
1	99.9999390	ε
2	99.9999390	ε
3	2.00000000	0
4	-99.9999390	ε
5	99.9999390	ε
6	2.00000000	0
7	99.9999390	ε
8	-99.9999390	ε
9	2.00000000	0
10	-99.9999390	ε
11	-99.9999390	ε
12	2.00000000	0
13	0.000000000E+00	0
14	0.000000000E+00	0
15	0.000000000E+00	0
16	0.000000000E+00	0

ITERATION 2

I=	X(I)=
1	99.9999390
2	99.9999390
3	2.00000000
4	-99.9999390
5	99.9999390
6	2.00000000
7	99.9999390
8	-99.9999390
9	2.00000000
10	-99.9999390
11	-99.9999390
12	2.00000000
13	0.238418434E-04
14	0.238418434E-04
15	0.238418434E-04
16	0.238418434E-04

ITERATION 3

I=	X(I)=
1	99.9999695
2	99.9999695
3	2.00000095
4	-99.9999390
5	99.9999237
6	2.00000000
7	99.9999237
8	-99.9999390
9	2.00000000
10	-99.9999390
11	-99.9999390
12	2.00000000
13	0.238418434E-04
14	0.238418434E-04
15	0.238418434E-04
16	0.238418434E-04

ITERATION 10

I=	X(I)=
1	99.9998627
2	99.9998627
3	1.99999809
4	-99.9999847
5	99.9999847
6	2.00000191
7	99.9999847
8	-99.9999847
9	2.00000191
10	-99.9998627
11	-99.9998627
12	1.99999809
13	0.119209217E-03
14	-0.953673734E-04
15	0.119209217E-03
16	-0.953673734E-04

ITERATION 20

I=	X(I)=
1	100.001541
2	100.001541
3	2.00004864
4	-99.9983063
5	99.9983063
6	1.99995136
7	99.9983063
8	-99.9983063
9	1.99995136
10	-100.001541
11	-100.001541
12	2.00004864
13	-0.321864872E-02
14	0.324249058E-02
15	-0.321864872E-02
16	0.324249058E-02

ITERATION 30

I=	X(I)=
1	99.9513550
2	99.9513550
3	1.99853516
4	-100.048492
5	100.048492
6	2.00146484
7	100.048492
8	-100.048492
9	2.00146484
10	-99.9513550
11	-99.9513550
12	1.99853516
13	0.975369811E-01
14	-0.975131392E-01
15	0.975369811E-01
16	-0.975131392E-01

ITERATION 40

I=	X(I)=
1	101.469406
2	101.469406
3	2.04430580
4	-98.5304413
5	98.5304413
6	1.95569420
7	98.5304413
8	-98.5304413
9	1.95569420
10	-101.469406
11	-101.469406
12	2.04430580
13	-2.94999886
14	2.95002270
15	-2.94999886
16	2.95002270

ITERATION 50

I=	X(I)=
1	55.5384216
2	55.5384216
3	0.659451008
4	-144.461411
5	144.461395
6	3.34054852
7	144.461395
8	-144.461411
9	3.34054852
10	-55.5383911
11	-55.5383911
12	0.659450054
13	89.2581329
14	-89.2581329
15	89.2581329
16	-89.2581329

FIG.10: *Decomposition 4B + I.* $D = \frac{100}{6}$, $X^{(0)} = X$.

seems to be of the order of 2.10^{-5}.

4.5.3. **Decomposition 2B + I.** The new ordering of the nodes is: 1,2,3,4,5,6,7/10,11,12,13,14,15,16/8,9. We choose $D = \frac{100}{6}$. For $B = {}^t(0,0,0,0,0,-202,-202,0,0,0,0,0,-202,-202,0,0)$, $X = {}^t(100,100,100,0,-100,2,2,-100,-100,100,0,-100,2,2,0,0)$. If we take $X^{(0)} = {}^t(0,0,0,0,0,100,100,0,0,0,0,0,100,100,0,0)$, the block Jacobi method converges theoretically in one iteration. Then take $X_\varepsilon^{(0)} = {}^t(0,0,0,\varepsilon,0,100,100,0,0,0,\varepsilon,0,100,100,0,0)$ and $\varepsilon = 0,1$. Even with $X^{(0)}$, the approximation is not better than 10^{-4} ; with $X_\varepsilon^{(0)}$ we have (almost) the same results.

The phenomenon described for decomposition 4B + I doesn't seem to occur.

4.5.4. **Some observations.** A modification of $X^{(0)}(i)$, i corresponding to a node representing the gradient on an axis of symmetry, has more influence if the node is on the axis of decomposition. The results indicates that decomposition 4B + I is less stable than decomposition 4B, and also than 2B + I.

5. **Conclusion.** When everything is symmetrical, the Jacobi method can converge in one iteration with a decomposition of dissection type, if an appropriate initial vector is choosen.

The decompositions in which the interface is separated give faster results when a not too high precision is requested ; but they cannot reach a very high precision, and in this case the results may deteriorate. This can be partly explained by the fact that after some outer iterations, the initialization vector of the Jacobi iterations has no more exactly the properties for a convergence in one iteration.

Decompositions for which the interface is separated seem to be less stable than the ones in which the interface is integrated to the domains.

The choice of the decomposition depends on what we need: a high precision or a fast result.

REFERENCES

(1) J.M. COLLART, C. FEDON-MAGNAUD, J.J. LAUTARD, Parallel Diffusion Calculation for the Phaeton on-line Multiprocessor Computer, Proceedings of the International Topical Meeting on Advances in Reactor Physics, Mathematics and Computation, Paris April 27-30, 1987.

(2) J.M. COLLART, C. FEDON-MAGNAUD, J.J. LAUTARD, The Phaeton Project, ENC 86 Transactions, 2, p 303, Geneva, June 1-6, 1986.

(3) F. COULOMB, C. FEDON-MAGNAUD, Mixed and Mixed Hybrid Elements for the Diffusion Equation, Proceedings of the International Topical Meeting on Advances in Reactor Physics, Mathematics and Computation, Paris April 27-30, 1987.

(4) R. DAUTRAY, J.L. LIONS, Analyse mathématique et calcul numérique pour les sciences et techniques, R. Dautray, Masson, 1985.

(5) J.J. DUDERSTADT, L.J. HAMILTON, Nuclear Reactor Analysis, J. Wiley & Sons, Inc.

(6) P.A. RAVIART, J.M. THOMAS, A Mixed Finite Element Method for 2nd Order Elliptic Problems, Mathematical Aspects of the Finite Element Method, Lecture Notes in Mathematics 606 Springer Verlag, 1977.

(7) J.M. THOMAS, Sur l'analyse numérique des méthodes d'éléments finis hybrides et mixtes, Thesis, Université P. et M. Curie, 1977.

CHAPTER 26

A Parallel Solver for the Linear Elasticity Equations on a Composite Beam

Philippe Destuynder*
François-Xavier Roux*

Abstract .

We report about parallel implementation on CRAY2 supercomputer of a parallel algorithm based on a domain decomposition method with Lagrange multiplier, for solving an ill conditioned three dimensional composite structural analysis problem, with as many as one million degrees of freedom .

We show that this method has very good features on both granularity and data dependancy viewpoints. We explain the practical differences between this method and the standard domain decomposition method with Gaussian elimination of the degrees of freedom inside the subdomains .

The tests performed prove that the choice of the local solver is very important to get an efficient global method. For the studied case, it is clear that solving the local problems with a direct method is the best solution, and we give some reasons why it will be the same for many other problems .

1. Presentation of a structural analysis problem for a composite beam .

We consider the linear elasticity equations for a composite beam made of a little more than one hundred stiff fibers (carbon or iron) bound by an uncompressible elastomer matrix .

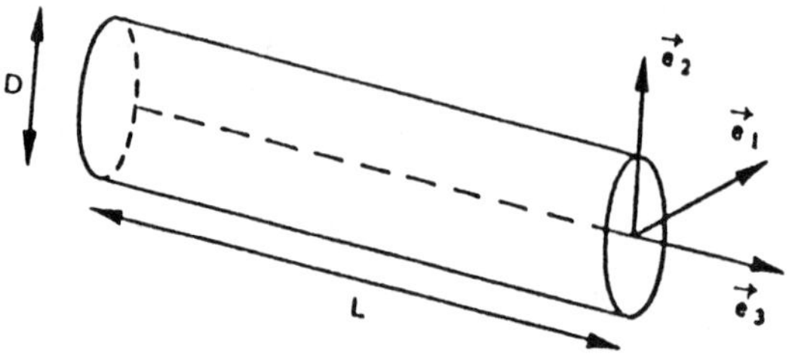

Fig. 1. Geometry of the beam .

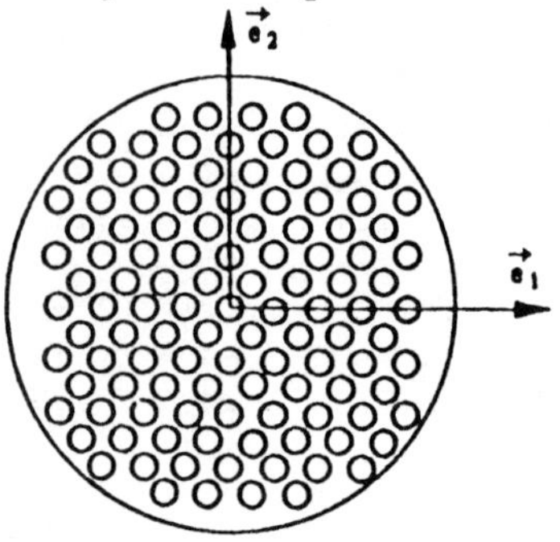

Beam cross section .

* O.N.E.R.A., Chatillon, France

Homogeneisation methods do not work for such a device with microscopic-scale discontinuity. But, due to the composite feature, the finite element mesh for solving the problem with discontinuous coefficients must be very refined to get a good representation of each substructure. This leads to a very large scale matrix, hence the problem can be solved only by iterative methods like the conjugate gradient method .

However, substructuring is very easy in the present case for the beam is obviously made of similar jointed composite "pencils". Hence, using a domain decomposition method for solving such a problem seems to be natural, because the decomposition is straightforward. Furthermore, all the subdomains are identical .

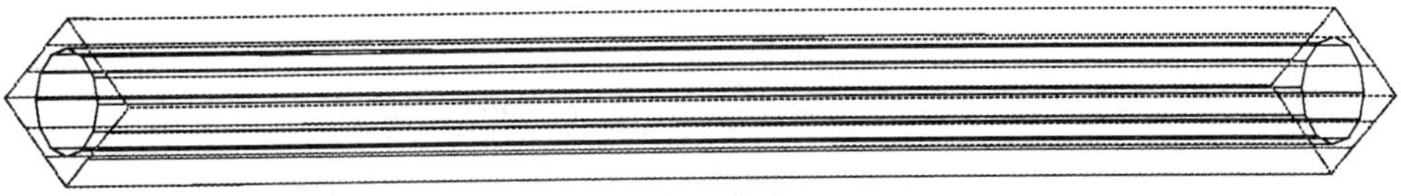

Fig. 2. A composite pencil .

It must be noticed that the problem we study is very ill conditioned for both geometrical and material reasons.

The material reasons are at first the composite feature. The Young moduli of the fiber have an order of magnitude 10^4 times the one of the elastomer. Secondly, the penalty method for the uncompressibility condition in the elastomer makes the condition number increase with the penalty parameter .

The geometrical reason is linked to the fact we try to solve a beam problem, with a ratio of the length upon the width equal to 6, that makes the pure bending problem very ill conditioned .

2. The primal hybrid variational principle .

The domain decomposition method we used for solving the problem is based on the so-called primal hybrid variational principle, and consists in introducing a Lagrange multiplier to remove the continuity condition on the interface (see for instance [1] or [2]).

Let $\mathbf{\Omega}$ be a bounded open subset of $\mathbf{R}^3$ with a smooth boundary Γ. The linear elasticity equations with homogeneous boundary conditions on Γ_0 a subset of Γ are :

$$\begin{cases} \mathbf{Au} = \mathbf{f} & \textit{in } \Omega \\ \mathbf{u} = 0 & \textit{on } \Gamma_0 \end{cases} \qquad \textit{with } (\mathbf{Au})_i = -\frac{\partial(a_{ijkh}\varepsilon_{kh}(\mathbf{u}))}{\partial x_j} \quad . \tag{1}$$

The usual variational form of this problem consists in finding $\mathbf{u}$ in $(\mathrm{H}^1(\Omega))^3$ and satisfying the boundary condition $\mathbf{u} = 0$ on Γ_0 wich minimizes the energy functional :

$$\mathbf{I}(\mathbf{v}) = \frac{1}{2}\mathbf{a}(\mathbf{v},\mathbf{v}) - (\mathbf{f},\mathbf{v}) \qquad \textit{with} \quad \mathbf{a}(\mathbf{u},\mathbf{v}) = \int_\Omega a_{ijkh}\varepsilon_{kh}(\mathbf{u})\varepsilon_{ij}(\mathbf{v})\, dx \ .$$

Let us consider a splitting of the domain $\mathbf{\Omega}$ into two open subsets Ω_1 and Ω_2 with smooth boundaries Γ_1 and Γ_2 . Let us assume that the boundary of the interface $\Sigma = \Gamma_1 \cap \Gamma_2$ is included in Γ_0 so that the traces on Σ of the displacements fields $\mathbf{u}$ satisfying the boundary condition $\mathbf{u} = 0$ on Γ_0 belong to the space $(\mathrm{H}_0^{1/2}(\Sigma))^3$.

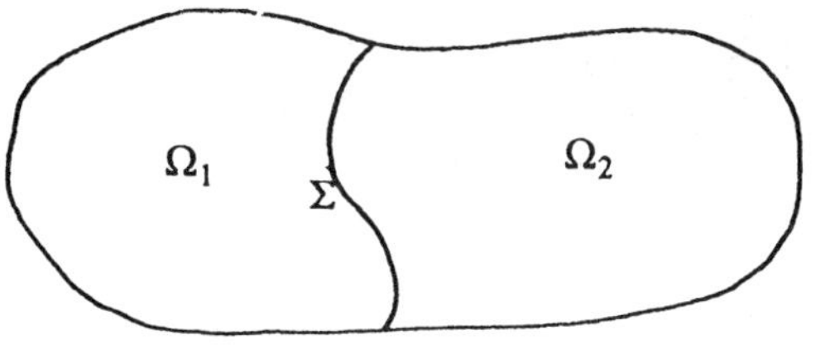

Solving the linear elasticity equation consists in finding two functions $\mathbf{u}_1$ and $\mathbf{u}_2$ defined on Ω_1 and Ω_2, satisfying the boundary conditions on Γ_1 and Γ_2 which minimize the energies $\mathbf{I}_1(\mathbf{v})$ and $\mathbf{I}_2(\mathbf{v})$ with the continuity constraint : $\mathbf{v}_1 = \mathbf{v}_2$ on Σ .

The weak form of the continuity constraint is :

$$(v_1 - v_2 , \mu)_\Sigma = 0 \quad \textit{for any } \mu \textit{ in } (H^{-1/2}(\Sigma))^3 \ .$$

One can show that the problem of minimization with constraint above is equivalent to finding the saddle-point of the Lagrangian :

$$L(v,\mu) = I_1(v_1) + I_2(v_2) + (v_1 - v_2 , \mu)_\Sigma \ . \tag{3}$$

This is equivalent to find the fields (u_1,u_2) in $(H^1(\Omega_1))^3 X (H^1(\Omega_2))^3$ and the Lagrange multiplier λ in $(H^{-1/2}(\Sigma))^3$ which satisfy :

$$L(u,\mu) \le L(u,\lambda) \le L(v,\lambda) \ ,$$

for any field $v = (v_1,v_2)$ in $(H^1(\Omega_1))^3 X (H^1(\Omega_2))^3$, and any μ in $(H^{-1/2}(\Sigma))^3$.

The left inequality imposes $(u_1 - u_2 , \mu)_\Sigma \le (u_1 - u_2 , \lambda)_\Sigma$ and so $(u_1 - u_2 , \mu)_\Sigma = 0$ for any μ in $(H^{-1/2}(\Sigma))^3$, thus the continuity constraint is satisfied by the solution of the saddle-point problem .

The right inequality implies $I_1(u_1) + I_2(u_2) \le I_1(v_1) + I_2(v_2)$ for any (v_1,v_2) in $(H^1(\Omega))^3$. It means that (u_1,u_2) minimizes the sum of the energies on Ω_1 and Ω_2 among the fields satisfying the continuity requirement. Hence u_1 and u_2 are the restrictions to Ω_1 and Ω_2 of the solution of the linear elasticity equation on Ω .

The classical variational interpretation of the saddle-point problem (3) leads to the equations :

$$\begin{cases} A_1 u_1 + B_1^* \lambda = f_1 & \textit{in } \Omega_1 \\ u_1 = 0 & \textit{on } \Gamma_0 \cap \Gamma_1 \\ A_2 u_2 - B_2^* \lambda = f_2 & \textit{in } \Omega_2 \\ u_2 = 0 & \textit{on } \Gamma_0 \cap \Gamma_2 \\ B_1 u_1 - B_2 u_2 = 0 & \textit{on } \Sigma \end{cases} \tag{4}$$

where A_1 and A_2 are the differential operators of the linear elasticity equations on Ω_1 and Ω_2, while B_1 and B_2 the trace operators onto Σ of functions belonging to $(H^1(\Omega_1))^3$ and $(H^1(\Omega_2))^3$.

The analysis of these equations shows that the Lagrange multiplier λ is in fact equal to the interaction force between the substructures along their common boundary .

On a mechanical point of view, it means that it is necessary to know the forces on the interface to obtain local independent displacements problem .

3. An iterative algorithm based on the hybrid finite element method .

A discretisation with Lagrangian finite elements for the displacements fields and for the interface constraints of the hybrid formulation (4) leads to the same equations where A_1, A_2, B_1 and B_2 are the matrices associated with the discrete linear elasticity and trace operators .

$$\begin{cases} A_1 u_1 + B_1^t \lambda = f_1 \\ A_2 u_2 - B_2^t \lambda = f_2 \\ B_1 u_1 - B_2 u_2 = 0 \end{cases} \tag{5}$$

By substitution in the equation (5) the problem can be written with respect to λ only :

$$\left[B_1 * A_1^{-1} * B_1^t + B_2 * A_2^{-1} * B_2^t \right] * \lambda = B_1 * A_1^{-1} * f_1 - B_2 * A_2^{-1} * f_2$$

Hence λ satisfies the following equation :

$$\mathbf{C} * \lambda = \mathbf{b} \tag{6}$$

where $\mathbf{C} = \begin{bmatrix} \mathbf{B}_1 , -\mathbf{B}_2 \end{bmatrix} * \begin{bmatrix} \mathbf{A}_1^{-1} & 0 \\ 0 & \mathbf{A}_2^{-1} \end{bmatrix} * \begin{bmatrix} \mathbf{B}_1^t \\ -\mathbf{B}_2^t \end{bmatrix}$ *and* $\mathbf{b} = \begin{bmatrix} \mathbf{B}_1 , -\mathbf{B}_2 \end{bmatrix} * \begin{bmatrix} \mathbf{A}_1^{-1} & 0 \\ 0 & \mathbf{A}_2^{-1} \end{bmatrix} * \begin{bmatrix} \mathbf{f}_1 \\ \mathbf{f}_2 \end{bmatrix}$

The the discrete displacements spaces $\mathbf{V}_i$, and the discrete Lagrange multiplier space $\mathbf{M}$ are chosen in order to satisfy the Babuska-Brezzi-Ladyzhenskaya condition :

$$\underset{|\mu|_{\mathbf{M}}=1}{Inf} \quad \underset{|\mathbf{u}_i|_{\mathbf{V}_i}=1}{Sup} \quad (\mathbf{B}_i * \mathbf{u}_i , \mu) = \beta > 0 \ .$$

Hence the $\mathbf{C}$ matrix is symmetric, positive definite, and so the equation (6) has only one solution .

We refer to [1] or [2] for the now classical error estimates for the hybrid formulations of elliptic differential equations .

The resolution by the conjugate gradient method of the interface problem (6) leads to a parallel algorithm with a very good granularity .

Let μ be a vector, computing the product $\xi = \mathbf{C} * \mu$ involves the following three steps .

Step one :

computation of the product , $\begin{bmatrix} \mathbf{v}_1 \\ \mathbf{v}_2 \end{bmatrix} = \begin{bmatrix} \mathbf{B}_1^* \\ -\mathbf{B}_2^* \end{bmatrix} * \begin{bmatrix} \mu \\ \mu \end{bmatrix}$,

that means computing two independent local vector-matrix products .

Step two :

computation of the product , $\begin{bmatrix} \mathbf{w}_1 \\ \mathbf{w}_2 \end{bmatrix} = \begin{bmatrix} \mathbf{A}_1^{-1} \\ \mathbf{A}_2^{-1} \end{bmatrix} * \begin{bmatrix} \mathbf{v}_1 \\ \mathbf{v}_2 \end{bmatrix}$,

that means computing the solution of two independent local sets of linear equations associated to local linear elasticity problems with Neuman boundary condition on the interface Σ .

Step three :
computation of the variation on the interface of the displacements fields $\mathbf{w}_1$ and $\mathbf{w}_2$,

$$\xi = \begin{bmatrix} \mathbf{B}_1 , -\mathbf{B}_2 \end{bmatrix} * \begin{bmatrix} \mathbf{w}_1 \\ \mathbf{w}_2 \end{bmatrix} = \mathbf{B}_1 * \mathbf{w}_1 - \mathbf{B}_2 * \mathbf{w}_2$$

The simplest choice for the space $\mathbf{M}$, consists in having on each interface vertex as many degrees of freedom for the multiplier as for the trace of the discrete displacements fields. In this case, the $\mathbf{B}_1$ and $\mathbf{B}_2$ operators are simply the operators of restrictions of the discrete displacements fields .

Then the condition, $\mathbf{B}_1\mathbf{u}_1 - \mathbf{B}_2\mathbf{u}_2 = 0$ on Σ, is equivalent to the condition : $\mathbf{u}_1 = \mathbf{u}_2$ for any degree of freedom located on Σ. So the first and third steps of the computation of the product by the dual matrix, are quite simple and involves only few operations .

In this case, it is straightforward to see that the dual matrix $\mathbf{C}$ is elliptic, and the solution ($\mathbf{u}_1$, $\mathbf{u}_2$) of the hybrid problem (5) is equal to the solution of the standard conforming problem on the whole domain .

Obviously, the main part of the computation for the resolution of the hybrid problem (5) by the conjugate gradient method consists in the step two of the matix-vector product, i.e. the resolution of local independent elasticity problems, which can be performed in parallel. The granularity is good, because data transfers involve only the boundary degrees of freedom, as the computations consist in the resolution of problems in the whole subdomains and require much more operations .

4. Comparisions with the Schur complement method .

The parallel algorithm associated with the hybrid method has the same kind of features as the Schur complement method, presented for instance in [3] or [4]. Both methods consist in solving a condensed problem on the interface, and when solving this problem by a gradient method, the main part of the computations consist in the resolution at each step of the outer iterative scheme of local independent problems in each subdomain .

The differences between the two algorithms lie in the facts that the Schur complement method is a conforming method, and that the local independent problems have Dirichlet boundary conditions for this method and Neuman boundary conditions for the hybrid method .

The first consequence is that the topology of the interface can be simpler with the hybrid method .

Two subdomains are neighbours for the hybrid method iff they have a common degree of freedom of the Lagrange multiplier. As the multiplier is associated with the weak formulation of the continuity constraint, $(\mathbf{v}_1 - \mathbf{v}_2 , \mu)_\Sigma = 0$ for any μ in $(\mathbf{H}^{-1/2}(\Sigma))^3$, its degrees of freedom have to be introduced only on interfaces between subdomains with a non zero integral .

Thus, for a decomposition of a two-dimensional problem into quadrangles, subdomains are neighbours for the hybrid method only if they have a common edge whereas a common vertex is enough for the conforming Schur complement method .

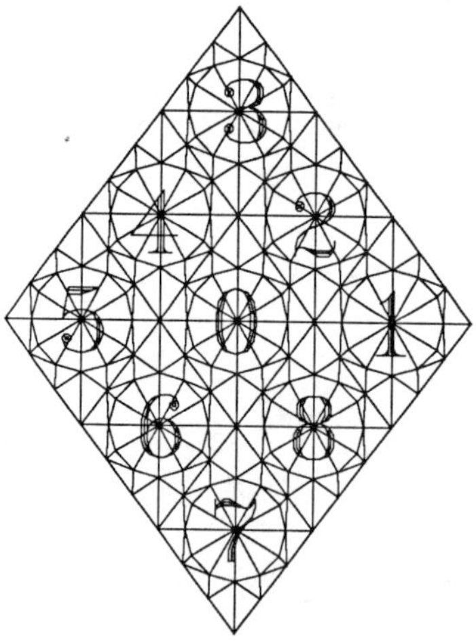

Fig. 3. Mesh of the bottom section for 9 pencils .

In the case of the decomposition of a cylinder into pencils, the topology is the same as for the two-dimensional problem for a cross section.

On the figure, we show the mesh on the bottom section of a nine pencils domain.

With the hybrid method, the pencil number 0 in the middle has only four neighbours, one for each edge, the subdomains 2,4,6 and 8, whereas all the eight surrounding pencils are neighbouring for the conforming domain decomposition method .

This feature makes the domain decomposition method with Lagrange multiplier more suitable for parallel machines with distributed memory. Furthermore, it would allow substructuring with real three-dimensional topology more easily than the Schur complement method .

The second consequence is linked with the boundary conditions of the local problems to be solved at each step of the outer gradient scheme .

With the Schur complement method, we get Dirichlet boundary conditions. So, the displacements degrees of freedom located on the interface are fixed. At the opposite, with the hybrid method, we have prescribed stresses on the interface, that means Neuman boundary conditions .

Except for the clamped bottom section, the boundary conditions we had to assume, for the beam problem we tackled, were prescribed stresses for the top section, and free boundary conditions for the beam sides .

So, when substructuring the domain into geometrically identical subdomains, each one made of one or more pencils, we get identical local matrices for the local problems with the hybrid methods. With the Schur complement method, when substructuring in a few subdomains, for instance nine subdomains like in the figure above, the matrices of the local problems can be all different, because the fixed degrees of freedom associated with the Dirichlet boundary conditions are not the same in each subdomain .

5. Results of the tests performed with the CRAY2 .

We performed tests for domains with the same features as the composite beam presented in the first section, with less fibers but with the same kind of geometry, in order to get condition numbers of the same order of magnitude but not too many degrees of freedom .

The first lesson with these tests concerns the choice of the solver for the local independent problems .

When trying to solve them with an iterative method, the conjugate gradient method, the algorithm we got was sometimes as many as three times more expensive than the standard global diagonal scaled conjugate gradient method for the complete domain .

The explanation lies in the fact that the ratio of the length upon the width of the subdomains, here they were pencils, was greater than for the complete domain, and so was the condition number, in such a way that the conjugate gradient method was not an efficient algorithm to solve the local problems .

This problem could appear in many other cases, because when substructuring a domain to use any domain decomposition method, one try to locate the interfaces in regions in which the solution is expected to be smooth, in order to keep the number of iterations needed for the outer iterative scheme as low as possible. This can lead to substructures with such an aspect that the local matrices are ill conditioned for geometrical reasons, particularly for structural analysis problems .

Finally, for ill conditioned problems it seems safer to use a direct local solver. Moreover, as each local problem has to be solved many times, the cost of the **L*U** decomposition of the matrices do not justify the use of iterative solvers .

The second lesson concerns the number of subdomains .

In all tests we performed we noticed that the number of outer iterations needed to solve the global problem grew much faster than the number of subdomains .

For instance, when trying to solve the same problem for 36 pencils with 9 subdomains made of 4 pencils, the number of outer iterations can be 10 times greater than for a decomposition in 4 subdomains, each made of 9 pencils, in such a way that the global time is 5 times longer, though each iteration of the hybrid method is less expensive with more numerous and smaller subdomains .

Hence, for solving the complete problem presented in the first section we took subdomains as large as possible, according to geometric constraints and memory requirements, that meant in practice subdomains made of nine fibers, with more than one million degrees of freedom .

6. Conclusions .

The tests we performed prove the ability of the domain decomposition method with

Fig. 4. Decomposition in 4 subdomains . Decomposition in 9 subdomains .

Lagrange multiplier to solve some large scale ill conditioned problems of structural analysis on parallel supercomputers .

With four subdomains, the speed-up we got on the CRAY2, with four processors and 256 megawords of memory, measured by the comparision of ellapsed times on a dedicated machine, was over 3.85 . And when using a vectorised direct local solver, each processor ran at more than one hundred megaflops, that leads to a global computation speed over 400 megaflops .

A remaining question is the problem of finding preconditioners for the dual interface problem, like the ones studied in [4] or [5] for the Schur complement method, to make this algorithm a really robust one .

References .

[1] P.A. Raviart and J.M. Thomas , *Primal hybrid finite element methods for 2nd order elliptic equations* , Math. of Comp. ,Vol. 31,number 138,pp. 391-413 .

[2] J.M. Thomas , Thesis , Univ. P & M Curie , Paris , 1977 (In French) .

[3] G.H. Golub and D. Meyers, *The use of preconditioning over irregular regions* , in Proceedings of the 6th International Conference on Computing Methods for Scientific Engineering, Versailles, France, December 12-16 1983 .

[4] P.E. Bjordstad and O.B. Wildlund , *Iterative methods for solving elliptic problems on regions partitioned into substructures* , SIAM J. Numer. Anal. ,Vol. 23, No. 6, December 1986 .

[5] T.F. Chan and D.C. Resasco , *A Framework for the Analysis and Construction of Domain Decomposition Preconditioners* , in SIAM Proceedings of the First International Symposium on Domain Decomposition Methods for Partial Differential Equations, R. Glowinski and G.H. Golub ,eds. ,Paris, 1987.

Domain Decomposition for Nonsymmetric Systems of Equations: Examples from Computational Fluid Dynamics

David E. Keyes*
William D. Gropp†

Abstract.
A block-preconditioned Krylov method which combines features of several previously developed techniques in domain decomposition and iterative methods for large sparse linear systems is described and applied to a few illustrative problems. The main motivation of the work is to examine the gracefulness of parallelization under the domain decomposition paradigm of the solution of systems of equations typical of finite-differenced fluid dynamical applications. Such systems lie outside of the realm of self-adjoint scalar elliptic equations for which most of the theory has been developed, and the present contribution is merely a first step in an attempt to approach them, raising several issues and settling none. However, results of tests run on an Encore Multimax with up to 16 processors show that even this first step has utility in the coarse-granularity parallelization of hydrocodes of practical importance.

1. Introduction

Interest in domain decomposition techniques for partial differential equations from a parallel computing perspective stems from their transformation of a large discrete problem defined on a spatial domain, irreducible in the matrix theory sense, into a fairly arbitrary and potentially large number of independent problems defined on simply-connected subdomains, at the cost of introducing a set of constraints at the interfaces of the subdomains. The interface equations contain all of the global coupling of the original problem; the subdomain problems are completely decoupled from each other once the interfacial degrees of freedom are assigned. The interfacial constraints, though complex in structure, involve lower-dimensional subsets of the unknowns of the original problem. With the development of sufficiently good approximations for the interface constraints, a preconditioned iterative scheme involving all of the unknowns of the problem can be constructed. The solution

*Department of Mechanical Engineering, Yale University, New Haven, CT 06520. The work of this author was supported in part by the National Science Foundation under contract number EET-8707109.

†Department of Computer Science, Yale University, New Haven, CT 06520. The work of this author was supported in part by the Office of Naval Research under contract N00014-86-K-0310 and the National Science Foundation under contract number DCR 8521451.

(or preconditioning) of the union of the subdomain problems is generally cheaper than the solution (or preconditioning) of the full domain problem due to their reduced size and bandwidth, in terms of which the appropriate operation counts may grow as badly as cubicly, depending upon the algorithm. This advantage can be traded off against the extra iterations needed due to the approximation of the interface equations. In the parallel context, additional advantages stemming from the independence and data locality of the subdomain problems enter into consideration.

In d-dimensional problems, the process of relegating the global coupling, inevitable at some level, to a set of lower-dimensional problems can, in principle, be recursively implemented d times. Thus, for example, a two-dimensional elliptic problem discretized with n subintervals on a side and decomposed into box-type subdomains with p subdomains on a side yields after two stages an irreducible system of $(p-1)^2$ equations for the points at the vertices of the boxes. This system is the Schur complement in the ambient (or global) matrix of the degrees of freedom defined at all other points [9]. Were the vertex degrees of freedom determined, $2p(p-1)$ decoupled sets of $(p-1)$ equations could be derived to yield the values at the points along the edges connecting them. Finally, given all of the interfacial degrees of freedom, p^2 decoupled sets of $(\frac{n}{p}-1)^2$ equations, simply sub-blocks of the original matrix, would yield the remaining values in the subdomain interiors. For general operators, the cost of deriving the exact lower-dimensional systems can greatly exceed the cost of direct banded Gaussian elimination on the original system. It follows that the success of domain decomposition hinges on the ability to efficiently approximate the lower-dimensional operators, by taking advantage of either known or "probed" structure.

Most published work to date on domain decomposition algorithms has concentrated on self-adjoint scalar elliptic equations, and several optimal algorithms are now known for this case, in the sense that the number of iterations required to solve the discretized PDE does not grow with the gridpoint density or the number of subdomains as these quantities are refined in proportion (see, *e.g.*, [5, 26]). These algorithms employ the conjugate gradient method, and are distinguished from each other primarily by the selection of solvers or preconditioners for the decoupled systems of equations for the subdomain interiors and for the coupled interface equation system. For certain constant coefficient problems, exact preconditioners can be obtained by means of Fourier analysis so that the iterations converge in a single step [4, 7]. Recently, Chan [6] has extended the class of problems for which good preconditioners are known to the scalar convective-diffusive case. However, problems involving several linearized convective-diffusive equations coupled to each other by source terms have received little attention in this context. The solution of such problems is an important computational kernel in implicit methods (for instance, Newton-like methods and linearized implicit time-stepping methods) commonly used for systems of nonlinear PDEs arising in science and engineering, and is often CPU-bound or memory-bound or both on the fastest and largest serial computers available. Furthermore, it is often the only computationally intensive part of such codes whose efficient parallelization is not straightforward, particularly when the distribution of data throughout the computer's memory hierarchy cannot be dictated exclusively by linear algebra considerations.

In this contribution, preconditionings of a "modified Schur complement" (MSC) type are applied to the solution of the large sparse linear systems arising at each stage of the application of Newton's method to a system of elliptic boundary value problems modeling a two-dimensional reacting fluid flow. The nomenclature derives from the fact that the preconditioner is built from blocks which are low-bandwidth approximations to actual Schur complements derived from various submatrices of the original global matrix. These low-bandwidth matrices are required to produce the same matrix-vector products as approximations to the true (dense) Schur complement when acting on a given set of trial vectors, and are obtainable by solving independent problems on subdomains. Since this type of requirement is sometimes imposed on incomplete factorizations, in which context the adjective "modified" has become solidly established, we employ it here as well. As a preconditioning technique for the interface equations, the modified Schur complement method exploits only the sparsity structure and clustering of large magnitudes around the diagonal of the actual operator, and no other properties like symmetry or constant coefficients. As a result, it performs suboptimally on many special problems, but generalizes with little difficulty to many other systems. In the absence of a general theory for the performance of this technique, experimental exploration of its convergence properties is required, and sample results are described herein (section 4), preceded by a description of the overall computational procedure (section 2), and its parallel implementation (section 3). We conclude in section 5 by listing some open issues in MSC-mediated domain decomposition.

2. Algorithmic Description

As sketched above, an iterative substructuring algorithm consists of an iterative procedure together with at least a two-level hierarchical preconditioner for the subdomain and interface systems. Our selection of a hybrid generalized minimum residual / incomplete LU-decomposition / modified Schur complement block algorithm is now motivated and described. The techniques woven into this hybrid are outlined in the following subsections. Fuller details are obtainable from the cited references.

2.1. Full Matrix Domain Decomposition

Domain decomposition enters our considerations at the level of the solution of a system of linear equations. In applications, this system is usually derived from the linearization of a nonlinear process, which gives the entire domain decomposition procedure the status of an inner iteration. To avoid any deeper nesting of iterations in applications, we iterate simultaneously on all of the unknowns in the linear system, in the sense that the subdomain problems are not individually iterated to convergence before their values are used to update the right-hand sides of the equations for the interfacial unknowns. This form of full matrix domain decomposition was advocated in [5] for problems in which no fast solver is known for the subdomain interior problems, and has been demonstrated therein to lead to an optimally convergent scheme for a class of self-adjoint strongly elliptic operators, provided (among other things) that spectrally equivalent subdomain preconditioners are employed.

This paradigm for domain decomposition is most easily illustrated in the decomposition into two strips of a rectangular region overlaid by a tensor-product grid. The single cut follows a line of gridpoints, which are ordered separately. For a 9-point operator on a grid with 16 interior subintervals in each direction (with Dirichlet boundary conditions eliminated) the resulting sparsity pattern for the operator A is indicated graphically in Fig. 1(a), and in matrix notation as follows:

$$A = \begin{pmatrix} A_{11} & 0 & A_{13} \\ 0 & A_{22} & A_{23} \\ A_{31} & A_{32} & A_{33} \end{pmatrix} . \tag{1}$$

Here, A_{33} renders the coupling between the points on the interface itself.

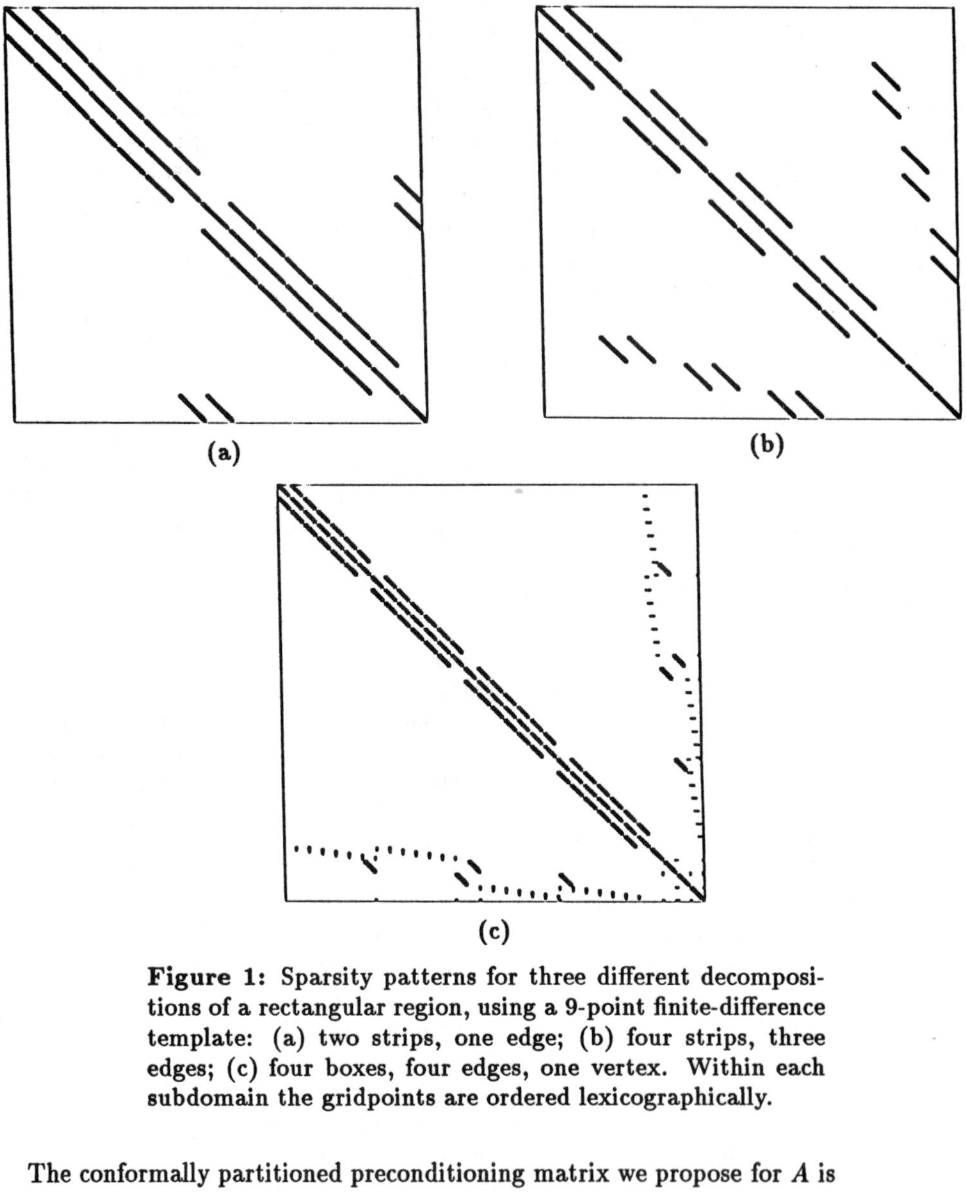

Figure 1: Sparsity patterns for three different decompositions of a rectangular region, using a 9-point finite-difference template: (a) two strips, one edge; (b) four strips, three edges; (c) four boxes, four edges, one vertex. Within each subdomain the gridpoints are ordered lexicographically.

The conformally partitioned preconditioning matrix we propose for A is

$$B = \begin{pmatrix} \tilde{A}_{11} & 0 & \tilde{A}_{13} \\ 0 & \tilde{A}_{22} & \tilde{A}_{23} \\ 0 & 0 & \tilde{C} \end{pmatrix}, \tag{2}$$

where $\tilde{C}$ approximates the Schur complement of A_{11} and A_{22} in A. The exact Schur complement, C, may be obtained from block-Gaussian elimination on A as:

$$C \equiv A_{33} - A_{31}A_{11}^{-1}A_{13} - A_{32}A_{22}^{-1}A_{23}. \tag{3}$$

The tilde-notation in the definition of B accommodates the replacement, if convenient, of the exact A_{ij} with approximations thereto. We assume throughout that the A_{ii} are invertible. (This is certainly a reasonable requirement for a discrete convective-diffusive operator.) Under this assumption, C is also invertible [9].

It has not been assumed above that A is symmetric. This provides the freedom to consider, without sacrifice of symmetry, a nonsymmetric B in (2) possessing instead the valuable property of block triangularity. Note that the inverse of B can be applied with one solve in each subdomain. Were A of the symmetric form

$$A = \begin{pmatrix} A_{11} & 0 & A_{13} \\ 0 & A_{22} & A_{23} \\ A_{13}^T & A_{23}^T & A_{33} \end{pmatrix} , \tag{4}$$

we would have considered a symmetric B, in order to take advantage of the two-term recurrence relationship of the conjugate gradient iteration for the preconditioned system:

$$B = \begin{pmatrix} \tilde{A}_{11} & 0 & \tilde{A}_{13} \\ 0 & \tilde{A}_{22} & \tilde{A}_{23} \\ \tilde{A}_{13}^T & \tilde{A}_{23}^T & \tilde{C} + \sum \tilde{A}_{i3}^T \tilde{A}_{ii}^{-1} \tilde{A}_{i3} \end{pmatrix} . \tag{5}$$

One application of the inverse of this symmetric B requires two solves in each subdomain [5], twice the work of the block triangular B.

2.2. Generalized Minimum Residual Method

Any algorithm intended for use in fluid dynamical applications must be robust with respect both to asymmetry, to allow for the presence of convective terms, and indefiniteness, to allow for the presence of linearized source terms whose coefficients oppose the algebraic sign of the diagonal term of the discrete convective-diffusive operator. (The latter can be particularly important in the modeling of chemically reacting flows in which at any point in the flow field some species may be created while others are consumed. The source terms in the transport equations for electrons and holes frequently employed in semiconductor device simulation can also give rise to indefiniteness.) The generalized minimum residual method (GMR) [21] is one such algorithm with which the authors have experience. Chebyshev iteration [17] would be an alternative suitable for cases in which the indefiniteness can be controlled (for instance, by the addition of a transient term to the continuous operator, which adds a term inversely proportional to some sufficiently small time step and of the correct sign to the diagonal of the discrete operator). All of our results to date employ GMR; however, we intend to experiment with adaptive Chebyshev iteration in the future because of its short recurrence relation.

Given a system of equations, $Mx = f$, M nonsingular, and an initial iterate, x_0, with initial residual, $r_0 = f - Mx_0$, GMR computes the solution x from finding $z \in K_m$ such that

$$(r_0 - Mz, v) = 0 ,$$

for all $v \in L_m$, and setting $x = x_0 + z$, where K_m and L_m are Krylov spaces based on r_0:

$$K_m \equiv \text{span}\{r_0, Mr_0, \ldots, M^{m-1}r_0\},$$

$$L_m \equiv \text{span}\{Mr_0, M^2r_0, \ldots, M^m r_0\}.$$

The solution x computed after m steps of GMR minimizes $\|r\|_2$ in the affine space $x_0 + K_m$. In a practical algorithm, an orthogonal basis for K_m is built up by means of a Gram-Schmidt or Householder process, which obviates the necessity of working with the normal equations. Suitable computer implementations of GMR have been given in [21] and [24], of which we use the former. Among the desirable properties of GMR are: (1) the only reference to M is in form of matrix-vector products, (2) the only matrix to be "inverted" is usually of order $m \ll \dim(M)$, (3) it cannot break down (in exact arithmetic) short of delivering the solution even for nonsymmetric systems with indefinite symmetric part, (4) it requires less storage and fewer operations per step than the mathematically equivalent GCR and ORTHODIR algorithms, and (5) the 2-norm of the residual is non-increasing and can be monitored without constructing intermediate solution iterates. The main disadvantage of GMR is the lack of a bounded recurrence relation, which causes the operation count and storage requirements to grow quadratically and linearly, respectively, in the iteration index. In many applications, restarting GMR after a predetermined number of steps successfully deals with this problem, but restarted GMR can also fail through stagnation.

2.3. Incomplete LU Subdomain Preconditioning

GMR is often uneconomical when left to act by itself on a general reaction-convection-diffusion operator. In an effort to control the work and storage required by GMR when A has a widely spread spectrum, we precondition the iterations by taking M in the formulae above to be AB^{-1}, for the A and B given in §2.1. This is "right" preconditioning, which first solves $M\tilde{x} = f$ for $\tilde{x}$, then $Bx = \tilde{x}$ for x. We adopt right over left preconditioning because in the latter the matrix B enters into the GMR residual convergence criterion in a direct way, making convergence comparisons between different preconditioning techniques difficult.

Approximate factorizations of the original matrix into triangular matrices, such as incomplete LU-decomposition (ILU) [18], are useful general purpose preconditioners for GMR and other Krylov methods. However, such factorizations can be bottlenecks in parallel implementations, because of the sequential nature of triangular factorizations and solves. Though wavefront-based or red-black reorderings of the standard sequential operations can alleviate this problem [20] in the context of sparse banded matrices, domain-decomposition approaches side-step it altogether by applying ILU within subdomains only.

In the present examples we employ the simplest of techniques appropriate to nonsymmetric systems, a Crout-Doolittle ILU(0), where the zero indicates that no fill-in outside the original sparsity pattern of A is allowed (see [25]). Thus, by its definition, the remainder matrix $R \equiv LU - A$ has (in the nine-point case) four nonzero diagonals, one on either side of the tridiagonal cluster about the main diagonal and one just inside each of the exterior tridiagonal clusters.

In a procedural language aided by compass-point subscript notation we have the following algorithm for a 9-point stencil in two dimensions:

$$
\begin{aligned}
u_0^i &\leftarrow a_0^i - l_{SW}^i u_{NE}^{i-n-1} - l_S^i u_N^{i-n} - l_{SE}^i u_{NW}^{i-n+1} - l_W^i u_E^{i-1} \\
u_E^i &\leftarrow a_E^i - l_S^i u_{NE}^{i-n} - l_{SE}^i u_N^{i-n+1} \\
u_{NW}^i &\leftarrow a_{NW}^i - l_W^i u_N^{i-1} \\
u_N^i &\leftarrow a_N^i - l_W^i u_{NE}^{i-1} \\
u_{NE}^i &\leftarrow a_{NE}^i \\
l_W^{i+1} &\leftarrow [a_W^{i+1} - l_{SW}^{i+1} u_N^{i-n} - l_S^{i+1} u_{NW}^{i-n+1}][u_0^i]^{-1} \\
l_{SE}^{i+n-1} &\leftarrow [a_{SE}^{i+n-1} - l_S^{i+n-1} u_E^{i-1}][u_0^i]^{-1} \\
l_S^{i+n} &\leftarrow [a_S^{i+n} - l_{SW}^{i+n} u_E^{i-1}][u_0^i]^{-1} \\
l_{SW}^{i+n+1} &\leftarrow [a_{SW}^{i+n+1}][u_0^i]^{-1}
\end{aligned}
\tag{6}
$$

Here a_0^i is the diagonal term in the ith row, a_E^i is the term in the first superdiagonal of the ith row, which corresponds in the natural ordering to the point due east of the diagonal entry, etc. There are n interior gridpoints in the most rapidly ordered direction. The entries $u_{()}^i$ and $l_{()}^i$ of the sparse upper and lower triangular factors overwrite partial rows and columns, respectively, of the original matrix as the factorization proceeds down the main diagonal. (l_0^i need not be computed because it is simply the identity.) Higher order factorizations, denoted ILU(k), $k > 0$, are possible in which more nonzero diagonals are permitted to accommodate the fill-in in U and L, relative to A. For $k = n - 2$ with the

9-point stencil (or $k = n-1$ with the 5-point stencil), the band fills in completely and the factorization becomes exact.

2.4. Modified Schur Complement Interface Preconditioning

For the interface equations, we use the modified Schur complement (MSC) technique. This was proposed in [7] and implemented and compared with several other preconditioners on symmetric problems in [14]. Like the ILU technique, MSC allows for a variable number of nonzero diagonals. In the two-dimensional two-subdomain example (1), we define MSC(k) by using for the matrix $\tilde{C}$ in (2) the following banded matrix:

$$\tilde{C}_k = \tilde{A}_{33} - E_k \ ,$$

where E_k is banded of semi-bandwidth k which satisfies as accurately as possible

$$E_k v_l = \sum_i (\tilde{A}_{3i} \tilde{A}_{ii}^{-1} \tilde{A}_{i3}) v_l \tag{7}$$

for a set of vectors v_l, $l = 1, 2, \ldots, L$.

For a nonsymmetric scalar system of equations, we set $L = 2k+1$ and use for the respective E_k the vectors:

$$\begin{aligned}
E_0 : v_1 &= [1,1,1,1,1,1,\ldots]^T \\
E_1 : v_1 &= [1,0,0,1,0,0,\ldots]^T \\
v_2 &= [0,1,0,0,1,0,\ldots]^T \\
v_3 &= [0,0,1,0,0,1,\ldots]^T \\
E_2 : v_1 &= [1,0,0,0,0,1,\ldots]^T \\
v_2 &= [0,1,0,0,0,0,\ldots]^T \\
v_3 &= [0,0,1,0,0,0,\ldots]^T \\
v_4 &= [0,0,0,1,0,0,\ldots]^T \\
v_5 &= [0,0,0,0,1,0,\ldots]^T \\
&\vdots
\end{aligned} \tag{8}$$

For a symmetric scalar system of equations, we set $L = k+1$ and use instead:

$$\begin{aligned}
E_0 : v_1 &= [1,1,1,1,1,1,\ldots]^T \\
E_1 : v_1 &= [1,0,1,0,1,0,\ldots]^T \\
v_2 &= [0,1,0,1,0,1,\ldots]^T \\
E_2 : v_1 &= [1,0,0,1,0,0,\ldots]^T \\
v_2 &= [0,1,0,0,1,0,\ldots]^T \\
v_3 &= [0,0,1,0,0,1,\ldots]^T \\
&\vdots
\end{aligned} \tag{9}$$

In the nonsymmetric case, there are $(2k+1)n - k(k+1)$ distinct elements in E_k of dimension n, and $(2k+1)n$ scalar equations in (7). Therefore, only the $k=0$ (simple row-sum preserving) case is well defined. $(2k+1)n - k(k+1)$ of the equations in (7) explicitly assign individual elements of E_k. The remaining $k(k+1)$ equations, for $k > 0$, involve none of the elements of E_k at all, but require a certain sum of elements from the matrix on the right-hand side of (7), each of which is at least $k+1$ diagonals away from the main diagonal, to vanish. This overdetermination is inconsistent, but not fatally so for a preconditioner, particularly if the Schur complement matrix being approximated has terms which decay rapidly away from the diagonal. It costs $2k+1$ solves on each subdomain to compute the right-hand side of (7).

In the symmetric case, there are $(k+1)n - k(k+1)/2$ distinct elements in E_k of dimension n and $(k+1)n$ scalar equations in (7). In this case, however, the overdetermination is consistent. The diagonal elements of E_k can be read off from explicit assignments and the remaining elements can be obtained in $O(kn)$ operations. It costs $k+1$ solves on each subdomain to compute the right-hand side of (7).

In either case, setting $k = n-1$ determines the Schur complement exactly (assuming that the tilde-quantities in (7) are identical to their non-tilde counterparts), in which case the overall algorithm converges in one iteration. This is, in effect, the approach used by Przemieniecki in [19], and requires as many subdomain solves as there are degrees of freedom on the interface. As a means of obtaining a specified finite level of precision in the final result (commensurate, for example, with the discretization error), taking k to be $O(n)$ is inefficient, and particularly so if the subdomain solves are themselves not exact so that more than one iteration is needed. Between the extremes of many cheap iterations and few expensive iterations determined by the index k in each of ILU(k) and MSC(k) will be an optimal trade-off. In subsequent sections the simple gridpoint-scale zero patterns in these v_l will need extensions to both coarser (§2.4.1) and finer (§2.5) scales.

2.4.1. Extension of MSC to Multiple Interfaces

The descriptions of MSC(k) above are for a simple interface between two subdomains, as depicted in Fig. 1(a). Generalizations to the compound interface cases shown in Fig. 1(b) (three edges) and Fig. 1(c) (four edges, with vertex) and finer decompositions can take a variety of forms. Their development is aided by a heuristic rather than formal approach to approximating the appropriate compound-interface generalization of the Schur complement (3), recognizing it to be, essentially, a discrete Green's function.

We note that the element C_{ij} represents the influence of the data at interfacial node j (source point) on the discrete residual at interfacial node i (field point). Varying each interface point separately in turn would enable filling in C column-by-column. In diffusive problems, we would expect this internodal influence to decay rapidly with physical separation. In mixed convective-diffusive problems, we would expect dependencies of even shorter range on the values at "downwind" nodes, and of somewhat longer range on the values at "upwind" nodes. Depending upon the magnitude of these influences we might want, in the interest of economy and efficient parallelism, to set large off-diagonal blocks of C corresponding to sufficiently distantly coupled degrees of freedom to zero, and to determine the remaining (assumed non-neglible) blocks by varying large numbers of source points simultaneously. This is analogous to the Curtis-Powell-Reid [10] technique for efficient sparse Jacobian estimation using vector function evaluations, except that we are prepared, in general, to accept much looser restrictions on which columns may be treated as corresponding to unrelated degrees of freedom and thus be evaluated simultaneously.

In the extreme limit of $k = 0$, we attempt to probe all columns of C simultaneously. As k grows, the resources for resolving more of the structure of C can be invested in different ways. This is done by the selection of the v_l in (7). The v_l listed above are appropriate in a purely diffusive problem with a spatially uniform diffusion coefficient and isolated interfaces (assuming that all the nodes on a given interface are ordered consecutively). By spreading out the active source points as evenly as possible, these v_l put a premium on resolving the influence of nearest neighbors *along* an interface. In a multiple-interface problem in which the subdomains possess high aspect ratio, it may be a better investment to isolate physically nearby interfacial degrees of freedom belonging to different edges than to isolate degrees of freedom distantly separated on the same edge. For the v_l above, taking k greater than the narrowest discrete dimension of the high-aspect-ratio subdomains would give rapidly diminishing returns, with a similar problem occuring for boxwise decompositions near the vertices. In such cases, we recommend assigning parity to the decomposition-defining cuts (in each dimension) and evaluating MSC blocks corresponding to odd and even parity in separate stages. In each stage, only one set of cuts bordering each subdomain is active; the elements of the v_l corresponding to all others are identically zero. For stripwise decompositions, this requires twice as many subdomain solves to compute $\tilde{C}$. For boxwise decompositions, four times the original preprocessing work is required.

2.4.2. Extension of MSC to Higher Levels of Complementation

The sparsity structure of the discrete operator A for the simplest decomposition admitting a two-level complementation is furnished in Fig. 1(c). The ambient matrix

$$A = \begin{pmatrix} A_{11} & 0 & 0 & 0 & A_{15} \\ 0 & A_{22} & 0 & 0 & A_{25} \\ 0 & 0 & A_{33} & 0 & A_{35} \\ 0 & 0 & 0 & A_{44} & A_{45} \\ A_{51} & A_{52} & A_{53} & A_{54} & A_{55} \end{pmatrix}, \tag{10}$$

where A_{55} renders the coupling between all the interfacial points (the union of the edge and vertex unknowns), is a bordered block diagonal matrix.

Eliminating A_{11} through A_{44} leaves a capacitance system which is *not* sparse in general, a situation which is fundamentally very different from that encountered in the ambient matrix. This poses problems in the parallel context, since it is desirable to construct a block triangular preconditioner for it in which each edge is handled independently.

Fortunately, cases are known in which bordered block diagonal approximations to this first Schur complement make acceptable preconditioners. The (symmetric) purely diffusive case is covered in [5], and the convectively-dominated case is discussed below in the absence of recirculation. Especially in such cases, though also in others, it may be worthwhile to recursively employ MSC techniques to approximate the Schur complement of the edge degrees of freedom in the matrix which approximates the first complement, and use the result to precondition the vertex equation system (which consists of a single degree of freedom in the example of Fig. 1(c)). It should be emphasized that the practical utility of higher-level MSC techniques has yet to be demonstrated. However, the fruitfulness of solving a globally coupled vertex system in the purely diffusive context [5], warrants attempted generalizations to other operators. Multigrid technology provides another, alternative path. The main advantage of the higher-level MSC approach will likely be the same as at the first level: it provides a path to constructing a preconditioner which becomes arbitrarily accurate as the bandwidth k is increased at all levels, independent of special operator features. Against this advantage must be traded off the expense of approximately solving exponentially more subdomain problems at each level of complementation.

2.5. Block-structured GMR/ILU/MSC domain decomposition

In most fluid dynamical applications there are several fields defined at each point in the domain, including the momenta and pressure (alternatively the streamfunction and vorticity), and possibly species concentrations, temperature, density, etc. The several unknowns at a single gridpoint couple strongly with one another, not only directly through source terms, but also indirectly through material-dependent transport properties in multicomponent mixtures. In the ordering of these fields into a discrete vector of unknowns, it is natural to preserve the locality of this coupling by grouping the r unknowns defined at each point together. This brings about a block-structured linearized operator in which the $r \times r$ blocks are dense, but distributed sparsely, typically in a 5- or 9-diagonal structure for two-dimensional problems or a 7- or 27-diagonal structure in three dimensions.

The ILU and MSC techniques can be generalized to accommodate multicomponent problems. In §2.3 the $a^i_{()}$, $u^i_{()}$ and $l^i_{()}$ must be reinterpreted as $r \times r$ blocks and the ones in (8) replaced with each of the r unit vectors in sequence, *e.g.*,

$$\begin{aligned} E_0 : v_{1,1} &= [e_1, e_1, e_1, e_1, e_1, e_1, \ldots]^T \\ v_{1,2} &= [e_2, e_2, e_2, e_2, e_2, e_2, \ldots]^T \\ &\vdots \\ v_{1,r} &= [e_r, e_r, e_r, e_r, e_r, e_r, \ldots]^T \end{aligned} \tag{11}$$

To resolve the intra-point coupling in the MSC method requires $(2k+1)r$ subdomain solves in the nonsymmetric case. Block-point preconditioning has been found superior in the single-domain context to the componentwise ILU schemes with the same storage requirements on the convective-diffusive-reactive operators considered in §4. The virtues of block-line preconditioning in the $r = 1$ case have also been explored in [8] and [23].

3. Parallel Implementation

There are potentially three penalties to be paid in distributing the GMR/ILU/MSC solution algorithm over an array of independent processors: synchronization overhead, communication overhead, and degradation of convergence. These penalties are measured indirectly through the speedup and efficiency figures-of-merit of a parallel implementation. The speedup is the ratio of the uniprocessor execution time of a given algorithm to that of the multiprocessor execution of the same algorithm. The efficiency is usually defined as the speedup divided by the number of processors. For many algorithms, these definitions are unnatural in the sense that one would never use the same algorithm in both uniprocessor and multiprocessor environments. (Usually *better* uniprocessor algorithms exist, so the parallel efficiency as defined above is inflated relative to its advantage.) We adopt measures in which the execution times are obtained from the most *natural* algorithm for each environment, namely, given p processors we employ exactly p subdomains.

The synchronization penalty arises if the processors have dependencies which force them them to wait for data which are not yet available. Even if the processors are programmed homogeneously this can happen at convergence checkpoints, for instance, if they have unequal amounts of work to do, or at points where reduction to a scalar of data distributed over all processors is required. The amount of work to be done in a processor is a function of the number of gridpoints in the subdomain assigned to it and the stencil of the discrete equations to be enforced at those gridpoints. The relative number of boundary gridpoints (which require somewhat less computational work than interior ones) decreases as the mesh is refined, and though their distribution between the processors becomes more uneven, only a small number of processors are thus periodically idled. If the gridpoints are allocated to the processors as evenly as possible within shape and contiguity constraints (which is *not* accomplished in our preliminary examples to follow) synchronization delays can be made relatively unimportant.

The communication penalty is the time spent gaining access to shared data even after it becomes available. The significance of this penalty depends on the amount of data to be shared, on its routing between memory and processors, on the amount of arithmetic which the algorithm must perform between fetches and writes, and on the communication-to-computation speed ratios of the hardware in question.

In an effort to increase the number and length of the independent threads which comprise a parallel computation, global coupling may be reduced in ways that degrade the convergence rate of the algorithm. In the context of domain decomposition, this penalty may arise as the ratio of interface to interior degrees of freedom is increased in refinements of the decomposition, since the approximations required to form diagonal preconditioner blocks for the former are often more severe.

In view of the above considerations as well as programming convenience, our parallel implementation consists of decomposing the logical tensor product computational domain into logically congruent strips or boxes of contiguous unknowns in two dimensions, (with the obvious generalizations in three dimensions) mapping these subdomains onto a network of processors, and programming the processors homogeneously. In this paper, our principal interest is in convergence rates, so we report on a bus-connected shared memory machine only: an Encore Multimax 320. The dependent variable arrays are placed in the shared memory and each processor is confined to roam over subranges of the array indices. The timings in the tables to follow include the parallel generation of the ILU and MSC blocks and the entire GMR iteration. To reduce the number of synchronization points inside each GMR iteration, a small QR factorization from which the coefficients of the Krylov basis vectors in the solution vector are derived is carried out redundantly in each processor. An analogous consideration led to the redundant solution of the equations for the vertex degrees of freedom in the parallel domain decomposition method described in [13]. Certain pre- and post-processing tasks, like the coefficient generation of the original operator and some spectral analysis, are done in serial and not included in the timings below.

4. Results for Model Problems

This section contains numerical results that display a few of the possibilities of MSC preconditioning, and, more generally, of the hybrid GMR/ILU/MSC algorithm. In the solution of linear systems arising from finite-differenced systems of conservation equations of the form

$$\frac{\partial u_r}{\partial t} + \vec{c}(\vec{u}) \cdot \nabla u_r - \nabla \cdot D_r(\vec{u}) \nabla u_r = f_r(\vec{u})$$

for $r = 1, \ldots, R$, combinations of the first, third, and fourth terms can be handled well with known methods [7], when $R = 1$. Comparisons on purely diffusive problems of MSC(k) for $k = 1, 2, 3$ with the optimal preconditioners may be found in [14]. Therefore, the cases $|\vec{c}| > 0$ and $R > 1$ are of particular interest. In the preliminary results contained herein, only stripwise decompositions (one level of complementation) are considered, and $k = 0$ in both MSC(k) and ILU(k).

4.1. Scalar Convection-Diffusion Problems

In the limit of unbounded mesh refinement and bounded convection velocity c, the nonsymmetric first term in the discrete operator for

$$c u_y - D \nabla^2 u = f \tag{12}$$

will be dominated by the diffusive term, and the optimal preconditioners such as the square root of the one-dimensional Laplacian (denoted $K^{1/2}$ in [14]) will be asymptotically superior to any MSC preconditioner. However, much practical CFD computation occurs in the opposite singularly perturbed limit. Especially in the early, coarse-grid stages of an adaptive grid calculation it is important to have methods which support the artificially diffusive upwind-differencing of the convective term.

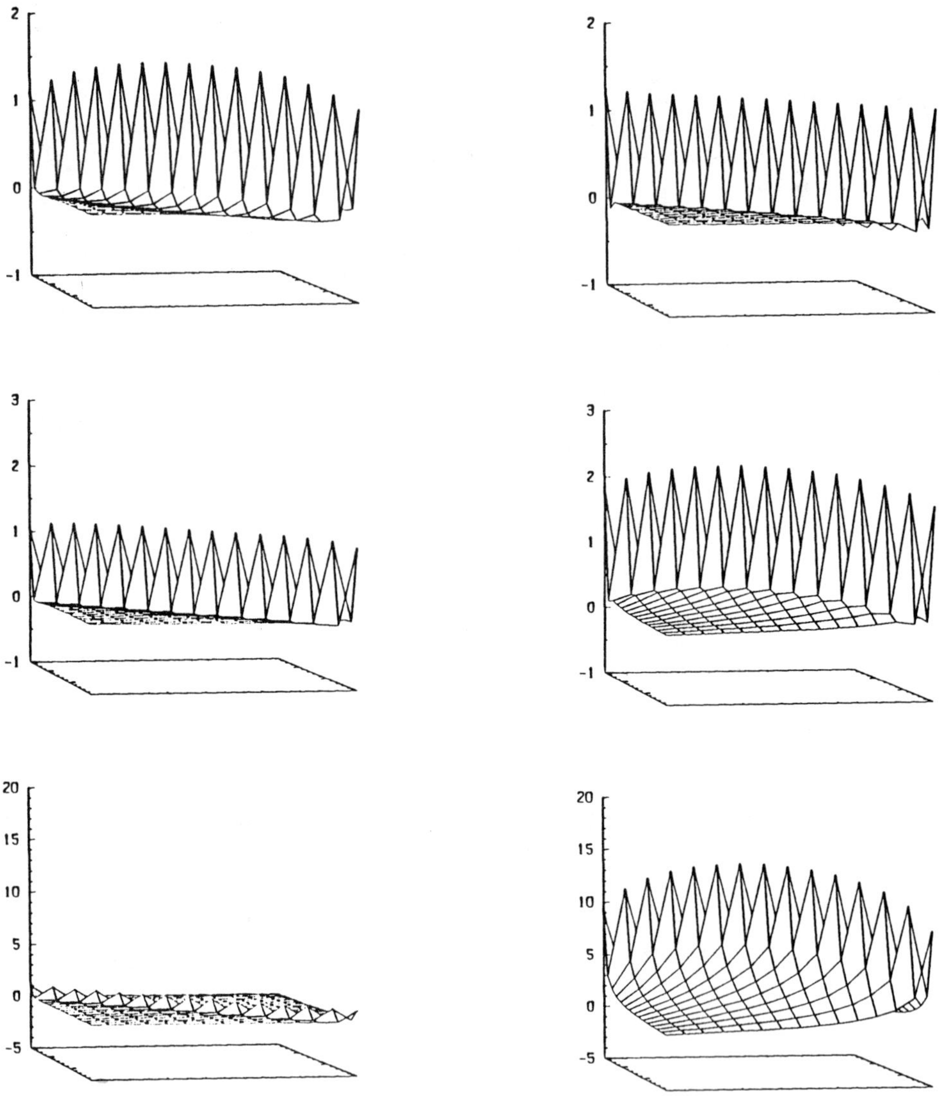

Figure 2: Element plots of the preconditioned capacitance system $\tilde{C}^{-1}C$ for the convective-diffusive operator (12) on the unit square with Dirichlet boundary conditions at $\mathrm{Re}_c = 0$ (top), $\mathrm{Re}_c = 2$ (middle), and $\mathrm{Re}_c = 20$ (bottom), using the modified Schur complement (left) and root-one-dimensional Laplacian (right). The horizontal axes are the row and column indices. Note that the vertical scale is different in each row of plots.

Fig. 2 consists of a set of plots of the matrix elements of the preconditioned interfacial system for the decomposition of Fig. 1(a), for the convective-diffusive operator of (12). To eliminate any interaction with ILU, the subdomain solves are carried out exactly in constructing the capacitance matrix. Two preconditioners, $\tilde{C}_D = K^{1/2}$ and $\tilde{C}_M = \text{MSC}(0)$, are considered at three cell Reynolds numbers, $\text{Re}_c \equiv |c|h/D = 0$, 2, and 20, respectively. For a good preconditioner, $\tilde{C}^{-1}C$ should be as close as possible to the identity. It is clear that the "straw man" $\tilde{C}_D$ fails as Re_c departs from 0, while $\tilde{C}_M$ improves. Another means of visualizing this improvement is the plot of the capacitance spectra in Fig. 3.The clustering of the eigenvalues near unity as Re_c becomes large is due to the easily physically visualizable fact that at least one of the factors in each of the triple products in equation (3) becomes small relative to the diagonal term A_{33} in this limit. High Re_c parabolizes the flow, making the matrices A_{32} and A_{13}, which transmit information upwind when $c > 0$, negligible. The orthogonality of the flow direction and the interface is not essential to this argument. For high enough Re_c, at least one in each pair of (A_{13}, A_{31}) and (A_{23}, A_{32}) is negligible compared to A_{33} at any flow orientation, including alignment with the interface. However, it is essential that the interface not cut a zone of fluid recirculation.

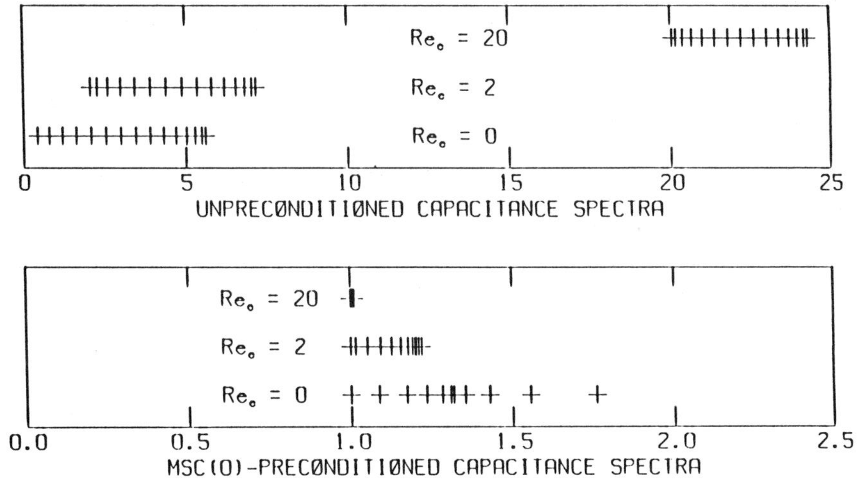

Figure 3: Spectra of the exact capacitance matrix for the convective-diffusive operator (above) and of the MSC(0)-preconditioned capacitance system (below), for three different cell Reynolds numbers.

The performance of MSC(0) in conjunction with approximate subdomain solves, is given in Table 1 for two different convective directions. (12) is the "vertical" convection case; for the "horizontal" we replace the first term with cu_x. Neumann conditions are used at the outflow boundary in either case and Dirichlet elsewhere on the perimeter of a unit square. All boundary degrees of freedom are incorporated into the matrix; none are eliminated *a priori*. Speedups of 5 to 10 are obtained on the largest problems using 16 processors. Note that in one convective direction the iteration count degrades at high aspect ratio, while in the other it actually improves slightly. As suggested by

		Vertical Convection			Horizontal Convection		
p	n	I	T	e	I	T	e
1	17	8	4.8	1.00	10	5.9	1.00
2	17	9	3.3	.73	11	3.9	.76
4	17	11	2.4	.50	11	2.6	.57
1	33	11	23.9	1.00	14	30.5	1.00
2	33	12	15.3	.78	15	19.0	.80
4	33	14	10.0	.60	15	10.7	.71
8	33	16	7.0	.43	15	6.5	.59
1	65	15	127.	1.00	21	186.	1.00
2	65	16	76.7	.83	22	109.	.85
4	65	17	43.1	.74	22	57.4	.81
8	65	20	28.7	.55	22	32.7	.71
16	65	26	24.7	.32	21	17.8	.65

Table 1: Iteration count I, CPU time T, and efficiency e for the convective-diffusive problem (12) at $\mathrm{Re}_c = 2.0$ with horizontal strips and either vertical convection or horizontal convection, as a function of number of processors p and spatial resolution n.

the different iteration counts in the undecomposed case $p = 1$, this anisotropy may be due as much to the difference in angle between the convection direction and that of the most rapid ordering of the unknowns in the ILU subdomain blocks (always horizontal) as to any interface orientation effect. Different flow orientations, interface orientations, and ILU orderings, leave a multitude of primitive combinations to be investigated. Note that neither combination considered here is superior to the other at all granularities of the decomposition. The relatively greater efficiency of the aligned case at high p is due to its poorer absolute $p = 1$ performance (compare the CPU times of the $n = 65$ data).

4.2. Vector Source-Diffusion Problems

As an example of a multicomponent problem with coupling through source terms, we consider a compact discretization of a sixth-order operator by means of three second-order operators. To be specific, we test

$$\begin{aligned} -\nabla^2 u_1 &= f \\ -\nabla^2 u_2 &= u_1 , \\ -\nabla^2 u_3 &= u_2 \end{aligned} \tag{13}$$

on a unit square with Dirichlet boundary conditions for all components. This introduces bidiagonal 3×3 diagonal blocks which the hybrid algorithm regards as fully dense. As with scalar Poisson problems, preconditioning techniques superior to MSC can be devised for such systems by exploiting their special structure.

Results are shown in Table 2 for the same problem decomposed into either horizontal or vertical strips. Speedups close to 9 are achieved on 16 processors. Differences in performance between each pair of cases in the same row of the table can be induced only by the ordering of the ILU preconditioning, which is fastest along the interface direction for horizontal strips, and fastest in the normal direction for vertical strips. Non-monotonic iteration count behavior characterizes both orientations; in fact, the decrease in iterations in going from 2 to 4 subdomains in the $n = 65$ case contributes to superlinear relative speedup (the execution time is better than halved).

		Horizontal Strips			Vertical Strips		
p	n	I	T	e	I	T	e
1	17	11	18.4	1.00	11	18.4	1.00
2	17	12	12.9	.71	12	13.0	.70
4	17	13	8.1	.57	12	7.6	.60
1	33	18	109.	1.00	18	109.	1.00
2	33	22	78.2	.70	22	78.8	.69
4	33	20	38.9	.70	20	39.5	.69
8	33	23	25.1	.54	21	24.0	.57
1	65	33	818.	1.00	33	818.	1.00
2	65	40	564.	.73	40	567.	.72
4	65	37	267.	.77	37	271.	.75
8	65	37	145.	.71	36	145.	.71
16	65	40	96.1	.53	36	88.5	.58

Table 2: Iteration count I, CPU time T, and efficiency e for the coupled Poisson problem (13) with horizontal strips or vertical strips, as a function of number of processors p and spatial resolution n.

4.3. Vector Reaction-Convection-Diffusion Problems

We conclude with examples of Jacobian matrices from a real problem describing a two-dimensional axisymmetric methane-air laminar diffusion flame under the flamesheet approximation for the chemical kinetics. The flamesheet exploits a large Damkohler number (ratio of diffusion to reaction time scales) by replacing a reaction zone of finite thickness with an interface (of unknown location) across which gradients of the temperature and species are discontinuous. The interface subdivides the physical domain Ω into an oxidizer-free zone Ω_F and a fuel-free zone Ω_O, in either of which the full composition and thermodynamic state of the gas mixture can be recovered from a single conserved scalar. The flamesheet is an economical means of computing initial iterates for detailed kinetics calculations [15], since it requires just three components per gridpoint (a variable density Stokes streamfunction ψ, the vorticity ω, and the conserved scalar S) instead of the dozens that would be required with full chemistry. The continuous system is of the following form:

Streamfunction and Vorticity Definitions:

$$\rho r v_r = -\frac{\partial \psi}{\partial z}, \quad \rho r v_z = \frac{\partial \psi}{\partial r}, \quad \omega = \frac{\partial v_r}{\partial z} - \frac{\partial v_z}{\partial r}$$

Streamfunction Equation:

$$\frac{\partial}{\partial z}\left(\frac{1}{r\rho}\frac{\partial \psi}{\partial z}\right) + \frac{\partial}{\partial r}\left(\frac{1}{r\rho}\frac{\partial \psi}{\partial r}\right) + \omega = 0$$

Vorticity Equation:

$$r^2\left[\frac{\partial}{\partial z}\left(\frac{\omega}{r}\frac{\partial \psi}{\partial r}\right) - \frac{\partial}{\partial r}\left(\frac{\omega}{r}\frac{\partial \psi}{\partial z}\right)\right] - \frac{\partial}{\partial r}\left(r^3\frac{\partial}{\partial r}\left(\frac{\mu}{r}\omega\right)\right) - \frac{\partial}{\partial z}\left(r^3\frac{\partial}{\partial z}\left(\frac{\mu}{r}\omega\right)\right) +$$

$$r^2 g\frac{\partial \rho}{\partial r} + r^2\left[\frac{\partial}{\partial r}\left(\frac{v_r^2+v_z^2}{2}\right)\frac{\partial \rho}{\partial z} - \frac{\partial}{\partial z}\left(\frac{v_r^2+v_z^2}{2}\right)\frac{\partial \rho}{\partial r}\right] = 0$$

Conserved Scalar Equation:

$$\frac{\partial}{\partial z}\left(S\frac{\partial\psi}{\partial r}\right) - \frac{\partial}{\partial r}\left(S\frac{\partial\psi}{\partial z}\right) - \frac{\partial}{\partial r}\left(r\rho D\frac{\partial S}{\partial r}\right) - \frac{\partial}{\partial z}\left(r\rho D\frac{\partial S}{\partial z}\right) = 0$$

State Relations:

$$\rho(\vec{x}) = \begin{Bmatrix} \rho_F(S), & \vec{x} \in \Omega_F \\ \rho_O(S), & \vec{x} \in \Omega_O \end{Bmatrix}$$

$$\mu(\vec{x}) = \begin{Bmatrix} \mu_F(S), & \vec{x} \in \Omega_F \\ \mu_O(S), & \vec{x} \in \Omega_O \end{Bmatrix}$$

$$D(\vec{x}) = \begin{Bmatrix} D_F(S), & \vec{x} \in \Omega_F \\ D_O(S), & \vec{x} \in \Omega_O \end{Bmatrix}$$

In practice (see *e.g.*, [22, 16]), these equations are solved by a modified Newton method and a pseudo-transient continuation scheme (in which the time-step is adaptively chosen, becoming infinite asymptotically to recover the quadratic convergence of Newton's method) on an adaptively refined (and severely nonuniform) grid.

p	n	I	T	e
1	17	9	18.3	1.00
2	17	13	16.1	.57
4	17	18	12.5	.37
1	33	14	100.	1.00
2	33	17	73.3	.68
4	33	20	47.3	.53
8	33	22	29.1	.43
1	65	18	507.	1.00
2	65	18	300.	.85
4	65	18	158.	.80
8	65	18	87.0	.73
16	65	17	52.2	.61

Table 3: Iteration count I, CPU time T, and efficiency e for Jacobians from the flamesheet problem with horizontal strips at various stages of spatial resolution n, as a function of number of processors p.

Table 3 reports results on Jacobians drawn from different stages of a calculation in which an initially adaptively selected grid was constrained to be refined by factors of two to avoid load-balancing considerations. There is no relation between problem difficulty at the three resolutions chosen. In fact, the first two are near the domain of convergence of Newton's method on their respective grids (near infinite time step) while the last has a large diagonal contribution from the pseudo-transient term. The ten-fold speedups on large problems would lead to worthwhile improvements in turnaround times on detailed kinetics problems, in which several supercomputer CPU hours can be spent on linear algebra alone. Though linear algebra does not necessarily account for the dominant share of the total CPU time in such models, it is the only part whose parallelization is non-trivial. Furthermore, linear solvers of simple relaxation type *do* contribute to the leading order in the communication overhead in parallelization implementations developed to date [16]. Domain-decomposed solvers are capable of doing more arithmetic between neighbor-neighbor interchanges and global convergence checks, at least up to moderate decomposition granularities.

5. Conclusions and Directions for Further Research

Included in the GMR/ILU/MSC framework is a family of schemes governed by bandwidth parameters which lie between the extremes of decoupled block-diagonal preconditioned GMR and domain-decomposed direct elimination. Though surpassed in efficiency by known methods in several contexts, they provide a means for the parallel solution of rather general linear systems. Improvements should be sought, however, in several areas:

1. *Tuning of order of approximation between levels.* It is clear that each level in the MSC hierarchy is limited in its attainable preconditioning ability by the levels beneath it from which the right-hand sides of (7) are computed. There would clearly be vanishing returns in cranking up k indefinitely in MSC while leaving it fixed in ILU. Further work is needed to establish theoretically on model problems and experimentally on representative real problems how the different tuning parameters should be coordinated.
2. *Locally operator-adaptive approximation.* In problems with different physical properties on different subdomains the requirements on the preconditioning blocks at the same level in the hierarchy may vary from subdomain to subdomain. Depending on a locally-averaged Reynolds number, for instance, a fast Poisson solver might be a chosen over ILU on some subdomain. Alternatively, the k in ILU(k) might be chosen differently within different subdomains. A mature technique should exploit such versatility, perhaps even adaptively, taking into consideration a load-imbalance penalty function.
3. *Other approximate Schur complements.* MSC is only one means of obtaining a compact approximation of the Schur complement. In problems with simplifying structure, other compact approximations may exist. For instance, noting the near translation independence along an interface in a Poisson problem, Golub and Mayers proposed a Toeplitz approximation to C in [12], requiring just n scalars to represent a matrix of n^2 elements. Generalizing along these lines, periodically sampled individual rows of C_{ij} might be assumed to hold over a surrounding row range. Dorr [11] has also recently proposed an extremely low-dimensional parameterization of the interdomain communication by means of Lagrange multipliers, in which it is affordable to explicitly construct and directly solve a generalized capacitance matrix. Multicomponent problems would require a "blocking" of these ideas.
4. *Efficient methods for nonunidirectional convection.* The restriction of the MSC to block diagonal form in the interest of clean parallelism may be too steep a price to pay in convergence rate for problems in which placing cuts through convective recirculation zones cannot be avoided. One is therefore left with the necessity of introducing global coupling at a lower level. Implementation issues related to the evaluation of and solution of the MSC systems with off-diagonal blocks have yet to be addressed.

It would also be of interest, considering the cost and complexity of constructing hierarchical MSC-based preconditioners, to ask how well three other classes of preconditioners do on the same problems: (1) non-hierarchical coupled preconditioners, (2) non-hierarchical decoupled preconditioners with overlap, and (3) non-hierarchical decoupled preconditioners without overlap. An example of the first would be GMR/ILU in which the incomplete factorization is carried out over the entire domain, rather than within subdomains. Since no interfacial blocks requiring approximation are introduced, this technique, appropriately blocked at the gridpoint level, will have a better iteration count than MSC-based techniques on most problems; however, its parallel efficiency can be so low in some problems that it is inferior even to unpreconditioned GMR [3] as a parallel algorithm. An example of the second Schwarz-type method would be GMR/ILU in which the incomplete factorization is carried out independently over overlapping subdomains, followed by some arbitration scheme for the common degrees of freedom. The disadvantages of such methods are the redundant effort expended on the common unknowns and the trade-off (in the parallel context) between sequential bottlenecks and the use of "lagged" data in the overlap regions. An example of the third method would be the use of

$$A = \begin{pmatrix} A_{11} & A_{12} \\ A_{21} & A_{22} \end{pmatrix}, \quad B = \begin{pmatrix} \tilde{A}_{11} & 0 \\ 0 & \tilde{A}_{22} \end{pmatrix} \tag{14}$$

where the $\tilde{A}_{ii}$ are ILU approximations, instead of (1) and (2). In this method, the creation of independent threads of computation in the preconditioner comes at the expense of severing the interdomain coupling altogether. It therefore parallelizes with virtually no overhead, but can suffer in iteration count as the granularity of the decomposition is refined and more and more of the coupling is discarded, the limit being block-point diagonal preconditioning. Some experiments with this technique have been reported in [1] and [2], and we have also tested it on all of the problems reported above, since it involves only minor modifications to the MSC-based code. For a given n and p, it is nearly always inferior in iteration count, but not always so in CPU time, since the cost of constructing and applying the preconditioning is less. Guidelines on when the interface coupling has sufficient incremental value to warrant (1)-(2) over (14) are lacking except in obvious special cases.

There is no consideration of complex domain geometry in the present work although ease of generalization to this case is an equally relevant motivation. The case of problems requiring too much memory to be managed by just one processor is another one in which the proposed technique is attractive, despite possible inefficiency relative to a global technique.

References.

[1] C. Ashcraft and R. Grimes, *On Vectorizing Incomplete Factorizations and SSOR Preconditioners,* Technical Report ETA-TR-41, Boeing Computer Services, December 1986.

[2] C. Ashcraft, *Domain Decoupled Incomplete Factorizations,* Technical Report ETA-TR-49, Boeing Computer Services, April 1987.

[3] D. J. Baxter, *personal communication,* 1987.

[4] P. E. Bjorstad and O. B. Widlund, *Iterative Methods for the Solution of Elliptic Problems on Regions Partitioned into Substructures,* Technical Report 136, Courant Institute of Mathematical Sciences, NYU, September 1984.

[5] J. H. Bramble, J. E. Pasciak and A. H. Schatz, *The Construction of Preconditioners for Elliptic Problems by Substructuring, I,* Math. Comp., 47(1986), pp. 103–134.

[6] T. F. Chan, *personal communication,* 1987.

[7] T. F. Chan and D. Resasco, *A Survey of Preconditioners for Domain Decomposition,* Technical Report 414, Computer Science Dept., Yale University, September 1985. In Proceedings of the IV Coloquio de Matemáticas del CINVESTAV, Workshop in Numerical Analysis and its applications, Taxco, Mexico, Aug. 18-24, 1985.

[8] P. Concus, G. H. Golub and G. Meurant, *Block Preconditioning for the Conjugate Gradient Method,* Technical Report 14856, Lawrence Berkeley Laboratory, July 1975.

[9] R. W. Cottle, *Manifestations of the Schur Complement,* Lin. Alg. Appl., 8 (1974), pp. 189–211.

[10] A. R. Curtis, M. J. Powell and J. K. Reid, *On the Estimation of Sparse Jacobian Matrices,* J. Inst. Math. Appl., 13 (1974), pp. 117–119.

[11] M. R. Dorr, Domain Decomposition via Lagrange Multipliers, *Second International Symposium on Domain Decomposition Methods,* 1988.

[12] G. H. Golub and D. Mayers, *The Use of Pre-Conditioning over Irregular Regions,* 1983. Lecture at Sixth Int. Conf. on Computing Methods in Applied Sciences and Engineering, Versailles, Dec. 1983.

[13] W. D. Gropp and D. E. Keyes, *Complexity of Parallel Implementation of Domain Decomposition Techniques for Elliptic Partial Differential Equations,* SIAM J. Sci. Stat. Comp., 9 (1988), pp. 312–326.

[14] D. E. Keyes and W. D. Gropp, *A Comparison of Domain Decomposition Techniques for Elliptic Partial Differential Equations and their Parallel Implementation,* SIAM J. Sci. Stat. Comp., 8 (1987), pp. s166-202.

[15] D. E. Keyes and M. D. Smooke, *Flame Sheet Starting Estimates for Counterflow Diffusion Flame Problems,* J. Comp. Phys., 73 (1987), pp. 267–288.

[16] ———, A Parallelized Elliptic Solver for Reacting Flows, A. K. Noor ed., *Parallel Computations and Their Impact on Mechanics,* ASME, 1987, pp. 375–402.

[17] T. A. Manteuffel, *The Tchebychev Iteration for Nonsymmetric Linear Systems,* Numer. Math., 28 (1977), pp. 307–327.

[18] J. A. Meierink and H. A. Van der Vorst, *Guidelines for the Usage of Incomplete Decompositions in Solving Sets of Linear Equations as they Occur in Practical Problems,* J. Comp. Phys., 44 (1981), pp. 134–155.

[19] J. S. Przemieniecki, *Matrix Structural Analysis of Substructures,* AIAA J., 1 (1963), pp. 138–147.

[20] Y. Saad and M. Schultz, *Parallel Implementation of Preconditioned Conjugate Gradient Methods,* Technical Report YALEU/DCS/RR–425, Computer Science Dept., Yale University, October 1985.

[21] ———, *GMRES: A Generalized Minimum Residual Algorithm for Solving Nonsymmetric Linear Systems,* SIAM J. Sci. Stat. Comp., 7 (1986), pp. 856–869.

[22] M. D. Smooke, *Solution of Burner-Stabilized Pre-Mixed Laminar Flames by Boundary Value Methods,* J. Comp. Phys., 48 (1982), pp. 72–105.

[23] R. R. Underwood, *An Approximate Factorization Procedure Based on the Block Cholesky Decomposition and its Use with the Conjugate Gradient Method,* Technical Report NEDO-11386, General Electric Co., Nuclear Energy Div., 1976.

[24] H. F. Walker, *Implementation of the GMRES Method Using Householder Transformations,* SIAM J. Sci. Stat. Comp., 9 (1988), pp. 152–163.

[25] J. W. Watts, III, *A Conjugate Gradient-Truncated Direct Method for the Iterative Solution of the Reservoir Simulation Pressure Equation,* Soc. Petrol. Engin. J., 21 (1981), pp. 345–353.

[26] O. B. Widlund, Iterative Substructuring Methods: Algorithms and Theory for Elliptic Problems in the Plane, R. Glowinski, G. H. Golub, G. A. Meurant and J. Periaux ed., *First International Symposium on Domain Decomposition Methods for Partial Differential Equations,* SIAM, 1988, pp. 113–128.

Multidomain Spectral Solution of Shock-Turbulence Interactions

David A. Kopriva*
M. Yousuff Hussaini†

Abstract. The use of a fitted-shock multidomain spectral method for solving the time dependent Euler equations of gasdynamics is described. The multidomain method allows short spatial scale features near the shock to be resolved throughout the calculation. Examples presented are of a shock- plane wave, shock-hot spot and shock-vortex street interaction.

1. Introduction. Spectral methods are global approximation methods in which solution unknowns are expanded in polynomials which are the eigenfunctions of a singular Sturm-Liouville problem. The primary advantage of these expansions is the rapid rate of convergence for problems with smooth solutions. However, the global nature of the approximation can also be a drawback. In particular, it is difficult to handle complicated geometries and to resolve locally important features.

Domain decomposition is one way to avoid the disadvantages of global approximation functions. The need for global mappings is eliminated when a computational domain is broken down into several smaller subdomains. It also be-

* Mathematics Department and Supercomputer Computations Research Institute Florida State University, Tallahassee, FL 32306. This research was supported in part by the Florida State University through time granted on its Cyber 205 supercomputer.

† Institute for Computer Applications in Science and Engineering(ICASE), NASA Langley Research Center, Hampton, VA 23665. Research was supported by the National Aeronautics and Space Administration under NASA Contract No. NAS1-18107 while in residence at ICASE.

comes easy to resolve important features of a solution since expansions of different orders can be used in different subdomains. The use of multidomain spectral methods for these purposes can be found, for example, in the papers by Kopriva [4,5].

In this paper, we use domain decomposition and grid refinement to resolve short spatial scale phenomena which are generated during a shock-plane wave, a shock-hot spot, and a shock-vortex street interaction in two spatial dimensions. For the shock-plane wave problem, we use Chebyshev- Fourier collocation within each subdomain. For the other problems, Chebyshev-Chebyshev collocation is used.

2. Shock Interaction Problem. We assume that the initial state of each shock interaction problem is a uniform gas, rapidly moving from left to right, which terminates in an infinite, normal shock. To the right of the shock, the gas is quiescent except for some specified fluctuation. This fluctuation might be a pressure, vorticity or entropy perturbation (or any combination of the three). As time progresses, the shock moves to the right and passes through the fluctuation. A result of this interaction is that the strength of the perturbation may be amplified or damped. A wave of one given type may also generate travelling waves of the other two types. The important feature of these waves from a numerical point of view is that while entropy and vorticity fluctuations move with the gas, sound waves move at the sound speed relative to the speed of the gas. This means that, spatially, there are two length scales associated with features generated by the shock: one corresponds to acoustic responses which will move far from the shock, the other is associated with entropy and vorticity responses which remain near the shock. A detailed discussion of the shock-turbulence interaction problem can be found in Zang, Hussaini and Bushnell [9].

The shock is handled by fitting it as a moving boundary. In the streamwise (x) direction the computational domain consists of the continually expanding region between the moving shock and a fixed upstream boundary at which inflow boundary conditions can be applied. In the computational domain, we model the gas by the inviscid Euler gas-dynamics equations. Because the shock is fitted, it is appropriate to write the equations in non-conservation form in terms of the logarithm of the pressure (P), velocity (u,v) , entropy (s) and temperature (T)

$$\begin{aligned} P_t + uP_x + vP_y + \gamma(u_x + v_y) &= 0 \\ u_t + uu_x + vv_y + TP_x &= 0 \\ v_t + uv_x + vv_y + TP_y &= 0 \\ s_t + us_x + vs_y &= 0 \end{aligned} \tag{1}$$

For the ratio of specific heats, γ, a value of 1.4 is used.

Previous numerical simulations of shock-turbulence interactions have included finite difference calculations by Zang, Hussaini and Bushnell [9]. Pao and Salas [6] also used a finite difference method to compute the related problem of the generation of sound waves by a shock-vortex interaction. Single-domain spectral solutions to the shock-turbulence and shock-vortex interactions were reported by Salas, Zang and Hussaini [7], Zang, Kopriva and Hussaini [8] and Hussaini, Kopriva, Salas and Zang [3]. In each case, the use of the expanding computational domain meant that the effective resolution of the grid and the accuracy of the solution decreased as the calculation progressed in time.

3. Multidomain Strategy. Figure 1 illustrates the use of domain decomposition to allow for grid refinement near the shock when the moving shock is fitted. The region between the fixed inflow boundary and the moving shock boundary is divided into a number of strips (subdomains). The interface positions are constant in the vertical direction, but vary in time to allow them to move with the shock. In this way, constant grid resolution can be maintained near the shock where the short scale effects of the interaction occur.

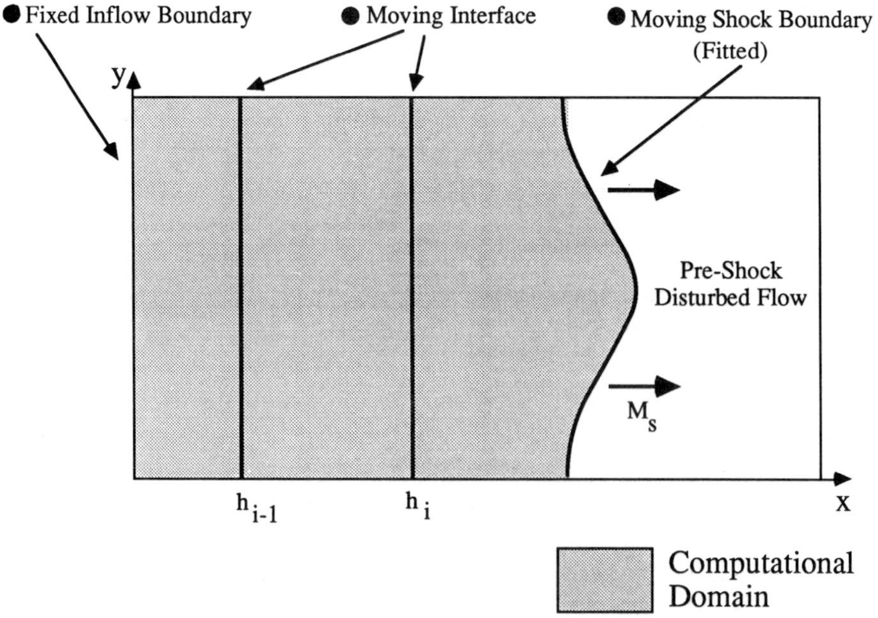

Figure 1. Diagram of the computational domain for the shock-turbulence interaction

Because the flow to the right of the shock is completely determined by a specified perturbation of a quiescent gas, we compute within a semi-infinite region between the shock and an arbitrarily placed upstream boundary $x = h_0$. For notational purposes, we will denote the shock position by $h_K(y,t)$. The extent of the region in the vertical direction is $-\infty < y < \infty$. We subdivide the region between the shock and the inflow boundary into K strips by placing interfaces at positions $x = h_i(y,t)$, $i = 1, 2, \ldots, K-1$. Each subdomain is then mapped onto a unit square. In the streamwise direction, we use the mapping

$$X = \frac{(x - h_{i-1}(y,t))}{(h_i(y,t) - h_{i-1}(y,t))} \tag{2}$$

The mapping in the vertical direction depends on whether the problem is periodic or non-periodic. The shock-plane wave interaction problem is periodic in the vertical direction. For that problem we use

$$Y = y/L \tag{3a}$$

where L is the vertical length scale. For the other problems, we use the mapping

$$Y = \frac{1}{2}\left(1 + \frac{y}{\sqrt{\alpha^2 + y^2}}\right) \tag{3b}$$

where α is a parameter which governs the clustering of the grid points near $y = 0$.

In the mapped coordinates on each subdomain, eq. 1 can be written as the system

$$Q_t + AQ_X + BQ_Y = 0 \tag{4}$$

where $Q = [P\ u\ v\ s]^T$. The coefficient matrices are

$$A = \begin{bmatrix} U & \gamma X_x & \gamma X_y & 0 \\ TX_x & U & 0 & 0 \\ TX_y & 0 & U & 0 \\ 0 & 0 & 0 & U \end{bmatrix} \quad \text{and} \quad B = \begin{bmatrix} V & \gamma Y_x & \gamma Y_y & 0 \\ TY_x & V & 0 & 0 \\ TY_y & 0 & V & 0 \\ 0 & 0 & 0 & V \end{bmatrix}$$

where the variables U and V represent the contravariant velocity components $U = X_t + uX_x + vX_y$ and $V = uY_x + vY_y$.

Within each subdomain, the solution, Q, is approximated by a grid function, Q_{ij}, at a finite number of collocation (grid) points (X_i, Y_j). In the X direction, the collocation points are the nodes of the Gauss-Lobatto-Chebyshev quadrature rule mapped onto [0,1]:

$$X_i = \frac{1}{2}(1 - \cos(i\pi/N)) \quad i = 0, 1, \ldots, N \tag{5}$$

In the vertical direction, the uniform grid

$$Y_j = j/M \quad j = 0, 1, \ldots, M \tag{6a}$$

is used for the plane wave interaction problem. For the others, the Gauss-Lobatto grid

$$Y_j = \frac{1}{2}(1 - \cos(j\pi/M)) \quad j = 0, 1, \ldots, M \tag{6b}$$

is used.

Each derivative of Q at the grid points is approximated by the corresponding derivative of the spectral interpolant which passes through the Q_{ij}'s. For problems which are periodic in the vertical direction, this interpolant is the Chebyshev-Fourier expansion

$$Q(X, Y, t) = \sum_{p=0, q=-M/2}^{N, M/2-1} \tilde{Q}_{pq}(t) T_p(2X - 1) e^{2\pi i q Y} \tag{7}$$

Problems which are periodic in neither direction require a Chebyshev-Chebyshev expansion

$$Q(X, Y, t) = \sum_{p=0, q=0}^{N, M} \tilde{Q}_{pq}(t) T_p(2X - 1) T_q(2Y - 1) \tag{8}$$

For details, see Canuto, Hussaini, Quarteroni and Zang [1].

Five types of boundary conditions are required. At the far right is the shock boundary which is fitted in the manner described by Pao and Salas [6]: An ordinary differential equation is derived for the motion of the shock and this is integrated along with the interior equations. At the inflow boundary on the left, a characteristic boundary condition is used. If the inflow boundary is subsonic, the velocity and entropy are fixed in time. The pressure is computed by integrating the compatibility equation

$$P_t = (U - aX_x)\left(\frac{\gamma}{a} u_X - P_X\right) \tag{9}$$

which is derived from the pressure and momentum equations and is written in terms of the sound speed, $a = \sqrt{\gamma T}$. If the boundary is supersonic, all variables are fixed in time.

The boundaries at infinity for the non-periodic problems are actually boundaries at large values of y. There, the velocities and the entropy are computed from the interior approximation. The pressure is determined by extrapolating along a characteristic projection normal to the boundary in the manner described by Gottlieb, Gunzburger and Turkel [2]. Periodic boundary conditions are handled trivially by the Fourier approximation.

At an interface, the approximation to the X derivative uses a weighted average of the derivatives from the left and right sides. We write

$$Q_t + A^L Q_X^L + A^R Q_X^R + BQ_Y = 0 \tag{10}$$

where Q_X^L, Q_X^R are the spectral derivative approximations from the left and the right. Since the problem is hyperbolic, it is necessary that information is propagated in the proper directions. Thus, a characteristic weighting is necessary and the left and the right coefficient matrices are written as

$$A^L = 1/2(A + |A|), \quad A^R = 1/2(A - |A|)$$

The matrix absolute value is defined as

$$|A| = Z|\Lambda|Z^{-1}$$

where Z is the matrix of the right eigenvectors of A and Λ is the diagonal matrix of eigenvalues. For the problem solved here, the matrix absolute value at an interface is quite simply

$$|A| = \frac{1}{2}\begin{bmatrix} (\lambda^+ + \lambda^-) & \gamma/a(\lambda^+ - \lambda^-) & 0 & 0 \\ a/\gamma(\lambda^+ - \lambda^-) & (\lambda^+ + \lambda^-) & 0 & 0 \\ 0 & 0 & 2\lambda^0 & 0 \\ 0 & 0 & 0 & 2\lambda^0 \end{bmatrix} \tag{11}$$

where $\lambda^\pm = U \pm aX_x$ and $\lambda^0 = U$ are the eigenvalues of A.

3. Applications.

3.1 Shock-Plane Wave Interaction. The first application that we consider is the interaction of a Mach 8 supersonic shock with a plane pressure wave. Ahead of the shock, the pressure perturbation is given by

$$p' = \beta e^{i(\mathbf{k}\cdot\mathbf{x} - \omega t)} \tag{12}$$

where the vector $\mathbf{k} = k(\cos(\theta), \sin(\theta))$ is the wavenumber vector, ω is the frequency and β is the amplitude. The angle of incidence, θ, is measured normal to the shock. See Zang et. al [9] for a detailed description of this problem.

To avoid overshoots associated with an abrupt start of the wave, the amplitude, β, is multiplied by the factor

$$s(t) = \begin{cases} 3(t/t_s)^2 - 2(t/t_s)^3 & 0 \le t \le t_s \\ 1 & t \ge t_s \end{cases} \tag{13}$$

The startup time, t_s, is chosen as 1/2 the time it takes the shock to encounter one full wavelength of the incident wave.

Linear theory predicts that the pressure wave will be amplified and diffracted by the shock. Also, a plane vorticity and entropy wave will be generated. We can define the pressure transmission coefficient as the ratio of the refracted to incident pressure wave amplitudes. The vorticity or entropy transmission coefficients are the ratio of the generated wave amplitude to the incident pressure amplitude.

Fig. 2 shows the pressure and vorticity transmission coefficients at $t = 0.2$ for a 10% ($\beta = 0.1$) amplitude pressure wave incident at 30° to the shock. Three subdomains were used, each with 16 horizontal and 8 vertical grid points. At the time indicated, the interfaces were at $x = 0.7$ and $x = 1.5$. The transmission coefficients were computed at each grid line in x from the first coefficient of the Fourier transform of the solution in the vertical direction. Compared with the computed solutions are the predictions of the linear theory.

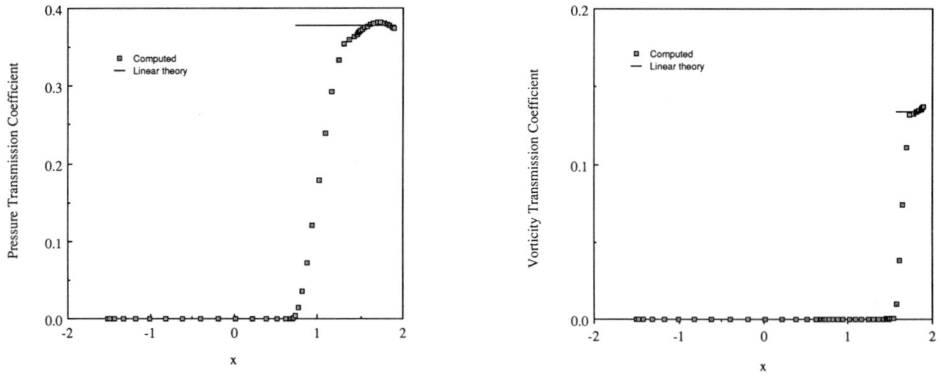

Figure 2. Dependence at $t = 0.21$ of acoustic (left) and vorticity (right) responses to an acoustic wave incident at 30° to a Mach 8 shock.

The computed responses clearly occur on two different length scales, the extent of the vorticity behind the shock being roughly one fourth that of the pressure. Nevertheless, it was possible to resolve the vorticity wave with eight grid points in the horizontal direction. This high resolution is necessary because the vorticity is computed from derivatives of the flow variables. Single domain calculations show noisy transmission coefficient profiles for the vorticity [8,9].

3.2 Shock-Hot Spot Interaction. A physically more complex problem is the interaction of a shock with a temperature spot. In this case, the temperature ahead of the shock is prescribed by

(a)

(b)

(c)

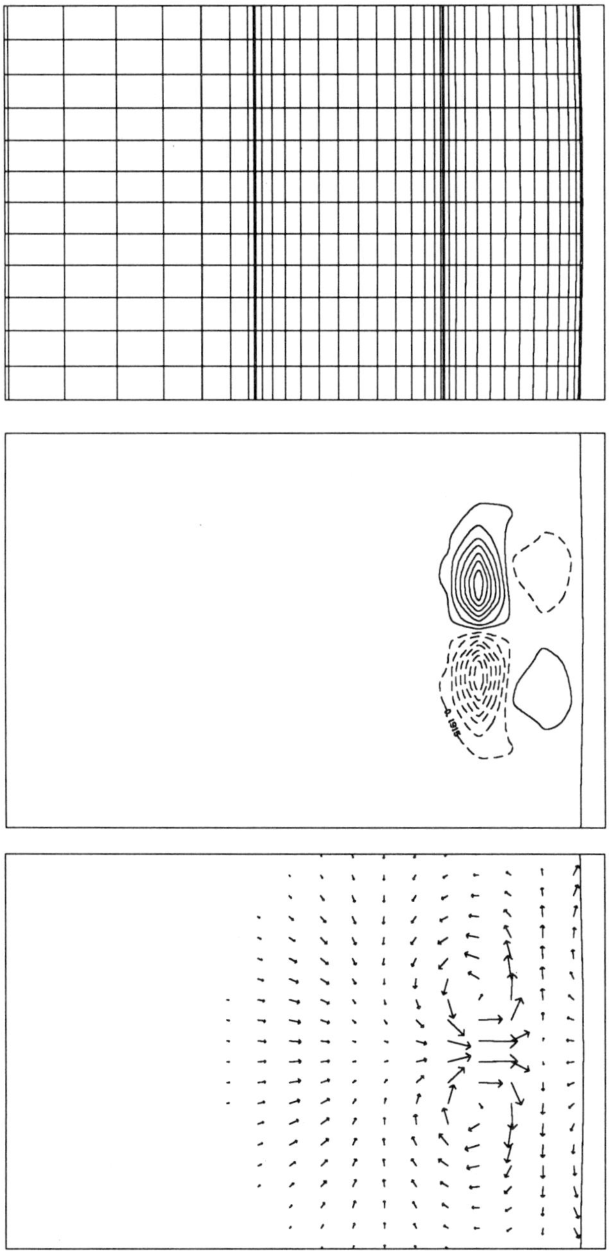

Figure 3. Grid (a), Vorticity contours (b) and velocity vectors (c) for the 25% hot spot interaction with a Mach 2 shock at time $t = 0.5$.

$$T(x, y) = 1 + \varepsilon e^{-\left[(x-x_o)^2+y^2\right]/2\sigma^2} \tag{14}$$

where ε is the maximum fractional temperature perturbation, $(x_o, 0)$ is the center of the spot and σ is its width. For this simulation, we chose $\varepsilon = 0.25$, $x_o = 0.5$, and $\sigma = 0.1$.

Figure 3 shows the grid, vorticity contours and velocity vectors at time, $t = 0.5$. Again, three subdomains were used, this time with 15 horizontal and 40 vertical grid points in each. What is actually shown are the results on a portion of the grid near the shock and hot spot covering the physical space rectangle $[-.25,\ 1.25] \times [-.5,\ .5]$. Notice the strong counter-rotating vortices which have been generated by the shock. A weaker set also appears to have been generated even closer to the shock.

3.3 Shock-Vortex Street Interaction. In this case, the initial perturbation ahead of the shock is a vortex field. The stream function for this field is

$$y = \frac{\kappa}{2\pi} \log\left[\cosh\left(\frac{2\pi}{c}\sqrt{r^2 + (y \pm b/2)^2} - \cos\left(\frac{2\pi}{c}(x \pm c/2)\right)\right)\right] \tag{15}$$

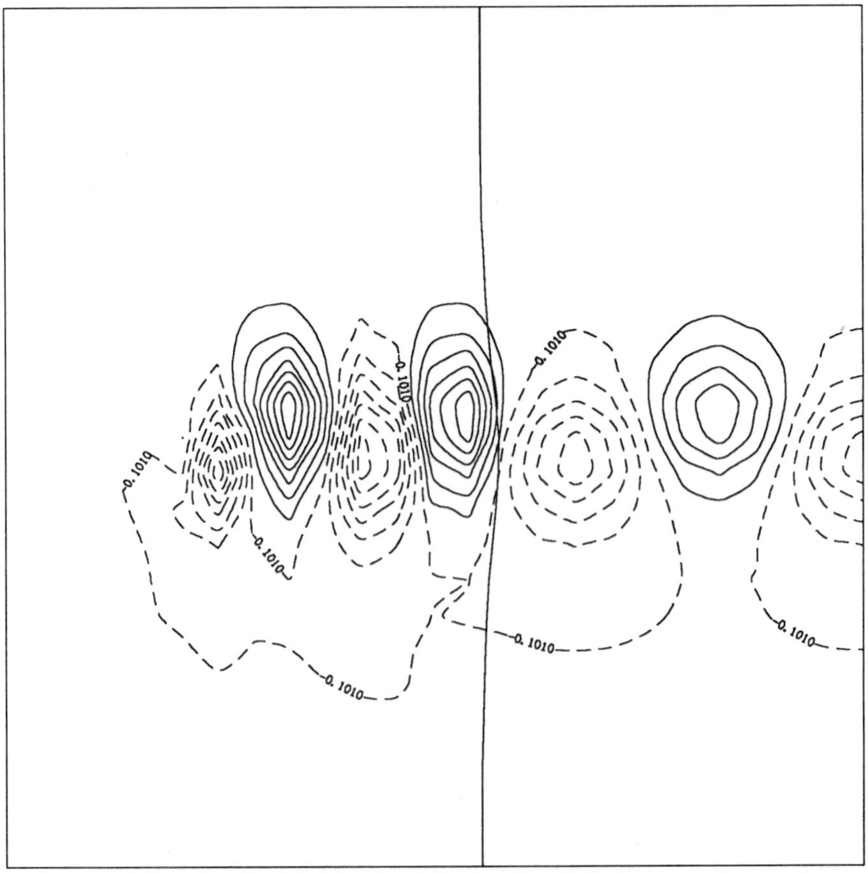

Figure 4. Vorticity contours at $t = 0.36$ for a Karman vortex street after interaction with a Mach 1.3 Shock.

The circulation, core radius, shock Mach number and vortex separation parameters used were $\kappa = 0.186$, $r = 0.1$, $M_s = 1.3$, $c = 0.33$ and $b = 0.048$ to correspond to the single domain calculation of Salas, Zang and Hussaini [7].

Figure 4 shows the vorticity contours of the shock-vortex street interaction at time, $t = 0.36$, for a three subdomain calculation. Within each subdomain a 15×40 grid was used. Notice the longitudinal compression and lateral elongation of the vortex field behind the shock. The results are also quite smooth. This is to be contrasted with the single domain calculations of [7] which required strong filtering every 160 time steps and still produced solutions which were not smooth. No artificial smoothing was required for this multidomain calculation.

4. Summary. The solution of three simple shock-turbulence interaction models has been described. Each problem has the characteristic that the length scales associated with the pressure and with the entropy/vorticity are substantially different, thus making it difficult to resolve them efficiently in the horizontal direction with a single Chebyshev grid. By using domain decomposition it was possible to resolve these two length scales.

REFERENCES

(1) C. Canuto, M. Y. Hussaini, A. Quarteroni and T. A. Zang, *Spectral Methods in Fluid Dynamics*, Springer-Verlag, New York, 1987.

(2) D. Gottlieb, M. Gunzburger and E. Turkel, *On Numerical Boundary Treatment of Hyperbolic Systems for Finite Difference and Finite Element Methods*, SIAM J. Numer. Anal., **19** (1982), pp. 671–682.

(3) M. Y. Hussaini, D. A. Kopriva, M. D. Salas,and T. A. Zang, *Spectral Methods for the Euler Equations: Part II – Chebyshev Methods and Shock Fitting*, AIAA Journal, **23** (1985) pp. 234–240.

(4) D. A. Kopriva, *A Spectral Multidomain Method for the Solution of Hyperbolic Systems*, Applied Numerical Math., **2** (1986), pp. 221–241.

(5) D. A. Kopriva, *Solution of Hyperbolic Equations on Complicated Domains with Patched and Overset Chebyshev Grids*, FSU-SCRI-87-35 preprint. Submitted to SIAM J. Sci. Stat. Comp.

(6) S. P. Pao and M. D. Salas, *A Numerical Study of Two-Dimensional Shock Vortex Interaction*, AIAA Paper 81-1205, 1981.

(7) M. D. Salas, T. A. Zang and M. Y. Hussaini, *Shock Fitted Euler Solutions to Shock-Vortex Interactions*, Proceedings of the 8th International Conference on Numerical Methods in Fluid Dynamics, E. Krause, ed., Springer-Verlag, New York, 1982, pp. 461–467.

(8) T. A. Zang, D. A. Kopriva and M. Y. Hussaini, *Pseudospectral Calculation of Shock Turbulence Interactions*, Proceedings of the 3rd International Conference on Numerical Methods in Laminar and Turbulent Flow, C. Taylor, ed., Pineridge Press, Swansea, Wales, 1983, pp. 210–220.

(9) T. A. Zang, M. Y. Hussaini, and D. M. Bushnell, *Numerical Computations of Turbulence Amplification in Shock Wave Interactions*, AIAA Journal, **22** (1984), pp. 13–21.

A Domain Decomposition Approach for the Computation of Incompressible Flow by the Pseudospectral Matrix Element Method and Its Parallel Implementation

Hwar C. Ku*
Richard S. Hirsh*
Thomas D. Taylor*
Allan P. Rosenberg*†

Abstract

A recently developed pseudospectral matrix element (PSME) method [1], which extended the formulation of the pseudospectral matrix (PSM) method to a multi-element scheme, has been applied by the authors to the solution of the incompressible, primitive variable, Navier-Stokes equations for complex geometries in two and three dimensions. This has been accomplished by a new implementation of the Schwarz alternating procedure (SAP), which allows computation of the flow field in such geometrical configurations. The SAP divides the computational domain into a number of overlapping subdomains, of simpler geometry with patched grid points, in which the solution is more easily obtained. With an iterative procedure between subdomains, the complete solution is found.

A new domain decomposition (DD) procedure follows the original PSM method and employs only the continuity equation as the subdomain pressure boundary condition (including the overlapped interfaces) that permits one to utilize the eigenfunction expansion of the resulting pressure operator to reduce the original multi-dimensional problem to a series of one-dimensional ones. Appropriate downstream boundary conditions have been developed for the PSME method in the domain decomposed solutions which makes this method very attractive for computing inflow-outflow problems on a truncated domain of fewer nodes.

The PSME method on the decomposed domains has been efficiently implemented in parallel without making any processor sit idle. This is accomplished by developing fully parallelizable code for each overlapping subdomain and solving for each in turn.

Computational results for both two- and three-dimensional flow over a backward facing step at different Reynolds numbers are presented in this paper. No pronounced three-dimensional effects are observed except in the boundary layer along in the spanwise direction.

* The Johns Hopkins University Applied Physics Laboratory, Laurel, Maryland 20707
† also at Sachs/Freeman Associates

1 Introduction

The solution of boundary-value problems for complex geometries has been successfully implemented by exchanging data among the different zones (or subdomains), i.e. solving the problem on each subdomain separately and then updating the boundary values on the overlapped interface. The data exchange can been done in a straight forward manner, this is what is usually called SAP [2]. Fuchs [3] pointed out that one major difficulty with the SAP when solving the incompressible Navier-Stokes equations in streamfunction-vorticity form for a rigid body confined in a flow region is that the value of streamfunction is unknown on the surface of the rigid body. In primitive variables the question arises of how the boundary conditions for the pressure are handled on the interatively solved subdomains even with the prescribed velocity components. Instead of directly imposing conditions on both the pressure and its first derivatives at the edge of each overlapped region [4], the consistent use of the continuity equation to generate the subdomain pressure boundary conditions (including the overlapped interfaces) should be preferable. This is mainly because the former definitely creates non divergence-free velocity fields along the imaginary overlapped interfaces while the latter still guarantees differential mass conservation there.

Flow over a backward step is always presented as the standard test problem for different numerical schemes. Some investigators [5,6] assumed that the fully developed velocity profile is established right at the step and some considered the upstream effect before the step. Recently Armaly [7] *et al.* have performed a careful experimental study of such a flow for a wide range of Reynolds number. Although great care was taken to ensure that the initial inlet flow was two-dimensional, in addition to the well-known entrance effects they also found that as the Reynolds number increased above 450 three-dimensional effects became gradually apparent downstream from the step. These results motivated us to do numerical experiments on three-dimensional flow over a backward step.

For the purpose of high-speed computation, the issue of how to map a numerical algorithm for the solution of each subdomain onto the specific architecture of parallel machines so as to achieve highly parallel performance remains an open question. There are roughly two approaches to this problem: (1) distribute processors proportional to the amount of cpu time required per subdomain or (2) allocate all the processors to a single subdomain and do different subdomains in sequence. The first one seems to be far reached due to the requirement of a complier smart enough to anticipate the cpu time difference among subdomains, otherwise some processors with their jobs done earlier have to be sit idle until other processors finish their jobs. On the contrary, there is no data communication constraint for the second one and it becomes more attractive if the computation of each subdomain can be highly parallelizeable. In this paper, authors would like to use this concept to apply PSME method to the solution of incompressible flows in complex geometries. Although originally designed for a shared memory parallel processor with a few computational elements, our code is written entirely in terms of the proposed Fortran 8x standard and can be run on any machine (parallel or not) for which a complier implementing the standard exists.

2 Calculation of derivatives by PSME method

For simplicity, the spatial domain is divided into NE elements each of which has $N+1$ collocations, $x_j = 1/2[(b^e - a^e)\cos\pi(j-1)/N + b^e + a^e]$ $(1 \le j \le N+1; 1 \le e \le NE)$. The derivatives of a function $f(x)$ in the interior of element e can be discretized as

$$f_x^e(x_i) = \frac{1}{L^e}\sum_{m=1}^{N+1} \hat{GX}_{i,m}^{(1)} f_m^e \tag{1a}$$

$$f_{xx}^e(x_i) = \frac{1}{(L^e)^2}\sum_{m=1}^{N+1} \hat{GX}_{i,m}^{(2)} f_m^e = \frac{1}{L^e}\sum_{m=1}^{N+1} \hat{GX}_{i,m}^{(1)} (f_x^e)_m \tag{1b}$$

here L^e is the length of eth element defined on the interval $[a^e, b^e]$; $f_m^e \equiv f^e(x_m)$ and $\hat{\mathbf{GX}}^{(1)}$, $\hat{\mathbf{GX}}^{(2)}$ are the invariant derivative matrices based on the domain [0,1].

The interfacial derivatives at the inter-element points, are approximated by weighting of the derivatives from each side, according to the relations

$$f_x|_{interface} = \alpha f_x^e + \beta f_x^{e+1} \tag{2a}$$

$$f_{xx}|_{interface} = \alpha f_{xx}^e + \beta f_{xx}^{e+1}, \quad \alpha + \beta = 1, \quad 1 \le e \le NE - 1 \tag{2b}$$

With the choice of $\alpha = L^e/(L^e + L^{e+1})$, the fraction of total length of two adjacent elements, Eqs. (2), in view of Eqs. (1), now become

$$f_x|_{interface} = \frac{1}{L^e + L^{e+1}} \sum_{m=1}^{N+1} (\hat{GX}^{(1)}_{N+1,m} f_m^e + \hat{GX}^{(1)}_{1,m} f_m^{e+1}) \tag{3a}$$

$$f_{xx}|_{interface} = \frac{1}{L^e + L^{e+1}} \sum_{m=1}^{N+1} (\hat{GX}^{(1)}_{N+1,m} (f_x^e)_m + \hat{GX}^{(1)}_{1,m} (f_x^{e+1})_m) \tag{3b}$$

In Eq. (3a), c^0 continuity is explicitly assumed whenever the calculation of interface values of the first derivative is required. However, c^0 continuity is only implicitly assumed for the second derivative calculation. We can expand Eq. (3b) using Eq. (1b) and impose c^1 continuity to get

$$\begin{aligned} f_{xx}|_{interface} = & \frac{1}{L^e + L^{e+1}} [\sum_{m=1}^{N+1} (BX^*_{N+1,m} f_m^e + BX^{**}_{1,m} f_m^{e+1}) + \\ & (\hat{GX}^{(1)}_{N+1,N+1} + \hat{GX}^{(1)}_{1,1}) f_x|_{interface}] \end{aligned} \tag{4}$$

where it has been implicitly assumed that $f_x^e(x_{N+1}) = f_x^{e+1}(x_1) = f_x|_{interface}$. As will be shown shortly, the explicit value of $f_x|_{interface}$ is never needed. This requirement is met by the finite element method employing variational or Galerkin procedures wisth trial functions that are c^0 across element boundaries, i.e., flux (first derivative) continuity is intrinsically satisfied through integration by parts. This is exactly the same for the PSME method because the second term in the bracket of Eq. (4) is automatically cancelled out since, $\hat{GX}^{(1)}_{1,1} = (2N^2 + 1)/3 = -\hat{GX}^{(1)}_{N+1,N+1}$. The elements of the modified matrices $\mathbf{BX}^*, \mathbf{BX}^{**}$ are

$$BX^*_{N+1,m} = \frac{1}{L^e} \sum_{n=1}^{N} \hat{GX}^{(1)}_{N+1,n} \hat{GX}^{(1)}_{n,m} \tag{5a}$$

$$BX^{**}_{1,m} = \frac{1}{L^{e+1}} \sum_{n=2}^{N+1} \hat{GX}^{(1)}_{1,n} \hat{GX}^{(1)}_{n,m} \tag{5b}$$

In order to set up a simple and efficient matrix operation for derivatives by the PSME method, an operator with global-type structure which combines each local element derivative can be constructed. Therefore, Eqs. (1a) and (3a) representing the first derivatives can be cast into the form

$$\mathbf{f}' = \mathbf{G}^{(1)} \mathbf{f} \tag{6}$$

where $\mathbf{G}^{(1)}$ has the diagonal form

$$\mathbf{G}^{(1)} = \begin{bmatrix} \boxed{A^{(1)}} & & & & & \\ & \boxed{A^{(2)}} & & & & \\ & & \boxed{A^{(3)}} & & & \\ & & & \ddots & & \\ & & & & \boxed{A^{(NE-1)}} & \\ & & & & & \boxed{A^{(NE)}} \end{bmatrix} \tag{7}$$

The hatched area in Eq. (7) arises from the Eq. (3a) at the element-element interface while the non-overlapped area is simply Eq. (1a) in the interior of each element. The blocks $\mathbf{A}^{(n)}, n = 1, ..., NE$ are of size $(N+1)^2$ with one point overlaped region at the corners, and the overall size of matrix $\mathbf{G}^{(1)}$ is $(NX+1)^2$.

In an analogous manner, second derivative of Eqs. (1b) and (3b) can be plugged into the form

$$\mathbf{f}'' = \mathbf{G}^{(2)} \mathbf{f} \tag{8}$$

where $\mathbf{G}^{(2)}$ has the same diagonal form as $\mathbf{G}^{(1)}$.

The proposed PSME method has been tested on a standard one-dimensional convection-diffusion problem [1] as well as stationary shock wave of Burger's equation [8]. Spectral accuracy is observed for both cases.

3   Governing equations

The three-dimensional flow over a backward step (shown in Fig. 1) with an expansion ratio 2 in the vertical direction and an aspect ratio of 8 in the spanwise direction can be represented in a primitive variable formulation. In cartesian coordinates, the time-dependent Navier-Stokes equations in dimensionless form can be written as

$$\frac{\partial u_i}{\partial t} + u_j \frac{\partial u_i}{\partial x_j} = -\frac{\partial p}{\partial x_i} + \frac{1}{\text{Re}} \frac{\partial^2 u_i}{\partial x_j^2} \tag{9a}$$

$$\frac{\partial u_i}{\partial x_i} = 0 \tag{9b}$$

Here u, v, w are the velocity components in the horizontal, vertical and spanwise direction, respectively, and Re is the Reynolds number, $\rho U S / \mu$, where ρ is the density, U is the maxium velocity of inflow, S is the step height, and μ is the viscosity.

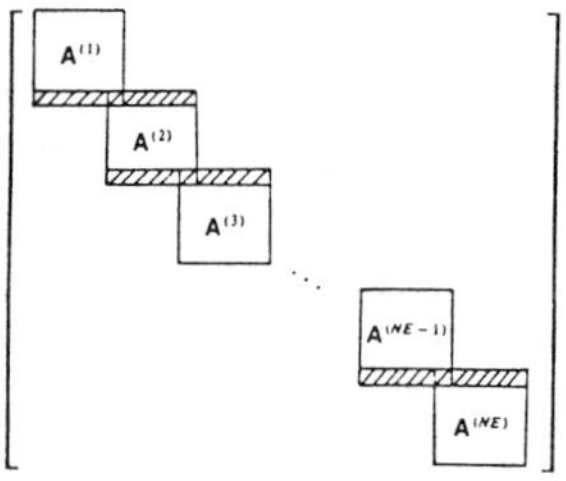

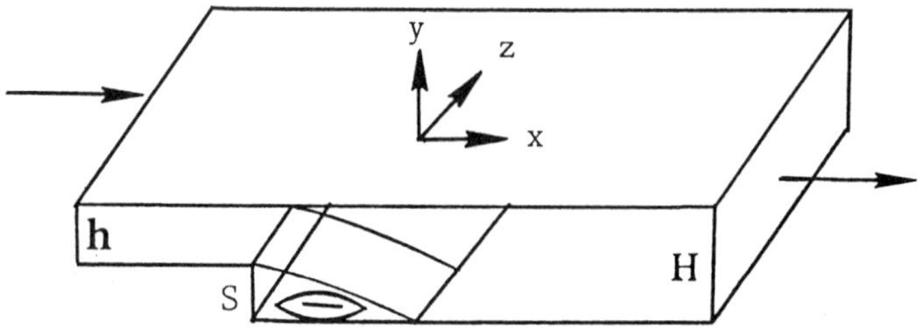

Figure 1: Three-dimensional configuration for flow over a backward step

Eqs. (9) are only solved for half of the domain due to the experimental results [7] of symmetry about the xy plane at $z = 0$, i.e., u, v, p are symmetric and w is anti-symmetric with respect to the central plane. Initially fully developed profiles are used for both the upstream and downstream regions. At the time $t > 0$, the boundary conditions are given by

$$v = w = 0, u = 4(y - y^2) \quad \text{at } x = -2 \tag{10a}$$

$$\frac{\partial^2 u}{\partial x^2} = \frac{\partial^2 v}{\partial x^2} = \frac{\partial^2 w}{\partial x^2} = 0 \quad \text{at } x = l \tag{10b}$$

$$u = v = w = 0 \quad \text{at } y = 1 \tag{10c}$$

$$u = v = w = 0 \quad \text{at } y = 0 \text{ for } x \leq 0 \tag{10d}$$

$$u = v = w = 0 \quad \text{at } y = -1 \text{ for } x \geq 0 \tag{10e}$$

$$u = v = w = 0 \quad \text{at } z = 4 \tag{10f}$$

$$\frac{\partial u}{\partial z} = \frac{\partial v}{\partial z} = \frac{\partial p}{\partial z} = 0, w = 0, \quad \text{at } z = 0 \tag{10g}$$

Appropriate outflow boundary conditions should have little influence on the upstream flow development. This is usually the case when the upstream continuously generates the disturbance which will be propagated into the downstream.

Note that different downstream boundary conditions are possible, for instance in the two-dimensional case

1. u, v both prescribed
2. $\frac{\partial u}{\partial x}=0,\ v=0$
3. $\frac{\partial v}{\partial x}=0,\ \frac{\partial^2 u}{\partial x^2}=0$
4. $\frac{\partial^2 u}{\partial x^2}=0,\ \frac{\partial^2 v}{\partial x^2}=0$

Conditions (1) & (2) seem too restrictive to be applied on the truncated domain, while conditions (3) & (4) yield the least effect upon the upstream flow development.

4 Primitive variable formulation

The method used for solving the Navier-Stokes equation is Chorin's [9] splitting technique. According to this scheme, the equations of motion, in tensor form, are

$$\frac{\partial u_i}{\partial t} + \frac{\partial p}{\partial x_i} = F_i \tag{11}$$

where $F_i = -u_j \partial u_i / \partial x_j + 1/\mathrm{Re}\,\partial^2 u_i / \partial x_j^2$.

The first step is to split the velocity into a sum of predicted and corrected value. The predicted velocity is determined by time integration of the momentum equations without the pressure term

$$\bar{u}_i^{n+1} = u_i^n + \Delta t F_i^n \tag{12}$$

The second step is developing the pressure and corrected velocity field that satisfies the continuity equation by using the following relationships

$$u_i^{n+1} = \bar{u}_i^{n+1} - \Delta t \frac{\partial p}{\partial x_i} \tag{13a}$$

$$\frac{\partial u_i^{n+1}}{\partial x_i} = 0 \tag{13b}$$

By taking the divergence operator to Eq. (13a) in order to satisfy the continuity equation, Eq. (13b), throughout the whole domain, and incorporating of prescribed velocity boundary conditions, it generates the pressure Poisson equations in the interior as well as the supplemental pressure equations at the boundaries. Neumann type boundary conditions rather than those derived from the continuity equation are used for the pressure in the z direction because the inflow is discontinuous in the z direction. This leads to non divergence-free velocity only at the solid wall in the z direction, while the continuity equation is still satisfied for the rest of the domain. This problem does not appear in the two dimensional case.

Following the eigenfunction expansion technique (see detailed procedure in Ref. [10], [12]), the original three-dimensional pressure equation is reduced to a simple one-dimensional matrix operator. Therefore, the overall solution of pressure can be obtained through the linear superposition of each eigenvalue and its associated eigenvectors.

5 Schwarz alternating procedure

The SAP for iterative solution of the incompressible Navier-Stokes equations in primitive variable form for the three-dimensional flow over a backward step is summarized as follows (see Figure 2):

1. First assume $u^{n+1}, v^{n+1}, w^{n+1}$ on □ABMN. Usually u^n, v^n, w^n would be a good initial guess.
2. Solve domain II ∪ III employing boundary conditions derived from the continuity of velocity field on □ABMN, i.e., $\partial u/\partial x = -\partial v/\partial y - \partial w/\partial z$ where the pressure Poisson equation is solved by an eigenfunction expansion technique.
3. With the solutions of $u^{n+1}, v^{n+1}, w^{n+1}$ on □NBEX from step (2), solve domain I ∪ II employing the same type boundary conditions as above for the pressure on □NBEX, i.e., $\partial v/\partial y = -\partial u/\partial x - \partial w/\partial z$, to update $u^{n+1}, v^{n+1}, w^{n+1}$ on □ABMN.
4. Repeat steps (2) & (3) until the convergence tolerance has been met for $u^{n+1}, v^{n+1}, w^{n+1}$ along □ABMN, □NBEX.

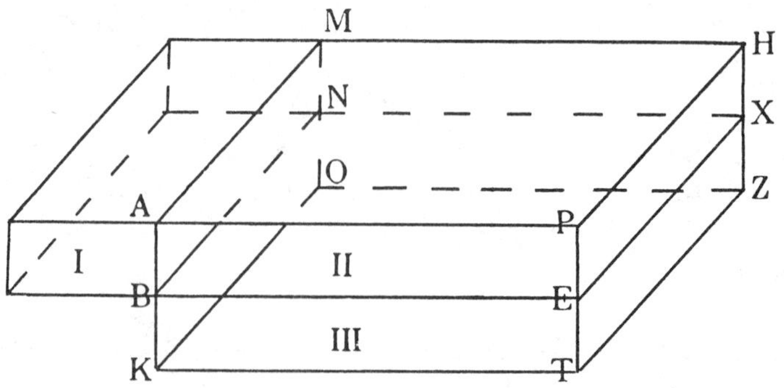

Figure 2: Three-dimensional configuration of domain decomposition

The novel features of this iteration scheme are: (i) for each iteration the divergence free condition is exactly satisfied everywhere except the boundary points in the direction where inflow carries the singular information; (ii) for every time step only a few iterations (usualy

$2 \sim 4$) are required to reach the converged solution for velocity components u, v, w; (iii) consistent mass conservation holds along $\overline{BN}$ despite a singular value for the vorticity and (iv) within each subdomain the eigenfunction expansion technique is still available to decompose the original three-dimensional problem into a simple one dimensional matrix operator.

6 Parallel implementation of N-S Equations

The method used to compute the solution of the Navier-Stokes equations in parallel performance is very important for the technique of domain decomposition with SAP. The basic idea is that in alternately performing steps (2) and (3) of section 5 all processors are used to update whichever subdomain II ∪ III or I ∪ II is current. Updating such a geometrically simple region can be performed efficiently in parallel [1] as we now sketch.

In our version of the time splitting approach for a single domain three steps account for most of the run time. These are computing the partial derivatives involved in updating the predicted velocity, transforming back and forth from physical to eigenfunctional space, and solving the sequence of reduced one-dimensional pressure equations in eigenfunction space.

The first two steps can be reduced to *dotproducts* and *matrix multiplications* between subsets of multimensional arrays - basic linear algebra operations defined in the proposed Fortran 8x standard and efficiently implementable on wide classes of vector, parallel and vector-parallel machines.

The key point of the last step is that instead of performing the whole backward substitution for each reduced pressure equation in turn, we perform one step of the backward substitution for all of the reduced pressure equations before going to the next step. Since the eigenfunction expansion has made each reduced equation independent, it is easy to parallelize each partial backward substitution.

Table I & II show some timings for the implementation of our program for the solution of both two- and three-dimensional flow over a backward step at Reynolds number 375 (four iterations of SAP) with $NX = 61$ (10 elements), $NY = 37$ (6 elements) at downstream, while NX=13 (2 elements), NY=19 (3 elements) atupstream and $NZ = 49$ (single element), running on an Alliant FX/8-series (eight processors) computer.

Table I. Cpu time in seconds per time step (2D)

number of processors	1	2	3	4	8
	4.07	2.10	1.48	1.18	0.74
speedup*	1	1.94	2.75	3.45	5.52

*: compared to 1 processor

Table II. Cpu time in seconds per time step (3D)

number of processors	1	2	3	4	8
	235.5	127.7	90.1	73.5	43.5
speedup*	1	1.84	2.61	3.20	5.42

*: compared to 1 processor

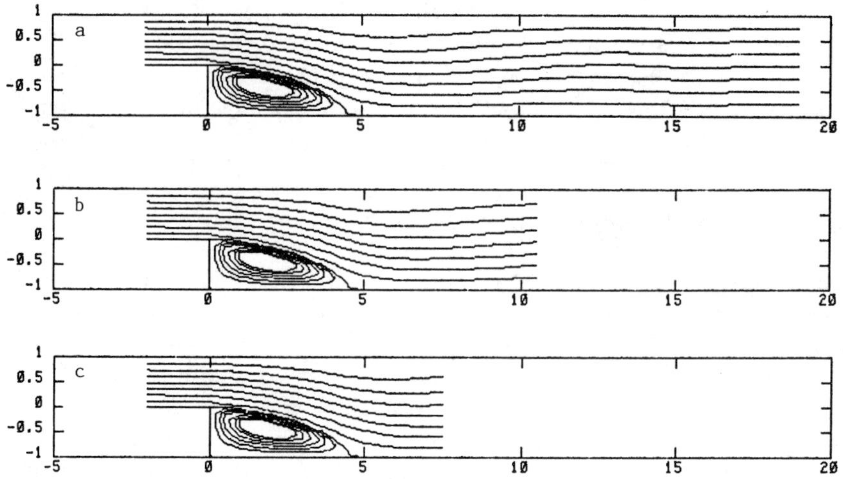

Fig. 3. Flow over a step, Re = 150, with downstream conditions at x/S = : a) 19.0, b) 10.5, c) 7.5

Note that using eight processors speeds the computation by a factor of 5.5, approximately 70% of the maximum possible value. It is worth emphasizing that our program is written entirely in terms of the proposed Fortran 8x standard and can be run on any machine for which a compiler implementing the standard exists.

7 Results and discussion

This section presents the results obtained by applying the present scheme, domain decomposition with SAP, to the two- and three-dimensional flow over a backward step. First, we will examine the effect of our downstream boundary conditions, the second derivative of velocity components vanishing in the streamwise direction, upon upstream flow development. Fig. 3 shows that the reattachment length at Re = 150 with downstream position X/S =19 (8 elements), 10.5 (5 elements) and 7.5 (4 elements) agrees very well with those found by Dirichlet boundary condition applied at infinity by the same method. Even the minimum streamfunction, i.e., ψ_{min}= -0.04421, -0.04434, -0.04425 for X/S = 19, 10.5 and 7.5, respectively, exhibits the same accuracy when compared to [11], ψ_{min}= -0.0436 by using 33*297 grid points. Fig. 4 provides streamline plots for the range of Reynolds number between Re =75 and Re = 375. As expected, with increasing Reynolds number, the downstream reattachment length of flow development increases up to Re = 375.

As Armaly *et al.* have pointed out, the deviation of the two-dimensional flow calculations from experimental results is due to three-dimensionality of the experimental flow observed above some certain Reynolds number. From the numerical results of three-dimensional flow, we find that the flow remains two-dimensional except in the boundary layers for Reynolds number, Re = 225, 300, 375, as is demonstrated by the plot of the spanwise location of the reattachment line shown in Fig. 5.

8 Conclusions

Domain decomposition with SAP has been used to solve the three-dimensional incompressible Navier-Stokes equations in primitive variable form in order to simulate flow over a

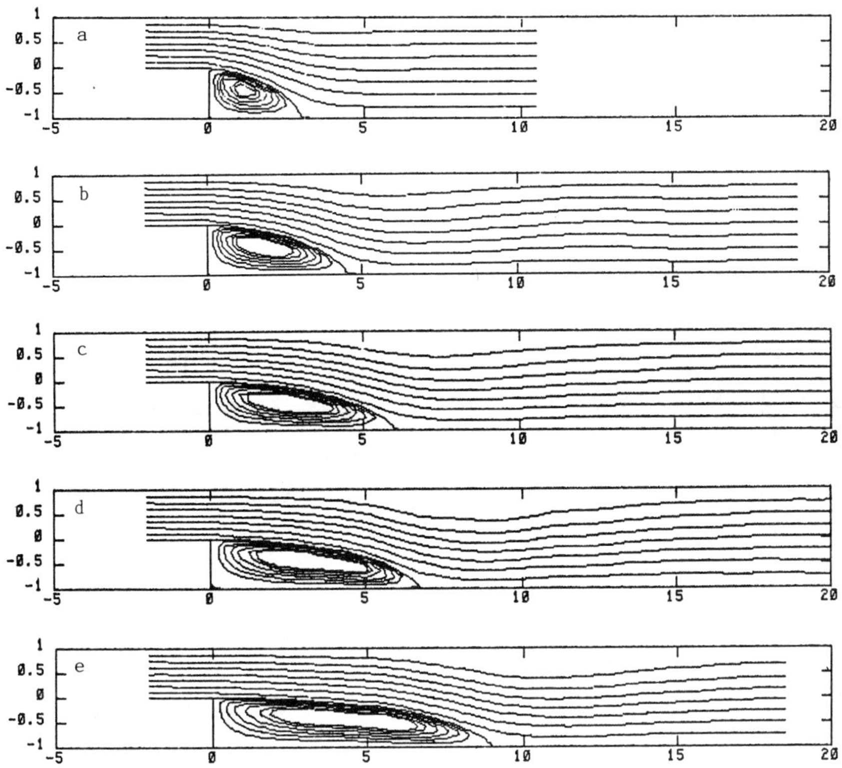

Fig. 4. Streamlines for flow over a step with Re = : a) 75, b) 150, c) 225, d) 300, e) 375

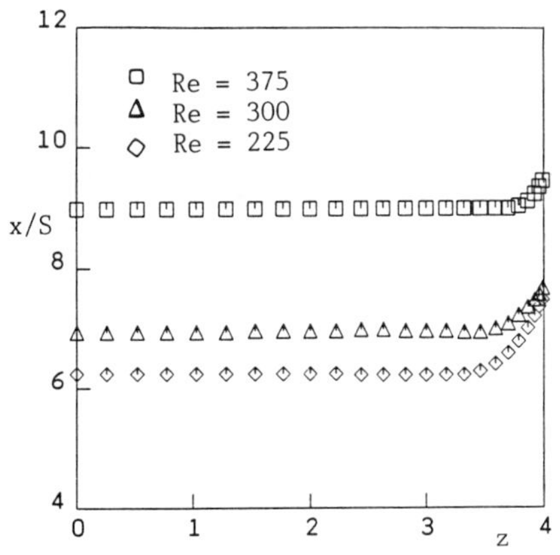

Fig. 5. Spanwise location of reattachment line with different Reynolds numbers

backward step. In the solution approach, the complex geometry is divided into a few subdomains, each of simple geometry, where the pressure solution can be easily obtained. With the continuity equation as the boundary conditions at the overlapped interfaces, an eigenfunction expansion technique can be applied to each subdomain to give the direct and fast solution of the three-dimensional pressure Poisson equations in one-dimensional form so that the parallel implementation of each subdomain could be easily achieved.

Numerical results computed by the PSME method are in good agreement with the two-dimensional experimental results. For Reynolds number up to 450, we found three-dimensional effects confined to the boundary layer region in the spanwise direction.

Acknowledgement

The authors would like to thank Mr. John Dutton of Alliant Computer Systems Corporation for his assistance in getting timings on the 4 and 8 processor machines. This work was partially supported by the Office of Naval Research under the Contract Number N00039 - 87 - C - 5301.

References

[1] H. C. Ku, R. S. Hirsh, T. D. Taylor and A. P. Rosenberg, submitted to *J. Comp. Phys.*

[2] L. W. Ehrlich, *SIAM J. Sci. Sta. Comput.* **7**, 989 (1986).

[3] L. Fuchs, in *Proceedings of the Seventh GAMM Conference on Numerical Methods in Fluid Mechanics*, edited by M. Deville (Vieweg and Sohn, Louvain-la-Neuve, 1987).

[4] Y. Morchoisne, *AIAA* paper No 81-0109 (1981).

[5] J. Kim and P. Moin, *J. Comput. Phys.* **59** 308 (1985).

[6] S. E. Rogers J. L. Chang and D. Kwak, *J. Comput. Phys.* **73** 364 (1987).

[7] B. F. Armaly, F. Durst, J. C. F. Pereira and B. Schönung, *J. Fluid Mech.* **127** 473 (1983).

[8] H. C. Ku, R. S. Hirsh and T. D. Taylor, in *Proceedings of the Seventh GAMM Conference on Numerical Methods in Fluid Mechanics*, edited by M. Deville (Vieweg and Sohn, Louvain-la-Neuve, 1987).

[9] A. J. Chorin, *Math. Comp.* **22**, 745 (1968).

[10] H. C. Ku, R. S. Hirsh and T. D. Taylor, *J. Comput. Phys.* **70**, 439 (1987).

[11] Y. D. Schkalle, F. Thiele, *Notes on Numerical Fluid Mech.* edited by K. Morgan, J. Periaux and F. Thomasset (Vieweg ans Sohn), 372 (1984).

[12] H. C. Ku, R. S. Hirsh, T. D. Taylor and A. P. Rosenberg, in preparation.

A Spectral Multi-Domain Technique Applied to High-Speed Chemically Reacting Flows

Michele G. Macaraeg*
Craig L. Streett†
M. Yousuff Hussaini††

Abstract. The first applications of a spectral multi-domain method for viscous compressible flow are presented. The method imposes a global flux balance condition at the interface so that high-order continuity of the solution is preserved. The global flux balance is imposed in terms of a spectral integral of the discrete equations across adjoining domains. Since the discretized equations interior to each domain are solved uncoupled from each other and since the interface relation has a block structure, the solution scheme can be adapted to the particular requirement in each sub-domain. The multiple scales associated with chemically reacting flows and transition important areas for hypersonic research, are motivating applications well-suited for the present spectral multi-domain technique. The following work will focus on both of these topics which are rapidly gaining widespread attention in computational fluid dynamics. The discretization techniques implemented for solution of these two problems are distinctly different: a nonstaggered multi-domain

*Computational Methods Branch, High-Speed Aerodynamics Division, NASA Langley Research Center, Hampton, Virginia.
†Theoretical Aerodynamics Branch, Transonic Aerodynamics Division, NASA Langley Research Center, Hampton, Virginia.
‡Institute for Computer Applications and Scientific Engineering, Hampton, Virginia.

discretization is utilized for the calculation of the chemically reacting flow, and the first implementation of a staggered multi-domain mesh is presented to accurately solve the stability equations for a viscous compressible fluid. The successful implementation of the latter discretization is strongly dependent on the interface condition of a multi-domain technique. The global-flux balance condition of the present method poses no problem for staggered mesh calculations.

1. Introduction. A number of spectral domain decomposition techniques have appeared in the literature and are becoming accepted tools for fluid dynamical calculations. For example, the spectral element method which applies finite element methodology using Galerkin spectral discretization in the variational formulation within elements is a popular technique [1,2]. This technique utilizes a split Galerkin-collocation discretization which restricts its application to convection-diffusion problems for incompressible flows. The spectral element method in practice is used in a manner similar to classical finite element techniques: Low-order internal discretization using many elements with no internal stretchings to improve resolution. The technique is most easily implemented if each element utilizes the same number of collocation points. Other domain decomposition techniques involve explicit enforcement of C^1 continuity across the interface [3,4]. It is not clear how well these techniques perform for strongly convection-dominated problems; the second author's experience with such techniques [5] has shown them to be not entirely satisfactory.

The spectral multi-domain technique of the present paper was developed with compressible flow applications in mind. The multiple scales associated with chemically reacting flows and transition, both features of hypersonic aerodynamics, was a further consideration in developing the multi-domain technique. The former issue will be addressed by incorporating a nonequilibrium chemistry model for air into the spectral multi-domain Navier-Stokes solution method. The application will focus on the chemical kinetics initiated as air passes through a fully resolved shock wave.

2. Spectral Multi-Domain Technique. Spectral collocation methods have proven to be an efficient discretization scheme for many aerodynamic (see e.g. [5-9]) and fluid mechanic (e.g. [10-13]) problems. The higher-order accuracy and resolution shown by these methods allows one to obtain engineering-accuracy solution on coarse meshes, or alternatively, to obtain solutions with very small error. There exist, however, drawbacks to spectral techniques which prevent their widespread usage. One drawback to these techniques has been the requirement that a complicated physical domain must map onto a simple computational domain for discretization. This mapping must be smooth if the high-order accuracy and exponential convergence rates associated with spectral methods are to be preserved [6]. Additionally, even smooth stretching transformations can decrease the accuracy of a spectral method, if the stretching is severe [9]. Such stretchings would be required to resolve the thin viscous region in an external aerodynamic problem, or the widely disparate scales which occur in chemically-reacting flows. Furthermore, problems with discontinuities in boundary conditions, very high-gradient regions or shocks cause oscillations in the spectral solution. The above situations are more the rule than the exception in hypersonic flows.

These restrictions are overcome in the present method by splitting the domain into regions, each of which preserve the advantages of spectral collocation, and allow the ratio of the mesh spacings between regions to be several orders of magnitude higher than allowable in a single domain [14]. Adjoining regions are interfaced by enforcing a global flux balance which preserves high-order continuity of the solution. This interface technique maintains spectral accuracy, even when mappings and/or domain sizes are radically different across the interface, provided that the discretization in each individual subdomain adequately resolves the solution there.

Spectral flux balance interface technique. A simple one-dimensional, two-region example will serve to illustrate the present method for interfacing two collocation-discretized regions. Given the second order, potentially nonlinear boundary-value problem:

$$\begin{aligned} &[F(U)]_x - \nu U_{xx} = S(U), && x \in [-1, 1], \\ &U(-1) = a, && U(1) = b, \end{aligned} \tag{1}$$

We wish to place an interface at the point $x = m$, and have independent collocation discretization in the regions $x^{(1)} \in [-1, m]$ and $x^{(2)} \in [m, 1]$. Even though the point $x = m$ is an interior point to the problem domain, simply applying a collocation statement there, utilizing a combination of the discretizations on either side, will not work; the resulting algebraic system is singular. This is because the spectral second-derivative operator has two zero eigenvalues; thus the patching together of two spectrally-discretized domains yields potentially four zero eigenvalues in the overall algebraic system. Two of these eigenvalues are accounted for by imposition of boundary conditions, and one by continuity of the solution at the interface leaving one zero eigenvalue in the system. To alleviate this difficulty, a global statement of flux balance is used. Rewriting (1) as:

$$[G(U)]_x = S(U) \tag{2}$$

where the flux is

$$G(U) = F(U) - \nu U_x, \tag{3}$$

then integrating (2) from -1 to 1 results in

$$G(U)\big|_{x=1} - G(U)\big|_{x=-1} + [G]\big|_{x=m} = \int_{-1}^{1} S(U)dx. \tag{4}$$

If the jump in flux at the interface, [G], is zero, then (4) may be written:

$$G(U)\big|_{x=-1} + \int_{-1}^{m} S(U)\, dx = G(U)\big|_{x=1} - \int_{m}^{1} (U)\, dx. \tag{5}$$

The statement of global flux balance across the two regions, along with the assumption that the solution is continuous, provides the condition necessary to close the equation set which results from spectral discretization of (1) in two regions. Note that the left side of (5) involves the discretization in the region $x^{(1)} \epsilon [-1, m]$ while the right side involves the region $x^{(2)} \epsilon [m, 1]$. Since spectral collocation discretization strongly couples all points in their respective regions, (5) couples all points in both discretizations.

Note also that no statement is made concerning whether or not (1) is advection- or diffusion-dominated. Equation (1) is considered a scalar equation here, although the above is extendable to a system.

3. Numerical Model of Nonequilibrium Shock Flow. The above technique will model the chemical kinetics and flow kinematics of a nonionized air mixture (O2, N2, NO, O, and N) passing through a fully resolved shock wave, thus alleviating the need for artificial viscosity. The governing equations are the quasi-one dimensional Navier-Stokes equations [15], and species conservation equations [16]. The quasi one-dimensional form is used to provide an artifice for controlling the shock location in the physical space for this otherwise indeterminate problem.

The conservation equations can be written as:

$$\frac{\partial U}{\partial t} + \frac{\partial F}{\partial x} = \frac{\partial V}{\partial x} + W \tag{6}$$

where the dependent variables are denoted by U, the convective flux by F, the dissipative flux by V, and the production rate by W. The equations are nondimensionalized by dividing the state and transport parameters by their dimensional free-stream values. Each of the quantities U, F, V, and W, have 8 components. These expressions are given explicitly in Appendix A.

The viscosity of each of the individual species is calculated from a curve fit relation [17]. Similarly, curve fits are used to obtain specific heats internal energies, and enthalpies [18,19]. The thermal conductivity of each specie is calculated from the Euken semi-empirical formula using the specie's viscosity and specific heat. Appropriate mixture rules are next used to obtain the transport properties of the mixture [20]. Experimental values of bulk viscosities, as obtained from acoustical interferometry and related experiments, are taken from Truesdell [21].

In the present work, the diffusion model is limited to binary diffusion with the binary diffusion coefficients specified by the Lewis number. The value of the Lewis number used is 1.4.

The temperature range under study will not exceed 8000 Kelvin, for conditions at an altitude of approximately 190,000 feet. Therefore, ionization reactions, which occur at roughly 9000 Kelvin, are not included. The chemical reactions utilized for the nonionized air mixture are impact dissociation and exchange reactions. The seventeen reactions included in the present study can be found in ref. 16, which also lists ionization reactions. The constants needed to evaluate reaction rates are given in ref. [16].

Initial conditions are obtained from a spectral multi-domain code for solution of the Navier-Stokes equations with equilibrium chemistry, written for the above problem. These governing equations may be found in ref 15. Transport properties are obtained in the manner previously discussed. The routines of ref. 18 generalized for air are used to obtain equilibrium concentrations.

Compressible multi-domain algorithm. The multi-domain discretization involves three independent subdomains, with the shock located in the center subdomain. Shock jump conditions are obtained by an iterative procedure to solve the Rankine-Hugoniot relations for real air.

A direct inversion of the coupled system is utilized to obtain a fully implicit method. The conserved variables are written in delta form, and a pseudo time iteration using backwards Euler is utilized to obtain the steady state solution as follows. Time-local linearization of equation (6) leads to the implicit form of the equation over the time step Δt:

$$\left\{I + \Delta t\left[\frac{\partial}{\partial x}\left(A + \frac{\partial B}{\partial x}\right) - \frac{\partial^2}{\partial x^2} B - S\right]\right\}\Delta U = \Delta t\left[W + \frac{\partial}{\partial x}(-F + V)\right] \tag{7}$$

where I is the unit matrix, and A, B, and S are Jacobian matrices: $A = \partial(F-V)/\partial U$, $B = \partial(F-V)/\partial U_x$, and $S = \partial W/\partial U$. These Jacobians are obtained analytically and are evaluated at the previous time step. Because of the large rank and ill-condition of the Jacobian matrix, iterative improvement of the Gaussian elimination solution was found to be required. Nonetheless the scheme required less than one second per time step on the Cray-2 at NASA Ames for typical discretizations used in this study.

Method verification. Validity of the multi-domain Navier-Stokes algorithm is demonstrated by comparison with experiment. A low-density wind tunnel study of shock-wave structure and relaxation phenomena in gases was conducted by Sherman [22]. The experiment measured shock wave profiles recorded in terms of the variation in the equilibrium temperature of a small diameter wire oriented parallel to the plane of the shock, as the wire is moved through the shock zone. Free stream Mach number is 1.98. For this test case, a Navier-Stokes spectral multi-domain calculation is performed for a perfect gas with temperature-dependent properties and a nonzero bulk viscosity corresponding to air [21]. A comparison with experimental temperatures normalized by the free stream temperature versus normalized distance is given in figure 1. The experimental data points are represented by the open symbols. The numerical results fall within a symbol width of the data. The multi-domain technique utilized three domains. The center domain, located between $x = -.15$, and $x = 0.3$, contains 21 points; the outer domains contain 11 points each. The computational domain spans -1 to 1. The unit Reynolds number of the flow is 80. A calculation for a unit Reynolds number of 1000 is given in figure 2, showing the ability of the method to accurately resolve strong gradients without numerical oscillations. The plot is of Mach number versus normalized distance. Three domains are again used; the center domain contains 17 points and the outer domains contain 11 points each with the interfaces located at -.15 and -.1.

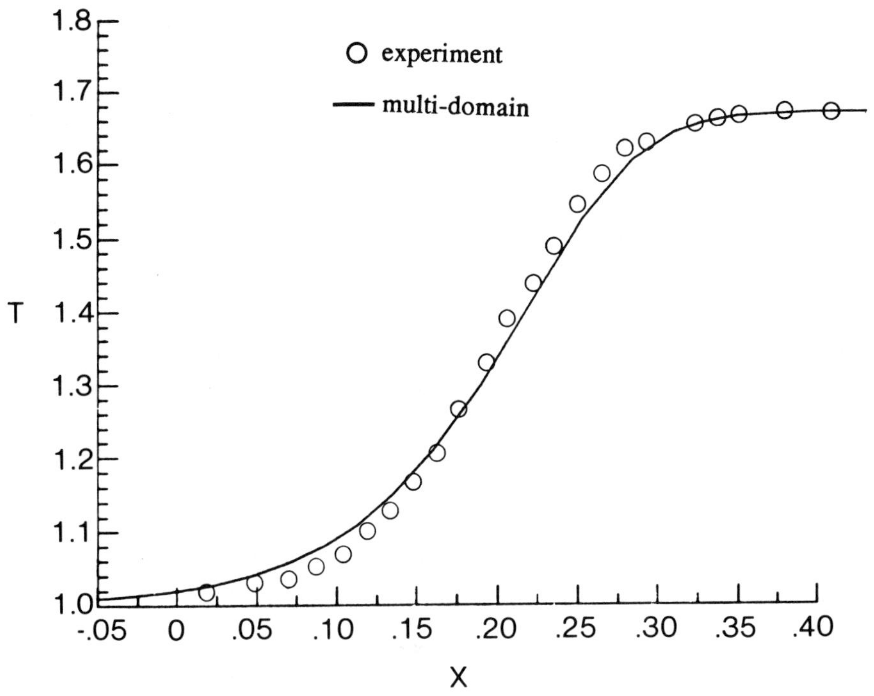

Figure 1. Comparison of multi-domain Navier-Stokes calculation with experimentally obtained temperatures; $M_\infty = 1.98$, Re = 80. Discretization 11/21/11, interfaces at -.15 and .3.

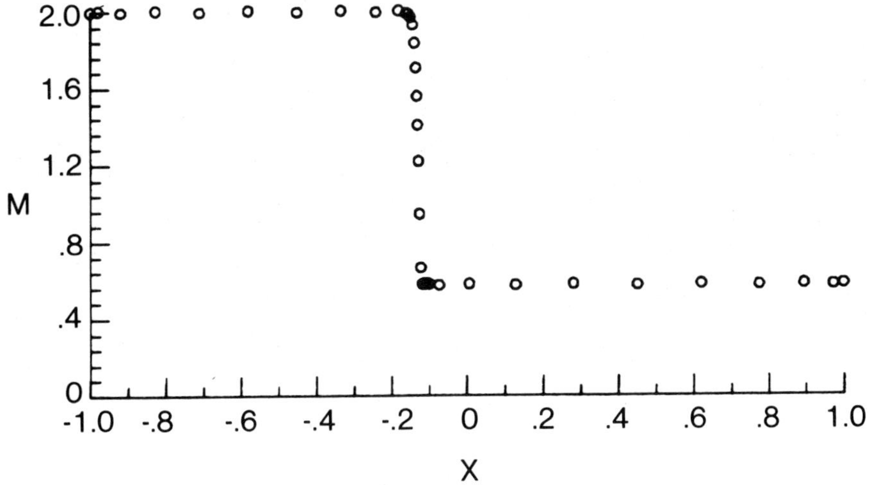

Figure 2. - Computed solution for Re = 1000; discretization 11/17/11, interface at -.15 and -.1.

Results. The method is used to calculate the chemical kinetics initiated as air passes through a hypersonic shock wave. The case to be discussed for nonequilibrium flow is $M_\infty = 11.0$, $T_\infty = 350K$, and $\rho_\infty = 6 \times 10^{-8} g/cm^3$. These conditions invoke primarily O_2 dissociations with N_2 dissociations just beginning. Temperatures are not yet high enough for ionization to occur, so electronic energy modes remain unexcited.

Typical discretizations used in this study were 15, 27 and 33 points in the upstream, middle and downstream domains, respectively. The backward-Euler implicit time-stepping algorithm typically required less than 2000 iterations to converge from an equilibrium starting solution, with at least an eight order of magnitude reduction in maximum residual.

A study of the effect of artificial visosity on the flow physics was carried out by adding the equivalent of second-order artificial viscosity to the momentum, energy, and species' concentration equations. The amount of artifical viscosity introduced was such that the shock was spread out to a thickness about three orders of magnitude wider than the fully-resolved no artificial viscosity solution. This width was chosen to represent the grid spacing of a typical shock-capturing computation on a full-scale configuration. Figures 3a and 3b show the Mach number and temperature profiles for the resolved-shock and smeared-shock cases respectively; note that the entire physical domain is shown in Fig. 3b, whereas only the near-shock region on a greatly expanded scale is plotted in Fig. 3a. For the resolved-shock case, endpoints are at -1 and 200, and interface points are at -.3 and .1. The interface locations for the smeared shock case are at 65 and 100, with endpoints at 0 and 270. The most important feature to note in comparing these profiles is the 20% reduction in the temperature overshoot as a result of the artificial viscosity. In a calcuation with ionization, such a reduction could prevent its onset; or similarly, a calculation with combustion chemistry may not reach threshold temperatures necessary for ignition due to this artificial damping. In addition, the high temperature zone following the passage of air through the shock persists for roughly two orders of magnitude downstream further than the resolved shock case. Computationally, this effect of artificial viscosity could cause a chemical reaction to produce more of a given species than what occurs in the true physics.

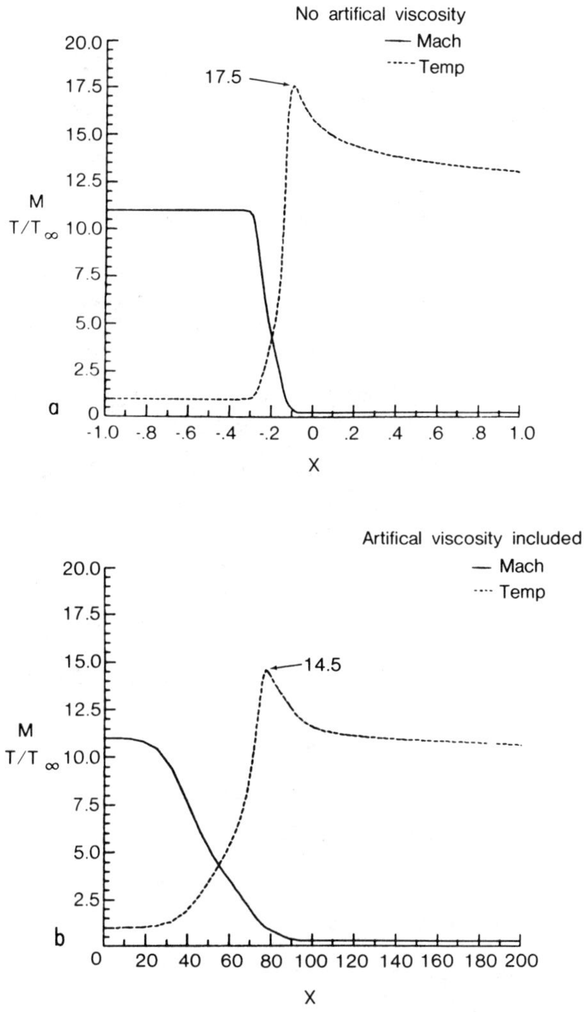

Figure 3. - Comparison of temperature and Mach number profiles for calculations; (a) without artificial viscosity, and (b) with artificial viscosity.

For the Mach 11 case considered here, although the path along which the chemistry relaxes is significantly altered in the near-shock region, the chemical end states from the computations with and without artificial viscosity are within 3-4% of each other. This can be seen in Figs. 4a,b which show the profiles for [N] and [NO] in the relaxation zone. This is not to say, however, that this situation will always occur especially for higher Mach numbers where electronic excitation (ionization) occurs. The reduction in the temperature overshoot could result in a large enough change in the relaxation path that the end state could be affected significantly.

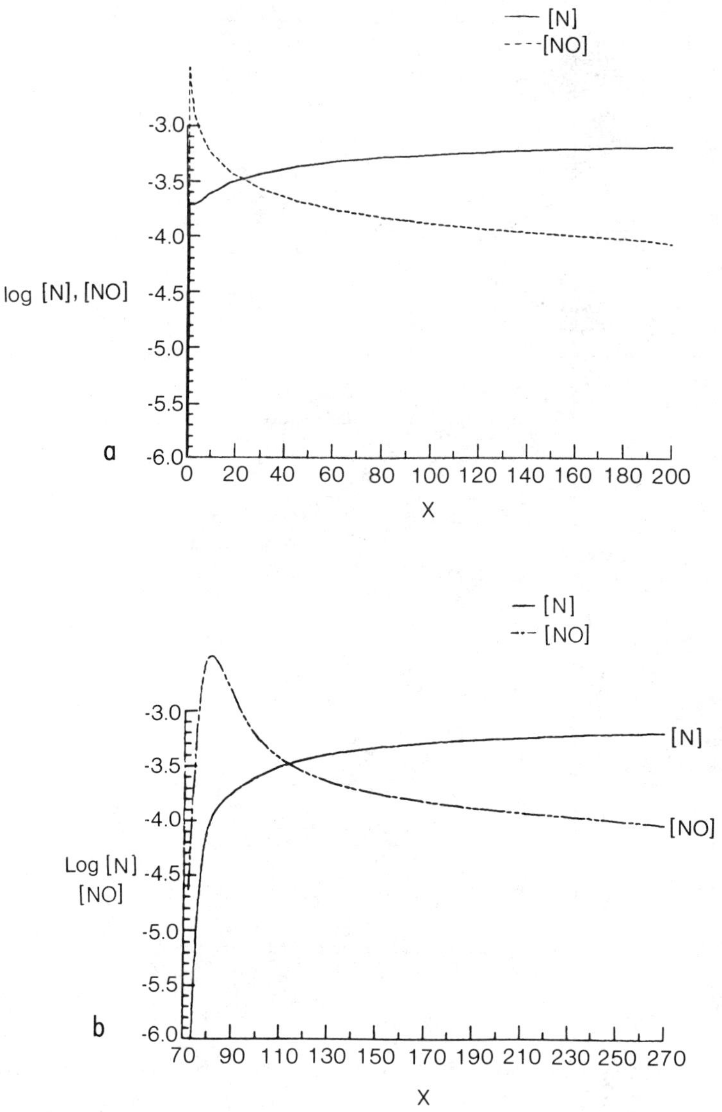

Figure 4. - Comparison of relaxation pathway for [N] and [NO] from calculations; (a) without artificial viscosity, and (b) with artificial viscosity.

4. Stability of Compressible Flows. The second focus is a careful study of the stability of high-speed boundary layers and free-shear flows. A newly developed spectral stability code (staggered pressure mesh) is presented for analysis of compressible viscous flow stability. An order of magnitude less number of points is needed for equivalent accuracy of growth rates compared to those calculated by a finite-difference formulation. Supersonic disturbances which are found to have highly irregular structures have been resolved by a spectral multi-domain discretization, which requires a factor of three fewer points than the single domain spectral stability code.

At the present time, there is no prospect of a unified theory of transition even in low speed flows (where some of the underlying mechanisms are relatively well-known), let alone in hypersonic flows. As a short-term goal, it is imperative then to obtain linear stability results. The implicit assumption is that supersonic/hypersonic transition has its origin in linear instability and is not overly sensitive to details of the disturbance environment (not that Morkovin's bypasses are inapplicable to the hypersonic regime). Clearly, as Mack [23] points out, there is a need to develop efficient, accurate and robust linear stability codes for use in a large number of design calculations.

This is the motivation behind the present work which treats the linear compressible stability equations by a spectral collocation method. Results are presented to substantiate these claims. Furthermore, the multi-domain version of this method presented here deals economically with complex flows which can include such features as multiple interior shear layers. This section first validates the method for the case of boundary layers and free shear layers, then goes on to present some new results in the case of free mixing layers.

Formulation of compressible stability equations. The basic equations governing the flow of a viscous compressible fluid are the Navier-Stokes equations. For this stability study, the equations in Appendix A, modified for a perfect gas, describes the system.

In this study all velocities are scaled by U_e, the x component of velocity at the edge of the boundary layer, and all lengths are scaled by δ^*, the displacement thickness of the velocity profile in the x-direction. The Reynolds number and Mach number are given by

$$Re = \frac{U_e \delta^*}{\nu_e} \tag{8}$$

$$M = \frac{U_e}{\sqrt{\gamma \, Re \, T_e}} \tag{9}$$

where ν_e and T_e are the kinematic viscosity and mean temperature in the freestream, and gamma is the ratio of specific heats. The results in the present paper consider σ = Prandtl number = .72 and $\bar{\mu}^*$ is evaluated by Sutherland's law.

If we assume that the base flow is a locally parallel boundary layer then

$$u(x,y,z,t) = U_o(y) + \hat{u}(y)e^{i(\alpha x+\beta z-\omega t)} \tag{10}$$

$$v(x,y,z,t) = \hat{v}(y)\ e^{i(\alpha x+\beta z-\omega t)} \tag{11}$$

$$w(x,y,z,t) = W_o(y) + \hat{w}(y)e^{i(\alpha x+\beta z-\omega t)} \tag{12}$$

$$p(x,y,z,t) = \hat{p}(y)\ e^{i(\alpha x+\beta z-\omega t)} \tag{13}$$

$$\tau(x,y,z,t) = T_o(y) + \hat{\tau}(y)e^{i(\alpha x+\beta z-\omega t)} \tag{14}$$

where U_o, W_o, and T_o represent the steady unperturbed boundary layer (mean flow), and quantities with tildas denote complex disturbance amplitudes. $V_o(y)$ is assumed zero since the flow is parallel, and $P_o(y)$ is zero since pressure is assumed constant across the boundary layer. α and β are x and z disturbance wave numbers, respectively, and ω is the complex frequency. Equations (10)-(14) are substituted into he nondimensional Navier-Stokes equations, the mean flow terms subtracted out, and the terms which are quadratic in the disturbance neglected. The resulting system is the linearized compressible Navier-Stokes equations for the disturbance quantities as given in Appendix B.

Solution technique. The equations in Appendix B constitute an eigenvalue problem for the complex frequency ω, once the disturbance wavenumbers α and β are specified. Discretization of these equations in the y-direction forms a generalized matrix eigenvalue problem, suitable for computer solution. Equations (B1)-(B7) are essentially an eighth order system; thus the eight boundary conditions (eq. (B7)) are sufficient for solution, and no boundary condition is applied to the disturbance pressure. Whatever discretization scheme is used must respect this arrangement if an accurate solution is to be expected.

The discretization scheme used here is a spectral collocation technique, using Chebyshev polynomials as basis functions. The nodes of the variables $\hat{v}$, $q^+ = \alpha\hat{u} + \beta\hat{w}$, $q^- = \alpha\hat{w} - \beta\hat{u}$, and $\hat{\tau}$ are located at the Gauss-Lobatto points (the extrema of the last retained Chebyshev polynomial [24]); the energy and momentum equations are collocated at these points. Thus discrete boundary conditions may be imposed for these variables at both end points of the domain. The pressure nodes are located at the Gauss points (the zeroes of the first neglected polynomial) of a Chebyshev series one order less than that used for the other variables; the continuity equation is collocated at these points. Since no Gauss points fall on the boundary, we are free of any requirement of providing an artificial numerical boundary condition for pressure, and we have the proper balance of number of equations and unknowns.

The farfield boundary of the discretized domain is placed at a finite distance, typically 20-100 δ^* from the wall or shear-layer centerline. Extensive sensitivity studies were performed to determine the effect of this finite domain truncation.

Stretching is employed in the discretization to improve resolution near the wall/centerline. In the case of the boundary layer, either of two stretching forms are used:

$$y_p = y_{max}(C_2 - 1)^{C_1} y_c/(C_2 - y_c^2)^{C_1} \tag{15}$$

or

$$y_p = y_{max}\, C_3 y_c/(1 + C_3 - y_c) \tag{16}$$

where y_p is the coordinate in the physical space $[0, y_{max}]$, y_c is the computational coordinate ε $[0,1]$, and C_1, C_2, and C_3 are adjustable constants. In this work, C_1 is either 4 or 6, and C_2 ranged from 1.2 to 2.0; C_3 is used between .01 and .03. For the shear layer, equation (15) is used as the stretching, yielding a physical space of $[-y_{max}, y_{max}]$ from $y_c \varepsilon [-1,1]$.

Using standard spectral collocation discretization formulas [24], matrix differentiation operators are formed for both the Gauss-Lobatto ($\hat{v}$, q+, q-, $\hat{\tau}$) and the Gauss ($\hat{p}$) grids, incorporating the selected stretching function. Mean flow quantities from the spectral boundary layer code of [25] are spectrally interpolated onto the new mesh, and derivatives of these quantities obtained using the differentiation operators. The generalized matrix eigenvalue problem which results from this discretization is of the form:

$$A_{GL}L_{GL}^2\vec{\phi} + B_{GL}L_{GL}(\vec{\phi} + I_G^{GL}P) + C_{GL}(\vec{\phi} + I_G^{GL}P) = \omega\lfloor D_{GL}L_{GL}(\vec{\phi} + I_G^{GL}P)$$

$$+ E_{GL}(\vec{\phi} + I_G^{GL}P)\rfloor \tag{17}$$

for the momentum and energy equations, and

$$B_G L_G \, I_{GL}^{G} \vec{\phi} + C_G\left(I_{GL}^{G} \vec{\phi} + P\right) = \omega E_G\left(I_G^{GL} \vec{\phi} + P\right) \qquad (18)$$

for the continuity equation, where A, B, C, D, and E are matrix coefficients derived from equations (B1)-(B5), L denotes a spectral differentiation operator, the unknown vector $\phi = (\hat{v}, q^+, q^-, \hat{\tau})^T$, and the vector P contains the disturbance pressure $\hat{p}$. Subscripts GL and G denote location at or operation on Gauss-Lobatto and Gauss point grids, respectively; I_G^{GL} and I_{GL}^{G} are spectral interpolation matrices from Gauss to Gauss-Lobatto points and vice-versa.

The unknown vectors $\vec{\phi}$ and P are collected into a single vector, and the matrix equations (17) and (18) are assembled into a large generalized matrix eigenvalue problem for input into a standard library routine for solution. A complex modified QZ algorithm [26] is used to obtain the eigenvalues of the system directly; this is referred to as a global search. The most unstable eigenvalue is then selected, and used as an input to an inverse Rayleigh method to purify the eigenvalue of the effects of round-off error and to obtain the solution eigenvectors. In all cases, the global and local (Rayleigh-iterated) eigenvalues agreed to better than eight decimal places.

Verification.

Boundary layer. For verification, calculations are performed for the stability analysis of compressible two-dimensional similarity boundary layer profiles. A spectral mean flow code modified for compressible flow is used for this purpose [25].

A Mach number boundary layer profile perturbed by a three dimensional disturbance at $M_\infty = 4.5$, $T_e = 520°R$, $Re = 10{,}000$, $\alpha = .6$, $\beta = 1.0392$ is first analyzed. A resolution study for the eigenvalue computations are given in Tables I and II for the single domain spectral code and COSAL (a finite-difference compressible stability code, [27]), respectively. Roughly 3 significant digits for values of growth rate are obtained with the spectral code using 81 points; in Table II COSAL is seen to require approximately 800 points for equivalent accuracy. Eigenfunctions for $\hat{u}$ are shown in Fig. 5a,b from each code. A multi-domain spectral discretization (MDSPD) with two domains obtains equivalent accuracy with the single domain spectral stability code with 1/3 the number of points, as illustrated in Table III. The savings is significant considering that the number of operations in the spectral stability code goes as the cube of the number of points.

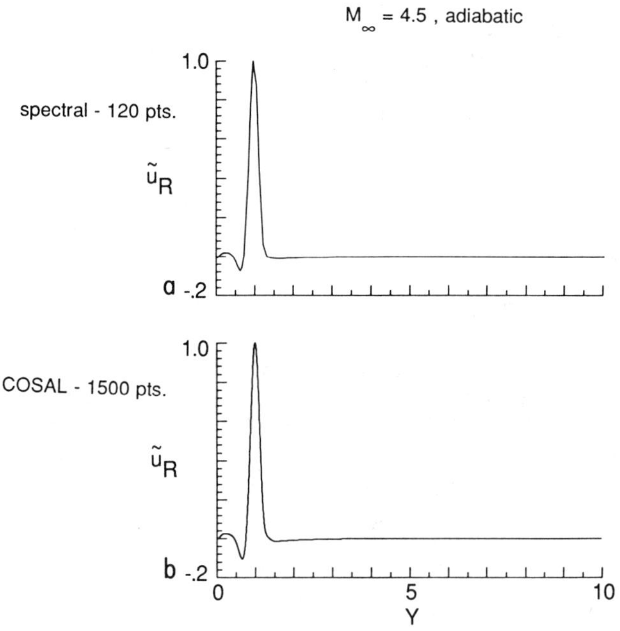

Figure 5. - $\hat{u}$-velocity fluctuation amplitude function for $\alpha = .6$, $\beta = 1.0392$, $M_\infty = 4.5$, Re = 10,000, $T_e = 520°R$. Calculation by:
a. Spectral stability code
b. COSAL

The reason for the increased efficiency with the MDSPD is that in addition to the significant structure at the wall (Fig. 5), the farfield boundary must be far enough out so that the exponential rate of decay associated with these disturbances is accurately captured. For this case, the outer extent is 30 displacement thickness units (y coordinate) from the wall at y = 0 to capture this exponential decay, otherwise the accuracy of the growth rate is effected. These issue put a severe demand on the stretching required for resolution with a single-domain discretization. The multi-domain utilized two domains with interface at 2. The inner domain has 25 points and is unstretched; the outer domain has 17 points and a tanh stretching.

Stability of a Viscous, Compressible Shear Layer. The spectral stability code is used to analize the stability of a viscous, compressible shear flow obtained from a spectral similarity solution obtained by modifying the boundary-layer code of ref. 25. The shear flow consists of two parallel gases: an injected stream into a quiescent gas. Issues of relevance in this study is understanding the impact of transition on fuel/air mixing efficiency in scramjet combustors. It has been observed experimentally that the mixing efficiency is decreased fourfold in the range of Mach 1 to 4. The cause of this trend is unkown. However, it is well known that turbulent mixing is many orders of magnitude faster than laminar. Ideally one would like to be able to manipulate the downstream evolution of shear layers to enhance mixing -- perhaps linear stability theory holds a clue. The initial stages of shear flow instabilities are driven by linear mechanisms which persist for a considerable distance downstream. Understanding the growth and propagation of the disturbance in these early stages will allow not only a better understanding of the onset of transition, but will allow initiation of the transition process in a numerical model so that the physics might be systematically studied. This study investigates a range of Mach numbers, gas temperatures and disturbance wave numbers. The mean flow in all cases involves a jet being injected into a quiescent gas.

It is observed that the disturbance eigenfunctions significantly tighten in structure as α increases which puts a greater demand on the resolution required for a single domain spectral discretization (SDSPD). A progression of eigenfunction plots for increasing α (phase angle $\theta = 60°$) is given in Figs. 6 and 7 illustrating this observation.

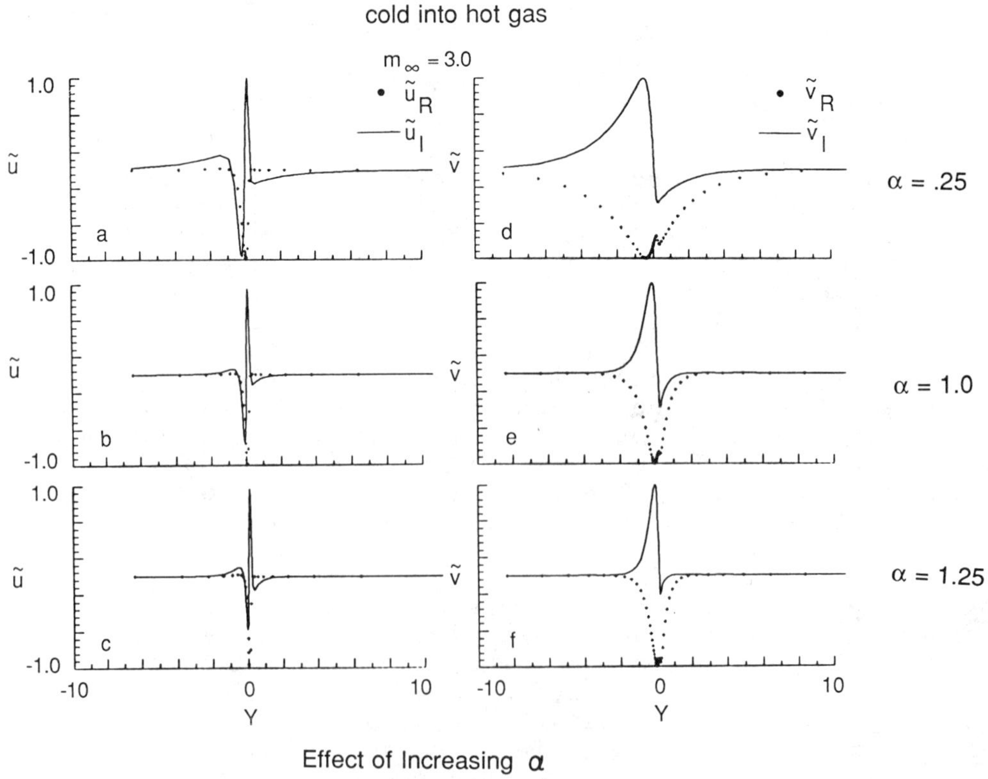

Figure 6. - Effect of increasing α on $\hat{u}$ and $\hat{v}$ eigenfunctions; $\theta = 60°$, $M_\infty = 3$.

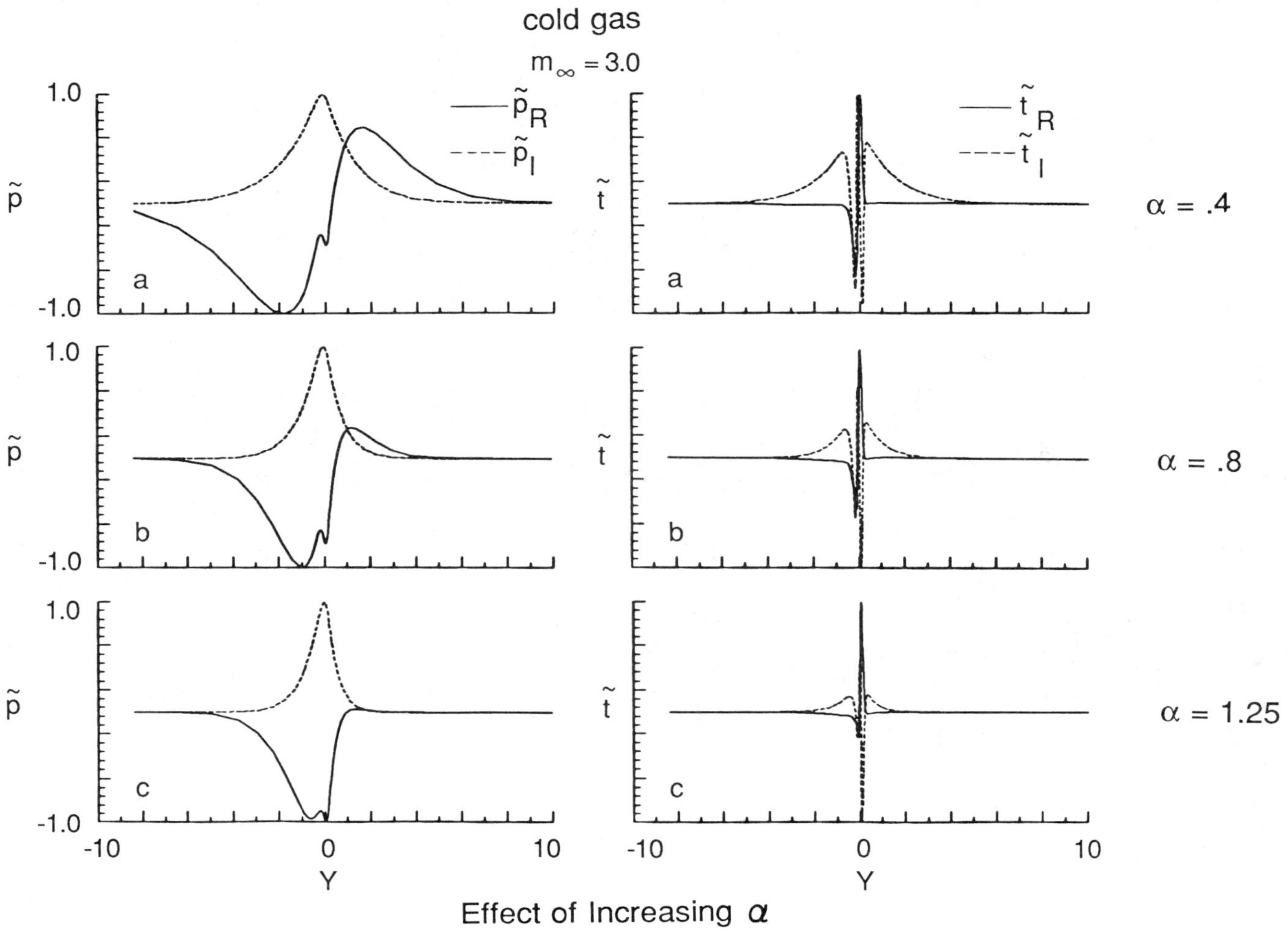

Figure 7. - Effect of increasing α on $\hat{p}$ and $\hat{t}$ eigenfunctions; $\theta = 60°$, $M_\infty = 3$.

Difficulties in resolution similarly occur for higher Mach number disturbances. The spectral stability code had difficulty resolving eigenfunctions beyond 3.75 for the cold injection cases and as early as Mach 3 for the hot injection case for disturbance wave angles of 60°. Restrictions on the allowable stretching for single domain spectral methods contributes to this difficulty. Examples of the MDSPD for cases requiring a very severe stretching will be given later.

The above studies involve three dimensional disturbances with a wave propagation angle of 60°. Preliminary

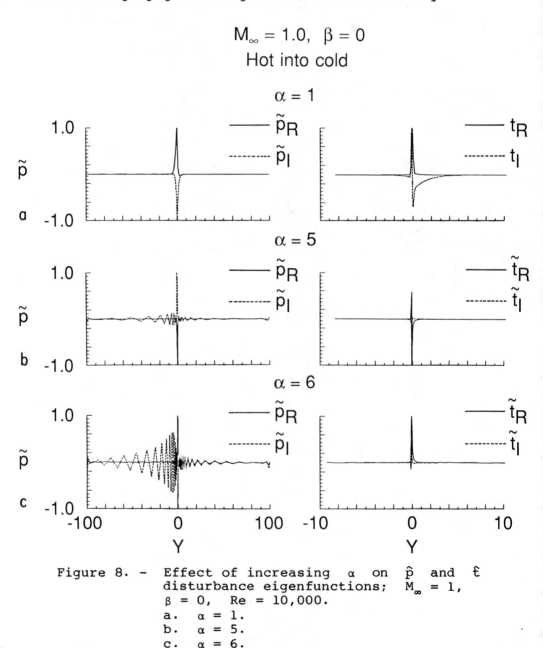

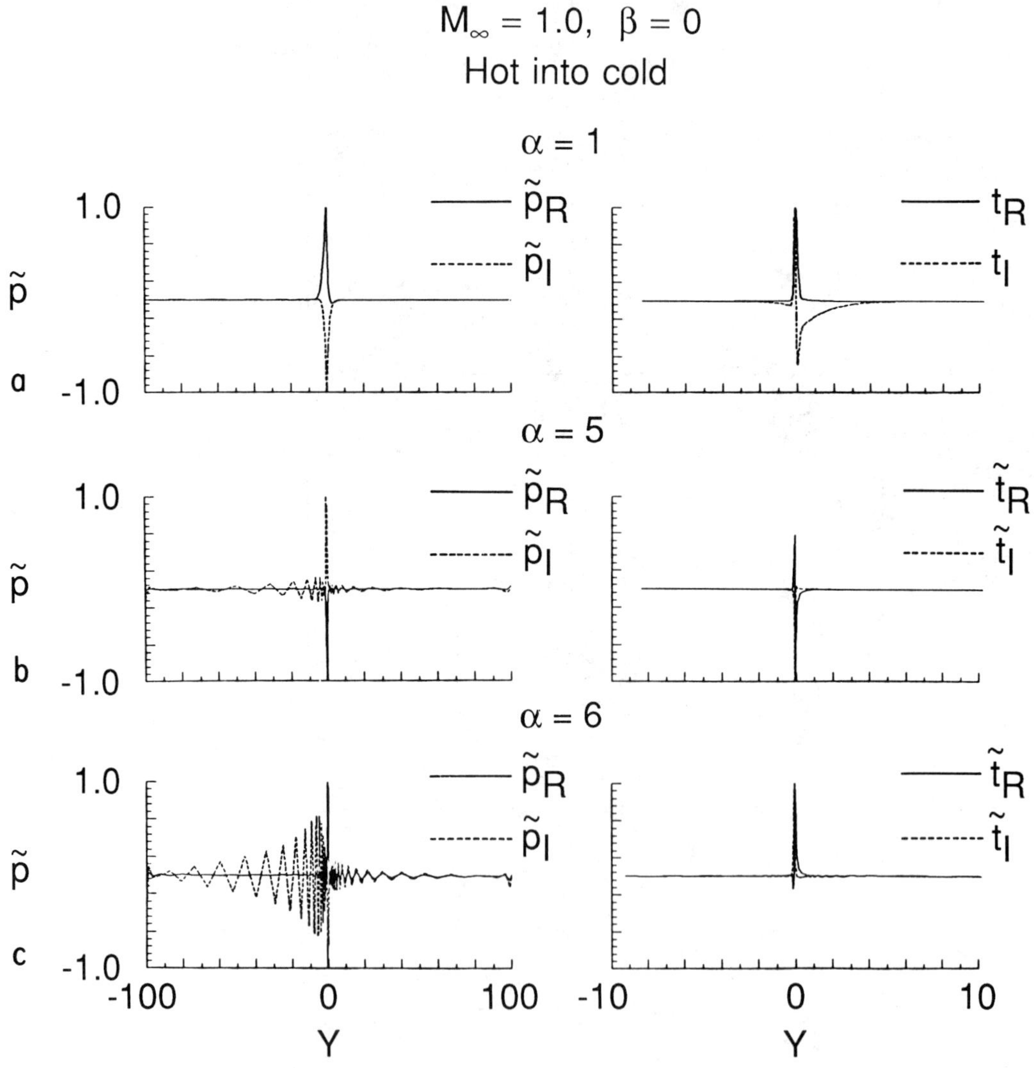

Figure 8. - Effect of increasing α on $\hat{p}$ and $\hat{t}$ disturbance eigenfunctions; $M_\infty = 1$, $\beta = 0$, Re = 10,000.
a. $\alpha = 1$.
b. $\alpha = 5$.
c. $\alpha = 6$.

results indicate that two dimensional disturbances ($\beta = 0$) exhibit similar trends; however, difficulties in resolving eigenfunctions with a SDSPD occur at lower Mach numbers. Again at higher α disturbance eigenfunctions tighten radically beyond the limit at which a single domain spectral method can resolve; the hardest temperature case to resolve is hot injection into a colder gas. This observation is demonstrated in Fig. 8 which plots a sequence of $\hat{p}$ and $\hat{t}$ eigenfunction for increasing α, $\beta = 0$. The mean flow corresponds to $M_\infty = 1$, Re = 10,000 with $\beta_T = .2$ ($\beta_T = T_{quiescent}/T_{injected}$). Note the extremely tight structures in $\hat{p}$ and $\hat{t}$ as α increases. Note that $\hat{t}$ is plotted on the interval [-10,10], the actual spatial extent is [-100,100]. Oscillations in $\hat{p}$ for $\alpha = 5$ and 6 are quite pronounced.
The MDSPD resolves these cases with relative ease. To illustrate this point, $\hat{p}$ and $\hat{v}$ eigenfunctions obtained from the MDSPD are displayed in Fig. 9 for the

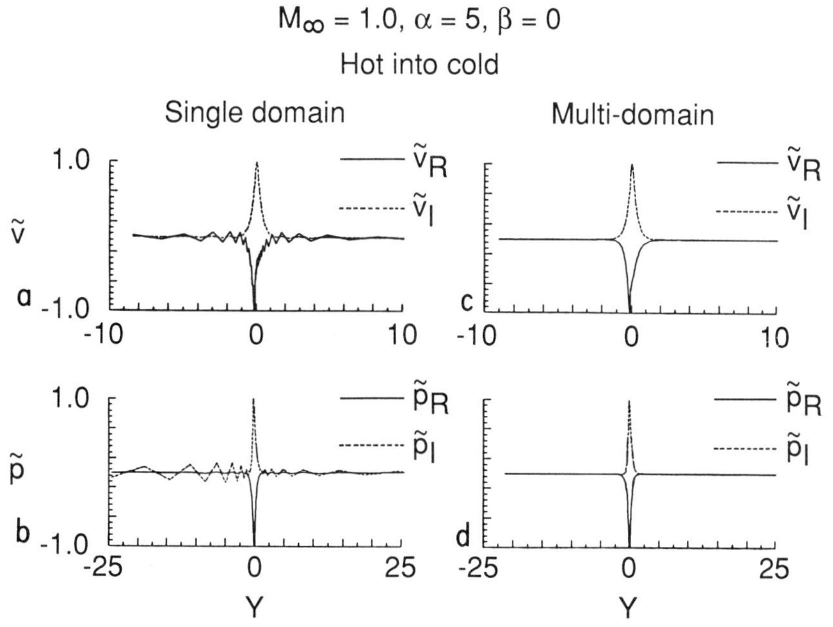

Figure 9. - $\hat{v}$ and $\hat{p}$ eigenfunctions obtained with spectral multi-domain discretization as opposed to single domain spectral discretization for case described in Fig. 8, $\alpha = 5$.

single domain: multi-domain:

a. $\hat{v}$ c. $\hat{v}$
b. $\hat{p}$ d. $\hat{p}$

Re = 10,000, $\alpha = 5$ case discussed previously. Adjacent to the multi-domain results are the single domain solution for $\hat{p}$ and $\hat{v}$. The multi-domain solution remains oscillation free. The number of points in each discretization is roughly 100, however, the multi-domain utilizes three domains, with 41 points in the center domain between -.5 and .5, 25 points in the left domain between -100. and -.5, and 37 points in the right domain between .5 and 100. The plot of the pressure disturbance for the entire spatial extent obtained from the multi-domain solution is given in Fig. 10, to illustrate the fineness of the structure which is resolved.

The preceding case ($\alpha = 5$) is found to be resolvable by a SDSPD but only after quite a bit of trial and error stretching of the mesh for a variety of resolutions. The important point is that the MDSPD is quite robust and gives accurate eigenvalues over a wide range of stretching parameters. This observation is illustrated in Table IV which lists a range of stretching parameters and corresponding phase speeds and growth rates for the preceding case. Both the SDSPD and MDSPD for this illustration utilized 99 points with the same outer extent. Note that for the SPSPD cases changing the

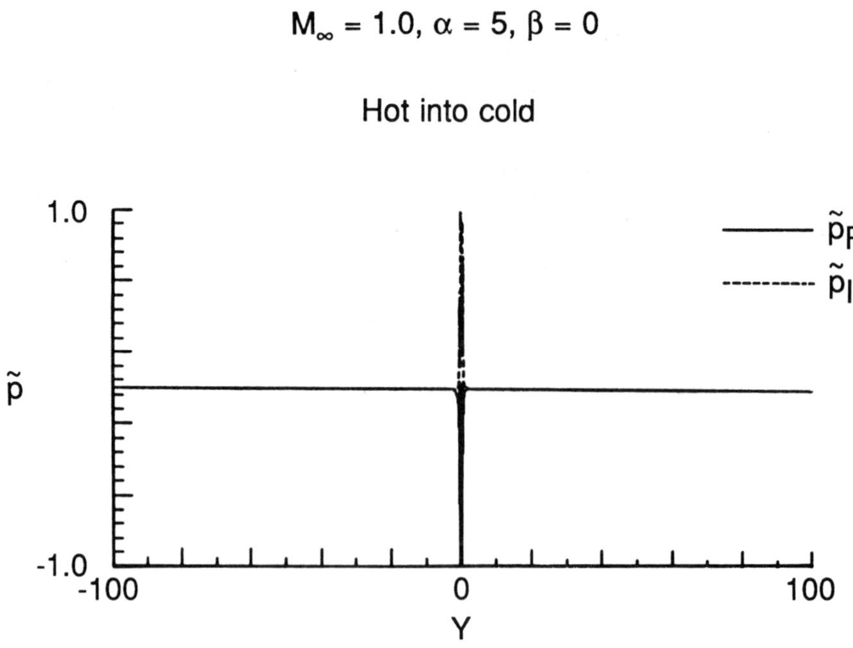

Figure 10. $\hat{p}$ disturbance eigenfunctions obtained by multi-domain stability code displayed on entire computational domain for case of Fig. 9. Discretization: center domain on [-.5,5], 41 pts.; left domain on [-100,-.5], 21 pts.; right domain on [.5,100], 35 pts.

stretching parameter by approximately 25% causes about a 10% change in phase speed and over a 50% change of value in growth rate. Contrast this sensitivity with the MDSPD cases. The stretching parameter is allowed to vary 100% in the center domain (three domains are utilized with interfaces at ±1). The phase speed has changed by less than 1% and the growth rate by only 6%. This robustness is extremely important since one usually has no idea of the value of the phase speed or growth rate. In addition, one needs to determine whether the disturbance mode is spurious--which is measured by its persistance over a wide range of resolutions.

As mentioned earlier, the disturbance eigenfunctions become increasingly complex as Mach number is increased. These higher Mach number cases are unresolvable by a SDSPD. To illustrate, a series of $\hat{p}$ and $\hat{v}$ eigenfunctions calculated by the SDSPD stability code are displayed in Fig. 11 and 12, respectively, for a disturbance wave angle of 60°. Note that at M_∞ = 3.5 the injected gas side of the disturbance begins to take on an oscillatory nature; at M_∞ = 3.75 these oscillations are more pronounced. It is well known that for supersonic disturbances the eigenfunction structure will be oscillatory [29]. (A supersonic disturbance occurs when the wave velocity of the disturbance relative to the local flow, in the direction of wave propagation, has a magnitude greater than the speed of sound.) The MDSPD is able to capture the structure of these supersonic modes with relative ease. Fig. 13 displays two unstable

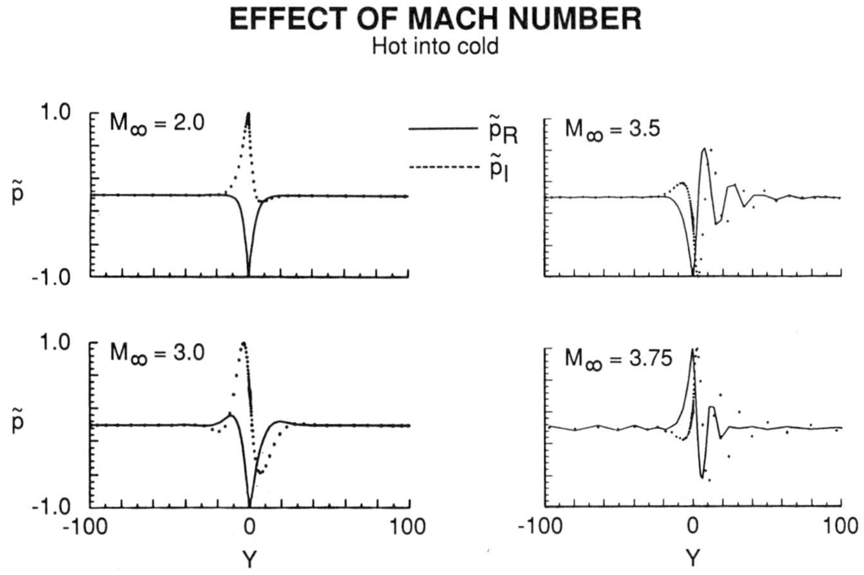

Figure 11.- $\hat{p}$ disturbance eigenfunctions for increasing Mach number and a SDSPD (θ = 60°).

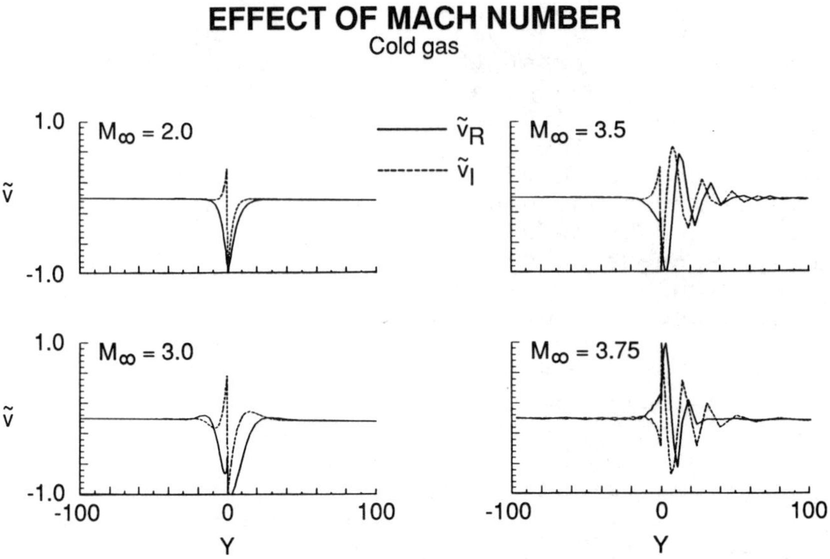

Figure 12.- $\hat{v}$ disturbance eigenfunctions for increasing Mach number and a SDSPD (θ = 60°).

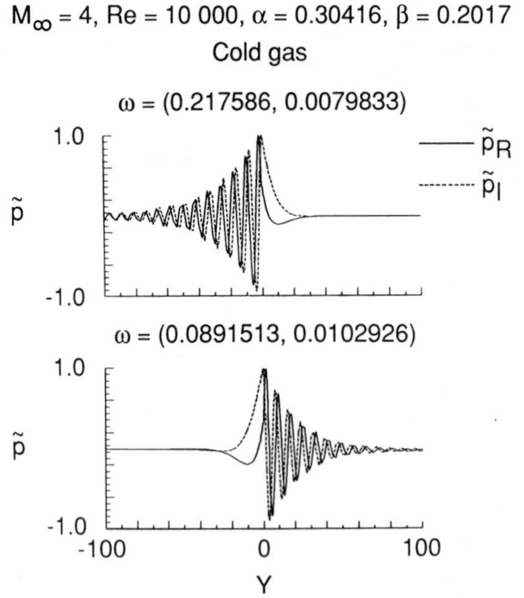

Figure 13.- $\hat{p}$ disturbance eigenfunctions for two unstable supersonic modes corresponding to a M_∞ = 4; ;Re = 10,000; θ = 60°. MDSPD with three domains: 41/105/41, interface at ±1.

supersonic modes which are associated with the instability of a Mach 4, $\beta_T = 1$, free shear flow. The disturbance wave numbers are $\alpha = .30416$, and $\beta_T = .2017$. The MDSPD involves three domains: 105 points on the oscillatory side, 41 points in the inner domain, and 41 points in the outer domain where the profile is smooth. Interface locations are located at ±1. Note further the level of complexity of the $\hat{t}$ eigenfunctions of this case. Fig. 14 is a plot of $\hat{t}$ for both modes on a full scale ±100 and greatly expanded scale ±1. The center structure is an added complexity to the structure which also requires adequate resolution, and further illustrated the necessity of a flexible discretization scheme like a MDSPD.

Conclusions. The present global flux balance spectral multi-domain method has demonstrated maintenance of exponential-order accuracy on a variety of advection- and diffusion-dominated test problems [14]. Extremely large differences in discretization across an interface through domain size, number of points and stretchings, have been shown to not disrupt this property of the present method. Additionally, this technique can be used to isolate certain types of coefficient, mapping, or boundary condition discontinuities.

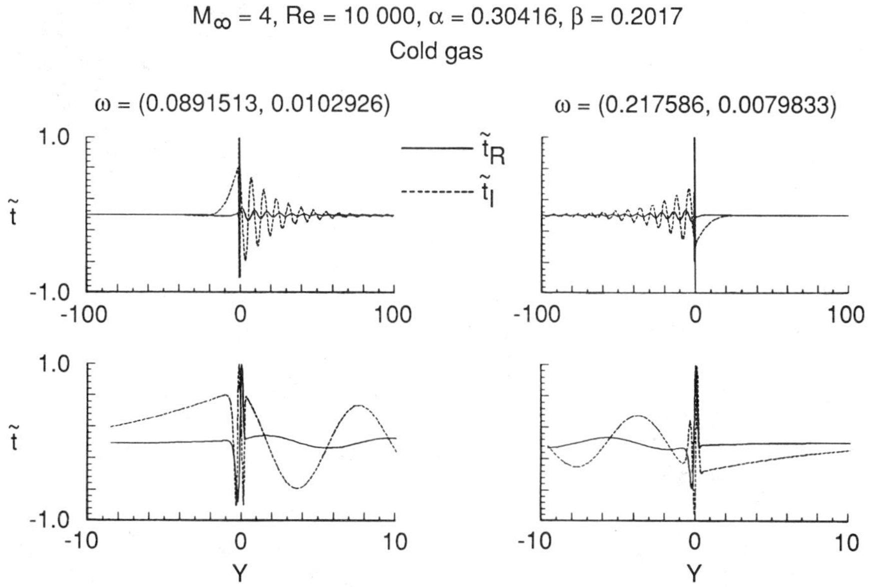

Figure 14.- $\hat{t}$ disturbance eigenfunctions for two unstable supersonic modes corresponding to a $M_\infty = 4$; Re = 10,000; $\theta = 60^0$, on a full scale (±100) and greatly expanded scale (±1). MDSPD: 41/105/41, interface at ±1.

These advantages have made possible the first compressible Navier-Stokes calculation by a spectral multi-domain technique. In addition, a Mach 11 shock calculation with nonequilibrium chemistry was performed to study the chemical kinetics initiated as air passes through a fully-resolved shock wave.

In addition, the first spectral collocation linear stability code for compressible flow is presented. The accurate discretization (staggered pressure grid) may be carried over to nonlinear simulations. Verification cases for high-speed boundary layers indicate an order of magnitude reduction in the number of points required for a single domain spectral discretization (SDSPD) to obtain equivalent accuracy in growth rates with a finite-difference formulation. In addition, a multi-domain spectral discretization (MDSPD) is found to require a factor of three less points man the (SDSPD), which is significant since the operation count of the spectral formulation goes as the cube of the number of points.

The first viscous stability analysis of a compressible shear flow is presented. The study indicates that for subsonic disturbances stability ofmixing layers is enhanced by viscosity, increasing Mach number and higher temperatures of the injected gas. Disturbances become supersonic at lower Mach numbers for 2-D disturbances ($M < 2$). 3-D disturbances become supersonic in the Mach number range of 3.5 to 4. The exact value of this Mach number depends upon the value of β_T the lower the value of β_T (cooling) the lower the Mach number at which the disturbance eigenfunctions become supersonic. The highly irregular structure of the supersonic modes are easily resolved by MDSPD stability code which is shown to be highly robust over a wide range of stretching parameters and resolutions.

Appendix A. Non-equilibrium one-dimensional Navier-Stokes equations

$$\frac{\partial U}{\partial \tau} + \frac{\partial F}{\partial x} = \frac{\partial V}{\partial x} + W \tag{A1}$$

Conservation variables

$$U_i = \rho\gamma_i, \quad 1 \leqslant i \leqslant NS, \text{ number of species} \tag{A2}$$

$$U_{NS+1} = \rho \tag{A3}$$

$$U_{NS+2} = \rho u \tag{A4}$$

$$U_{NS+3} = \rho E \tag{A5}$$

Convective fluxes

$$F_i = uU_i, \quad 1 \leqslant i \leqslant NS \tag{A6}$$

$$F_{NS+1} = \rho u \tag{A7}$$

$$F_{NS+2} = \rho u^2 + P \tag{A8}$$

$$F_{NS+3} = (E + P)u \tag{A9}$$

Viscous fluxes

$$V_i = \frac{Le}{Pr\ Re}\ \mu \frac{\partial \gamma_i}{\partial x}, \quad 1 \leqslant i \leqslant NS \tag{A10}$$

$$V_{NS+1} = 0 \tag{A11}$$

$$V_{NS+2} = \frac{(\lambda + 2\mu)}{Re} \frac{\partial \rho u}{\partial x} \tag{A12}$$

$$V_{NS+3}$$

$$= \frac{(\lambda + 2\mu)}{Re}\ u\ \frac{\partial \rho u}{\partial x} + \frac{\beta\ k}{(\beta - 1)\ Pr\ Re} \frac{\partial}{\partial x} \left[(\beta - 1)\left\{E - \frac{u^2}{2}\right\}\right]/z$$
$$+ \left[\frac{1 - \beta}{z}\right]\frac{Le\ m_\infty}{Pr\ Re}\left[E - \frac{\rho u^2}{2}\right] \sum_{i=1}^{NS} h_i \frac{\partial}{\partial x}\gamma_i \tag{A13}$$

where

m_∞ = freestream molecular weight

Le = Lewis number
z = compressibility: $P = z\ \rho\ T$

h_i = enthalpy of species "i"

$\beta = h/e$

Source production terms

$$W_i = \sum_{r=1}^{NR} (\beta_{i,r} - \alpha_{i,r}) \left[R_r^f - R_r^b\right] \tag{A14}$$

$$R_r^f = k_r^f \prod_{J=1}^{NS} (\rho\gamma_j)^{\alpha_{j,r}} \quad R_r^b = k_r^b \prod_{j=1}^{NS} (\rho\gamma_j)^{\beta_{j,r}} \tag{A15}$$

$$k_r^f = A_1\ T^{A_2} \exp(-A_3/T) \quad k_r^b = B_1\ T^{\beta_2} \exp(-B_3/T) \tag{A16}$$

NR = number of reactions

$\alpha_{i,r}$, $\beta_{i,r}$ = stoichiometric coef. for forward and backward reactions, resp.

$W_{NS+1} = W_{NS+2} = W_{NS+3} = 0$

Appendix B. Linearized compressible Navier-Stokes stability equations.

$$D^2(\alpha\hat{u} + \beta\hat{w}) + \frac{1}{\mu_o}\frac{d\mu_o}{dT_o}T_o' D(\alpha\hat{u} + \beta\hat{w}) + i(\lambda - 1)(\alpha^2 + \beta^2)(\alpha\hat{u} + \beta\hat{w}) - \left[\frac{Re}{\mu_o T_o}(\alpha U_o' + \beta W_o') - \frac{i}{\mu_o}\frac{d\mu_o}{dT_o}T_o'(\alpha^2 + \beta^2)\right]\hat{v} - \frac{iRe}{\mu_o}(\alpha^2 + \beta^2)\hat{p} = 0 \quad (B1)$$

$$D^2\hat{v} + \frac{i(\lambda-1)}{\lambda}D(\alpha\hat{u} + \beta\hat{w}) - \left[\frac{iRe}{\mu_o T_o\lambda}(\alpha U_o + \beta W_o - \omega) + \frac{\alpha^2 + \beta^2}{\lambda}\right]\hat{v} + \frac{i}{\lambda\mu_o}\frac{d\mu_o}{dT_o}(\alpha U_o' + \beta W_o')(\alpha\hat{w} - \beta\hat{u}) = 0 \quad (B2)$$

$$D\hat{v} + i(\alpha\hat{u} + \beta\hat{w}) - \frac{T_o'}{T_o}\hat{v} + i\gamma M^2(\alpha U_o + \beta W_o - \omega)\hat{p} - \frac{i}{T_o}(\alpha U_o + \beta W_o - \omega)\hat{\tau} = 0 \quad (B3)$$

$$D^2\hat{\tau} + 2(\gamma - 1)M^2\sigma\frac{(\alpha V_o' + \beta W_o')}{\alpha^2 + \beta^2}D(\alpha\hat{u} + \beta\hat{w}) + \frac{2i}{\mu_o T_o'}\frac{d\mu_o}{dT_o}D\hat{\tau}$$
$$+2(\gamma-1)M^2\sigma\frac{(\alpha W_o' - \beta U_o')}{(\alpha^2+\beta^2)}D(\alpha\hat{w} - \beta\hat{u}) - \left[\frac{Re}{\mu_o T_o}T_o' - 2i(\gamma-1)M^2\sigma(\alpha U_o' + \beta W_o')\right]\hat{v}$$
$$+\frac{iRe\sigma}{\mu_o}(\sigma-1)M^2(\alpha U_o + \beta W_o - \omega)\hat{p} - \left[\frac{iRe\sigma}{\mu_o T_o}(\alpha U_o + \beta W_o - \omega) + (\alpha^2 + \beta^2)\right.$$
$$\left. -(\gamma-1)\frac{\sigma M^2}{\mu_o}\frac{d\mu_o}{dT_o}(U_o'^2 + W_o'^2) - \frac{1}{\mu_o}\frac{d^2\mu_o}{dT_o^2}(T_o')^2 - \frac{1}{\mu_o}T_o''\frac{d\mu_o}{dT_o}\right]\hat{\tau} = 0 \quad (B4)$$

$$D^2(\alpha\hat{w} - \beta\hat{u}) + \frac{1}{\mu_o}\frac{d\mu_o}{dT_o}(\alpha W_o' - \beta U_o')D\tau' + \frac{1}{\mu_o}\frac{d\mu_o}{dT_o}T_o' D(\alpha\hat{w} - \beta\hat{u})$$
$$- \frac{Re}{\mu_o T_o}(\alpha W_o' - \beta U_o')\hat{v} + \left[\frac{T_o'}{\mu_o}\frac{d^2\mu_o}{dT_o^2}(\alpha W_o' - \beta U_o') + \frac{1}{\mu_o}\frac{d\mu_o}{dT_o}(\alpha W_o'' - \beta U_o'')\right]\hat{\tau}$$

$$-\left[\frac{iRe}{\mu_o T_o}(\alpha U_o + \beta W_o - \omega) + (\alpha^2 + \beta^2)\right](\alpha\hat{w} - \beta\hat{u}) = 0 \qquad \text{(B5)}$$

Where primed quantities indicate differentiation with respect to y. The equation of state is where Φ_o is the mean flow density and $\hat{\phi}$ the complex density disturbance. The boundary conditions are

$$(\hat{v},\ \alpha\hat{u} + \beta\hat{w},\ \alpha\hat{w} - \beta\hat{u},\ \hat{\tau}) = 0 \quad @ \quad y = 0$$

$$\text{(B7)}$$

$$(\hat{v},\ \alpha\hat{u} + \beta\hat{w},\ \alpha\hat{w} - \beta\hat{u},\ \hat{\tau}) \to 0 \quad \text{as } y \to \infty$$

Table I. Calculation of temporal eigenvalues for different grids using spectral stability code.

($M_\infty = 4.5$, $T_e = 520°R$, $Re = 10{,}000$, $\alpha = .6$, $\beta = 1.0392$)

# pts	ω
45	$(.531703936,\ 1.835681668 \times 10^{-2})$
51	$(.495671735,\ 3.892834157 \times 10^{-3})$
63	$(.495719127,\ 3.903565212 \times 10^{-3})$
65	$(.495936719,\ 3.890656057 \times 10^{-3})$
73	$(.495844598,\ 3.977529728 \times 10^{-3})$
81	$(.495811969,\ 3.933484970 \times 10^{-3})$
95	$(.495825367,\ 3.933674615 \times 10^{-3})$
120	$(.495824177,\ 3.935165423 \times 10^{-3})$
151	$(.495824195,\ 3.935192776 \times 10^{-3})$
200	$(.495824195,\ 3.935192907 \times 10^{-3})$

Table II. Calculation of temporal eigenvalues for different grids using COSAL

($M_\infty = 4.5$, $T_e = 520°R$, $R_e = 10{,}000$, $\alpha = .6$, $\beta = 1.0392$)

# pts	ω
211	$(.485935455,\ 3.720663427 \times 10^{-3})$
513	$(.495990048,\ 3.688194736 \times 10^{-3})$
1025	$(.495983122,\ 3.683247436 \times 10^{-3})$
1500	$(.495981891,\ 3.682367822 \times 10^{-3})$
1500 (with Richardson extrapolation)	$(.495980770,\ 3.681512173 \times 10^{-3})$

Table III. Calculation of temporal eigenvalues for different grids using multi-domain spectral stability code.

($M_\infty = 4.5$, $T_e = 520^0 R$, Re = 10,000, $\alpha = .6$, $\beta = 1.0392$)

2 Domains (total # of points)	ω	
25/17 (42)	.495824525	$3.935144920 \times 10^{-3}$
31/21 (52)	.495824188	$3.935209430 \times 10^{-3}$
37/21 (58)	.495824191	$3.935210299 \times 10^{-3}$

Table IV. Effect of stretching parameter: single vs. multi-domain spectral discretization

($M_\infty = 1.0$, Re = 10,000, $\beta_T = .2$, $\alpha = 5$, $\beta = 0$)

Single-domain, 99 pts. on [-100, 100]

Stretching Parameter	ω	
1.1	1.47179	.272236
1.2	1.49883	.159461
1.3	1.49806	.158226
1.4	1.57814	.266635

Multi-Domain, 17/65/17 (99 pts) on [-100/-1/1/100].

Stretching Parameter	ω	
1.6	1.497268	.167511
2.0	1.496913	.159498
2.2	1.497146	.158781
4.0	1.496359	.156986

References

[1] A.T. Patera, A spectral element method for fluid dynamics: laminar flow in a channel expansion. J. Comput. Physics 54 (1984).

[2] N. Ghaddar, A.T. Patera and B. Mikic, Heat transfer enhancement in oscillatory flow in a grooved channel, AIAA Paper 84-0495.

[3] B. Metivet and Y. Morchoisne, Multi-domain spectral techniques for viscous flow calculations, in: Proceedings of the 4th Conference on Numerical Methods in Fluid Dynamics, Oct. 1981.

[4] H.H. Migliore and E.G. McReynolds, Multi-element collocation solution for convective dominated transport, C. Taylor, J. Johnson, and W. Smith, Eds., Numerical Methods in Laminar and Turbulent Flow, 1983.

[5] D.L. Gottlieb, L. Lustman and C.L. Streett, Sectral methods for two-dimensional shocks, ICASE Report No. 82-83, Nov. 1982.

[6] C.L. Streett, A spectral method for the solution of transonic potential flow about an arbitrary two-dimensional airfoil, AIAA Paper No. 83-1949-CP, Paper presented at the AIAA 16th Computational Fluid Dynamics Conference, Danvers, MA, July 13-15, 1983.

[7] M.Y. Hussaini, C.L. Streett and T. Zang, Spectral methods for partial differential equations, NASA CR-172248, Aug. 1983.

[8] C.L. Streett, T.A. Zang and M. Hussaini, Spectral multigrid methods with applications to transonic potential flow, Comput. Physics 56 (1984).

[9] C.L. Streett, T.A. Zang, and M.Y. Hussaini, Spectral methods for solution of the boundary-layer equation, AIAA Paper 84-0170, Paper presented at the AIAA 22nd Aerospace Sciences Meeting, Reno, NV, Jan. 9-12, 1984.

[10] M.G. Macaraeg, Numerical model of the axisymmetric flow in a heated, rotating spherical shell, PhD Dissertation, University of Tennessee Space Institute, TX 1-541-627, June 1984.

[11] M.G. Macaraeg, The effect of power law body forces on a thermally-driven fluid between concentric rotating spheres, J. Atmospheric Sci. 62 (2) (1986).

[12] M.G. Macaraeg, A mixed pseudospectral/finite difference method for the axisymmetric flow in a heated, rotating spherical shell, J. Comput. Physics 43 (3) (1986).

[13] M.G. Macaraeg, A mixed pseudospectral/finite difference method for a thermally-driven fluid in a nonuniform gravitational field, AIAA Paper 85-1662, Present at the AIAA 18th Fluid Dynamics and Plasmadynamics and Laser Conference, Cincinnati, OH, July 15-18, 1985.

[14] M.G. Macaraeg, and C.L. Streett, Improvements in Spectral Collocation Discretization Through a Multiple Domain Technique, Applied Numerical Mathematics, vol. 2, no. 2, 1986.

[15] W.C. Davy, C.K. Lombard, and M.J. Green, Forebody and Base Real-Gas Flow in Severe Planetary Entry by a Factored Implicit Numerical Method, AIAA Paper 81-0282.

[16] C. Park, On Convergence of Computation of Chemically Reacting Flows, AIAA Paper 85-0247.

[17] F.G. Blottner, Nonequilibrium Laminar Boundary Layer Flow of Ionized Air, General Electric Report R645D56, 1964.

[18] W.D. Erickson, and R.K. Prabhu, Rapid Combustion of Chemical Equilibrium Composition: An Application to Hydrocarbon Combustion, AIChE Journal, vol. 32, 1986.

[19] Janaf Tables, Thermochemical Data - Dow Chemical Co., 1977.

[20] R.B. Bird, W.E. Stewart, and E.N. Lightfoot, Transport Phenomena, 1966.

[21] C. Truesdell, Precise Theory of the Absorption and Dispersion of Forces Plan Infinitesimal Waves According to the Navier-Stokes Equations, J. of Rational Mechanics and Analysis, vol. 2, no. 4, 1953.

[22] F.S. Sherman, A Low Density Wind-Tunnel Study of Shock Wave Structure and Relaxation Phenomena in Gases, NACA TN-3298, 1955.

[23] Mack, L. M., "Boundary Layer Stability Theory," NASA CR-131591 (Jet Propulsion Lab.), November 1969.

[24] Streett, C. L., "Spectral Methods and their Implementation to Solution of Aerodynamic and Fluid Mechanic Problems," presented at the Sixth International Symposium on Finite Element in Flow Problems, Antibes, France, June 1986.

[25] Streett, C. L., T. A. Zang, and M. Y. Hussaini, "Spectral Methods for Solution of the Boundary-Layers Equations," AIAA paper no. 84-0170, presented at the 22nd AIAA Aerospace Sciences Meeting, Reno, Nevada, January 1983.

[26] Wilkinson, J. H.: The Algebraic Eigenvalue Problem. Oxford, 1965.

[27] Malik, M. R., "COSAL - A Black-Box Compressible Stability Code for Transition Prediction in Three-Dimensional Boundary Layers," NASA CR-165925, May 1982.

[28] Drummond, J. P., "Numerical Simulation of a Supersonic, Chemically Reacting Mixing Layer," PhD dissertation, George Washington University, May 1987.

[29] Gropengiesser, H.: Study of the Stability of Boundary Layers and Compressible Fluids. NASA TT F-12,786, 1969.

CHAPTER 31

Nonconforming Mortar Element Methods: Application to Spectral Discretizations

Yvon Maday*†
Cathy Mavriplis†
Anthony T. Patera†

ABSTRACT

Spectral element methods are p-type weighted residual techniques for partial differential equations that combine the generality of finite element methods with the accuracy of spectral methods. We present here a new *nonconforming* discretization which greatly improves the flexibility of the spectral element approach as regards automatic mesh generation and non-propagating local mesh refinement. The method is based on the introduction of an auxiliary "mortar" trace space, and constitutes a new approach to discretization-driven domain decomposition characterized by a clean decoupling of the local, structure-preserving residual evaluations and the transmission of boundary and continuity conditions. The flexibility of the mortar method is illustrated by several nonconforming adaptive Navier-Stokes calculations in complex geometry.

* Laboratoire d'Analyse Numerique de l'Universite Pierre et Marie Curie, Paris, France

† Department of Mechanical Engineering, Massachusetts Institute of Technology, Cambridge, Massachusetts 02139

1 Introduction

Spectral element methods [22,25,27] are weighted residual techniques for the approximation of partial differential equations that combine the rapid convergence rate of spectral methods [6,14] with the generality of finite element techniques [8,12,29]. The spectral element discretization, coupled to fast order-independent iterative solvers [21,28,32], yields numerical algorithms which have proven computationally efficient on both serial and parallel processors [11,10]. Although the spectral element method is, by construction, applicable in complex geometries [16,18,27], the large indestructible geometric unit associated with high-order brick elements leads to a certain lack of flexibility as regards automatic mesh generation, adaptive mesh refinement, and the treatment of moving boundaries. In this paper we present a new method, the "mortar element method", which largely eliminates this rigidity by allowing for nonconforming matching between subdomains.

The "mortar element method" represents a new domain decomposition approach [7,13] in which there is a clean decoupling of local-structure-preserving internal residual evaluations and the transmissions of boundary (or continuity) conditions. The method is not based on Lagrange-multiplier interface constraints e.g. [9], but rather on the explicit construction of the appropriate nonconforming space of approximation through the introduction of a new intermediary mortar trace space. The explicit-space approach is more appropriate for fast iterative solution than the Lagrange-multiplier methods, as it avoids the necessity of solving a coupled, potentially ill-conditioned problem. Although we develop the mortar methods here for spectral element discretizations, they are also appropriate in the h-type finite element context [4], in which they constitute an extension and generalization of classical nonconforming methods [8,9,29,31].

We present here the "mortar element method" in its simplest form for the solution of two-dimensional second-order elliptic and saddle problems. The emphasis is on the numerical formulation, implementation, and demonstration of the technique, and the illustration of the flexibility of the nonconforming paradigm; theoretical support for the method is given in [4], in which the optimality of the discretization is proven. The outline of the paper is as follows. In Section 2 we present the basic discretization for the Poisson equation in terms of the function spaces over which the standard variational form is to be tested. In Section 3 we present the associated nonconforming bases and the resulting set of discrete equations. Conjugate gradient iterative solution of the mortar discretization is described, illustrating the strong domain decomposition nature of the residual evaluation procedure. In Section 4 the extension of the method to the solution of the Stokes and Navier-Stokes problem is presented. Lastly, in Section 5 we give several numerical examples.

2 Spectral Element Nonconforming "Mortar" Spaces

2.1 Problem Formulation

We consider first the solution of a Poisson equation on a domain Ω of $\mathbf{R}^2$: Find $u(x,y)$ such that

$$-\nabla^2 u = f \quad \text{in } \Omega, \tag{1a}$$

$$u = 0 \quad \text{on } \partial\Omega, \tag{1b}$$

where $\partial\Omega$ is the boundary of Ω, and f is the prescribed force. We suppose that Ω is rectangularly decomposable, that is, that there exist rectangular subdomains Ω^k, $k = 1, ..., K$ such that

$$\overline{\Omega} = \bigcup_{k=1}^{K} \overline{\Omega}^k, \quad \forall k,l,\ k \neq l,\ \Omega^k \cap \Omega^l = \emptyset. \tag{2}$$

The problem (1a,1b) is well posed in $X = H_o^1$ in the sense that the following weak formulation of the problem admits only one solution: Find $u \in X$ such that

$$(\nabla u, \nabla v) = < f, v >, \quad \forall v \in X. \tag{3}$$

Here $(.,.)$ represents the L^2 inner product, and $< .,. >$ denotes the duality pairing between X and its dual space. For the definition of standard spaces, norms and inner products we refer the reader to [1].

For the Galerkin numerical approximation of problem (1a,1b), we test the variational form (3) with respect to a family of discrete finite dimensional spaces X_h, where h denotes a discretization parameter: Find $u_h \in X_h$ such that

$$(\nabla u_h, \nabla v_h) = < f, v_h >, \quad \forall v_h \in X_h. \tag{4}$$

In the case of a conforming approximation, for which $X_h \subset X$, the convergence and convergence rate of u_h towards u is determined essentially by stability (ellipticity and continuity) and approximation theory (infimum of $||u - v_h||_{1,\Omega}$ over all $v_h \in X_h$, where $||\cdot||_{1,\Omega}$ refers to the H^1 norm over Ω). In the case of nonconforming approximations, for which $X_h \not\subset X$, we must also consider the consistency error, which measures the deviation of the approximation space X_h from the proper space X [8,29].

To date, spectral element approximations [22] have been based on domain decompositions that satisfy (2) as well as the additional constraint that the intersection of two adjacent elements is either an entire edge or a vertex; this second constraint is derived from the conforming assumption, and is also present in the finite element method. In the spectral element context this constraint can be prohibitively restrictive due to the large geometric units involved. Although relaxing the conforming constraint clearly introduces a new source of error, it has the potential advantage of greatly increasing the flexibility of the numerical method as regards mesh generation and adaptive refinement procedures. This increase in flexibility improves not only the efficiency of the algorithm, but also the tractability of calculations involving moving and sliding meshes [15]. Furthermore, the nonconforming approach achieves generality at no cost in loss of local structure, an important consideration as regards optimal solvers.

We present here a spectral method based on nonconforming approximations in which the consistency errors are commensurate with the approximation errors. To present the nonconforming spectral element space X_h we first describe the anatomy of the discretization. The K rectangular subdomains of (2) are now identified as spectral elements, and the (x, y) coordinate system is chosen so as to be aligned with the edges of the Ω^k. These edges are denoted $\Gamma^{k,l}$, $l = 1, ..., 4$, such that $\partial\Omega^k = \bigcup_{l=1}^{4} \overline{\Gamma}^{k,l}$. We next introduce the set of "mortars" γ^p, where

$$\gamma^p = \text{int}(\overline{\Omega}^k \cap \overline{\Omega}^l) \tag{5a}$$

for some k and l, or

$$\gamma^p = \text{int}(\overline{\Omega}^m \cap \partial\overline{\Omega}) \tag{5b}$$

for some index m, where p is an arbitrary enumeration $p = 1, ..., M$ of all (k, l) and m such that $\text{int}(\overline{\Omega}^k \cap \overline{\Omega}^l)$ or $\text{int}(\overline{\Omega}^m \cap \partial\overline{\Omega})$ is not empty. The intersection of all closures of all γ^p defines a set of vertices $\mathcal{V}$ composed of all elements

$$v^q = \overline{\gamma}^m \cap \overline{\gamma}^n \tag{6}$$

where q is an arbitrary enumeration $q = 1, ..., V$ of all couples (m, n) for which $(\overline{\gamma}^m \cap \overline{\gamma}^n)$ is not

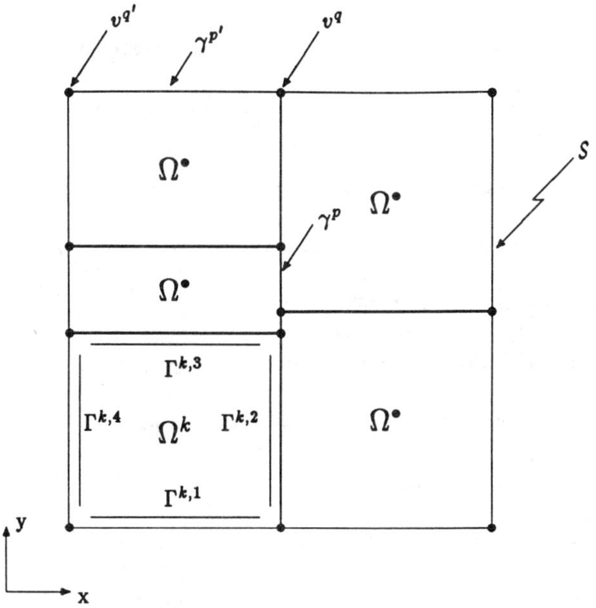

Figure 1: Subdomains and Mortars of a Nonconforming Decomposition

empty. (The set $\mathcal{V}$ is equal to the set of all the vertices of the Ω^k by definition of the mortars). Lastly, we define the skeleton S of the mortar system by

$$S = \bigcup_{p=1}^{M} \overline{\gamma}^p = \bigcup_{k=1}^{K} \partial\Omega^k. \tag{7}$$

The geometry of the nonconforming decomposition is shown graphically in Figure 1.

In order to define the nonconforming space X_h, we first require an auxiliary mortar space W_h

$$W_h = \{\phi \in C^0(S),\ \forall p = 1, ..., M,\ \phi_{|_{\gamma^p}} \in \mathbf{P}_N(\gamma^p),\ \phi_{|\partial\Omega} = 0\} \tag{8}$$

where $\mathbf{P}_N(\gamma^p)$ is the space of all polynomials on γ^p of degree $\leq N$. The nonconforming space is then given by

$$X_h = \{v \in L^2(\Omega),\ \forall k = 1,...,K,\ v_{|_{\Omega^k}} \in \mathbf{P}_N(\Omega^k) \text{ such that } \exists \phi \in W_h \text{ for which:}$$

$$\forall q = 1,...,V,\ \forall k = 1,...,K, \text{ such that } v^q \text{ is a vertex of } \Omega^k,\ v_{|_{\Omega^k}}(v^q) = \phi(v^q); \tag{9a}$$

$$\text{and } \forall l = 1,...,4,\ \forall k = 1,...,K,\ \forall \psi \in \mathbf{P}_{N-2}(\Gamma^{k,l}),\ \int_{\Gamma^{k,l}} (v_{|_{\Omega^k}} - \phi)\psi ds = 0 \ \}. \tag{9b}$$

Here $\mathbf{P}_N(\Omega^k)$ denotes the space of all polynomials on Ω^k of degree $\leq N$ in each spatial direction; the spectral element discretization parameter is the couple $h = (K, N)$. For a conforming approximation X_h is the standard spectral element space; here, and elsewhere in this paper, we assume the reader is familiar with the conforming spectral element method [22].

Let us summarize the properties of the approximation space X_h. First, as regards the uniqueness of the solution, we note that the uniqueness of the mortar element $\phi \in W_h$ is not of major importance as long as its image $u_h \in X_h$ is unique; it is u_h, not the mortar element, that must be close to u. The uniqueness of the discrete solution u_h follows from the ellipticity of the Laplacian form $(\nabla u_h, \nabla v_h)$, $\forall u_h \in X_h$, $\forall v_h \in X_h$ with respect to the following "broken $H^1(\Omega)$ norm",

$$\|v_h\|_{X_h} = \left[\sum_{k=1}^{K} \|(v_h)_{|_{\Omega^k}}\|^2_{1,\Omega^k}\right]^{1/2}, \quad \forall v_h \in X_h. \tag{10}$$

Although the proof of ellipticity is quite involved (see [4]), an elementary proof of uniqueness can be readily derived. To wit, we note that if u_h and u'_h are two solutions of (4), we get

$$0 = (\nabla v_h, \nabla v_h) = \sum_{k=1}^{K} \int_{\Omega^k} [\nabla (v_h)_{|_{\Omega^k}}]^2 \text{ with } v_h = u_h - u'_h,$$

and thus v_h is piecewise constant. Using the fact that the elements of X_h vanish over $\partial\Omega$ and are continuous at the vertices of $\mathcal{V}$, it follows that $v_h \equiv 0$ and thus $u_h = u'_h$.

Although uniqueness of $\phi \in W_h$ is not necessary, it is nevertheless true that spurious (or parasitic) modes in ϕ correspond to unprofitable work, and can potentially cause problems in the

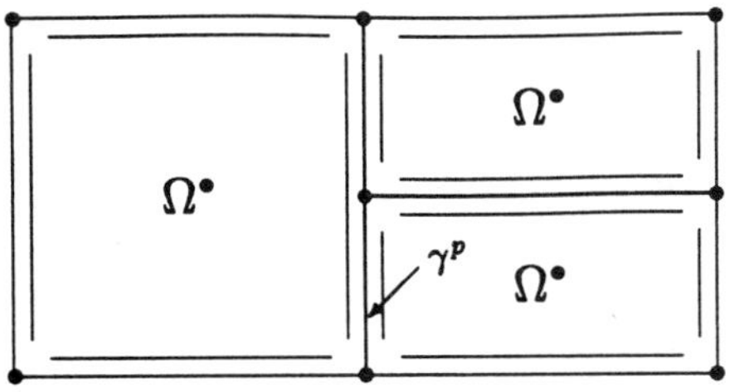

Figure 2: Nonconforming Discretization Derived from the Refinement of a Conforming Approximation

subsequent solution of the discrete system (see Section 3.2). There is one situation in which the uniqueness of ϕ follows easily; this is the case where for each γ^p there exists an element Ω^k that accepts γ^p as an entire edge (see Figure 2). This arises, for instance, from a refinement of a mesh which is initially conforming. In this paper we shall consider only this "refinement" case; development and analysis of the general problem of Figure 1 is more involved, and is relegated to future publications. For the "refinement" case uniqueness of ϕ results from the fact that the mortar element ϕ coincides exactly with the restriction $(v_h)_{|_{\Omega^k}}$ over γ^p. To show this we note that, by construction, the elements ϕ and $v_h^k = (v_h)_{|_{\Omega^k}}$ coincide at the endpoints of γ^p. This implies that $\phi - v^k_{h|_{\gamma^p}}$ is a polynomial of the local coordinate $\tilde{s}$,

$$(\phi - v^k_{h|_{\gamma^p}})(\tilde{s}) = (1 - \tilde{s}^2)\Phi(\tilde{s}), \tag{11}$$

where Φ is a polynomial of degree $\leq N - 2$. Here, and in what follows, $\tilde{s} = \tilde{x}$ (or $\tilde{s} = \tilde{y}$) for a horizontal (or vertical) mortar, where $\tilde{x}$ (or $\tilde{y}$) is a mortar-local variable which scales x (or y) such that γ_p corresponds to $]-1,1[$ (similarly, $\hat{s} = \hat{x}$ (or $\hat{s} = \hat{y}$) for a horizontal (or vertical) edge, where $\hat{x}$ (or $\hat{y}$) is an element-local variable on $]-1,1[$ which scales x (or y) to the appropriate $\Gamma^{k,l}$). From the orthogonality of $\phi - v^k_{h|\gamma^p}$ to all elements of $\mathbf{P}_{N-2}(\gamma^p)$ (9b), it follows that Φ is necessarily zero, and thus ϕ is exactly the trace of one piece of v_h. The uniqueness of the solution u_h to problem (4) thus yields the uniqueness of the corresponding mortar element.

Let us consider now the consistency error. The scheme (4) based on the definition (9a,9b) of X_h is optimal in that the consistency error is maintained small by the combination of the L^2 condition (9b) and the vertex condition (9a). In essence, the L^2 condition ensures that the jump in functions is small in the interior of internal boundaries, whereas the vertex condition ensures exact continuity at cross points where the normal derivative has more than one sense. We note that the superiority of the L^2- (versus pointwise-) matching of $v_{|_{\Omega^k}}$ and ϕ has been demonstrated previously [2]. The mortar methods are different from previously proposed nonconforming L^2 approximations in that the latter are mortarless master-slave spaces, whereas the current approach is democratic; this allows for very simple implementation in arbitrary topologies.

Lastly, the approximation properties of the space X_h are similar to those of past nonconforming approximations. For example, for the case of a square domain decomposed into several elements, as a first result one can use the best *global* polynomial approximation as a bound for approximation errors. The combination of stability, consistency and approximation result in an optimal scheme, the details, and degree of locality of which, are described in [4]. We note that for the special case of infinitely smooth solutions, u_h approaches u exponentially fast as $N \to \infty$ for fixed K (spectral convergence).

3 Representation and Discrete Equations

3.1 Bases

Although the spaces W_h and X_h appear quite complicated, they have a simple basis and evaluation procedure which yields an efficient domain decomposition algorithm. In this section we discuss the basis, and in the following section we describe residual evaluation.

To begin, we write for the space W_h,

$$\phi_{|_{\gamma^p}} = \sum_{j=0}^{N} \phi_j^p h_j^N(\tilde{s}), \quad \forall p \in \{1, ..., M\} \tag{12}$$

where we assume that all indices increase with increasing x, y. Here the h_j^N are Lagrangian interpolants defined by

$$h_j^N \in \mathbf{P}_N(]-1,1[), \quad h_j^N(\xi_i) = \delta_{ij}, \quad \forall i,j \in \{0,...,N\}^2 \tag{13}$$

where the $\xi_i (= \xi_i^N)$ are the $N+1$ Gauss-Lobatto Legendre points defined by the zeroes of $L_N'(z)(1-z^2)$, and L_N is the Legendre polynomial of order N [30] so that

$$h_j^N(z) = -\frac{1}{N(N+1)L_N(\xi_i)} \frac{(1-z^2)L_N'(z)}{z-\xi_i} \quad z \in]-1,1[, \quad \forall j \in \{0,...,N\}.$$

The definition (12) is not sufficient given the requirement that $\phi \in W_h$ must be $C^0(S)$; to indicate the continuity condition, we resort to diagrammatic methods. The mortar conventions are described in Table 1a, with the basis for W_h shown in Figure 3b for the nonconforming mesh of Figure 3a.

We next construct a representation for $v \in X_h$ in terms of the mortar. To begin, we write

$$v_{|_{\Omega^k}} = \sum_{i=0}^{N} \sum_{j=0}^{N} v_{ij}^k h_i^N(\hat{x}) h_j^N(\hat{y}), \quad \forall k \in \{1, ..., K\} \tag{14}$$

where the h_i^N are defined in (13). The internal degrees-of-freedom, v_{ij}^k, $i,j \in \{1,...,N-1\}^2$, are clearly free, however the boundary degrees-of-freedom are constrained through (9a,9b). Based on

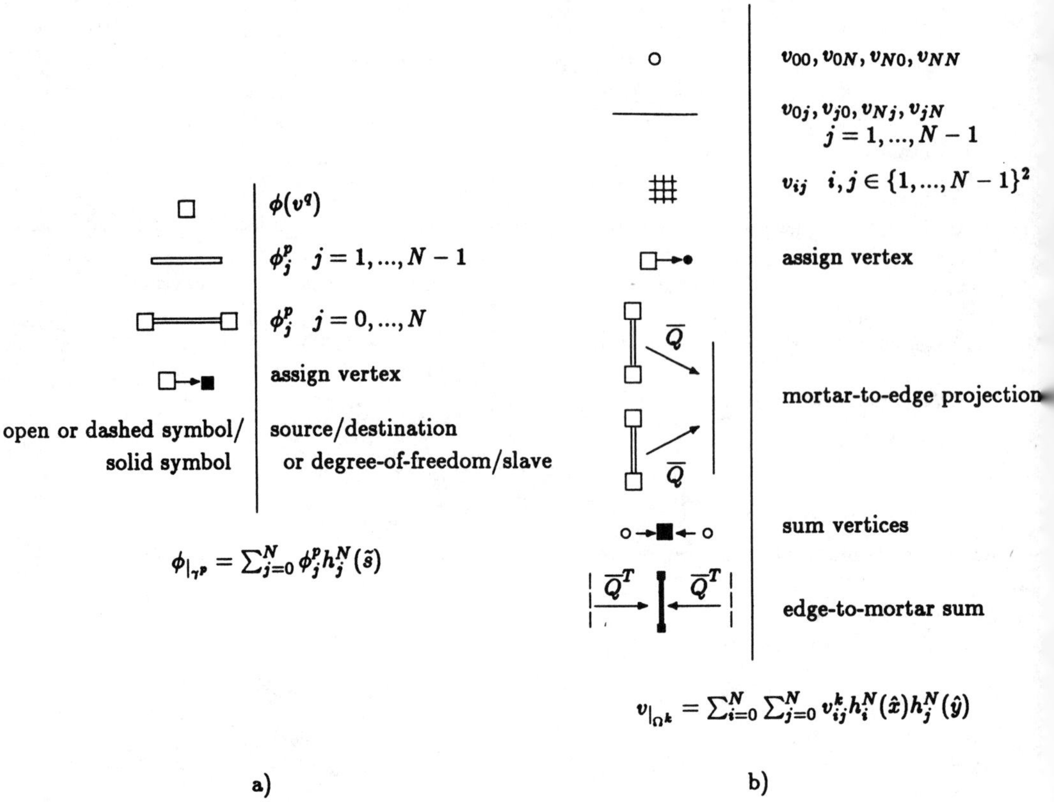

Table 1: Symbols for Diagrammatic Basis Representation

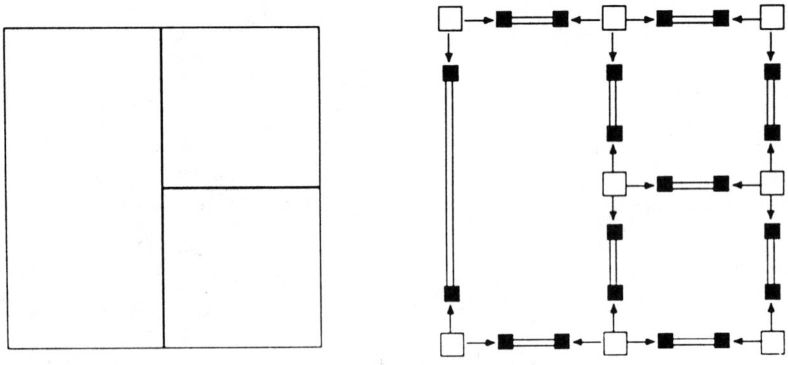

a) b)

Figure 3: a) Nonconforming Mesh and b) Associated Mortar Basis

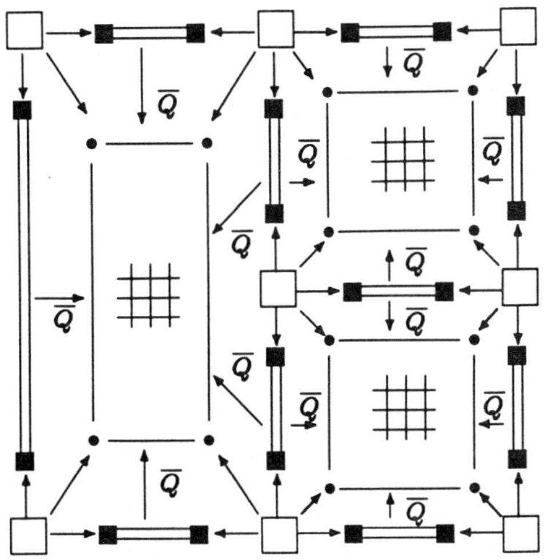

Figure 4: Diagrammatic Representation of the Basis for X_h on Nonconforming Mesh of Fig. 3a

the diagrammatic conventions of Table 1 the admissible v are given by Figure 4, where $\overline{Q}$ derives from the projection (9b). In order to construct $\overline{Q}$ we require a basis for ψ, which we choose as

$$\psi|_{\Gamma^{k,l}} = \sum_{q=1}^{N-1} \beta_q \eta_q^{N-2}(\hat{s}) \tag{15}$$

where

$$\eta_q^{N-2}(z) = (-1)^{N-q} \frac{L'_N(z)}{\xi_q - z} \qquad z \in]-1,1[, \quad q \in \{1,...,N-1\}. \tag{16}$$

To calculate the projection of (9b) we then perform (here exact) piecewise Gauss-Lobatto quadrature on $N+1$ points on the element edges and mortar segments, giving

$$\sum_{j=1}^{N-1} \overline{B}_{ij} v_j = \sum_{j=0}^{N} \overline{P}_{ij} \phi_j, \quad \forall i \in \{1,...,N-1\} \tag{17}$$

where for the destination edge $\Gamma^{k,l}$ (v_j) and source mortar γ^p (ϕ_j)

$$\int_{\Gamma^{k,l}} v\psi \longrightarrow \overline{B}_{ij} = \frac{|\Gamma^{k,l}|}{2}(-1)^{N-i}(-L''_N(\xi_i))\rho_i \delta_{ij}, \quad \forall i,j \in \{1,...,N-1\}^2 \tag{18}$$

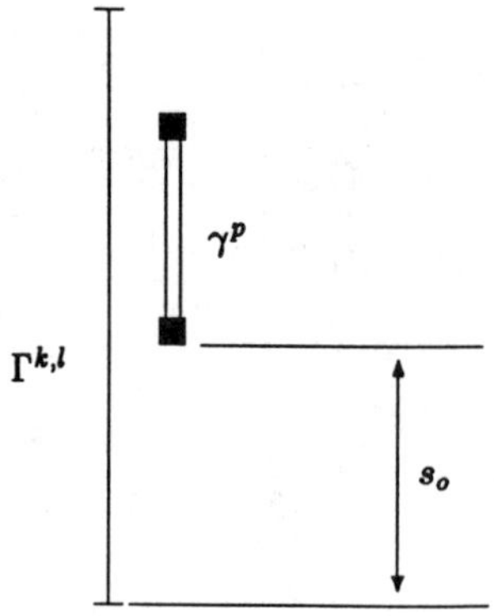

Figure 5: Definition of Mortar Offset s_o

and

$$\int_{\Gamma^{k,l}} \phi|_{\gamma^p}\psi \longrightarrow \overline{P}_{ij} = \frac{|\gamma^p|}{2}\eta_i^{N-2}\left(2\frac{s_o}{|\Gamma^{k,l}|} - 1 + (\xi_j+1)\frac{|\gamma^p|}{|\Gamma^{k,l}|}\right)\rho_j - \begin{cases} \frac{|\Gamma^{k,l}|}{2}\eta_i^{N-2}(-1)\rho_0\delta_{0,j} & \text{if } s_o = 0 \\ \frac{|\Gamma^{k,l}|}{2}\eta_i^{N-2}(1)\rho_N\delta_{N,j} & \text{if } s_o + |\gamma^p| = |\Gamma^{k,l}| \end{cases} \quad (19)$$
$$\forall i \in \{1,...,N-1\} \ \forall j \in \{0,...,N\}.$$

Here s_o is the offset of the mortar γ^p from the edge $\Gamma^{k,l}$, as shown in Figure 5; the endpoint terms of (19) derive from the vertex-pinning condition of (9a). Finally we arrive at

$$\overline{Q}_{ij} = [\overline{Q}] = [\overline{B}]^{-1}[\overline{P}], \quad \forall i \in \{1,...,N-1\}, \forall j \in \{0,...,N\}. \quad (20)$$

Note that by proper choice of the basis for ψ we can explicitly form the matrix $\overline{Q}$, that is, we are able to directly invert the diagonal inner product $\overline{B}$. The alternating-sign term in η_q^{N-2} assures that the entries of $\overline{B}$ are positive.

Although in practice we shall evaluate $v|_{\Omega^k}$ from the diagram without forming the global linear projection operator, it is nevertheless of theoretical interest to remark that the diagram is equivalent to

$$\begin{pmatrix} v^k_{ij}|_{interior}, \ \forall k \\ v^k_{ij}|_{interface}, \ \forall k \end{pmatrix} = \begin{pmatrix} [I] & 0 \\ 0 & [\overline{Q}] \end{pmatrix} \begin{pmatrix} v^k_{ij}|_{interior}, \ \forall k \\ \phi(v^q), \ \forall q, \ \phi^p_j, \ \forall p, \ \forall j \in \{1,...,N-1\} \end{pmatrix} \tag{21a}$$

or

$$\underline{v}^* = Q\underline{v}. \tag{21b}$$

We denote the vector $\underline{v}$ as the *algebraic* basis, in that this variable represents the finite-dimensional approximation space with an equivalent number of degrees-of-freedom; the proper *functional* basis corresponds to the images of $\underline{v}^T = (1,0,0...), (0,1,0,....), ..., (0,...,0,1)$ in $\underline{v}^*$ through the transformation Q, acting on the local bases $h_i h_j$ as described by (14).

3.2 Discrete Equations

Armed with the variational forms of Section 2 and the bases of Section 3.1, it is now a simple matter to construct the discrete equations. In particular, we note that our basis construction (21) allows us to express admissible elemental degrees-of-freedom $\underline{v}^*$ in terms of $\underline{v}$. This, in turn, permits us to construct the global discrete equations directly from local structure-preserving elemental equations, which is at the heart of the discretization-driven domain decomposition approach.

We first construct the decoupled elemental matrices and inhomogeneity,

$$blk(\hat{A}^k) = \begin{pmatrix} (\nabla h_p h_q, \nabla h_i h_j)^{k=1} & 0 & 0 \\ 0 & (\nabla h_p h_q, \nabla h_i h_j)^{k=2} & 0 \\ & & \\ 0 & 0 & (\nabla h_p h_q, \nabla h_i h_j)^{k=K} \end{pmatrix} \tag{22}$$

$$blk(\hat{f}^k) = \begin{pmatrix} (h_p h_q, f)^{k=1} \\ (h_p h_q, f)^{k=2} \\ \\ \cdot \\ \\ (h_p h_q, f)^{k=K} \end{pmatrix}, \qquad \forall i, j, p, q \in \{0, ..., N\}^4.$$

The kth block of $blk(\hat{A}^k)$ represents the Neumann Laplace operator on the elemental domain Ω^k. We now recognize that not all elemental $h_i h_j$ are possible, and that not all $h_p h_q$ are admissible; indeed, the admissible degrees-of-freedom follow from the Q transformation of (21). We thus arrive, rather simply, at the fully discrete equations:

$$Q^T blk(\hat{A}^k) Q \underline{u} = Q^T blk(\hat{f}^k). \tag{23}$$

We note that independent of the size of the mortar nullspace (Q right nullspace), (23) is solvable. A sufficient condition for a unique mortar function is that $Q^T Q$ be invertible; in the conforming cases $Q^T Q$ is simply the multiplicity of a node (that is, the number of elements in which it appears).

Equation (23) illustrates that the global Laplace operator can be thought of as a local operator "mortared" together by the Q^T, Q operations; indeed, the Q^T operator is the algebraic form of the standard direct stiffness procedure (here extended to nonconforming elements). In the implementation of iterative procedures the Q, Q^T are, of course, never explicitly formed, but rather are evaluated; diagrammatic evaluation of Q^T (direct stiffness summation) is shown in Figure 6 in terms of the diagram conventions defined in Table 1. The domain decomposition decoupling afforded by the implicit construction of the image basis through $\underline{u}^*$ allows for efficient parallel implementation following the methods described in [11] for conforming techniques.

Although the emphasis in the current paper is on the mortar discretization, the bases and

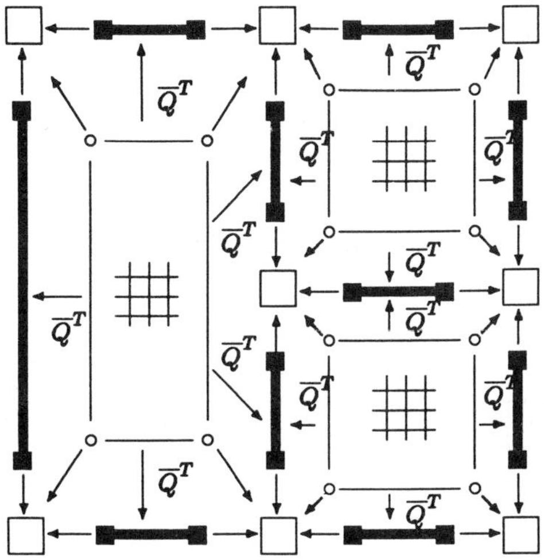

Figure 6: Direct Stiffness Summation Q^T of Residuals on Nonconforming Mesh

evaluation procedure have been tailored to admit efficient iterative solution, and it is therefore appropriate to briefly indicate how the method is used in conjunction with (for example) conjugate gradient iteration. To solve (23) we write

$$\underline{u}_0; \qquad \underline{r}_0 = Q^T blk(\hat{f}^k) - Q^T blk(\hat{A}^k) Q \underline{u}_0; \qquad \underline{q}_0 = \underline{r}_0 \tag{24}$$

$$a_m = (\underline{r}_m, \underline{r}_m)/(\underline{q}_m, Q^T blk(\hat{A}^k) Q \underline{q}_m)$$

$$\underline{u}_{m+1} = \underline{u}_m + a_m \underline{q}_m$$

$$\underline{r}_{m+1} = \underline{r}_m - a_m Q^T blk(\hat{A}^k) Q \underline{q}_m$$

$$b_m = (\underline{r}_{m+1}, \underline{r}_{m+1})/(\underline{r}_m, \underline{r}_m)$$

$$\underline{q}_{m+1} = \underline{r}_{m+1} + b_m \underline{q}_m,$$

where m refers to iteration number, $\underline{r}_m$ is the residual, $\underline{q}_m$ the search direction and $(.,.)$ is the usual discrete inner product. All evaluations are performed through the diagrams of Figures 4 and 6. The $blk(\hat{A}^k)$ operations are entirely local at the elemental level, with all transmission and coupling through Q. The local $\hat{A}^k$ calculations are the standard conforming spectral element tensor product evaluations, as the mortar decoupling allows all local structure to remain intact despite global irregularity (e.g. non-propagating mesh refinement).

4 The Stokes and Navier-Stokes Problems

In this section we consider the extension of the nonconforming mortar method to the solution of the two-dimensional steady Stokes problem in a rectangularly-decomposable domain Ω,

$$\begin{aligned} -\nu\nabla^2\mathbf{u} - \nabla p &= \mathbf{f} \\ \text{div}\mathbf{u} &= 0, \end{aligned} \tag{25}$$

with homogeneous Dirichlet velocity boundary conditions $\mathbf{u} = 0$ on $\partial\Omega$. Here $\mathbf{u}$ is the velocity, p is the pressure, $\mathbf{f}$ is the forcing vector, and ν is the kinematic viscosity. The associated variational problem is: Find $(\mathbf{u}, p) \in (H_o^1(\Omega)^2, L_o^2(\Omega))$ such that

$$\begin{aligned} \nu(\nabla\mathbf{u}, \nabla\mathbf{w}) - (p, \text{div}\mathbf{w}) &= (\mathbf{f}, \mathbf{w}), \quad \forall\mathbf{w} \in H_o^1(\Omega)^2 \\ (q, \text{div } \mathbf{u}) &= 0, \quad \forall q \in L_o^2(\Omega), \end{aligned} \tag{26}$$

where $L_o^2(\Omega)$ is the space of L^2 functions of zero mean.

The discrete formulation of the problem consists of choosing two discrete approximation spaces, one for the velocity field and one for the pressure. It is shown in [3,22,24] for the conforming spectral element approximation that choosing *both* of these spaces to be polynomials of degree less than or equal to the *same* degree N leads to an ill-posed problem, in which spurious pressure modes arise [5,12]. The existence of such modes is in contradiction with the verification of the "inf-sup" condition [5]. As regards our nonconforming methods for the Stokes problem, our starting point is the conforming staggered mesh method defined in [23] for which the "inf-sup" condition is satisfied. The correct nonconforming extension is to use the velocity space $(X_h)^2$ defined in (9a,9b), and for the pressure the space $M_h = \{\phi \in L^2, \ \phi_{|_{\Omega^k}} \in \mathbf{P}_{N-2}(\Omega^k)\}$ associated with the conforming approximation. In essence, the fact the pressure is L^2 implies that it need not be modified when the constraints on the velocity are relaxed.

With these spaces we arrive at the following nonconforming discretization:

Find $(\mathbf{u}_h, p_h) \in ((X_h)^2, M_h)$ such that

$$\begin{aligned} \nu(\nabla \mathbf{u_h}, \nabla \mathbf{w_h}) - (p_h, \operatorname{div}\mathbf{w_h}) &= <\mathbf{f}, \mathbf{w_h}>, \quad \forall \mathbf{w_h} \in (X_h)^2 \\ (q_h, \operatorname{div}\, \mathbf{u_h}) &= 0, \quad \forall q_h \in M_h \end{aligned} \tag{27}$$

from which uniqueness, stability, and spectral error properties follow from the results of previous sections and [4], suitably modified within the Stokes context as described in [23,24]. (We note that, as elsewhere in this paper, we do not dwell on quadrature issues which are, by now, standard practice.) We then choose a basis for M_h

$$p_{|_{\Omega^k}} = \sum_{i=1}^{N-1} \sum_{j=1}^{N-1} p_{ij}^k g_i^{N-2}(\hat{x}) g_j^{N-2}(\hat{y}), \quad \forall k \in \{1, ..., K\} \tag{28}$$

where the g_j^{N-2} are the $N-2$th order Gauss-Legendre interpolants, that is, those polynomials of $\mathbf{P}_{N-2}$ such that $g_j^{N-2}(\varsigma_i^{N-2}) = \delta_{ij}$, where ς_i^{N-2} are the $N-2$ zeroes of L_{N-2} [30]. We thus arrive at the discrete saddle problem

$$\begin{aligned} Q^T blk(\hat{A}^k) Q\underline{u} - Q^T blk(\hat{\mathbf{D}}^k)^T \underline{p} &= Q^T blk(\hat{\mathbf{f}}^k) \\ blk(\hat{\mathbf{D}}^k) Q\underline{u} &= 0. \end{aligned} \tag{29}$$

Here $\underline{p}$ is the algebraic basis for p analogous to $\underline{u}$ of (21b), $blk(\hat{A}^k), Q^T, Q, blk(\hat{\mathbf{f}}^k)$ are defined as in

(18-22), and $\hat{\mathbf{D}}$ is the gradient operator given by

$$blk(\hat{\mathbf{D}}^k) = \begin{pmatrix} (g_p^{N-2}g_q^{N-2}, \nabla h_i h_j)^{k=1} & 0 & . & 0 \\ 0 & (g_p^{N-2}g_q^{N-2}, \nabla h_i h_j)^{k=2} & & 0 \\ & & . & \\ & & & . \\ 0 & 0 & & (g_p^{N-2}g_q^{N-2}, \nabla h_i h_j)^{k=K} \end{pmatrix} \tag{30}$$

$$\forall i,j \in \{0,...,N\}^2, \quad \forall p,q \in \{1,...,N-1\}^2.$$

Extension to Navier-Stokes is straightforward given the lower-order nature of the convective terms.

As in the pure elliptic discretization, (29) is amenable to iterative solution. We currently use a semi-implicit procedure for Navier-Stokes, in which the nonlinear terms are treated explicitly, and the Stokes subproblem is handled with a Uzawa nested iteration [20]; conforming multigrid techniques [28] are currently being extended to the nonconforming case. In addition to the staggered mesh Stokes treatment, elliptic-splitting methods appropriate for higher Reynolds number flows are also used [17,19]; these discretizations represent sequences of elliptic operations (23), and thus their extension to the nonconforming case follows from Sections 2 and 3.

5 Numerical Examples

In this section we illustrate various aspects of our nonconforming method by a number of examples. The central point is the flexibility and ease-of-implementation afforded by a nonconforming approach based on a consistent, non-context-dependent matching. As our first example we consider the Helmholtz problem

$$-\nabla^2 u + \lambda^2 u = f \quad \text{on} \quad \Omega =]0,1[\times]0,1[\tag{31}$$

$$u = e^{\frac{\lambda}{\sqrt{2}}((x-1)+(y-1))} \quad \text{on} \quad \partial\Omega$$

where f is chosen such that the exact solution in Ω is given by $u = e^{\frac{\lambda}{\sqrt{2}}((x-1)+(y-1))}$. In Figure 7a we show a high-resolution conforming mesh $h = (K = 16, N = \bullet)$; in Figure 7b we show a nonconforming mesh $h = (K = 10, N = \bullet)$, in which the local structure-preserving mesh refinement is illustrated. In Figure 8 we plot the error in the X_h norm of (10) for both solutions as a function of N (K fixed) for $\lambda = 50$. This example demonstrates the rapid (here exponential) convergence of the spectral element approach, and the superior resolution properties of the nonconforming discretization, which achieves the same accuracy as the conforming approximation with significantly fewer degrees-of-freedom.

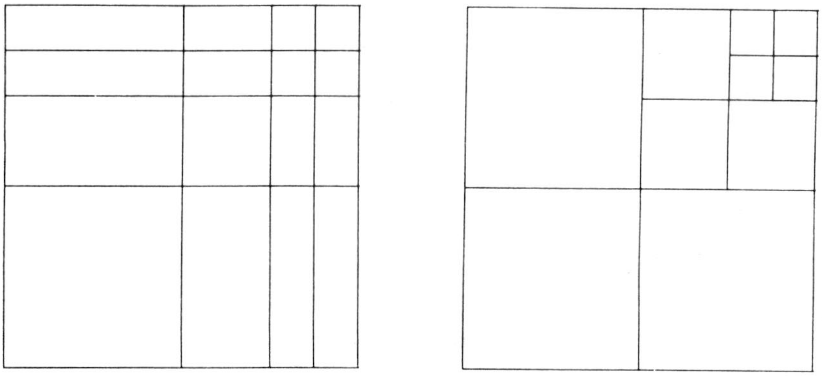

a) b)

Figure 7: High Resolution a) Conforming and b) Nonconforming Meshes for Helmholtz Problem

As our second example we demonstrate the utility of nonconforming methods in constructing appropriate meshes; we consider the labyrinth channel of Figure 9, in which the two meshes for the

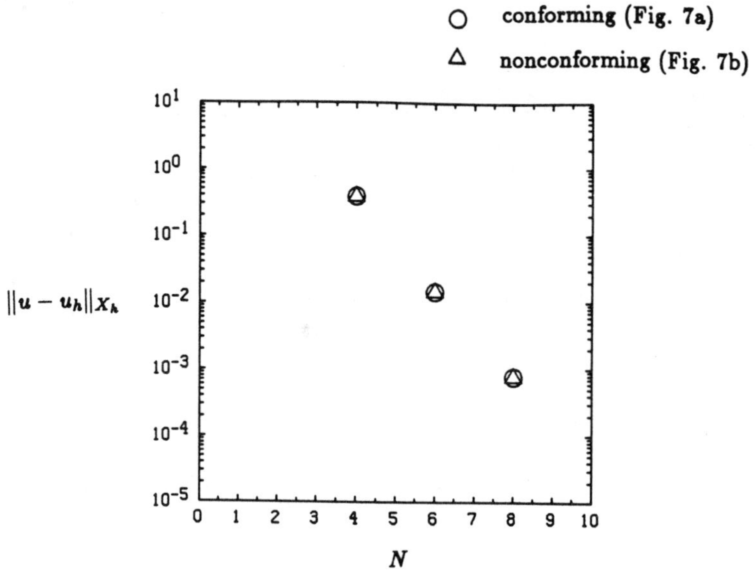

Figure 8: Convergence of the Broken H^1 Error for Helmholtz Problem

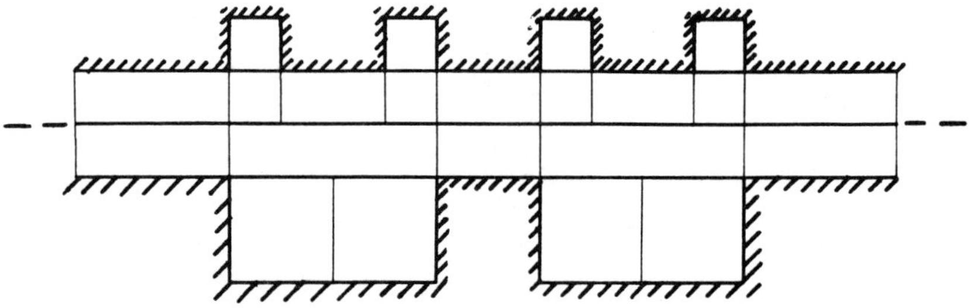

Figure 9: Nonconforming Labyrinth Channel Mesh, $h = (K = 22, N = 9)$

two sides of the channel are constructed "separately", and subsequently merged by mortar. The boundary conditions are given as: a parabolic velocity profile at inflow, no slip on the channel walls, and outflow (constant pressure) at the exit. In Figure 10 we show streamlines for the steady Stokes flow calculated by the discretization (29) and the nested conjugate gradient Uzawa method; notable are the continuity at element boundaries and the lack of spurious pressure modes. The mesh in Figure 9 can be thought of as one instance of a sliding channel calculation; nonconforming methods, with appropriate extension (as in Figure 1), should prove to be powerful techniques for moving boundary problems when used in conjunction with arbitrary-Lagrangian-Eulerian techniques [15].

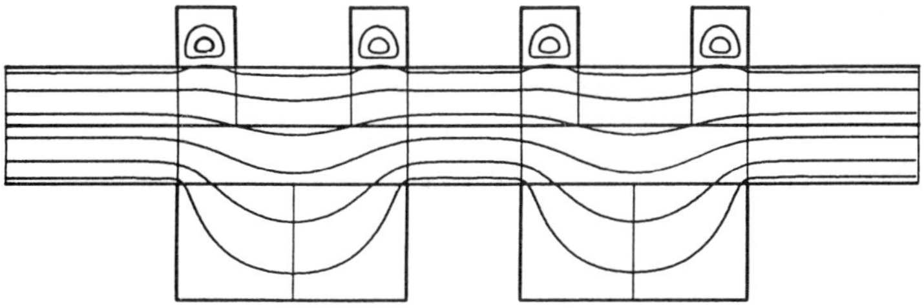

Figure 10: Stokes Solution for the Labyrinth Channel

Lastly, we consider a moderate Reynolds number flow past a wedge [17,26] in a channel; the utility of the nonconforming methods in generating an appropriately refined mesh is apparent in the mesh shown in Figure 11. Note that we relax here the constraint, introduced for simplicity of exposition in previous sections, that the elements be rectangular; treatment of general curved elements represents a simple extension of the methods described in Sections 2-4. In Figure 12 we show the short time solution of the startup vortex near the tip of the wedge, for a Reynolds number $R = 500$ at a time $\tau = \frac{tV}{H} = .085$ on the mesh $h = (K = 16, N = 9)$ of Figure 11. We prescribe a slug velocity profile at inflow, no slip boundary conditions on the walls, and outflow (constant pressure) at the exit. Here $R = \frac{VH}{\nu}$, where V is the channel average velocity, H the channel width, and ν is the kinematic viscosity, and t is time. The high resolution in the vicinity of the wedge allows for a detailed description of the startup vortex.

Acknowledgements

This work was supported by Avions Marcel Dassault-Bréguet Aviation, the ONR and DARPA under Contract N00014-85-K-0208, the ONR under Contract N00014-88-K-0188, and the NSF under Grant ASC-8806925. This work was partially done while the first author was in residence at ICASE.

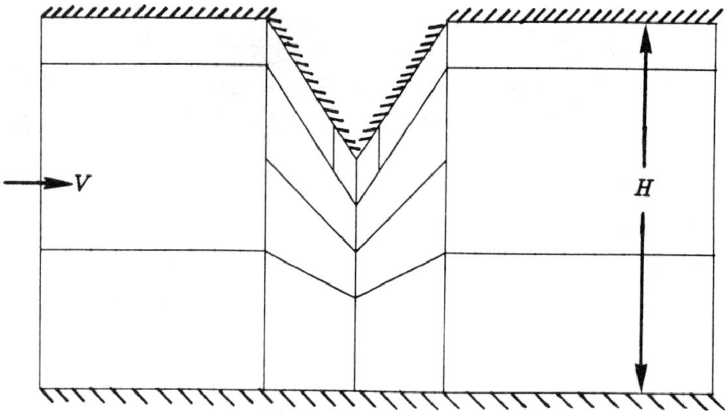

Figure 11: Nonconforming Mesh for Startup Flow Past a Wedge

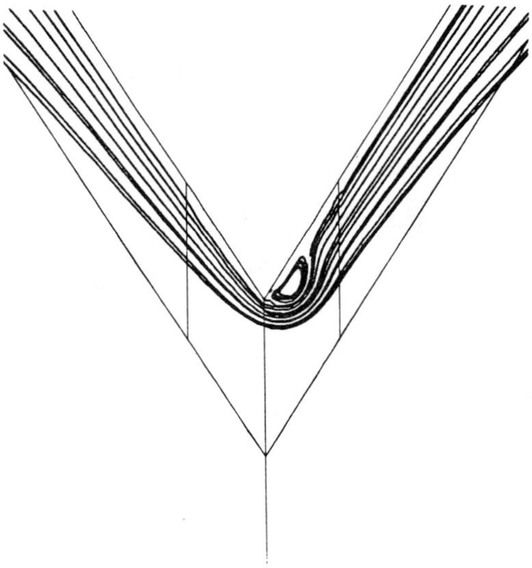

Figure 12: Navier-Stokes Solution for Flow Past a Wedge at $R = 500$, $\tau = .085$

References

[1] R. Adams. *Sobolev Spaces.* Academic Press, 1975.

[2] C. Bernardi, N. Débit, and Y. Maday. Coupling spectral and finite element methods for the Laplace equation. *Mathematics of Computation*, to appear (also ICASE report no. 87-70 and Publication du Laboratoire d'Analyse Numérique de l'Université Pierre et Marie Curie, no. 87031).

[3] C. Bernardi, Y. Maday, and B. Métivet. Calcul de la pression dans la résolution spectrale du problème de Stokes or Computation of the pressure in the spectral approximation of the Stokes problem. *La Recherche Aérospatiale (and English edition)*, 1–21, 1987.

[4] C. Bernardi, Y. Maday, and A.T. Patera. A new nonconforming approach to domain decomposition: the mortar element method. In H. Brezis and J. L. Lions, editors, *Nonlinear Partial Differential Equations and Their Applications, Collège de France Seminar*, Pitman, to appear.

[5] F. Brezzi. On the existence, uniqueness and approximation of saddle-point problems arising from Lagrange multipliers. *Rairo Anal. Numer.*, 8 R(2), 129, 1974.

[6] C. Canuto, M. Y. Hussaini, A. Quarteroni, and T. A. Zang. *Spectral Methods in Fluid Dynamics.* Springer-Verlag, 1987.

[7] T. Chan, editor. *Proceedings of the Second International Conference on Domain Decomposition Methods for Partial Differential Equations*, SIAM, Philadelphia, 1988.

[8] P. Ciarlet. *The Finite Element Method.* North Holland, 1978.

[9] M. Dorr. Domain decomposition via Lagrange multipliers. *submitted to Numerische Mathematik*, 1988.

[10] P. F. Fischer, L. W. Ho, G. E. Karniadakis, E. M. Rønquist, and A. T. Patera. Recent advances in parallel spectral element simulation of unsteady incompressible flows. *Proceedings of the Symposium on Advances and Trends in Computational Structural Mechanics and Computational Fluid Dynamics, Washington, D.C.*, 1988.

[11] P. F. Fischer and A. T. Patera. Parallel spectral element solution of the Stokes problem. *submitted to Journal of Computational Physics*, 1988.

[12] V. Girault and P. A. Raviart. *Finite Element Methods for the Navier-Stokes Equations.* Springer, 1986.

[13] R. Glowinski, G. Golub, G. Meurant, and J. Périaux, editors. *Proceedings of the First International Conference on Domain Decomposition Methods for Partial Differential Equations*, SIAM, Philadelphia, 1987.

[14] D. O. Gottlieb and S.A. Orszag. *Numerical Analysis of Spectral Methods.* SIAM, Philadelphia, 1977.

[15] L. W. Ho. *A Spectral Element Method for Free-Surface and Moving Boundary Flows.* PhD thesis, Massachusetts Institute of Technology, 1989.

[16] G. E. Karniadakis. Numerical simulation of heat transfer from a cylinder in crossflow. *International Journal of Heat and Mass Transfer*, **31**(1), 107, 1988.

[17] G. E. Karniadakis. Spectral element simulations of laminar and turbulent flows in complex geometries. *Numerical Applied Mathematics*, to appear, 1988.

[18] G. E. Karniadakis, B. B. Mikic, and A. T. Patera. Minimum-dissipation transport enhancement by flow destabilization: Reynolds' analogy revisited. *Journal of Fluid Mechanics*, **192**, 365, 1988.

[19] K.Z. Korczak and A.T. Patera. An isoparametric spectral element method for solution of the Navier-Stokes equations in complex geometry. *Journal of Computational Physics*, **62**, 361, 1986.

[20] Y. Maday, D.I. Meiron, A.T. Patera, and E.M. Rønquist. Iterative saddle problem decomposition methods for the steady and unsteady Stokes equations. *submitted to Journal Computational Physics*, 1988.

[21] Y. Maday and R. Muñoz. Spectral element multigrid. II. theoretical justification. *submitted to Journal of Scientific Computing*, 1988.

[22] Y. Maday and A.T. Patera. Spectral element methods for the Navier-Stokes equations. In A.K. Noor, editor, *State-of-the-art surveys in computational mechanics*, ASME, New York, 1988.

[23] Y. Maday, A.T. Patera, and E.M. Rønquist. The $\mathbf{P}_n - \mathbf{P}_{N-2}$ Legendre spectral element method for the multi-dimensional Stokes problem. in preparation.

[24] Y. Maday, A.T. Patera, and E.M. Rønquist. Optimal Legendre spectral element methods for the Stokes semi-periodic problem. *submitted to SIAM Journal of Numerical Analysis.*

[25] A.T. Patera. A spectral element method for fluid dynamics: laminar flow in a channel expansion. *Journal of Computational Physics*, **54**, 468, 1984.

[26] D. I. Pullin and A. E. Perry. Some flow visualization experiments on the starting vortex. *Journal of Fluid Mechanics*, **97**, 239, 1980.

[27] E. M. Rønquist. *Optimal Spectral Element Methods for the Unsteady Three-Dimensional Incompressible Navier-Stokes Equations*. PhD thesis, Massachusetts Institute of Technology, 1988.

[28] E.M. Rønquist and A.T. Patera. Spectral element multigrid. I. Formulation and numerical results. *Journal of Scientific Computing*, **2**(4), 389, 1987.

[29] G. Strang and G. Fix. *An Analysis of the Finite Element Method.* Prentice-Hall, 1973.

[30] A.H. Stroud and D. Secrest. *Gaussian Quadrature Formulas*. Prentice Hall, Englewood Cliffs, New Jersey, 1966.

[31] J. M. Thomas. *Sur l'analyse numérique des méthodes d'éléments finis hybrides et mixtes.* Thèse d'Etat, Université Pierre et Marie Curie, 1987.

[32] T.A. Zang, Y.S. Wong, and M.Y. Hussaini. Spectral multigrid methods for elliptic equations. *Journal of Computational Physics*, **48**, 485, 1982.

Two Domain Decomposition Techniques for Stokes Problems*

Joseph E. Pasciak†

Abstract. I will develop two domain decomposition techniques for Stokes problems in this talk. The first uses a reformulation of the saddle point system developed in [4] and reduces the derivation of domain decomposition algorithms for Stokes to the definition of domain decomposition preconditioners for second order problems. The second applies domain decomposition directly to Stokes and gives rise to a saddle point system for the velocity nodes on the subdomain boundaries and the mean values of the pressure on the subdomains. This system is solved iteratively.

1. Introduction. In this talk, I will discuss two domain decomposition techniques for the iterative solution of the discrete systems which arise in finite element approximation to Stokes problems. Specifically, we consider the velocity-pressure formulation of the Stokes equations where the divergence constraint is treated by a Lagrange multiplier technique and the pressure variable corresponds to the multiplier. The discrete systems which arise are of the saddle point type.

In Section 2, we review some properties of saddle point systems and discuss a reformulation of the saddle point system developed in [4]. This reformulation provides a framework for the development of iterative methods for saddle point problems. The rate of convergence of the resulting iterative methods can be estimated in terms of a corresponding 'inf-sup' condition. Moreover, preconditioning can be incorporated into the scheme.

* This manuscript has been authored under contract number DE-AC02-76CH00016 with the U.S. Department of Energy. Accordingly, the U.S. Government retains a non-exclusive, royalty-free license to publish or reproduce the published form of this contribution, or allow others to do so, for U.S. Government purposes. This work was also supported in part under the Air Force Office of Scientific Research, Contract No. ISSA86-0026.

† Brookhaven National Laboratory, Upton, New York 11973

Section 3 defines the model Stokes problem and gives the corresponding weak formulation. The finite element approximation is then defined in terms of the weak formulation.

Section 4 defines the first domain decomposition technique for Stokes. Using the reformulation of Section 2, the task of developing rapidly convergent algorithms for the full saddle point problem is reduced to the development of effective preconditioners for second order problems. Domain decomposition algorithms for Stokes result from the use of standard domain decomposition preconditioners developed earlier in, for example, [5,6,7,8,9].

In Section 5, we develop iterative algorithms for Stokes problems by directly applying domain decomposition to the discrete Stokes systems. We develop iterative algorithms for the solution of the original Stokes system which require the solution of discrete Stokes problems on subdomains at each iterative step. The work in [11] provided insight for the development of this technique.

To present the ideas most clearly, I will only consider the simplest applications and approximation techniques. Many generalizations are possible and will be addressed elsewhere.

2. Iterative methods for saddle-point systems. We consider two techniques used to develop iterative methods for saddle point systems in this section. The first technique is well known and the second was developed in [4]. We include this discussion for completeness and continuity of exposition since the techniques will be used extensively in later sections of this paper.

Let H^1 and H^2 be Hilbert spaces and consider the problem

$$M\begin{pmatrix} X \\ Y \end{pmatrix} = \begin{pmatrix} f \\ g \end{pmatrix} \tag{2.1}$$

where $X, f \in H^1$ and $Y, g \in H^2$. We study operators M of the form,

$$M = \begin{pmatrix} A & B \\ B^* & 0 \end{pmatrix}. \tag{2.2}$$

We assume that A is a positive definite, symmetric operator on H^1 with a bounded inverse and that B and B^* are adjoints with respect to the inner products in H^1 and H^2. We further assume that $A^{-1}B$ and $B^*A^{-1}B$ are bounded. We shall use the notation $(\cdot,\cdot)$ and $\|\cdot\|$ to denote the inner products and norms on H^1 and H^2.

Saddle point problems of the form (2.1) arise in many applications. For example, such systems must be solved for finite element Lagrange multiplier approximations to Dirichlet and interface problems [2,3], velocity-pressure formulations of the equations of Stokes and elasticity [10], and mixed finite element methods [15].

Applying block Gaussian elimination to (2.1) implies that the solution of (2.1) satisfies

$$\begin{pmatrix} A & B \\ 0 & B^*A^{-1}B \end{pmatrix} \begin{pmatrix} X \\ Y \end{pmatrix} = \begin{pmatrix} f \\ B^*A^{-1}f - g \end{pmatrix}. \tag{2.3}$$

Thus, (2.1) is solvable if and only if $B^*A^{-1}B$ is invertible (A, B and B^* need not be bounded). But $B^*A^{-1}B$ is symmetric and non-negative. Hence, $B^*A^{-1}B$ is solvable if and only if it is definite. A straightforward computation gives

$$(B^*A^{-1}Bu, u) = \sup_{\theta\in H^1} \frac{(Bu,\theta)^2}{(A\theta,\theta)} \tag{2.4}$$

and hence solvability of (2.1) will follow if we can verify

$$\sup_{\theta\in H^1} \frac{(Bu,\theta)^2}{(A\theta,\theta)} \geq c_0 \|u\|^2 \tag{2.5}$$

holds for some positive constant c_0. Inequality (2.5) is equivalent to the classical L-B-B (Ladyzhenskaya-Babuška-Brezzi) condition. In addition to being a sufficient condition for the solvability of (2.1), the constant c_0 in (2.5) will be an ingredient in determining convergence rates for the iterative methods to be subsequently discussed.

By (2.3), we see that the solution of (2.1) can be computed by by first solving

$$B^*A^{-1}BY = B^*A^{-1}f - g \tag{2.6}$$

and then back solving (2.3) for X, i.e. $X = A^{-1}(f - BY)$. For our applications, $B^*A^{-1}B$ is a full matrix and expensive to compute. One alternative is to iteratively solve (2.6), e.g. apply conjugate gradient iteration. The rate of convergence for this iteration is related to the condition number K of $B^*A^{-1}B$. From the above discussion, we clearly have that $K \leq c_1/c_0$ where c_0 satisfies (2.5) and c_1 satisfies the reverse inequality,

$$\sup_{\theta\in H^1} \frac{(Bu,\theta)^2}{(A\theta,\theta)} \leq c_1 \|u\|^2 . \tag{2.7}$$

One gets a rapidly convergent algorithm for the computation of Y if the condition number K is not too large. This is the first iterative technique for solving (2.1) to be considered.

One problem with the iterative technique just developed is that it requires the evaluation of the action of A^{-1} at each step in the iteration. In many applications, the action of A^{-1} is more expensive to compute than that of a suitable preconditioner. The next technique was developed and analysed in [4] and leads to a rapidly convergent algorithm for solving (2.1) which utilizes the preconditioner for A^{-1} without requiring the computation of the action of A^{-1}.

Let A_0 be a good preconditioner for A^{-1}. This means that the evaluation of the action of A_0^{-1} is much more economical than that of A^{-1} and that A_0 satisfies inequalities of the form

$$\alpha_0(Av,v) \leq (A_0v,v) \leq \alpha_1(Av,v) \qquad \text{for all } v \in H^1$$

with α_1/α_0 not too large. By scaling A^{-1}, we may assume that $\alpha_1 < 1$. Again, applying block matrix manipulations to (2.1) gives

$$\begin{pmatrix} A_0^{-1}A & A_0^{-1}B \\ B^*A_0^{-1}(A-A_0) & B^*A_0^{-1}B \end{pmatrix} \begin{pmatrix} X \\ Y \end{pmatrix} = \begin{pmatrix} A_0^{-1}f \\ B^*A_0^{-1}f - g \end{pmatrix} \tag{2.8}$$

which we rewrite

$$\tilde{M} \begin{pmatrix} X \\ Y \end{pmatrix} = \tilde{f} \tag{2.9}$$

with the obvious definitions of $\tilde{M}$ and $\tilde{f}$. It is straightforward to see that $\tilde{M}$ is a symmetric operator in the inner product

$$\left[\begin{pmatrix} U \\ V \end{pmatrix}, \begin{pmatrix} W \\ X \end{pmatrix} \right] = ((A-A_0)U, W) + (V, X). \tag{2.10}$$

Moreover, it was shown in [4] that $\tilde{M}$ is positive definite in this inner product and is well conditioned provided that c_1/c_0 and α_1/α_0 are not too large. The second iterative technique for solving (2.1) applies conjugate gradient in the $[\cdot,\cdot]$ inner product to the reformulated (well conditioned) problem (2.8).

3. The model Stokes problem. In this section, we describe the model Stokes problem and its finite element discretization. Let Ω be a domain in N dimensional Euclidean space for $N = 2$ or $N = 3$. The velocity-pressure formulation of the steady-state Stokes problem is: Find $\mathbf{u}$ and P satisfying

$$\begin{aligned} -\Delta \mathbf{u} - \nabla P &= \mathbf{F} \text{ in } \Omega, \\ \nabla \cdot \mathbf{u} &= 0 \text{ in } \Omega, \\ \mathbf{u} &= 0 \text{ on } \partial\Omega, \\ \int_\Omega P &= 0. \end{aligned} \tag{3.1}$$

Here, $\mathbf{u}$ is a vector valued function defined on Ω and P is a scale valued function defined on Ω. The first equation is, of course, a vector equality at each $x \in \Omega$ and Δ denotes the componentwise Laplace operator.

We restrict ourselves to the model problem (3.1) for simplicity. Applications to problems with variable coefficients and the equations of linear elasticity are similar.

We consider a weak formulation of problem (3.1). Let $(\cdot,\cdot)$ denote the $L^2(\Omega)$ inner product and $\|\cdot\|$ the denote the corresponding norm applied either to scalar or vector functions. Let $H_0^1(\Omega)$ be the Sobolev space of functions defined on Ω which vanish (in an appropriate sense) on $\partial\Omega$ and which along with their first derivatives are square integrable on Ω. Define $\mathbf{H} \equiv H_0^1(\Omega) \times H_0^1(\Omega)$ and let $\|\cdot\|_1$ denote the corresponding norm. Let $\Pi = L^2(\Omega)$ and $\Pi/1$ denote the functions in Π with zero mean value on Ω. Multiplying (3.1) by functions in $\mathbf{H}$ and Π and integrating by parts when appropriate, it is easy to see that the solution $(\mathbf{u}, P)$ satisfies

$$\begin{aligned} D(\mathbf{u},\mathbf{v}) + (P, \nabla \cdot \mathbf{v}) &= (\mathbf{F}, \mathbf{v}) \qquad \text{for all } \mathbf{v} \in \mathbf{H}, \\ (\nabla \cdot \mathbf{u}, q) &= 0 \qquad \text{for all } q \in \Pi/1. \end{aligned} \tag{3.2}$$

Here, D is the Dirichlet form defined by

$$D(\mathbf{w},\mathbf{v}) \equiv \sum_{i=1}^{N} \int_{\Omega} \nabla w_i \cdot \nabla v_i \, dx.$$

Clearly, (3.2) is of the form of (2.1). The corresponding operator A is unbounded but has a bounded inverse. Moreover, it is well known that the corresponding inf-sup condition:

$$\sup_{\theta \in \mathbf{H}} \frac{(p, \nabla \cdot \theta)^2}{A(\theta,\theta)} \geq C_0 \|p\|^2 \qquad \text{for all } p \in \Pi/1 \tag{3.3}$$

holds for some positive constant C_0. It then follows that there is a unique solution $(\mathbf{u}, P)$ in $\mathbf{H} \times \Pi/1$ to (3.2).

To approximately solve (3.2), we introduce a collection of pairs of approximation subspaces $\mathbf{H}_h \subset \mathbf{H}$ and $\Pi_h \subset \Pi$ indexed by h in the interval $0 < h < 1$. We will assume that the inf-sup condition holds for the pair of spaces; i.e. we assume that there is a constant c_0 which does not depend upon h such that

$$\sup_{\theta \in \mathbf{H}_h} \frac{(p, \nabla \cdot \theta)^2}{A(\theta,\theta)} \geq c_0 \|p\|^2 \qquad \text{for all } p \in \Pi_h/1. \tag{3.4}$$

Many subspace pairs satisfying (3.4) have been studied and their approximation properties are well known [10,14,16].

The approximations to the functions $(\mathbf{u}, P)$ are defined by replacing the spaces in (3.2) by their discrete counterparts. Specifically, the approximations are defined as the functions $\mathbf{u}_h \in \mathbf{H}_h$ and $P_h \in \Pi_h/1$ satisfying

$$\begin{aligned} D(\mathbf{u}_h,\mathbf{v}) + (P_h, \nabla \cdot \mathbf{v}) &= (\mathbf{F},\mathbf{v}) \qquad \text{for all } \mathbf{v} \in \mathbf{H}_h, \\ (\nabla \cdot \mathbf{u}_h, q) &= 0 \qquad \text{for all } q \in \Pi_h/1. \end{aligned} \tag{3.5}$$

Existence and uniqueness for the solution of (3.5) follows from (3.4) and the discussion in Section 2.

We conclude this section with an example of a pair of approximation subspaces. For simplicity of exposition, we shall only describe these spaces when Ω is the unit square. Generalizations to certain more complex domains are possible.

Let $n > 0$ be given. We start by breaking the square into $2n \times 2n$ subsquares and define $h = 1/2n$ (see Figure 3.1). Let $x_i \equiv ih$ and $y_j = jh$ for $i, j = 1, \ldots, 2n$. We partition the subsquares into pairs of triangles using one of the subsquares diagonals (for example, the diagonal going from the bottom right corner to the upper left corner of the subsquare). Let H_h be the collection of functions which vanish on the boundary of the square and are piecewise linear and continuous on this triangulation. The subspace $\mathbf{H}_h$ is defined to be $H_h \times H_h$.

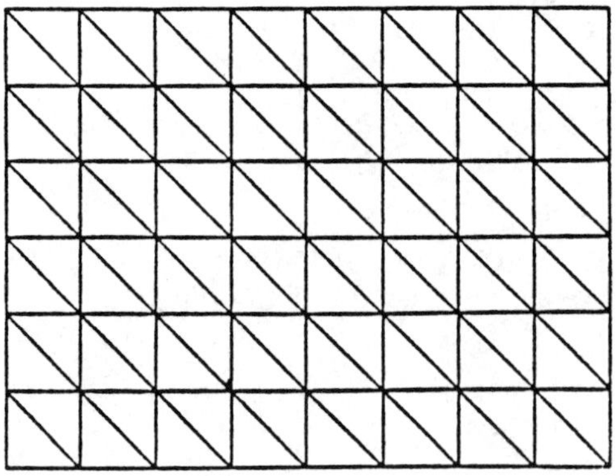

Figure 3.1: *The triangular mesh.*

To define the space Π_h, we first consider the space $\tilde{\Pi}_h$ which is defined to be the space of functions which are piecewise constant on the subsquares (see Figure 3.2). It is interesting to note [12] that the subspace pair $\{\mathbf{H}_h, \tilde{\Pi}_h/1\}$ is not stable in L^2, i.e. the inf-sup condition fails to hold for the subspace pair. To get a stable pair, we shall consider a somewhat smaller subspace of $\tilde{\Pi}_h$. Let θ_{kl} for $k, l = 1, \ldots, 2n$ be the function which is one on the subsquare $[x_{k-1}, x_k] \times [y_{l-1}, y_l]$ and vanishes elsewhere. We define the functions $\phi_{i,j} \in \tilde{\Pi}_h$, for $i, j = 1, \ldots, n$, by (see also, Figure 3.2)

$$\phi_{ij} \equiv \theta_{2i-1,2j-1} - \theta_{2i,2j-1} - \theta_{2i-1,2j} + \theta_{2i,2j}. \tag{3.6}$$

We then define Π_h by

$$\Pi_h \equiv \{Q \in \tilde{\Pi}_h | (Q, \phi_{ij}) = 0 \text{ for } i, j = 1, \ldots, n\}.$$

An estimate of the form of (3.4) holds with c_0 independent of h for the subspace pair $\{\mathbf{H}_h, \Pi_h\}$ [12]. Furthermore, the exclusion of the functions of the form (3.6) does not result in a change in the order of approximation for the space (we obviously still have the subspace of constants on the mesh of size $2h$).

REMARK: The exclusion of functions of the form (3.6) poses no difficulty in practice. In fact, it only affects the definition of the corresponding B^* in a trivial way. By definition, $B^*\mathbf{v} \equiv Q$ where $Q \in \Pi_h$ solves

$$(Q, R) = (\nabla \cdot \mathbf{v}, R) \qquad \text{for all } R \in \Pi_h/1.$$

It is easy to see that Q is the L^2 orthogonal projection (into $\Pi_h/1$) of the function $\tilde{Q} \in \tilde{\Pi}_h$ satisfying

$$(\tilde{Q}, R) = (\nabla \cdot \mathbf{v}, R) \qquad \text{for all } R \in \tilde{\Pi}_h. \tag{3.7}$$

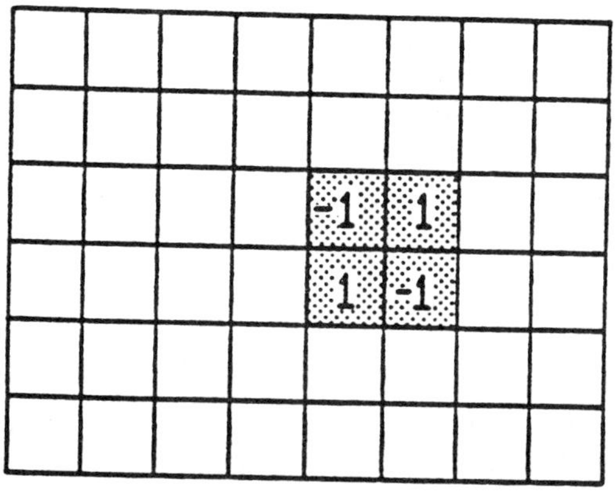

Figure 3.2. *The rectangular mesh used for* $\tilde{\Pi}_h$*; the support (shaded) and values for a typical* ϕ_{ij}.

This projection is a trivial local operation since the supports of the functions $\{\phi_{ij}\}$ are disjoint. Furthermore, the computation of $\tilde{Q}$ is straightforward since the gram matrix for (3.7) is diagonal (with the obvious choice of basis).

4. The first domain decomposition technique for Stokes. We consider the direct application of the second iterative technique of Section 2 to the saddle point problem corresponding to the Stokes discretization. As we shall see, all that is required is effective preconditioners for second order problems. Thus, domain decomposition algorithms for Stokes result from standard domain decomposition preconditioners for second order problems.

Let us introduce some operator notation. Let $A : \mathbf{H}_h \mapsto \mathbf{H}_h$ be defined by

$$(Av, w) = D(v, w) \qquad \text{for all } w \in \mathbf{H}_h. \tag{4.1}$$

Clearly, (4.1) defines a symmetric positive definite operator on $\mathbf{H}_h$. We define $B : \Pi_h/1 \mapsto \mathbf{H}_h$ by

$$(Bp, \mathbf{w}) = (p, \nabla \cdot \mathbf{w}) \qquad \text{for all } \mathbf{w} \in \mathbf{H}_h.$$

Its adjoint, $B^* : \mathbf{H}_h \mapsto \Pi_h/1$ is then defined by

$$(B^*\mathbf{w}, q) = (\nabla \cdot \mathbf{w}, q) \qquad \text{for all } q \in \Pi_h/1,$$

and is nothing more than the divergence followed by L^2 projection into $\Pi_h/1$. The discrete solution pair $(\mathbf{u}_h, P_h)$ satisfies (2.1).

The operator A involves two componentwise operators corresponding to the standard discrete Dirichlet operator on H_h. Consequently, it can be preconditioned componentwise by domain decomposition preconditioners for the Dirichlet problem. Results concerning the development of domain decomposition preconditioners for the general second order problems have been given in [5,6,7,8,9]. One example, described in [6], develops a domain decomposition preconditioner for the second order problem in R^2 and gives rise to a problem which, even though not well conditioned, has a condition number growth bounded by $c(1+\ln^2(d/h))$. Here, d is roughly the size of the subdomains. In this case, the condition number of (2.8) also will grow like $c(1+\ln^2(d/h))$.

5. A direct domain decomposition approach. In this section, we shall directly apply domain decomposition to the Stokes problem. We shall develop algorithms for solving the discrete system (3.5), which only require the solution of smaller discrete Stokes systems on the subdomains and some type of reduced system. In this case, the reduced system will involve the values of $\mathbf{u}_h$ on the boundary of the subdomains and the mean value of the pressure on the subdomains.

We assume that $\bar{\Omega}$ has been partitioned into a number of subdomains $\bar{\Omega} = \cup_{i=1}^m \bar{\Omega}_i$. We require that the boundary of the subdomains $(\Gamma \equiv \cup_{i=1}^m \partial\Omega_i)$ align with the mesh in $\mathbf{H}_h$ and Π_h. We then define

$$\mathbf{H}_h^i = \{\phi \in \mathbf{H}_h | \text{support}(\phi) \subset \Omega_i\} \tag{5.1}$$

and

$$\Pi_h^i = \{\phi \in \Pi_h | \text{support}(\phi) \subset \Omega_i\}. \tag{5.2}$$

We shall assume that the inf-sup condition holds for each subspace pair, i.e.

$$\sup_{\theta \in \mathbf{H}_h^i} \frac{(q, \nabla \cdot \theta)^2}{A(\theta,\theta)} \geq c_0 \|q\|_{\Omega_i}^2 \qquad \text{for all } q \in \Pi_h^i/1 \tag{5.3}$$

and that the function which is one on Ω_i and vanishes in the remainder of Ω is an element in Π_h. Note that, since the functions in $\mathbf{H}_h$ are continuous, the subspace pair $(\mathbf{H}_h^i, \Pi_h^i)$ can be used to approximate the Stokes problem with zero boundary conditions on the subdomains.

Because of (5.3), local Stokes problems on the subdomains are solvable. The first step is to solve these local problems and reduce the problem to one which implicitly involves fewer degrees of freedom. To do this, we let $(\mathbf{v}_h^i, Q_h^i)$ be the solution of

$$\begin{aligned} A(\mathbf{v}_h^i, \mathbf{w}) + (Q_h^i, \nabla \cdot \mathbf{w}) &= (\mathbf{F}, \mathbf{w}) \qquad \text{for all } \mathbf{w} \in \mathbf{H}_h^i, \\ (\nabla \cdot \mathbf{v}_h^i, q) &= 0 \qquad \text{for all } q \in \Pi_h^i/1. \end{aligned} \tag{5.4}$$

We set $\mathbf{v}_h = \sum \mathbf{v}_h^i$, $Q_h = \sum Q_h^i$ and define $\mathbf{w}_h = \mathbf{u}_h - \mathbf{v}_h$ and $R_h = P_h - Q_h$. Then, $\mathbf{w}_h$ and R_h satisfy

$$\begin{aligned} A(\mathbf{w}_h, \mathbf{v}) + (R_h, \nabla \cdot \mathbf{v}) &= F(\mathbf{v}) \qquad \text{for all } \mathbf{v} \in \mathbf{H}_h, \\ (\nabla \cdot \mathbf{w}_h, q) &= G(q) \qquad \text{for all } q \in \Pi_h/1. \end{aligned} \tag{5.5}$$

The functionals F and G vanish for functions in $\mathbf{H}_h^i$ and $\Pi_h^i/1$ respectively. Thus, the functions $\mathbf{w}_h$ and R_h lie in a subspace of $\mathbf{H}_h \times \Pi_h/1$ with significantly lower dimension. We shall parameterize this subspace and then derive equations for the parameters corresponding to the solution $\mathbf{w}_h$ and R_h.

We shall parameterize the solution $(\mathbf{w}_h, R_h)$ in terms of parameters $\sigma \in \mathbf{H}(\Gamma)$ and $\lambda \in \Pi_0$ where

$$\mathbf{H}(\Gamma) \equiv \{\phi|_\Gamma, \phi \in \mathbf{H}_h\}$$

and

$$\Pi_0 \equiv \{\phi \in \Pi_h/1 \text{ such that } \phi \text{ is constant on } \Omega_i \text{ for each } i\}.$$

To do this, we define the operators $S : \mathbf{H}(\Gamma) \mapsto \Pi_h$ and $T : \mathbf{H}(\Gamma) \mapsto \mathbf{H}_h$ satisfying the following:

(1) $S(\gamma)|_{\Omega_i} \in \Pi_h^i/1$,
(2) $T(\gamma)|_\Gamma = \gamma$,
(3) $D(T(\gamma), \phi) + (S(\gamma), \nabla \cdot \phi) = 0 \quad$ for all $\phi \in \mathbf{H}_h^i$,
(4) $(\nabla \cdot T(\gamma), q) = 0 \quad$ for all $q \in \Pi_h^i/1$.

It is not difficult to show that the above conditions uniquely define S and T. Moreover, if $\sigma = \mathbf{w}_h|_\Gamma$ and $\lambda \in \Pi_0$ is the function which has the same mean values on the subdomains as R_h then

$$\mathbf{w}_h = T(\sigma) \text{ and } R_h = S(\sigma) + \lambda. \tag{5.6}$$

Thus, (5.6) gives a parameterization of $\mathbf{w}_h$ and R_h in terms of the parameters (σ, λ) in $\mathbf{H}(\Gamma) \times \Pi_0$. Note that given a value of γ, the evaluation of $S(\gamma)$ and $T(\gamma)$ essentially only involves the solution of discrete Stokes problems on the subdomains.

We next give equations for the determination of σ and λ satisfying (5.6). To do this, we define a quadratic form $E : (\mathbf{H}(\Gamma) \times \Pi_0)^2 \mapsto R^1$ given by

$$\begin{aligned} E((\gamma_1, \delta_1), (\gamma_2, \delta_2)) = D(T(\gamma_1), T(\gamma_2)) &+ (\delta_1, \nabla \cdot T(\gamma_2)) \\ &+ (\nabla \cdot T(\gamma_1), \delta_2). \end{aligned} \tag{5.7}$$

It is not difficult to see that, given local bases, $\{\phi_i\}$ for $\mathbf{H}(\Gamma)$ and $\{\psi_i\}$ for Π_0, we can compute the data $\tilde{F}$ satisfying

$$E((\sigma, \lambda), (\phi_i, \psi_j)) = \tilde{F}(\phi_i, \psi_j) \tag{5.8}$$

from the functionals F and G in (5.5). This can be carried out only using a few local operations per basis function, without knowing σ, λ and without computing $T(\phi_i)$.

From the definition of E, it is clear that (5.8) gives rise to a symmetric indefinite system of the form (2.1) which can be used to compute σ, λ satisfying (5.6). The form $D(T(\gamma_1), T(\gamma_2))$ corresponds to the operator A in (2.1). The form $(\delta_1, \nabla \cdot T(\gamma_2))$ corresponds to B, etc. Stability properties for the above system are given in the following theorem which will be proven in a subsequent paper under reasonable assumptions on the domain subdivision.

THEOREM. *There are positive constants $\alpha_0, \alpha_1, c_0, c_1$, independent of d and h, such that*

$$\alpha_0 D(T(\gamma),T(\gamma)) \leq \sum_{i=1}^{m} |\gamma|^2_{1/2,\partial\Omega_i} \leq \alpha_1 D(T(\gamma),T(\gamma)), \tag{5.9}$$

and

$$c_0 \|\delta\|^2 \leq \sup_{\gamma\in\mathbf{H}(\Gamma)} \frac{(\delta, \nabla\cdot T(\gamma))^2}{D(T(\gamma),T(\gamma))} \leq c_1 \|\delta\|^2. \tag{5.10}$$

where $|\cdot|_{1/2,\partial\Omega_i}$ denotes the Sobolev semi-norm of order 1/2 on $\partial\Omega_i$ (c.f. [13]).

Inequalities (5.10) imply that the 'inf-sup' condition corresponding to form E on the subspace pair $(\mathbf{H}(\Gamma), \Pi_0)$ is well conditioned independently of h. The boundary form $D(T(\gamma),T(\gamma))$ is not well conditioned but is equivalent to a sum of seminorms on the boundaries of the subdomains. The corresponding form,

$$\ll \gamma_1, \gamma_2 \gg_{1/2} \equiv \sum_{i=1}^{m} < \gamma_1, \gamma_2 >_{1/2,\partial\Omega_i}$$

has been well studied in the development of domain decomposition preconditioners for second order problems. In fact, each domain decomposition technique developed in [1,5,6,7,8,9] gives rise to a computationally effective domain decomposition preconditioner for $\ll \cdot,\cdot \gg_{1/2}$. Thus, we can solve (5.8) by using the second iterative technique of Section 2, with preconditioner A_0 corresponding componentwise to the boundary part of a second order method developed in [1,5,6,7,8,9]. For example, we can use the technique presented in [6]. This means the preconditioner for the boundary velocities will involve inverting the $l_0^{1/2}$ operator on the edge segments and the solution of a coarse grid problem with the number of unknowns equal to the number of 'cross-points' in the subdomain subdivision. The resulting symmetric positive definite reformulation of (5.8) will have a condition number bounded by $C(1+\ln^2(d/h))$.

It is possible to implement the above technique in such a way that each Stokes subdomain problem need be solved only once per step in the iterative algorithm for the solution of σ, λ. Once these parameters are solved to satisfactory accuracy, $\mathbf{w}_h$ and R^h can be computed with one more set of subdomain solves.

6. Conclusion. We have provided two domain decomposition techniques for solving Stokes problems. The technique of Section 4 applies an algebraic reformulation to the discrete Stokes equations and consequently can utilizes any available domain decomposition preconditioner for the second order problem. The technique of Section 5 applies domain decomposition directly to the discrete Stokes equations.

Although the technique of Section 5 is theoretically interesting, we feel that the one described in Section 4 will probably lead to the most flexible and computationally effective algorithms. The first reason for this is that, with the second order preconditioners, one has a much greater flexibility in the form of the subproblems which are to be solved. In contrast, since the method of Section 5 is a dimension reduction technique, the given Stokes problem must be solved on the subdomains. In addition, there are a large number of techniques available for developing 'fast-solvers' for second order problems while there are few (if any) fast Stokes-solvers available.

REFERENCES

(1) P.E. Bjørstad and O.B. Widlund, *Iterative methods for the solution of elliptic problems on regions partitioned into substructures*, SIAM J. Numer. Anal. **23** (1986), 1097–1120.

(2) J.H. Bramble, *The Lagrange multiplier method for Dirichlet's problem*, Math. Comp. **37** (1981), 1–12.

(3) J.H. Bramble and J.E. Pasciak, *A boundary parametric approximation to the linearized scalar potential magnetostatic field problem*, Appl. Numer. Math. **1** (1985), 493–514.

(4) J.H. Bramble and J.E. Pasciak, *A preconditioning technique for indefinite systems resulting from mixed approximations of elliptic problems*, Math. Comp. **50** (1988), 1–18.

(5) J.H. Bramble, J.E. Pasciak and A.H. Schatz, *An iterative method for Elliptic problems on regions partitioned into substructures*, Math. Comp. **46** (1986), 361–369.

(6) J.H. Bramble, J.E. Pasciak and A.H. Schatz, *The construction of preconditioners for elliptic problems by substructuring, I*, Math. Comp. **47** (1986), 103–134.

(7) J.H. Bramble, J.E. Pasciak and A.H. Schatz, *The construction of preconditioners for elliptic problems by substructuring, II*, Math. Comp. **49** (1987), 1–16.

(8) J.H. Bramble, J.E. Pasciak and A.H. Schatz, *The construction of preconditioners for elliptic problems by substructuring, III*, Math. Comp. (to appear).

(9) J.H. Bramble, J.E. Pasciak and A.H. Schatz, *The construction of preconditioners for elliptic problems by substructuring, IV*, Math. Comp., (submitted).

(10) V. Girault and P. Raviart, "Finite Element Approximation of the Navier-Stokes Equations," Lecture Notes in Math. # 749, Springer-Verlag, New York, 1981.

(11) R. Glowinski and M.F. Wheeler, *Domain decomposition methods for mixed finite element approximation*, in "Proceedings, 1'st Inter. Conf. on Domain Decomposition Methods," SIAM, Philadelphia, 1988, pp. 144–172.

(12) C. Johnson and J. Pitkäranta, *Analysis of some mixed finite element methods related to reduced integration*, Math. Comp. **38** (1982), 375–400.

(13) J.L. Lions and E. Magenes, "Problèmes aux Limites non Homogènes et Applications," Dunode, Paris, 1968.

(14) J.C. Nedelec, *Elements finis mixtes incompressibles pour l'equation de Stokes dans* R^3., Numer. Math. **39** (1982), 97–112.

(15) P.A. Raviart and J.M. Thomas, *A mixed finite element method for 2-nd order elliptic problems*, in "Mathematical Aspects of Finite Element Methods, Lecture

Notes in Mathematics," #606 (Eds. I. Galligani and E. Magenes) Springer-Verlag, New York, 1977, pp. 292–315.

(16) L.R. Scott and M. Vogelius, *Conforming finite element methods for incompressible and nearly incompressible continua*, Inst. for Phys. Sci. and Tech., Univ. of Maryland, Tech. Rep. BN-1018.

Domain Decomposition Algorithms for the Stokes Equations

Alfio Quarteroni*

Abstract We show that the capacitance (or Schur complement) matrix C associated with a multidomain finite element approximation to the Stokes equations is symmetric and positive definite. We then propose several preconditioners which are symmetric, positive definite, and spectrally equivalent to C. Their analysis is first carried out for decomposition by two subdomains, then is extended to cover the case of strips (M adjoint subdomains, see Fig.5.1) and boxes (four subdomains sharing an internal vertex, see Fig.5.2). As stated in section 1, despite most of this paper is concerned with finite element approximation of the Stokes equations, the arguments here developed can be applied to different kind of approximations (e.g., those based on spectral methods), as well as to different kind of boundary value problems.

1. Finite dimensional approximations to boundary value problems. Let V and Q be two Hilbert spaces, with norm $\|\cdot\|$ and $|\cdot|$ respectively. We consider the problem:

$$(1.1) \qquad \begin{cases} \text{find } u \in V, p \in Q \text{ s.t.} & \\ a(u,v) + b(v,p) = f(v) & \forall v \in V \\ b(u,q) = g(q) & \forall q \in Q \end{cases}$$

where $a : V \times V \to \mathbb{R}$ and $b : V \times Q \to \mathbb{R}$ are two bilinear and continuous forms, and f and g are two linear functionals defined on V and Q respectively. We assume that there exist two strictly positive constants α, β such that:

$$(1.2) \qquad \forall v \in V \quad a(v,v) \geq \alpha \|v\|^2 \ ; \qquad \forall q \in Q \quad \sup_{v \in V} b(v,q)/\|v\| \geq \beta |q|$$

These properties ensure that problem (1.1) is well posed.

* Dipartimento di Matematica, Università Cattolica di Brescia, 25121 Brescia, and Istituto di Analisi Numerica del C.N.R., 27100 Pavia - ITALY

We recall that (1.1) is the general setting for the variational formulation of the incompressible Stokes equations in $\Omega \subset \mathbb{R}^n$. In such case, u denotes the velocity field, p is the pressure,

$$a(u,v) = \upsilon \int_\Omega \nabla u \cdot \nabla v, \qquad b(v,q) = -\int_\Omega q \operatorname{div} v,$$

where $\upsilon > 0$ is the viscosity, $V=[H^1_0(\Omega)]^n$ and $Q = L^2(\Omega)/\mathbb{R}$ (see, e.g., [6]). The second equation of (1.1) (with $g \equiv 0$) enforces the divergence free constraint on u. Second and fourth order elliptics equations can also be cast into the same setting (with $f \equiv 0$) by letting either u or p be the primitive variable, and the others be ∇u or Δu, respectively (such formulation is suitable in view of approximation by mixed finite elements, see [4]). However, in order to keep our presentation plain, from now on we will specifically refer to (1.1) as to the variational form of the Stokes equations.

Let now V_h and Q_h be two finite dimensional subspaces of V and Q respectively. We introduce the following approximation to (1.1):

(1.3)
$$\begin{cases} \text{find } u_h \in V_h\ ,\ p_h \in Q_h & \text{s.t.} \\ a(u_h,v) + b(v,p_h) = f(v) & \forall\, v \in V_h \\ b(u_h,q) = g(q) & \forall\, q \in Q_h \end{cases}$$

We assume that the following *inf-sup condition* holds:

(1.4)
$$\forall\, q \in Q_h \qquad \sup_{v \in V_h} b(v,q)/\| v \| \geq \beta_h\, | q |$$

where $\beta_h > 0$ might depend on h. Under this assumption, the problem (1.3) has a unique solution, and the following error bound holds (see [3]):

(1.5)
$$\| u - u_h \| + | p - p_h | \leq C(\beta_h) \{ \inf_{v_h \in V_h} \| u - v_h \| + \inf_{q_h \in Q_h} | p - q_h | \}$$

Remark 1.1 Finite element approximations to the Stokes equations that satisfy (1.4) with β_h independent of the finite element mesh size h are numerous (see, e.g., [6] and [4]). In such cases, the bound (1.5) yields optimal convergence estimates. Some Fourier-Legendre spectral approximations that satisfy (1.4) with β_h independent of the polynomial degree of the spectral solutions are also known ([7] and [11]). More generally however, spectral collocation approximation (and finite element approximation numerical integration), using can be cast in the framework (1.3) provided two discrete bilinear forms a_h and b_h are used instead of a and b. Further, for spectral Chebyshev approximation, the bilinear form intervening in the momentum equation is actually different than that used in the continuity equation. For these general cases, the inequality (1.5) becomes much more complicated (see, e.g., [6], [1]), and the forthcoming discussion and relative results should be modified accordingly.

In matrix notation, the problem (1.3) can be written as

(1.6)
$$\begin{cases} \mathbf{Au} + \mathrm{B}^T\mathbf{p} = \mathbf{f} \\ \mathbf{Bu} = \mathbf{g} \end{cases}$$

where: **u** (resp. **p**) is the vector of the values of u (resp p) at the grid points,

$$a_{ij} = a(\varphi_j, \varphi_i) \ , \ b_{lm} = b(\varphi_m, q_l) \ , \ f_i = f(\varphi_i) \ , \ g_l = g(q_l)$$

and $\{\varphi_i\}$ (resp $\{q_l\}$) is the Lagrange basis of V_h (resp Q_h) relative to the grid points of the finite dimensional approximation.

2. The domain decomposition formulation and the associated capacitance matrix. We assume now that (1.1) is a boundary value problem set in an open domain Ω whose boundary is $\partial\Omega$. We make the assumption that Ω is partitioned into two disjoint subdomains Ω_1 and Ω_2 whose common boundary will be denoted by Γ. Despite most part of the forthcoming discussion applies to general approximations of the form (1.3), from now on we will explicitly refer to finite element approximations only. In this framework we require that each element of the decomposition does not cross Γ, i.e., it is contained in either Ω_1 or Ω_2. Then, for k = 1, 2 we denote by $V_{h,k}$ (resp $Q_{h,k}$) the space of the restrictions of the elements of V_h (resp Q_h) to Ω_k and by Φ_h the space of restrictions to Γ of the elements of V_h. Finally, we denote by $V^*_{h,k}$ the subspace of $V_{h,k}$ of those functions that vanish on Γ. It is proven in [9] that the single-domain finite element problem (1.3) is equivalent to the following multi-domain problem:

find $u_{h,k} \in V_{h,k}$, $p_{h,k} \in Q_{h,k}$, $k = 1, 2$ s.t.

(2.1) $\quad a_1(u_{h,1}, v) + b_1(v, p_{h,1}) = f_1(v) \qquad \forall v \in V^*_{h,1}$

(2.2) $\quad b_1(u_{h,1}, q) = g_1(q) \qquad \forall q \in Q_{h,1}$

(2.3) $\quad u_{h,1} = u_{h,2} \qquad$ on Γ

(2.4) $\quad a_1(u_{h,1},\rho_1\varphi) + b_1(\rho_1\varphi,p_{h,1}) + a_2(u_{h,2},\rho_2\varphi) + b_2(\rho_2\varphi,p_{h,2}) = f_1(\rho_1\varphi) + f_2(\rho_2\varphi) \ \forall\varphi\in\Phi_h$

(2.5) $\quad a_2(u_{h,2},v) + b_2(v,p_{h,2}) = f_2(v) \qquad \forall v\in V^*_{h,2}$

(2.6) $\quad b_2(u_{h,2},q) = g_2(q) \qquad \forall q\in Q_{h,2}$

Here $\rho_k\varphi \in V_{h,k}$ is the interpolant extension of $\varphi \in \Phi_h$ to Ω_k, i.e., $\rho_k\varphi = \varphi$ on Γ, $\rho_k\varphi = 0$ at each finite element node internal to Ω_k, while a_k, b_k, f_k and g_k are the restrictions of a, b, f and g, respectively, to Ω_k. Actually, one has $u_{h,k} = u_{h|\Omega k}$, $p_{h,k} = p_{h|\Omega k}$, k = 1, 2, provided the pressures space Q_h is made of discontinuous functions across the interelement boundaries. In the case where $Q_h \subset C^\circ(\Omega)$, (2.1)-(2.6) is no more equivalent to (1.3).
The matrix representation of (2.1)-(2.6) is as follows. Denote, for each k=1,2, by $\{\varphi^k_i\}$, $\{q^k_j\}$ and $\{\psi_m\}$ the finite element Lagrange bases of $V^*_{h,k}$, $Q_{h,k}$ and Φ_h, respectively, and by $\mathbf{U}^\circ_k$, $\mathbf{P}_k$ and $\mathbf{U}_3$ the vectors of the corresponding finite element unknowns.
Then (2.1)-(2.6) is equivalent to the linear system of Fig.2.1, where, for k = 1, 2:

$$A_{kk}(N_k \times N_k) : (A_{kk})_{ij} = a_k(\varphi^k_j, \varphi^k_i) \ , \ B_{kk}(M_k \times N_k) : (B_{kk})_{ij} = b_k(\varphi^k_j, q^k_i)$$

(2.7) $$A_{k3}(N_k \times N_3) : (A_{k3})_{ij} = a_k(\rho_k \psi_j, \varphi^k_i) \ , \ B_{k3}(M_k \times N_k) : (B_{k3}) = b_k(\rho_k \psi_j, q^k_i)$$

$$A^k_{33}(N_3 \times N_3) : (A^k_{33})_{ij} = a_k(\rho_k \psi_j, \rho_k \psi_i), \ \ A_{33} = A^1_{33} + A^2_{33}$$

$$(\mathbf{F}_k)_i = f_k(\varphi^k_i), \ \ (\mathbf{F}_{k3})_i = f_k(\rho_k \psi_i), \ \ \mathbf{F}_3 = \mathbf{F}_{13} + \mathbf{F}_{23}$$

$$\begin{bmatrix} A_{11} & B_{11} & A_{13} & 0 & 0 \\ B^T_{11} & 0 & B_{13} & 0 & 0 \\ A^T_{13} & B^T_{13} & A_{33} & A^T_{23} & B^T_{23} \\ 0 & 0 & A_{23} & A_{22} & B_{22} \\ 0 & 0 & B_{23} & B^T_{22} & 0 \end{bmatrix} \begin{bmatrix} \mathbf{U}^\circ_1 \\ \mathbf{P}_1 \\ \mathbf{U}_3 \\ \mathbf{U}^\circ_2 \\ \mathbf{P}_2 \end{bmatrix} = \begin{bmatrix} \mathbf{F}_1 \\ \mathbf{0} \\ \mathbf{F}_3 \\ \mathbf{F}_2 \\ \mathbf{0} \end{bmatrix}$$

Fig.2.1 The finite element system

If we define for i = 1, 2

(2.8)
$$K_i = \begin{pmatrix} A_{ii} & B_{ii} \\ B^T_{ii} & 0 \end{pmatrix} \qquad C_{i3} = \begin{pmatrix} A_{i3} \\ B_{i3} \end{pmatrix} \qquad J_i = (A^T_{i3} \quad B^T_{i3}) \ ,$$

by block elimitation we deduce from the system in Fig.2.1the following Schur complement system with respect to the vector of the interface unknowns $\mathbf{U}_3$:

(2.9)
$$C\mathbf{U}_3 = \mathbf{G} \quad , \quad \textit{with} \quad C = A^1_{33} - J_1 K_1^{-1} C_{13} + A^2_{33} - J_2 K_2^{-1} C_{23}$$

The right hand side of (2.9) is $\mathbf{G} = \sum_{i=1}^{2} \mathbf{F}_{i3} - J_i K_1^{-1} (\mathbf{F}_i, \mathbf{0})^T$. The matrix C is the *capacitance (or Schur complement) matrix.* Note that the complement is taken with respect to the interface values of the velocity only, and not to those of the pressure.

3. Functional interpretation of the capacitance matrix For any element φ of Φ_h, and for k = 1, 2, we look for $w_k(\varphi) \in V_{h,k}$, $\pi_k(\varphi) \in Q_{h,k}$ such that

(3.1)
$$\begin{cases} a_k(w_k(\varphi), v) + b_k(v, \pi_k(\varphi)) = 0 & \forall\, v \in V^*_{h,k} \\ b_k(w_k(\varphi), q) = 0 & \forall\, q \in Q_{h,k} \\ w_k(\varphi) = \varphi & \text{on } \Gamma \end{cases}$$

From now on we will refer to $(w_k(\varphi), \pi_k(\varphi))$ as to the finite element *Stokes extension* of φ to Ω_k, for k = 1, 2. Now set:

(3.2)
$$\mathcal{A}(\varphi,\psi) := \sum_{k=1}^{2} [a_k(w_k(\varphi), \rho_k\psi) + b_k(\rho_k\psi, \pi_k(\varphi))]$$

In view of (2.7)-(2.9), it is easy to show that

(3.3)
$$[C\mathbf{x}_\varphi, \mathbf{x}_\psi] = \mathcal{A}(\varphi, \psi) \qquad \forall \varphi, \psi \in \Phi_h$$

where $\mathbf{x}_\varphi$ is a vector whose N_3 components are the values of φ at the gridpoints on Γ, $\mathbf{x}_\psi$ is defined similarly, and $[\cdot, \cdot]$ is the euclidean inner product of $\mathbb{R}^{N3}$.

Remark 3.1 For a Stokes problem, from (3.2) we deduce that $\forall \varphi \in \Phi_h$ the capacitance matrix C associates to the vector $\mathbf{x}_\varphi$ the vector $C\mathbf{x}_\varphi$ whose N_3 components are the values at the interface nodes of $\sigma_1(\varphi) + \sigma_2(\varphi)$, where

$$\sigma_k(\varphi) := \partial w_k(\varphi)/\partial n_k - \pi_k(\varphi)\, n_k \tag{3.4}$$

(n_k = outward normal direction to $\partial\Omega_k$) is the *normal stress* on Γ associated with the Stokes extension of φ to Ω_k. Hence, equation (2.4) amounts to require the "natural" condition that the normal stress of the solution be continuous across Γ.

Proposition 3.1 *The following equality holds*

$$\mathcal{A}(\varphi, \psi) = a_1(w_1(\varphi), w_1(\psi)) + a_2(w_2(\varphi), w_2(\psi)) \quad \forall \varphi, \psi \in \Phi_h \, . \tag{3.5}$$

Proof Using the definition (3.2) we have:

$$\mathcal{A}(\varphi,\psi) = \sum_{k=1}^{2} [a_k(w_k(\varphi),w_k(\psi)) + a_k(w_k(\varphi), \rho_k\psi - w_k(\psi)) + b_k(\rho_k\psi - w_k(\psi), \pi_k(\varphi)) + b_k(w_k(\psi), \pi_k(\varphi))]$$

By the second equation of (3.1) (with $\varphi=\psi$), the last term of the sum is zero. Moreover, the sum of the second and third term in the bracket is zero due to the first equation of (3.1) (with $v=\rho_k\psi - w_k(\psi) \in V^*_{h,k}$).

Corollary 3.1 *If the bilinear form a $(\cdot,\cdot)$ is symmetric, then the capacitance matrix C is symmetric and positive definite.*

Proof. Denoting as above by $\{\psi_j\}$ the Lagrange basis of Φ_h, from (3.2) one obtains

$$C_{ij} = \mathcal{A}(\psi_j, \psi_i) \qquad 1 \le i, j \le N_3 \tag{3.6}$$

C is therefore the matrix associated with the form $\mathcal{A}(\cdot,\cdot)$. Since the forms $a_k(\cdot,\cdot)$ are symmetric, the symmetry of C follows from (3.5). Furthermore, C is positive definite since the forms $a_k(\cdot,\cdot)$ are coercive, as stated by the first inequality of (1.4).

4. An optimal preconditioner for the capacitance matrix We introduce the reduced bilinear form on Φ_h

$$\mathcal{B}(\varphi,\psi) := a_2(w_2(\varphi), w_2(\psi)) \qquad \forall \varphi, \psi \in \Phi_h \tag{4.1}$$

whose associated matrix is (see (2.9))

$$B = A^2_{33} - J_2\, K_2^{-1}\, C_{23} \tag{4.2}$$

It is shown in [9], Lemma 6.1, that for k = 1, 2

$$(4.3) \qquad \| H_k(\varphi) \|_k \le \| w_k(\varphi) \|_k \le (1 + \beta_{h,k}^{-1}) \| H_k(\varphi) \|_k \qquad \forall \varphi \in \Phi_h$$

where $\beta_{h,k}$ is the constant of the inf-sup condition (1.4) on Ω_k, $\| \cdot \|_k$ is the norm induced by the form $a_k(\cdot,\cdot)$, and $H_k(\varphi)$ is the *harmonic extension* of φ to Ω_k, i.e.,

$$(4.4) \qquad H_k(\varphi) \in V_{k,h} \ : \ a_k(H_k(\varphi), v) = 0 \qquad \forall v \in V^*_{k,h} \ , \ H_k(\varphi) = \varphi \text{ on } \Gamma$$

It is shown in [8] and [2] that if the finite element decomposition of Ω is quasi-uniform, then $\| H_1(\varphi)\|_1$ is uniformely equivalent to $\| H_2(\varphi) \|_2$, i.e.,

$$(4.5) \qquad C_1 \| H_1(\varphi) \|_1 \le \| H_2(\varphi) \|_2 \le C_2 \| H_1(\varphi) \|_1 \qquad \forall \varphi \in \Phi_h$$

where C_1, C_2 are two constants independent of h. Using (4.3) and (4.5) we get:

$$(4.6) \qquad C_1(1 + \beta_{h,1}^{-1})^{-1} \| w_1(\varphi) \|_1 \le \| w_2(\varphi) \|_2 \le C_2(1 + \beta_{h,2}^{-1}) \| w_1(\varphi) \|_1$$

It follows (as already noticed in [9]) that if the inf-sup conditions hold in Ω_k with constants $\beta_{h,k}$ uniformely bounded from below by a constant independent of h, then $\| w_1(\varphi) \|_1$ and $\| w_2(\varphi) \|_2$ are uniformely equivalent. Hence, in view of (3.5) and (4.1) we conclude that there exists two constants K_1 and K_2 independent of h s.t.

$$(4.7) \qquad K_1 \mathcal{B}(\varphi, \varphi) \le \mathcal{A}(\varphi, \varphi) \le K_2 \mathcal{B}(\varphi, \varphi) \qquad \forall \varphi \in \Phi_h$$

Thus the matrix B is *spectrally equivalent* to C, i.e.

(4.8) *the condition number of* $B^{-1} C$ *is independent of h*

Since B is symmetric, positive definite and spectrally equivalent to C, it can be used as an optimal preconditioner for conjugate gradient (or other) iterations on the capacitance system (2.9). If used with Richardson iterations, it gives rise to the generalization to Stokes equations of the Dirichlet-Neumann algorithm for elliptic equations (see [12] and [5], [8]). Clearly, the same kind of conclusion holds taking the matrix $B = A^1_{33} - J_1 K_1^{-1} C_{13}$.

5. Generalization to many subdomains We extend now the previous arguments to decompositions with several subdomains. For simplicity of exposition we will consider cartesian decompositions of "strips" and "boxes" only. For either case we will determine the associated capacitance matrix as well as several preconditioners.

5.1 Strips We consider first a domain Ω divided into M adjoining, non intersecting subdomains Ω_i. The common boundary between Ω_i and Ω_{i+1} is denoted by Γ_i (see Fig.5.1). We will assume that M is even.
The finite element multidomain problem relative to the current situation can still be defined as in (2.1)-(2.6), provided now Ω_1 denotes the set of the odd subdomains, Ω_2 that of the even ones, $\Gamma := \bigcup\{\Gamma_i, i=1,...,M-1\}$ and $\Phi_h := \prod\{\Phi_h(\Gamma_i), i=1,...,M-1\}$, where $\Phi_h(\Gamma_i)$ is

the finite element space V_h restricted to Γ_i. The capacitance matrix C relative to the current situation can be defined by means of some auxiliary matrices. For each i, denote by $\{\psi^i_m\}$ the Lagrange basis of $\Phi_h(\Gamma_i)$ at the finite element nodes $\{\xi^i_m\}$ of Γ_i. Then we define the four matrices $S^i_1,\dots,S^i_4$ as follows:

$$(S^i_1)_{km} = a_i(w_i(\psi^i_m), w_i(\psi^i_k)) \tag{5.1}$$

$$(S^i_2)_{km} = a_i(w_i(\psi_m^{i-1}), w_i(\psi_k^{i-1})) \tag{5.2}$$

$$(S^i_3)_{km} = a_i(w_i(\psi^i_m), w_i(\psi_k^{i-1})) \tag{5.3}$$

$$(S^i_4)_{km} = a_i(w_i(\psi_m^{i-1}), w_i(\psi^i_k)) \tag{5.4}$$

where, for each $\psi \in \Phi_h(\Gamma_j)$, $(j = i\text{-}1,i)$, $w_i(\varphi)$ is the velocity field of the finite element Stokes extension of φ on Ω_i, with $w_i(\varphi) = \varphi$ on Γ_j, and $w_i(\varphi) = 0$ on $\partial\Omega_i\backslash\Gamma_j$.
If for each $\varphi \in \Phi_h(\Gamma_j)$ we denote by $\mathbf{x}_\varphi$ the vector whose components are the values of φ at the finite element nodes on Γ_j, we have

$$[S^i_4\, \mathbf{x}_\varphi, \mathbf{x}_\psi] = a_i(w_i(\psi), w_i(\varphi)) \qquad \forall\, \varphi \in \Phi_h(\Gamma_{i-1}),\ \forall\, \psi \in \Phi_h(\Gamma_i)\ .$$

Thus, S^i_4 is the algebraic representation of the finite element approximation of the Steklov-Poincaré operator $\mathcal{S} : [H^{1/2}_{00}(\Gamma_{i-1})]^2 \to [H^{-1/2}(\Gamma_i)]^2$ that associates to a vector function φ defined on Γ_{i-1} the normal stress on Γ_i of the solution to a Stokes problem in Ω_i whose right hand side is zero, and whose velocity field is equal to φ on Γ_{i-1} and is zero on $\partial\Omega_i\backslash\Gamma_{i-1}$. The other matrices defined in (5.1)-(5.3) have a similar meaning. The capacitance matrix C is block-tridiagonal and reads as

$$C = \begin{bmatrix} S^1_1 + S^2_2 & S^2_3 & & & \\ S^2_4 & S^2_1 + S^3_2 & S^3_3 & & \\ & & \cdots & & \\ & & S_4^{M-2} & S_1^{M-2} + S_2^{M-1} & S_3^{M-1} \\ & & & S_4^{M-1} & S_1^{M-1} + S_2^{M} \end{bmatrix} \tag{5.5}$$

Its associated bilinear form is:

$$\mathcal{A}(\varphi,\psi) = \sum_{i=1}^{M} a_i(w_i(\varphi), w_i(\psi)) \qquad \forall\, \varphi, \psi \in \Phi_h\ . \tag{5.6}$$

Then C is symmetric and positive definite, provided a $(\cdot,\cdot)$ is symmetric.
Following what is proposed in [10] for multidomain spectral approximations to elliptic problems, we define now some preconditioners for the capacitance matrix C given in (5.5). The first preconditioner we consider has the following block diagonal structure:

$$
(5.7)\qquad B = \begin{bmatrix} S^2{}_2 & S^2{}_3 & & & & \\ S^2{}_4 & S^2{}_1 & & & & \\ & & S^4{}_2 & S^4{}_3 & & \\ & & S^4{}_4 & S^4{}_1 & & \\ & & & & S_2{}^M & S_3{}^M \\ & & & & S_4{}^M & S_1{}^M \end{bmatrix}
$$

This is precisely the generalization to the case of several subdomains of the matrix (4.2) (note that only the even subdomains are considered).The bilinear form associated with (5.7) is:

$$
(5.8)\qquad \mathcal{B}(\varphi,\psi) = \sum_{i\ \mathrm{even}} a_i(w_i(\varphi), w_i(\psi)) \qquad \forall\ \varphi, \psi \in \Phi_h
$$

(i between 2 and M in the sum) whence the matrix B is symmetric and positive definite. The equivalence between $\mathcal{A}$ and $\mathcal{B}$ can be established by proving that

$$
(5.9)\qquad \sum_{i\ \mathrm{even}} \| w_i(\varphi) \|_i^2 \ \textit{is equivalent to} \ \sum_{i\ \mathrm{odd}} \| w_i(\varphi) \|_i^2 \ , \quad \forall\ \varphi \in \Phi_h \ .
$$

By (4.5) (case of two subdomains) and the argument used in [10], proof of theorem 4.1, we can show that the equivalence claimed in (5.9) holds with two constants independent of h but possibily depending on M^2. Therefore, we conclude that

$$
(5.10)\qquad \textit{condition number of}\ B^{-1} C \le K M^2 \ \cdot\ K\ \textit{independent of}\ h\ .
$$

To the same conclusion we can arrive taking in (5.8) the summation on all odd (rather than even) integers. The matrix (5.7) modifies accordingly.

A different block diagonal preconditioner (with M-1 blocks) is:

$$
(5.11)\qquad B = \begin{bmatrix} S^2{}_2 & & & & \\ & S^3{}_2 & & & \\ & & S^4{}_2 & & \\ & & & \ddots & \\ & & & & S_2{}^M \end{bmatrix}
$$

Its associated bilinear form is

$$
(5.12)\qquad \mathcal{B}(\varphi,\psi) = \sum_{i=1}^{M} a_i(w_i^-(\varphi), w_i^-(\psi)) \qquad \forall\ \varphi, \psi \in \Phi_h\ .
$$

where $w^-_i(\varphi)$ is the first component of a finite element Stokes extension such that $w^-_i(\varphi) = \varphi$ on Γ_{i-1} and $w_i(\varphi) = 0$ on Γ_i. By the same kind of arguments used above one can prove that the matrix B given in (5.11) is symmetric and positive definite, and that (5.10) still holds.

A symmetric situation occurs if we take the matrix B associated with a bilinear form like (5.12) with $w^+_i(\varphi)$ and $w^+_i(\psi)$ instead of $w^-_i(\varphi)$ and $w^-_i(\psi)$, respectively.

Finally, consider the *lower* bidiagonal preconditioner

$$(5.13)\qquad B = \begin{bmatrix} S^1_1 & & & & \\ S^2_4 & S^2_1 & & & \\ & S^3_4 & S^3_1 & & \\ & & & \ddots & \\ & & & S_4^{M-1} & S_1^{M-1} \end{bmatrix}$$

whose associated form is:

$$(5.14)\qquad \mathcal{B}(\varphi,\psi) = \sum_{i=1}^{M} a_i(w_i(\varphi), w_i^+(\varphi)) \qquad \forall\, \varphi, \psi \in \Phi_h\ .$$

Its *upper*, bidiagonal counterpart is:

$$(5.15)\qquad B = \begin{bmatrix} S^2_2 & S^2_3 & & & \\ & S^3_2 & S^3_3 & & \\ & & \ddots & & \\ & & & S_2^{M-1} & S_3^{M-1} \\ & & & & S_2^M \end{bmatrix}$$

and its associated form reads as

$$(5.16)\qquad \mathcal{B}(\varphi,\psi) = \sum_{i=1}^{M} a_i\,(w_i\,(\varphi), w_i^-\,(\psi)) \qquad \forall\, \varphi, \psi \in \Phi_h\quad .$$

Either (5.14) and (5.16) (and consequently, the matrices (5.13) and (5.15)) fail to be symmetric. However, they are still positive, and (5.10) still holds.

Remark 5.1 Using iterative methods with block diagonal preconditioners for the capacitance system yields algorithms whose parallelism degree (i.e., the number of *independent* subproblems to be solved at each step) is equal to the number of diagonal blocks of the preconditioner.

5.2 Boxes (see Fig.5.2) We consider now the case of a box, i.e. of a domain Ω decomposed into four subdomains sharing a common internal vertex. We denote by Γ_i the common interface between Ω_i and Ω_{i+1}, i=1,...,4 (we identify Ω_5 with Ω_1).
As in the previous subsection, we introduce the capacitance matrix for the current case by means of auxiliary interface operators. We recall that ψ^r_m is the Lagrange function associated with the node ξ^r_m of Γ_r. We are not considering here neither the node corresponding to the interior vertex, nor the one belonging to $\partial\Omega$.

If N_s is the number of the interior nodes of Γ_s, we set for each i=1,...,4:

(5.17) $$(S^i_{rt})_{km} = a_i\,(w^i\,(\psi^r_m),\, w^i\,(\psi^t_k)) \quad k=1,\dots,N_r\ ,\quad m=1,\dots,N_t\ .$$

i is the index of the subdomain, r and t those of the interfaces, while k and m denote rows and columns of the matrix. The matrix S^i_{rt} is the algebraic representation of the finite element approximation of a Steklov-Poincaré operator $\mathcal{S}$. Precisely, $\mathcal{S}$ associates to a vector function φ defined on Γ_r the normal stress on Γ_t of the solution to a Stokes problem in Ω_i with zero right hand side, and with a velocity field that vanishes on $\partial\Omega_i\backslash\Gamma_r$, and coincides with φ on Γ_r. We will conventionally denote the interior vertex by Γ_0, and the corresponding node by 0. This allows us to extend the definition of the operators (5.17) to cover the case in which at least one of the indices r,t is equal to zero. In this way S^i_{0t} is a columm vector of lenght N_t, $(S^i_{r0})^T$ is a vector of lenght N_r, while S^i_{00} is a scalar. S^i_{0t} associates to the Lagrange finite element function the point 0 the normal pertaining to stress tensor (taken in the usual variational sense) of the corresponding Stokes extension (in Ω_i) at all internal nodes of the interface Γ_t. Simmetrically, S^i_{r0} associates to every Lagrange function on Γ_r the value at the point 0 of the stress tensor associated with the corresponding finite element Stokes extension in Ω_i. In terms of the above matrices and vectors the capacitance matrix C associated to the current multidomain finite element problem is (the interface unknowns are ordered as: $\mathbf{U}_1,\mathbf{U}_2,\mathbf{U}_0,\mathbf{U}_3,\mathbf{U}_4$, with $\mathbf{U}_i \in \Gamma_i$)

(5.18) $$C = \begin{bmatrix} S^1_{11}+S^2_{11} & S^2_{21} & \sigma_{10} & 0 & S^1_{41} \\ S^2_{12} & S^2_{22}+S^3_{22} & \sigma_{20} & S^3_{32} & 0 \\ \sigma_{01} & \sigma_{02} & \sigma_{00} & \sigma_{03} & \sigma_{04} \\ 0 & S^3_{23} & \sigma_{30} & S^3_{33}+S^4_{33} & S^4_{43} \\ S^1_{14} & 0 & \sigma_{40} & S^4_{34} & S^4_{44}+S^1_{44} \end{bmatrix}$$

where for convenience of notation we have set: $\sigma_{i0} = S^i_{0i} + S^{i+1}_{0i}$, $\sigma_{0i} = S^i_{i0} + S^{i+1}_{i0}$, and $\sigma_{00} = \sum_{i=1}^{4} S^i_{00}$ (as usual, a super index equal to five should be identified with 1). Its associated bilinear form is:

(5.19) $$\mathcal{A}(\varphi,\psi) = \sum_{i=1}^{4} a_i\,(w_i\,(\varphi),\, w_i\,(\psi)) \qquad \forall\ \varphi,\psi \in \Phi_h\ ,$$

whence C is positive definite and symmetric, provided a $(\cdot,\cdot)$ is symmetric .
A block diagonal preconditioner which is the counterpart of (5.7) can be obtained from (5.18) by disregarding all matrices and vectors S^i_{rt} with a super index i odd. Such a

preconditioner is still symmetric and positive definite, however it is neither block diagonal nor spectrally equivalent to C. Actually, using the results of [12] we can show that the condition number of the corresponding preconditioned matrix grows like $K(1+lg(H/h))^2$, where K is a positive constant, H is the maximum size of each subdomain, and h is, as usual, the finite element mesh size.

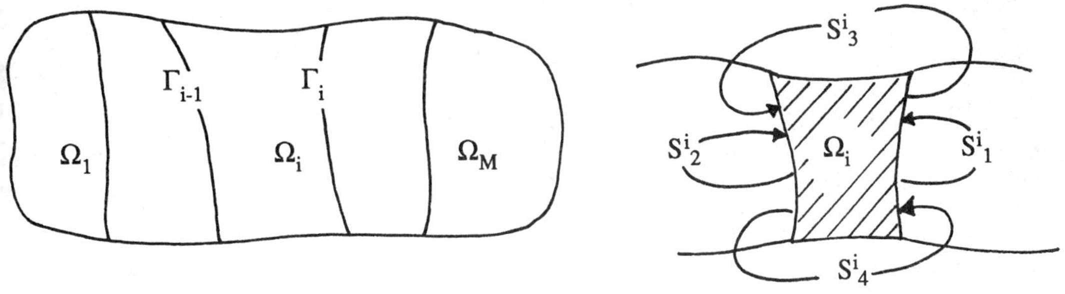

Fig.5.1 A strip and its associated interface operators

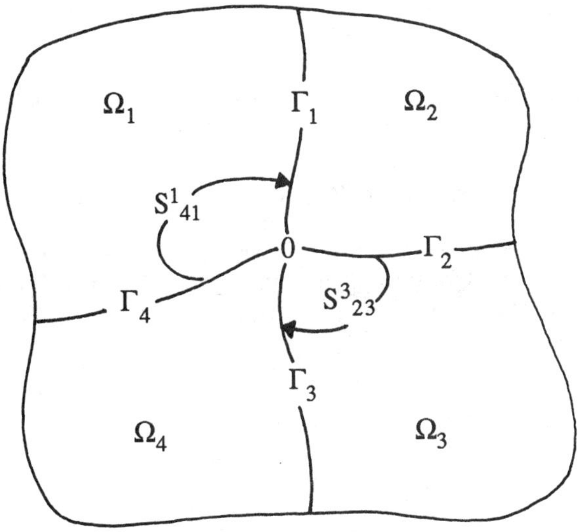

Fig.5.2 A box and its associated interface operators

REFERENCES

(1) C.BERNARDI, C.CANUTO and Y.MADAY, Generalized Inf-Sup Conditions for Chebyshev Spectral Approximation of the Stokes Problem, to appear in SIAM J. Numer. Anal., 1988.

(2) G.BJORSTADT and O.WIDLUND, Iterative methods for the solution of elliptic problems on regions partitioned into substructures, SIAM J. Numer. Anal., 23 (1986), pp.1097-1120.

(3) F.BREZZI, On the existence, uniqueness and approximation of saddle-point problems arising from lagrangian multipliers, RAIRO Numer. Anal., 8 (1974), pp.129-151.

(4) F.BREZZI and M.FORTIN, Book, in preparation.

(5) D.FUNARO, A.QUARTERONI and P.ZANOLLI, An iterative procedure with interface relaxation for domain decomposition methods, to appear in SIAM J. Numer. Anal., 1988.

(6) V.GIRAULT and P.A.RAVIART, Finite Element Methods for the Navier-Stokes Equations: Theory and Algorithms, Springer, Berlin, 1986.

(7) Y.MADAY, A.PATERA and E.RONQUIST, A well posed optimal spectral element approximation for the Stokes problem, NASA CR-178343, 1987.

(8) L.D.MARINI and A.QUARTERONI, An iterative procedure for domain decomposition methods: a finite element approach, in Domain Decomposition Methods for Partial Differential Equations, I, R.Glowinski et Al., Eds., SIAM, Philadelphia, 1988, pp.129-143.

(9)L.D.MARINI and A.QUARTERONI, A relaxation procedure for domain decomposition methods using finite elements, I.A.N.-C.N.R. Publication n.577, Pavia, 1987.

(10) A.QUARTERONI and G.SACCHI-LANDRIANI, Domain Decomposition Preconditioners for the Spectral Collocation Method, I.A.N.-C.N.R. Publication n.594, Pavia, 1987.

(11) H.VANDEVEN, Doctoral Thesis, Université P. et M. Curie, Paris 1987.

(12) O.B.WIDLUND, Iterative substructuring methods: algorithms and theory for elliptic problems in the plane, in Domain Decomposition Methods for Partial Differential Equations, I, R.Glowinski et Al., Eds., SIAM, Philadelphia, 1988, pp.113-128.

Adaptive Implicit-Explicit and Parallel Element-By-Element Iteration Schemes*

T. E. Tezduyar†
J. Liou†
T. Nguyen††
S. Poole††

Abstract

Adaptive implicit-explicit (AIE) and grouped element-by-element (GEBE) iteration schemes are presented for the finite element solution of large-scale problems in computational mechanics and physics. The AIE approach is based on the dynamic arrangement of the elements into differently treated groups. The GEBE procedure, which is a way of rewriting the EBE formulation to make its parallel processing potential and implementation more clear, is based on the static arrangement of the elements into groups with no inter-element coupling within each group. Various numerical tests performed demonstrate the savings in the CPU time and memory.

1. Introduction

For large-scale problems in computational mechanics and physics, solution (by direct methods) of the linear equation systems involving massive global matrices and the storage of such matrices generate substantial demand on the computational resources in terms of the CPU time and memory. For most problems of practical interest, especially in three dimensions, this demand becomes too heavy to accommodate even for today's most generous computers. Matrices of this volume usually arise from the implicit time-integration of spatially discretized time-dependent problems or from spatial discretization of equations with no (real or pseudo) temporal derivatives. For example, the incompressible Navier-Stokes equations in the vorticity-stream function formulation involve a time-dependent convection-diffusion equation for the vorticity and a Poisson's equation for the stream function. In this paper we describe, in the context of the vorticity - stream function formulation, alternative solution techniques which reduce or eliminate the need for the storage and factorization of large global matrices.

The implicit-explicit algorithm proposed by Hughes and Liu [6,7] (for solid mechanics and heat conduction problems) involves the static allocation of the implicit and

---

* This research was sponsored by NASA-Johnson Space Center under contract NAS-9-17892 and by NSF under grant MSM-8796352.

† Department of Aerospace Engineering and Mechanics, and Minnesota Supercomputer Institute, University of Minnesota, Minneapolis, MN 55455.

‡ IBM Corporation, Houston, TX 77056.

explicit elements based on the stability limit of the explicit algorithm. The adaptive implicit-explicit (AIE) scheme was first presented, in its preliminary stage, in Tezduyar and Liou [13]. It is based on the dynamic grouping of the elements into the implicit and explicit subsets as dictated by the element level stability and accuracy considerations. For this purpose the algorithm monitors the element level Courant number and some measure of the dimensionless wave number. The element level Courant number is based on the local convection and diffusion transport rates whereas the dimensionless wave number reflects the local smoothness of the "previous" solution and the spatial discretization. The "previous" solution can come from the previous time step, or nonlinear iteration step, or pseudo-time iteration step, whichever is appropriate. In this approach we can have "implicit refinement" where it is needed. Elsewhere in the domain, computations are performed explicitly thus resulting in substantial memory and CPU time savings. The savings can be further increased by performing, as often as desired, an equation renumbering at the implicit zones to obtain optimal bandwidths. Compared to the adaptive schemes based on grid-moving or element-subdividing the AIE method involves minimal bookkeeping and no geometric constrains.

It is important to note that the dynamic grouping of the AIE scheme does not necessarily have to be with respect to implicit and explicit elements. The letter "I" and "E" in "AIE" could be referring to any two procedures. In fact, there is no reason for the number of possible choices to be limited to two. For a given element, the rationale for favoring one procedure over another one could be based on any factor, such as the cost efficiency, the type of spatial or temporal discretization, the differential equations used for modelling, etc. In this paper, as an example, we implement the AIE scheme also in conjunction with the flux corrected transport (FCT) method [9,15].

It has been more than a decade since the first CRAY-1 computer was delivered to Los Alamos National Laboratories. The evolution of vector architecture in the past few years is evident. With the inception of new vectorizing compiler technology, the job of the application developers has become less architecture specific but rather more numerical and algorithmic in nature. Since the introduction of the CRAY-XMP, a multi-processor vector computer, in 1982, parallel processors and parallel processing have become more popular among the computer vendors in the Engineering and Scientific market sector. The fact that IBM introduced the Vector Facility for the 3090 multi-processors series is the ultimate acknowledgement that vector processing is an accepted standard mode of operation and parallel processing is eminent in the conventional camp.

The element-by-element (EBE) factorization scheme is very well suited for vectorization and parallel processing. It was proposed by Hughes, Levit, and Winget [5] for problems in transient heat conduction, solid mechanics, and structural mechanics. The EBE implementations in computational fluid dynamics were first reported, in the context of compressible Euler equations, by Hughes, Levit, Winget, and Tezduyar [8]. Applications to convection-diffusion equations and the incompressible flow problems can be seen in Tezduyar and Liou [13]. The EBE scheme selects a sequential product of the element level matrices as the preconditioning matrix to be used with the iterative solution of large linear equations systems arising from the finite element formulation. For symmetric and positive-definite matrices this preconditioner can be used with the conjugate-gradient method [3]. In the EBE approach the need for the formation, storage, and factorization of bulky global matrices is eliminated. The element level matrices can be either recomputed or stored. In the case the element level matrices are stored, the storage needed is still only linearly proportional to the number of elements.

The grouped element-by-element (GEBE) scheme is a variation of the regular EBE scheme. It has some common grounds with the operator splitting and domain decomposition methods [3,4]. The GEBE preconditioner is a sequential product of the element group matrices with the condition that no two elements in the same group can be "neighbors"; this is achieved by a simple element grouping algorithm which is applicable to arbitrary meshes. In our parallel computations, before we start processing a new group we first have to finish processing the current one. To minimize the overhead associated with this synchronization, we try to minimize the number of groups. Within each group, to increase the vector efficiency of the computations, elements are processed in chunks of

64 elements (or whatever the optimum chunk size for the given vector environment is). When parallel processing is invoked, each of the available CPUs work on one chunk at a time until all chunks are finished.

2. Problem Statement and the Finite Element Formulation

We consider the problems governed by the convection-diffusion equation and the two-dimensional incompressible Navier-Stokes equations. For the Navier-Stokes equations, we use the vorticity - stream function formulation which consists of a time-dependent transport equation for the vorticity ω, and a Poisson's equation relating the stream function ψ to the vorticity. Since the procedures for the spatial and temporal discretizations for the convection-diffusion equation are similar to those for the transport equation of the vorticity - stream function formulation, only the latter will be discussed in this section.

The field equations for the vorticity - stream function formulation of the two-dimensional incompressible Navier-Stokes equations are given as

$$\frac{\partial \omega}{\partial t} + \mathbf{u} \bullet \nabla \omega = \nu \nabla^2 \omega, \tag{2.1}$$

$$\nabla^2 \psi = -\omega, \tag{2.2}$$

where $\mathbf{u}$ is the velocity and ν is the kinematic viscosity. The incompressibility condition is automatically satisfied by the following definition of the stream function:

$$\frac{\partial \psi}{\partial x_2} = u_1, \tag{2.3a}$$

$$\frac{\partial \psi}{\partial x_1} = -u_2. \tag{2.3b}$$

The flow domain can have several internal boundaries corresponding to possible obstacles (holes) in the flow field. In such cases the unknowns of the problem are: the value of the vorticity at all interior points (ω_*), the value of the vorticity at all boundary points where both components of the velocity are specified (ω_{Gq}) (this includes all the external and internal solid boundaries), and the value of the stream function at all interior points and on the internal boundaries (ψ_{*q}).

The differential equations given by (2.1) and (2.2), together with the initial condition for the vorticity and the suitable boundary conditions for the vorticity and the stream function, can be translated, via a proper variational formulation [11], into a set of ordinary differential equations. That is

$$\tilde{\mathbf{M}}(\mathbf{d}_{*q})\, \mathbf{a}_* + \tilde{\mathbf{C}}(\mathbf{d}_{*q})\mathbf{v}_* = \tilde{\mathbf{F}}(\mathbf{d}_{*q}, \mathbf{v}_{Gq}, \mathbf{a}_{Gq}), \tag{2.4a}$$

$$\mathbf{v}_*(0) = (\mathbf{v}_*)_0, \tag{2.4b}$$

$$\mathbf{K}\, \mathbf{d}_{*q} = \mathbf{F}_{II}(\mathbf{v}_*, \mathbf{v}_{Gq}), \tag{2.5a}$$

$$\mathbf{M}\, \mathbf{v}_{G\,q} = \mathbf{F}_{III}(\mathbf{d}_{*q}, \mathbf{v}_*), \tag{2.5b}$$

where $\mathbf{d}_{*q}$, $\mathbf{v}_*$, $\mathbf{a}_*$, $\mathbf{v}_{Gq}$, and $\mathbf{a}_{Gq}$ are the vectors of nodal values of ψ_{*q}, ω_*, $\frac{\partial \omega_*}{\partial t}$, ω_{Gq}, and $\frac{\partial \omega_{Gq}}{\partial t}$, respectively.

Remark 2.1 Particular attention must be paid to the boundary conditions for the vorticity on the (external and internal) solid surfaces and for the stream function on the internal boundaries. Discussion of these boundary conditions and the derivation of the proper variational formulations can be found in [11].

Remark 2.2 Equation systems (2.4a) and (2.5) are derived from the variational formulations corresponding to (2.1) and (2.2), respectively. The initial condition for (2.1) is represented by (2.4b).

Remark 2.3 The matrices **K** and **M** are symmetric and positive-definite. The matrix **M** is topologically one-dimensional because it is associated with the vector $\mathbf{v}_{Gq}$ corresponding to the (unknown) value of the vorticity on the boundaries. The matrices $\tilde{\mathbf{M}}$ and $\tilde{\mathbf{C}}$ are functions of the stream function because in the variational formulation of (2.1) we employ a Petrov-Galerkin method [10] with weighting functions dependent upon the velocity field.

Remark 2.4 Equations (2.4) and (2.5) are solved by employing a predictor/multi-corrector algorithm which, in nature, is a block-iteration scheme [12]. In fact the way we have written these equations reflects the way the block-iteration scheme works. In the blocks corresponding to (2.4a),(2.5a), and (2.5b), we update $\mathbf{v}_*$, $\mathbf{d}_{*q}$, and $\mathbf{v}_{Gq}$, respectively. To move from time step n to time step n+1 we perform iterations as outlined below:

block 1:

$$\bar{\mathbf{M}}^i_{n+1}\ (\Delta\mathbf{a}_*)^i_{n+1} = \mathbf{R}^i_{n+1}\,, \tag{2.6a}$$

where

$$\bar{\mathbf{M}}^i_{n+1} = \tilde{\mathbf{M}}((\mathbf{d}_{*q})^i_{n+1}) + \alpha\,\Delta t\,\tilde{\mathbf{C}}((\mathbf{d}_{*q})^i_{n+1}), \tag{2.6b}$$

and

$$\mathbf{R}^i_{n+1} = \tilde{\mathbf{F}}((\mathbf{d}_{*q})^i_{n+1}, (\mathbf{v}_{Gq})^i_{n+1}, (\mathbf{a}_{Gq})^i_{n+1}) - (\tilde{\mathbf{M}}((\mathbf{d}_{*q})^i_{n+1})\,(\mathbf{a}_*)^i_{n+1} + \tilde{\mathbf{C}}((\mathbf{d}_{*q})^i_{n+1})\,(\mathbf{v}_*)^i_{n+1}\,, \tag{2.6c}$$

with updates

$$(\mathbf{v}_*)^{i+1}_{n+1} = (\mathbf{v}_*)^i_{n+1} + \alpha\,\Delta t\,(\Delta\mathbf{a}_*)^i_{n+1}, \tag{2.6d}$$

$$(\mathbf{a}_*)^{i+1}_{n+1} = (\mathbf{a}_*)^i_{n+1} + (\Delta\mathbf{a}_*)^i_{n+1}. \tag{2.6e}$$

Here i is the iteration count, Δt is the time step, and α is a parameter which controls the stability and accuracy of the time integration.

block 2:

$$\mathbf{K}(\mathbf{d}_{*q})^{i+1}_{n+1} = \mathbf{F}_{II}((\mathbf{v}_*)^{i+1}_{n+1},(\mathbf{v}_{Gq})^i_{n+1}). \tag{2.7}$$

block 3:

$$\mathbf{M}(\mathbf{v}_{G\,q})^{i+1}_{n+1} = \mathbf{F}_{III}((\mathbf{d}_{*q})^{i+1}_{n+1}, (\mathbf{v}_*)^{i+1}_{n+1}). \tag{2.8}$$

The iterations continue until a predetermined convergence criterion is met.

3. The Adaptive Implicit-Explicit (AIE) Scheme

In our block-iterative procedure, at every iteration we need to solve three equation systems : (2.6a),(2.7), and (2.8). The cost involved in (2.8) is not major (see Remark 2.3) and therefore we solve it with a direct method. We are mainly interested in minimizing the CPU time and memory demands of equations (2.6a) and (2.7) which, after dropping all the subscripts and superscripts, can be rewritten as follows:

$$\bar{\mathbf{M}}\,\Delta\mathbf{a} = \mathbf{R}, \tag{3.1}$$

$$\mathbf{K}\,\mathbf{d} = \mathbf{F}. \tag{3.2}$$

We need to remember that **K** is symmetric and positive-definite but $\bar{\mathbf{M}}$, in general, is not. We propose to employ the AIE scheme for both of these equations.

<u>the AIE scheme in the context of the vorticity transport equation</u>

Let $\mathcal{E}$ be the set of all elements, e=1,2,..., n_{el}. The assembly of the global matrix $\bar{\mathbf{M}}$ in equation (3.1) can be expressed as follows:

$$\bar{\mathbf{M}} = \sum_{e\in\mathcal{E}} \bar{\mathbf{M}}^e, \tag{3.3}$$

where $\bar{\mathbf{M}}^e$ is the element contribution matrix which is obtained by permutating the element level matrix $\bar{\mathbf{m}}^e$. It should be noted that $\bar{\mathbf{M}}^e$ has the same dimensions as the global matrix $\bar{\mathbf{M}}$ but only as many number of nonzero entries as $\bar{\mathbf{m}}^e$ (e.g. 4×4 for a two-dimensional quadrilateral element with a scalar unknown).

Let $\mathcal{E}_I$ and $\mathcal{E}_E$ be the subsets of $\mathcal{E}$ corresponding to the implicit and explicit elements, respectively, such that

$$\mathcal{E} = \mathcal{E}_I \cup \mathcal{E}_E, \tag{3.4a}$$

$$\varnothing = \mathcal{E}_I \cap \mathcal{E}_E. \tag{3.4b}$$

Consequently, from equation (3.3) we get

$$\bar{\mathbf{M}} = \sum_{e\in\mathcal{E}_I} \bar{\mathbf{M}}^e + \sum_{e\in\mathcal{E}_E} \bar{\mathbf{M}}^e. \tag{3.5}$$

The AIE scheme is based on modifying $\bar{\mathbf{M}}$ by replacing $\bar{\mathbf{M}}^e$ ($\forall$ e $\in$ $\mathcal{E}_E$)with its lumped mass matrix part; that is

$$\bar{\mathbf{M}} = \sum_{e\in\mathcal{E}_I} \bar{\mathbf{M}}^e + \sum_{e\in\mathcal{E}_E} \mathbf{M}_L^e. \tag{3.6}$$

The grouping given by (3.4) is done dynamically (adaptively) based on the stability and accuracy considerations.

The stability criterion is in terms of the element Courant number $C_{\Delta t}$ which is defined as

$$C_{\Delta t} = \frac{\|\mathbf{u}\|\,\Delta t}{h}\ , \tag{3.7}$$

where h is the "element length" [10]. Any element with its Courant number greater than the stability limit of the explicit method should belong to the implicit group $\mathcal{E}_I$.

For the accuracy considerations we use a quantity which is meant to be a measure of the dimensionless wave number [10]. For this purpose, in [13] the "jump" in the solution across an element was defined as

$$j^e(\omega) = \max_a (\omega_a^e) - \min_a (\omega_a^e), \tag{3.8}$$

where a is the element node number and ω_a^e is the value of the dependent variable ω at node a of the element e. This definition reflects the magnitude of the variation of the solution across an element. It was proposed in [13] that the elements with jumps greater than a predetermined amount should belong to the group $\mathcal{E}_I$.

In our numerical tests involving the one-dimensional propagation of various triangular profiles we observed that the accuracy of the solution is affected not only by the jump in the solution but also by the derivative of the flux of the dependent variable. Therefore, as an alternative criterion we propose to employ the product of the jump and the derivatives of the flux; that is

$$\sigma^e = j^e(\omega) \left(\left| \frac{\partial(u_1\omega)}{\partial x_1} \right| + \left| \frac{\partial(u_2\omega)}{\partial x_2} \right| \right)^e. \tag{3.9}$$

A global scaling is needed for this quantity and we propose the following procedure:

$$\sigma_{\mathcal{E}}^e = \frac{\sigma^e}{\sigma_{max}^e}, \tag{3.10}$$

where

$$\sigma_{max}^e = \max_{e \in \mathcal{E}} \sigma^e . \tag{3.11}$$

In this case the elements with $\sigma_{\mathcal{E}}^e$ greater than a predetermined value belong to group $\mathcal{E}_I$.

Implementation of the AIE scheme is quite straightforward. A global search is performed on the entire set of elements. With this search, based on the two criteria given by equations (3.7) and (3.10), the elements are grouped into the subsets $\mathcal{E}_I$ and $\mathcal{E}_E$. Compared to having all elements treated implicitly, this grouping results in substantial savings in the CPU time and memory.

Remark 3.1 It is important to note that the grouping does not have to be with respect to implicit and explicit elements. In the acronym "AIE", the letter "I" can refer to the "I-elements" subject to some "I-procedure" whereas the letter "E" can refer to the "E-elements" subject to some other "E-procedure". Let us suppose that we are concerned with the computational cost but we also believe that "you get what you pay for". Then we might have an "I-procedure" which is sophisticated but costly and an "E-procedure" which is less sophisticated but also less costly. We can employ the "I-procedure" where it is needed and the "E-procedure" elsewhere. Even if the computational cost is not an issue, in certain regions of the domain we just might have a reason to prefer the "I-procedure" over the "E-procedure", or vice versa. The reasons could be based on the type of interpolation functions, spatial discretization, differential equations used for modelling, etc. The AIE concept is based on making decisions at the element level, dynamically, and according to some desired and reliable criteria. Later in this section we will give an example for an AIE scheme which is not based on the implicit-explicit grouping.

Remark 3.2 In the AIE approach one can have a high degree of refinement throughout the mesh and raise the implicit flag only for those elements which need to be treated implicitly.

Remark 3.3 The savings in the CPU time and memory can be maximized by performing, as often as desired, an equation renumbering at the "implicit zones" to obtain optimal bandwidths. Bandwidth optimizers are already available for finite element applications, especially in the area of structural mechanics.

Remark 3.4 Time-dependent convection-diffusion of a passive scalar can be treated as a special case of equation (2.1) with the velocity field **u** given. The only equation system that needs to be solved is (3.1).

the AIE scheme in the context of the flux-corrected transport (FCT) method

It was stated in Remark 3.1 that the AIE scheme does not have to be based on implicit-explicit grouping. It can be based on any element level, dynamic, and rational selection between an "I-procedure" and an "E-procedure". Here we give an example by employing the AIE scheme in conjunction with the FCT method described in [9,15]. We apply this method to the time-dependent convection-diffusion of a passive scalar (see Remark 3.4).

The FCT method given in [9,15] is based on the correction of the flux obtained with a lower-order scheme with the flux obtained with a higher-order scheme, while limiting the maximum allowable correction. The balance law is satisfied at the element level. In our case, as the lower-order scheme we use the one-pass explicit SUPG method and as the higher order scheme the multi-pass explicit SUPG method. With sufficient number of iterations (typically three) the multi-pass explicit method produces solutions indistinguishable from those obtained by the implicit method. In the utilization of our AIE scheme we define the "E-procedure" to be the one-pass explicit SUPG method alone and the "I-procedure" to be the FCT method based on the lower- and higher-order schemes described above.

Remark 3.5 It is not our intention here to make any statement about what lower- and higher-order schemes should be used with the FCT method or about whether the FCT method should be preferred over some other method. We just needed to pick an example.

For the AIE scheme we employ the "v-form" of the predictor/multi-corrector algorithm described by equation (2.6). That is

$$\mathbf{M}_L\,(\Delta \mathbf{v}_*)^i_{n+1} = \bar{\mathbf{R}}^i_{n+1}, \tag{3.12a}$$

where $\mathbf{M}_L$ is the (diagonal) lumped mass matrix version of $\tilde{\mathbf{M}}$ and

$$\bar{\mathbf{R}}^i_{n+1} = \Delta t\ \tilde{\mathbf{F}}_{n+1} - (\tilde{\mathbf{M}} + \alpha \Delta t \tilde{\mathbf{C}})\,(\mathbf{v}_*)^i_{n+1}$$
$$+ (\tilde{\mathbf{M}} - (1-\alpha)\Delta t\,\tilde{\mathbf{C}})\,(\mathbf{v}_*)_n. \tag{3.12b}$$

Note that these expressions are quite simpler due to the fact that this is a linear problem and that no stream function is involved.

The nodal value increment vector $(\Delta \mathbf{v}_*)^i_{n+1}$ and the residual vector $\bar{\mathbf{R}}^i_{n+1}$ can be written as sums of their element level contribution vectors

$$(\Delta \mathbf{v}_*)^i_{n+1} = \sum_{e \in \mathcal{E}} (\Delta \mathbf{v}^e_*)^i_{n+1}, \tag{3.13}$$

$$(\bar{\mathbf{R}}^i_{n+1}) = \sum_{e \in \mathcal{E}} (\bar{\mathbf{R}}^e)^i_{n+1}. \tag{3.14}$$

The AIE scheme employed is outlined below:

1. given $(\mathbf{v}_*)_n$
   $i = 0$
2. $(\Delta \mathbf{v}^e_*)^i_{n+1} = \mathbf{M}_L^{-1} (\bar{\mathbf{R}}^e)^i_{n+1} \quad \forall\, e \in \mathcal{E}$ (3.15)
3. $(\mathbf{v}_*)^{i+1}_{n+1} = (\mathbf{v}_*)^i_{n+1} + \sum_{e \in \mathcal{E}} (\Delta \mathbf{v}^e_*)^i_{n+1}$ (3.16)
4. $i = i+1$
5. $(\Delta \mathbf{v}^e_*)^i_{n+1} = \mathbf{M}_L^{-1} (\bar{\mathbf{R}}^e)^i_{n+1} \quad \forall\, e \in \mathcal{E}_I$ (3.17)
6. $(\mathbf{v}_*)^{i+1}_{n+1} = (\mathbf{v}_*)^i_{n+1} + \sum_{e \in \mathcal{E}_I} c^e (\Delta \mathbf{v}^e_*)^i_{n+1}$ (3.18)
7. if $i < n_{it}$ then go to 4
8. $n = n+1$ and goto 1

where c^e is a parameter [9,15] determined by the maximum correction allowed for the element e and n_{it} is the number of iterations per time step for the higher-order solution.

Limiting the steps 5 and 6 to the group $\mathcal{E}_I$ results in substantial savings in the computational cost involved.

<u>solution of the discrete Poisson's equation</u>

We use the preconditioned conjugate gradient method to solve the equation system given by (3.2). In conjunction with the AIE scheme employed for equation (3.1) we define our preconditioner matrix **P** to be

$$\mathbf{P} = \sum_{e \in \mathcal{E}_I} \mathbf{K}^e + \sum_{e \in \mathcal{E}_E} \mathrm{diag}(\mathbf{K}^e), \tag{3.19}$$

where $\mathbf{K}^e$ is the element contribution matrix of **K**. If $\mathcal{E}_E = \varnothing$ then $\mathbf{P} = \mathbf{K}$ and the solution technique becomes a direct one. If $\mathcal{E}_I = \varnothing$ then the method becomes a Jacobi iteration [2]. Other definitions for **P** are of course possible; the one given by (3.19) is simple to implement and reflects our belief that there should be a correlation between the solution procedures for (3.1) and (3.2).

<u>Remark 3.6</u> The incompressible Navier-Stokes equations in the velocity-pressure formulation, upon spatial and temporal discretizations, can also translate to a set of equations given by (3.1) and (3.2). In this case equations (3.1) and (3.2) would

correspond, respectively, to the momentum equation and the incompressibility constraint. The latter one would consist of a discrete Poisson's equation for pressure.

Remark 3.7 Another example for the type of problems which can translate to equations (3.1) and (3.2) is the electrophoresis separation process (see [1]). In the context of a block-iteration scheme, the time-dependent transport equation for the chemical species and the equation for the electric potential would transform to equations (3.1) and (3.2), respectively.

4. The Grouped Element-by-Element (GEBE) Preconditioned Iteration Method

The linear equation systems given by (3.1) and (3.2) are both in the form

$$\mathbf{A}\,\mathbf{x} = \mathbf{b} \tag{4.1}$$

In this section we describe our parallel GEBE-preconditioned iteration method for solving (4.1).

element grouping

We group the set of elements $\mathcal{E}$ in such a way that

1. $\mathcal{E} = \bigcup_{K=1}^{N_{pg}} \mathcal{E}_K,$ (4.2)
2. $\emptyset = \mathcal{E}_J \cap \mathcal{E}_K$ for $J \neq K,$ (4.3)
3. no two elements which belong to the same group can be neighbors (we define being neighbors as having at least one common node),

where N_{pg} is the number of such (parallelizable) groups.

Corresponding to this grouping, the matrix $\mathbf{A}$ can be written as

$$\mathbf{A} = \sum_{K=1}^{n_{pg}} \mathbf{A}_K, \tag{4.4}$$

where the "group matrices" are given as

$$\mathbf{A}_K = \sum_{e \in \mathcal{E}_K} \mathbf{A}^e\ ,\ K = 1,2,\dots, N_{pg}. \tag{4.5}$$

Remark 4.1 Within each group, since there is no inter-element coupling, computations performed in an element-by-element fashion (for example such operations performed on a group matrix) do not depend on the ordering of the elements.

Remark 4.2 In our parallel computations, before we start with a new group we first have to finish with the current one. To minimize the overhead associated with this synchronization, we try to minimize the number of groups. For example, in a two-dimensional problem with rectangular domain and 100×100 mesh, there will be 4 groups with each group having 2500 elements and arranged in a checkerboard style. A simple element grouping algorithm for arbitrary meshes is described below:

initialization:

$e=1$	(start with the first element)
$N_{pg} = 1$	(create the first group)
$e \in \mathcal{E}_{N_{pg}}$	(put the element in that group)

next element:

$e = e+1$	(take the next element)
if $e > n_{el}$ then return	(if there is no such element return)
$K = 0$	(otherwise get ready to search for an eligible group among the existing groups, starting with the first existing group)

next existing group:

$K = K+1$	(take the next existing group)
if e has no neighbors in $\mathcal{E}_K$ then	(check if the element can be put in that group)
$e \in \mathcal{E}_K$	(if so, do it)
go to the next element	
else if $K < N_{pg}$ then	(otherwise check if there is a next existing group)
go to next existing group	(if so, try that next group)
end if	

create a new group:

$N_{pg} = N_{pg} + 1$	(otherwise create a new group)
$e \in \mathcal{E}_{N_{pg}}$	(and put the element in that group)
go to the next element	

Remark 4.3 Within each group, to increase the vector efficiency of the computations performed, elements are processed in chunks of 64 elements (or whatever the optimum chunk size is for the given vector environment). For the example of Remark 4.2 each group would have 40 chunks (with an assumed chunk size of 64 elements).

GEBE-preconditioned iterative solution

Equation (4.1) is solved iteratively as follows:

$$\mathbf{P}\,\Delta\mathbf{y}_m = \mathbf{r}_m, \tag{4.6}$$

where the residual vector is defined as

$$\mathbf{r}_m = \mathbf{b} - \mathbf{A}\,\mathbf{x}_m, \tag{4.7}$$

and the two-pass GEBE-preconditioning matrix is given as

$$\mathbf{P} = \mathbf{D}^{-1} \left(\prod_{K=1}^{N_{pg}} \mathbf{E}_K \prod_{K=N_{pg}}^{1} \mathbf{E}_K \right) \mathbf{D}^{-1}, \tag{4.8}$$

with the scaling matrix

$$\mathbf{D} = \frac{(\Delta\theta)^{1/2}}{\mathbf{W}^{1/2}} . \tag{4.9}$$

Here $\Delta\theta$ is a free parameter and for $\mathbf{W}$, depending on the properties of $\mathbf{A}$, we can pick one of the following two choices:

$$\mathbf{W} = \text{diag}\,\mathbf{A}, \qquad ([5]) \tag{4.10}$$

and

$$\mathbf{W} = \mathbf{M}_L . \qquad ([8]) \tag{4.11}$$

Note that equation (4.9) needs $\mathbf{W}$ to be positive-definite. While the choice of (4.10) does not guarantee this when $\bar{\mathbf{M}}$ is not positive-definite, the alternative choice of (4.11) does. For problems in fluid mechanics $\bar{\mathbf{M}}$ is not in general positive-definite. A certain type of streamline-upwind/Petrov-Galerkin method, however, in some cases result in $\bar{\mathbf{M}}$ being symmetric and positive-definite.

The matrix $\mathbf{E}_K$ is defined as

$$\mathbf{E}_K = (\mathbf{I} + \frac{1}{2}\mathbf{D}\mathbf{B}_K\mathbf{D}), \qquad K=1,2,\ldots,N_{pg}, \tag{4.12}$$

where $\mathbf{I}$ is the identity matrix. Considering that there is no inter-element coupling within each group, $\mathbf{E}_K$ can also be written as follows:

$$\mathbf{E}_K = \prod_{e\in\mathcal{E}_k} (\mathbf{I} + \frac{1}{2}\mathbf{D}\mathbf{B}^e\mathbf{D}), \qquad K=1,2,\ldots,N_{pg}. \tag{4.13}$$

For the element matrix $\mathbf{B}^e$ we consider the following two choices:

$$\mathbf{B}^e = \mathbf{A}^e, \tag{4.14}$$

and

$$\mathbf{B}^e = \mathbf{A}^e - \frac{\mathbf{W}^e}{\Delta\theta} \qquad \text{(Winget regularization [2])}. \tag{4.15}$$

<u>Remark 4.4</u> The grouped EBE approach is just a way of rewriting the regular EBE formulation. The rewriting is based on arranging the elements into parallelizable groups. We can say that the GEBE formulation is "parallel-ready".

<u>Remark 4.5</u> The expression given by (4.8) depends on the ordering of the groups. This is the only expression that depends on any ordering. The expression given by (4.13) does not depend on the element ordering. We can therefore conclude that the EBE schemes depend on the ordering of the groups but not on the ordering of the elements in a strict sense. The corollary is that the number of possible ordering combinations is $(N_{pg})!$ but not $(n_{el})!$. There may be some degree of non-uniqueness in the grouping of the elements but we assume this non-uniqueness to be a minor one (nonexistent for structured meshes).

The updated value of $\mathbf{x}$ is computed according to the following formula:

$$\mathbf{x}_{m+1} = \mathbf{x}_m + s\,\Delta\mathbf{y}_m, \tag{4.16}$$

where s is the search parameter determined, depending on the properties of $\mathbf{A}$, as follows:

$$s = \frac{\Delta\mathbf{y}_m \bullet \mathbf{r}_m}{\Delta\mathbf{y}_m \bullet \mathbf{A}\Delta\mathbf{y}_m} \qquad \text{(if } \mathbf{A} \text{ is symmetric positive-definite)}, \tag{4.17a}$$

or

$$s = \frac{(\mathbf{A}\Delta\mathbf{y}_m)\bullet\mathbf{r}_m}{\|\mathbf{A}\Delta\mathbf{y}_m\|^2} \qquad \text{(if } \mathbf{A} \text{ is not symmetric positive-definite)}. \tag{4.17b}$$

The latter expression for s is obtained by minimizing $\|\mathbf{r}_{m+1}\|^2$ with respect to s.

Remark 4.6 If **A** is symmetric positive-definite, the preconditioned conjugate-gradient algorithm can be used with the preconditioner given by equation (4.8). This is what we do to solve equation (3.2).

5. Remarks on the Vectorization and Parallel Processing

In a multi-processor vector machine each processor is an individual vector processor. The single vector processor can gain high performance through overlapping multiple operations using segmented functional units. When a program is restructured or vectorized to facilitate this overlapping, the speed can be improved by an order of magnitude. Performance improvements from vectorization can be augmented through the use of parallel processing in a multi-processor machine. Parallel processing allows a single program to use more than one CPU to do the required work. This results in a reduction of the elapsed time needed to do the job. The amount of reduction is proportional to the percentage of the program that can be executed independently. In our implementation we have, so far, focus on the vectorization and parallel processing of element level matrix/vector calculations, factorization and back substitution, and global residual calculations.

All the above calculations have been vectorized in this work. The structure of the vectorized code contains an inner loop over elements that belong to particular chunks of elements with identical calculations. An important implementation point is the restructuring of the code to enable compiler optimization. A specific example is the partitioning of the element level matrix/vector calculations into multiple loops. Of course this introduces additional storage requirements as we need temporary arrays to hold the intermediate computational results across loops. Another example is the deployment of CVMxx, CRAY compiler extension for vectorizing IF construct.

Additionally, all calculations are parallel processed using Microtasking, a compiler directive driven preprocessor for parallel processing on the CRAY. Microtasking is implemented with comment card directives. Before compiling the code, a preprocessor expands the directives into Fortran code which includes calls to the Microtasking library. It is important to note that if the preprocessor is not used, the code is portable to other machines since the Microtasking directives appear as Fortran comment cards. The code is currently being used for benchmarking on additional architectures (e.g., IBM 3090/VF). This is the subject of a future report.

If there are n_{CPU} CPUs available for parallel processing, we would like to distribute the work among all CPUs in such a way that large task granularity, good load balance and synchronization are satisfactorily achieved. The parallel performance gain must be considered against the deterioration of the vectorization performance.

In the implementation of the parallel GEBE procedure, within each group, elements are sub-grouped into n_{CHUNK} chunks with each chunk containing n_{VR} (e.g. 64) elements. The last chunk in each group may contain less than n_{VR} elements. We choose n_{CHUNK} and n_{VR} in such a way that n_{CHUNK} is a multiple of n_{CPU}. When parallel processing is invoked (with a compiler directive in the outer loop over n_{CHUNK}), each of the available n_{CPU} CPUs will work on one chunk at a time until all chunks are finished. Because each chunk contains the same amount of work (except for the last chunk) and the overall number of chunks can evenly be distributed to n_{CPU} CPUs, load balancing is ensured.

6. Numerical Tests

We have tested the AIE and GEBE methods on problems governed by the convection-diffusion and the two-dimensional incompressible Navier-Stokes equations.

one-dimensional pure advection of a cosine wave

The purpose of this simple test is to provide a conceptual illustration of the way the AIE scheme works. A cosine wave of unit amplitude is being advected from the left to the right with an advection velocity of 1.0. A uniform mesh is being used and the element Courant number is 0.8.

The AIE scheme is based on the selection between the implicit and the one-pass explicit versions of the streamline-upwind/Petrov-Galerkin (SUPG) procedure. The threshold value for the criterion given by equation (3.8) is set to 1%. Figure 1 shows the solution and the corresponding distribution of the implicit elements at time steps 0, 50, and 100. Compared to the implicit method, the AIE scheme results in a 79% reduction in the CPU time and 81% in the memory needed for the global matrix; the solutions obtained by the implicit and the AIE schemes are indistinguishable.

one-dimensional pure advection of a discontinuity

The conditions in this case are nearly identical to those in the previous case. This time the profile which is being advected is a discontinuity which initially spans four elements. The AIE scheme is based on the selection between the one-pass explicit SUPG method and the FCT method implemented as described in Section 3. The threshold value given by (3.8) is set to 0.01%. Figure 2 shows the solution and the corresponding distribution of the FCT elements, at time steps 0, 50, and 100. Compared to the "full" FCT approach the savings in the CPU time is 42%.

two-dimensional pure advection of a plateau

This test is for evaluating the performance of the AIE scheme for two-dimensional problems with moving sharp fronts. A 40×40 mesh is chosen in a 1.0×1.0 computational domain. All the boundary conditions are the Dirichlet type. The advection velocity is in the diagonal direction and has unit magnitude. The time step is 0.01.

The AIE scheme is based on the selection between the implicit and the one-pass explicit versions of the discontinuity-capturing SUPG procedure [14]. The threshold value for the criterion given by (3.8) is 1%. Figure 3 shows the solution and the distribution of the implicit elements at time steps 0, 20, and 40. Compared to the implicit method, the AIE scheme results in a 48% reduction in the CPU time and 79% in the memory needed for the global matrix.

flow past a circular cylinder at Reynolds number 100

In this problem both the AIE and the GEBE schemes were tested. We used a mesh with 1940 elements and 2037 nodal points. The dimensions of the computational domain, normalized by the cylinder diameter, are 16 and 8 in the flow and the cross flow directions, respectively. Figure 4 shows the finite element mesh and its element-grouped version ("GEBE mesh") for the parallel computations. The free-stream velocity is 0.125 and the initial condition for the vorticity is zero everywhere in the domain. The time step is 1.0.

The AIE scheme employed to solve the vorticity transport equation and the Poisson's equation is exactly as described in Section 3. For the vorticity transport equation, both the implicit and the explicit methods are SUPG based and involve several iterations due to the nonlinearity of the equation. In this problem the Courant number for some elements is above the stability limit of the one-pass explicit algorithm (we estimate the maximum Courant number to be somewhere around 1.5-1.6). Therefore the stability criterion given by equation (3.7) also becomes active in the implicit-explicit grouping. In fact our numerical experiments show that the computations do not converge when the entire set of elements belong to the explicit group. For the accuracy considerations, the threshold value of the criterion given by (3.10) is set to 1%. For solution of the discrete Poisson's equation, the iterations continue until the residual norm goes below 10^{-8} (we realize that this is rather an overkill).

Both the AIE and the GEBE methods give the expected solution for this problem: initially a symmetric flow pattern with two attached eddies growing in the wake of the cylinder, then the symmetry breaks, and as time goes by the vortices are formed alternately at the upper and lower downstream vicinity of the cylinder and carried along the Karman vortex street. Figures 5,6, and 7 show the solution obtained by the AIE scheme and the corresponding distributions of the implicit elements at time 680, 1500, and 2000.

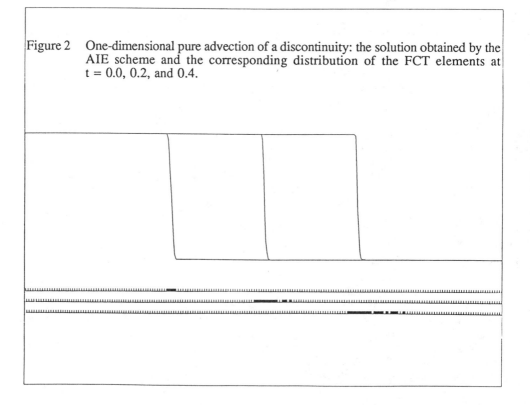

Figure 1 One-dimensional pure advection of a cosine wave: the solution obtained by the AIE scheme and the corresponding distribution of the implicit elements at t = 0.0, 0.2, and 0.4.

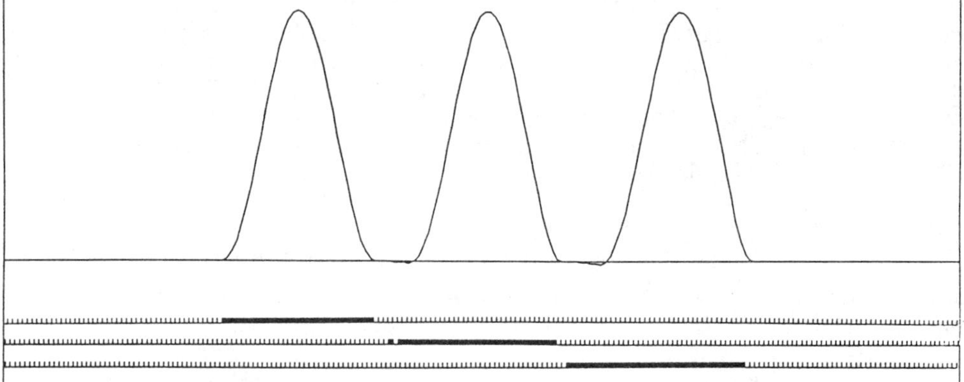

Figure 2 One-dimensional pure advection of a discontinuity: the solution obtained by the AIE scheme and the corresponding distribution of the FCT elements at t = 0.0, 0.2, and 0.4.

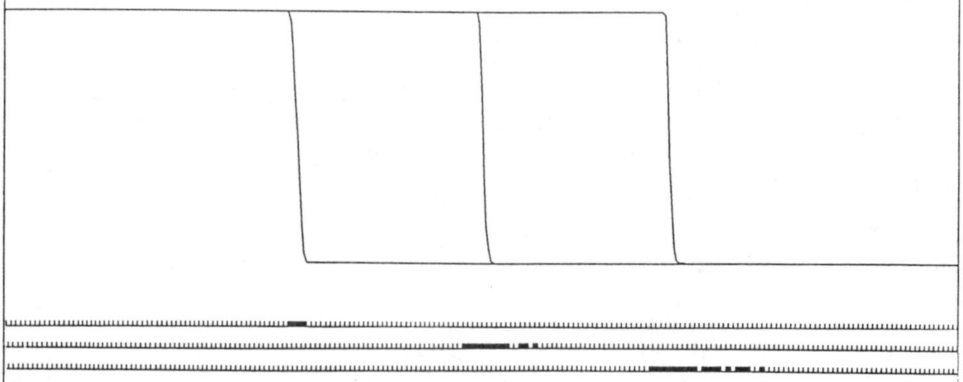

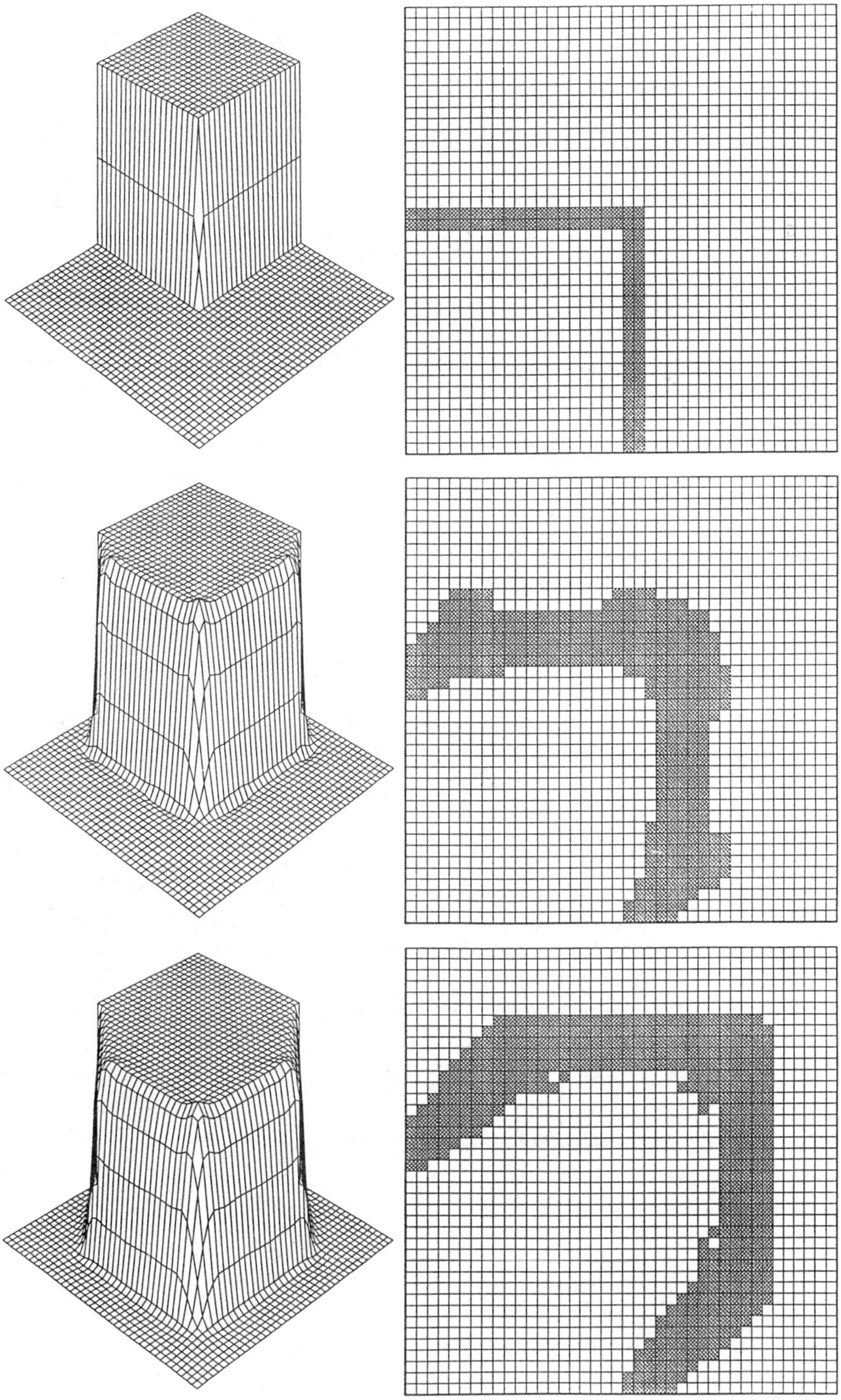

Figure 3 Two-dimensional pure advection of a plateau: the solution obtained by the AIE scheme and the corresponding distribution of the implicit elements at t = 0.0, 0.2, and 0.4.

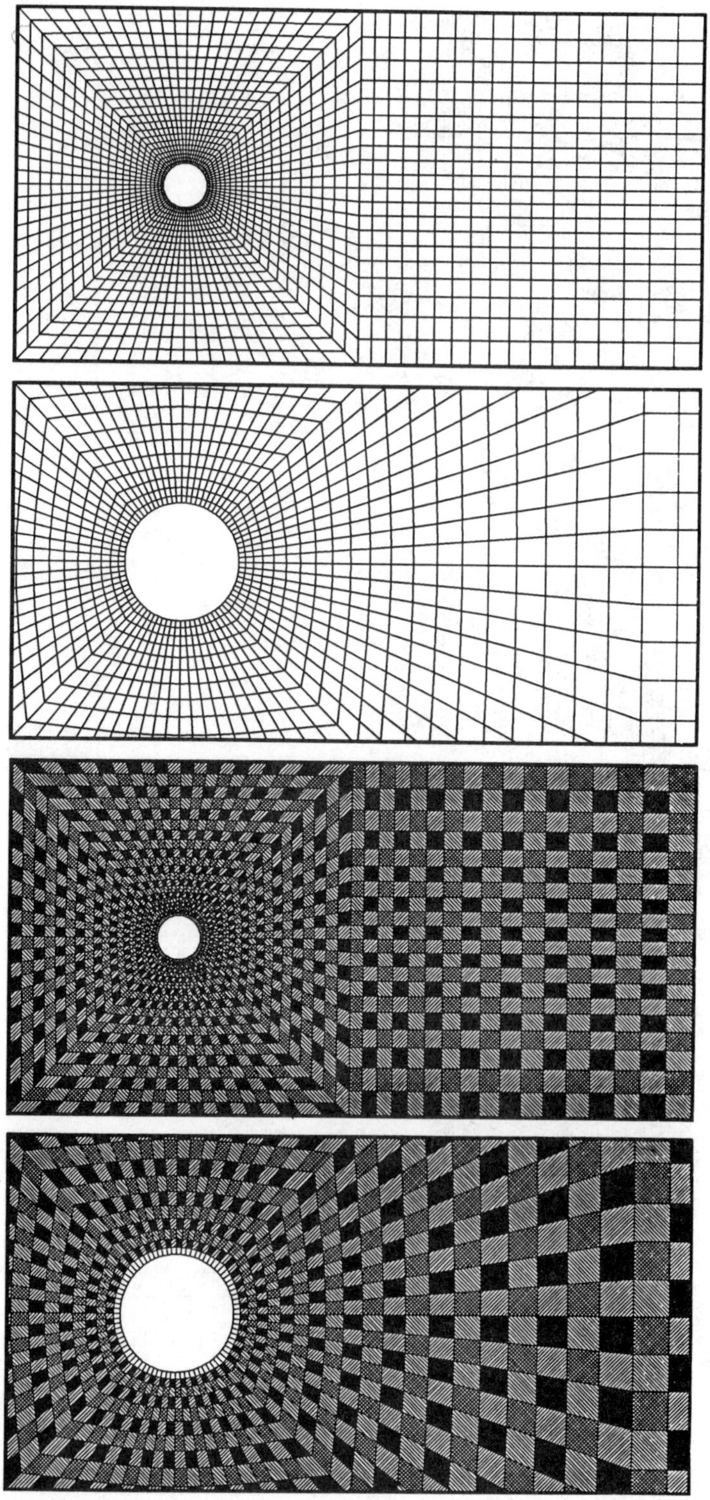

Figure 4 Flow past a circular cylinder at Reynolds number 100: the finite element mesh and its element grouped version ("GEBE mesh") for the parallel computations.

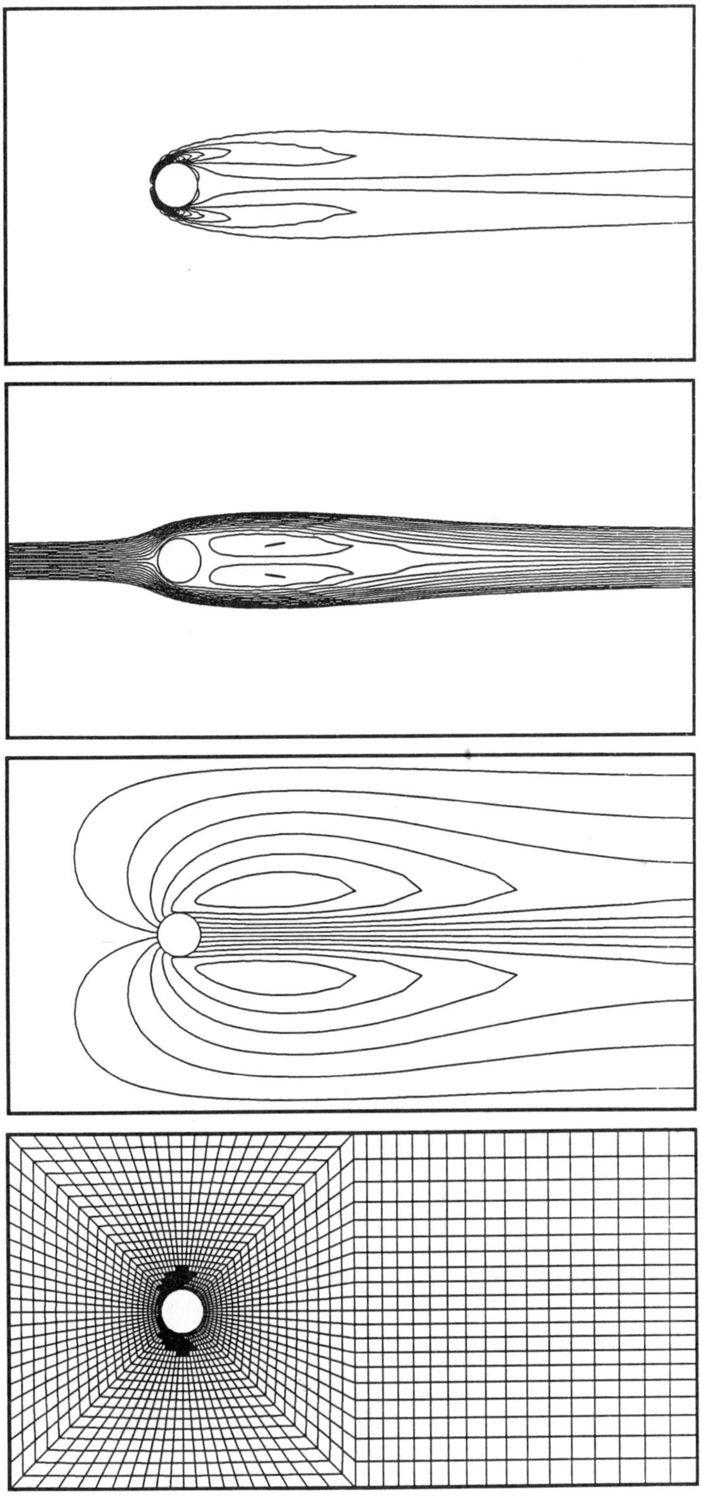

Figure 5 Flow past a circular cylinder at Reynolds number 100: the symmetrical solution obtained by the AIE scheme at t = 680; from top to bottom: vorticity, streamlines, relative streamlines, and distribution of the implicit elements.

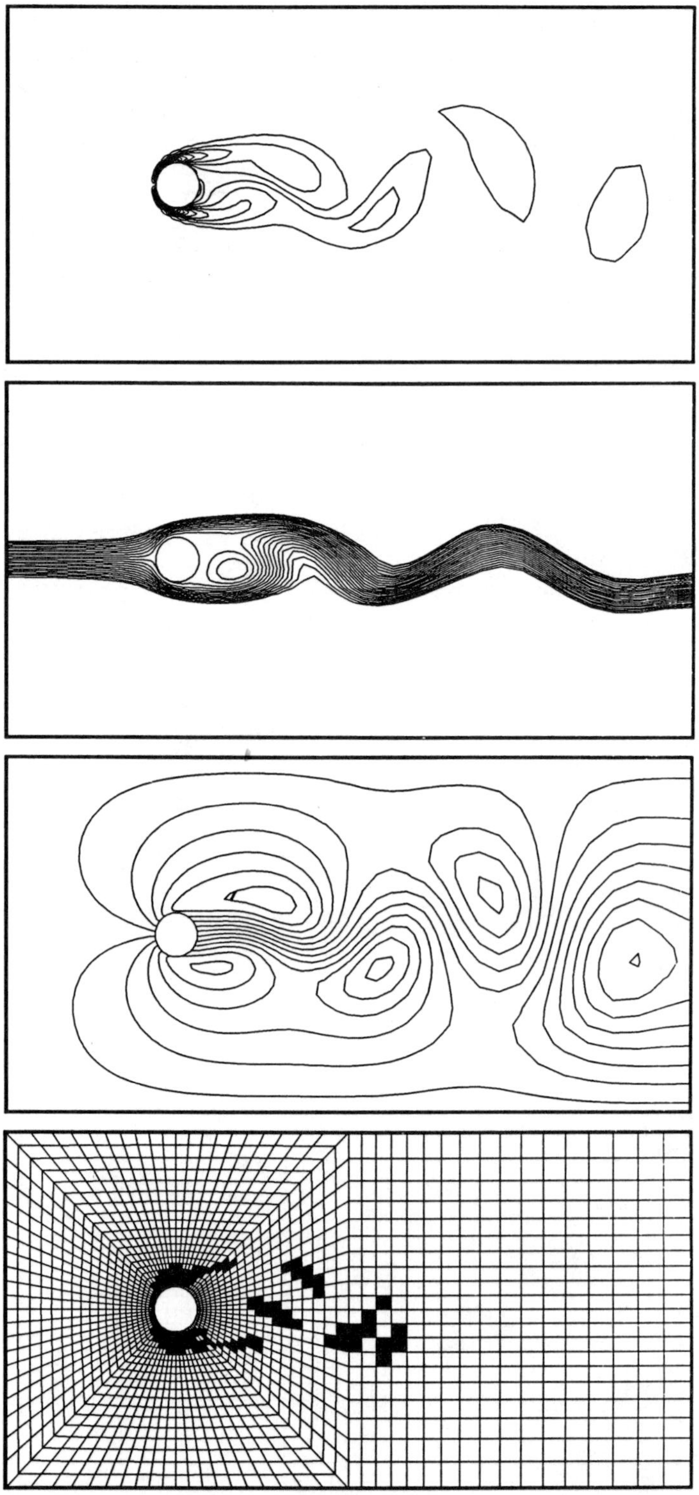

Figure 6 Flow past a circular cylinder at Reynolds number 100: the solution obtained by the AIE scheme at t = 1500; from top to bottom: vorticity, streamlines, relative streamlines, and distribution of the implicit elements.

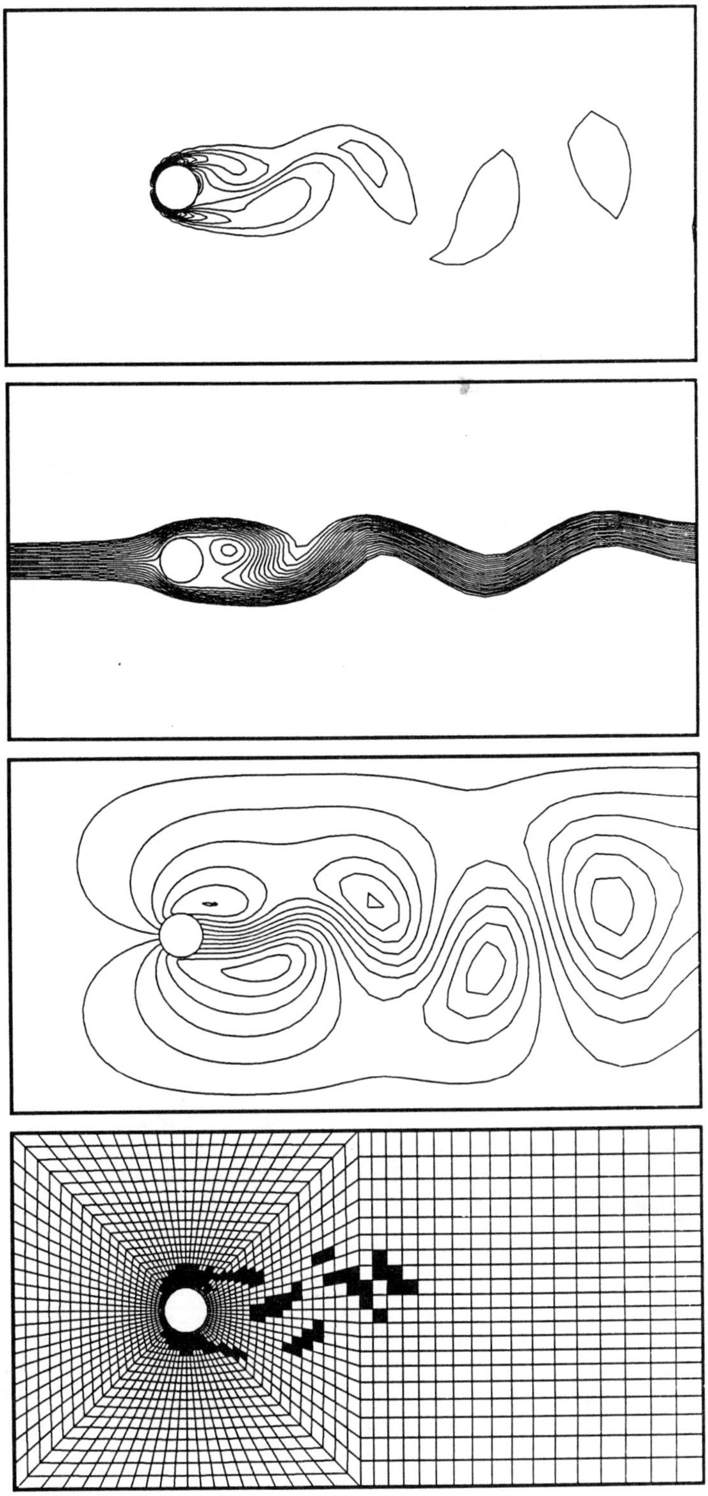

Figure 7 Flow past a circular cylinder at Reynolds number 100: the solution obtained by the AIE scheme at t = 2000; from top to bottom: vorticity, streamlines, relative streamlines, and distribution of the implicit elements.

The AIE method reduces the memory needed for the global matrices by 85%. The CPU time however is more than that of the implicit scheme. This is because, for the solution of the vorticity transport equation, the nonlinear iterations continue until a certain convergence criterion is met; this convergence criterion is the same whether it is the implicit or the AIE scheme. That is, for the solution of the vorticity transport equation, compared to the implicit method, the AIE scheme is not giving up any accuracy even in the explicit zones.

In the case of the GEBE computations, one has to note that for the first layer of elements around the cylinder computations can not be fully parallel. This is because of the way the problem is formulated to determine the unknown stream function value corresponding to this internal boundary. The tests for the GEBE computations were performed on a CRAY-XMP/416 : 4 CPUs, 8.5 ns clock, 128 Megabytes of memory, COS 1.16 operating system and products (Cray Research Mendota Heights data center).

The benchmark test lasted for 10 time steps with several non-linear iterations per time step. The vector performance reached 100 Mflops for the element level matrix/vector computations, 70 Mflops for the factorization and back substitution, and 25 Mflops for the global residual calculation. The overall program performance was at 48 Mflops. We expect to obtain higher performance rates as we vectorize the program further and filter out the initial setup/overhead cost. Parallel processing on the element level matrix/vector computations rendered speed ups of 3.5-3.8. The speed up on the factorization/back substitution and global residual calculation is around 2-3. This is due to the fact that the task granularity is small because of the relatively small size of the test problem.

7. Concluding Remarks

In this paper we have presented some efficient solution techniques for large equation systems emanating from the finite element formulation of fluid mechanics problems.

In the adaptive implicit-explicit (AIE) method the elements are dynamically grouped into implicit and explicit subsets based on the element level stability and accuracy criteria. This dynamic grouping idea can be applied in any context where there are some reasons for preferring a given procedure over another one. As an example we implemented the AIE concept in conjunction with the flux-corrected transport method.

The grouped element-by-element (GEBE) iteration scheme, which is a variation of the regular EBE scheme, is based on defining the preconditioner to be a sequential product of the element group matrices. To facilitate efficient parallel processing, the number of groups is minimized with the condition that there can be no inter-element coupling within each group.

We applied these methods to various test problems and demonstrated that substantial savings in the CPU time and memory can be achieved.

In our future studies we plan to experiment with the combinations of the AIE and GEBE schemes. For example we can use the GEBE scheme to solve the discrete Poisson's equation for the stream function (or the pressure), employ the AIE scheme to solve the time-dependent transport equation for the vorticity (or the momentum), but use, again, the GEBE scheme at the implicit zones of the implicit-explicit distribution.

Acknowledgement

We are grateful for the CRAY time provided by Cray Research Inc.

References

(1) D.K. Ganjoo, W.D. Goodrich, and T.E. Tezduyar, A New Formulation for Numerical Simulation of Electrophoresis Separation Processes, UMSI 88/37, April, 1988, University of Minnesota Supercomputer Institute.

(2) R. Glowinski, Numerical Methods for Nonlinear Variational Problems, Springer-verlag, New York, 1984.

(3) R. Glowinski, Q.V. Dinh, and J. Periaux, *Domain Decomposition Methods for Nonlinear Problems in Fluid Dynamics*, Computer Methods in Applied Mechanics and Engineering, 40 (1983), pp. 27-109.

(4) T.J.R. Hughes, R. M. Ferencz, *Fully Vectorized EBE Preconditioners for Nonlinear Solid Mechanics: Applications to Large-Scale Three-Dimensional Continuum, Shell and Contact/Impact Problems*, First International Symposium on Domain Decomposition Methods for Partial Differential Equations, R. Glowinski, G.H. Golub, G.A. Meurant, and J. Periaux (eds.), SIAM, 1988, pp. 261-280.

(5) T.J.R. Hughes, I. Levit and J.Winget, *An Element-by-Element Solution Algorithm for Problems of Structural and Solid Mechanics*, Computer Methods in Applied Mechanics and Engineering, 36 (1983), pp. 75-110.

(6) T.J.R. Hughes and W.K. Liu, *Implicit-Explicit Finite Elements in Transient Analysis: Stability Theory*, Journal of Applied Mechanics, 45 (1978), pp. 371-374.

(7) T.J.R. Hughes and W.K. Liu, *Implicit-Explicit Finite Elements in Transient Analysis: Implementation and Numerical Examples*, Journal of Applied Mechanics, 45 (1978), pp. 375-378.

(8) T.J.R. Hughes, J. Winget, I. Levit, and T.E. Tezduyar, *New Alternating Direction Procedures in Finite Element Analysis Based upon EBE Approximation Factorizations*, in Computer Methods for Nonlinear Solids and Mechanics, S.N. Atluri, and N. Perrone (eds.), AMD Vol. 54, ASME, New York, 1983, pp. 75-110.

(9) R. Lohner, K. Morgan, J. Peraire, and M. Vahdati, *Finite Element Flux-Corrected Transport (FEM-FCT) for the Euler and Navier-Stokes Equations*, Finite Elements in Fluids, 7 (1987), R.H. Gallagher, R. Glowinski, P.M. Gresho, J.T. Oden, and O.C. Zienkiewicz (eds.), John Wiley & Sons Ltd., pp. 105-121.

(10) T.E. Tezduyar and D.K. Ganjoo, *Petrov-Galerkin Formulations with Weighting Functions Dependent upon Spatial and Temporal Discretization: Application to Transient Convection-Diffusion Problems*, Computer Methods in Applied Mechanics and Engineering, 59 (1986), pp. 47-71.

(11) T.E. Tezduyar, R. Glowinski, and J. Liou, *Petrov-Galerkin Methods on Multiply-Connected Domains for the Vorticity - Stream Function Formulation of the Incompressible Navier-Stokes Equations*, to appear in the International Journal of Numerical Methods in Fluids.

(12) T.E. Tezduyar, R. Glowinski, J. Liou, T. Nguyen, and S. Poole, *Block-Iterative Finite Element Computations for Incompressible Flow Problems* , to appear in the Proceedings of the 1988 International Conference on Supercomputing, Saint-Malo, France, July 1988.

(13) T.E. Tezduyar and J. Liou, *Element-by-Element and Implicit-Explicit Finite Element Formulations for Computational Fluid Dynamics* , First International Symposium on Domain Decomposition Methods for Partial Differential Equations, R. Glowinski, G.H. Golub, G.A. Meurant, and J. Periaux (eds.), SIAM, 1988, pp. 281-300.

(14) T.E. Tezduyar and Y.J. Park, *Discontinuity-Capturing Finite Element Formulations for Nonlinear Convection-Diffusion-Reaction Equations*, Computer Methods in Applied Mechanics and Engineering, 59 (1986), pp. 307-325.

(15) S.T. Zalesak, *Fully Multidimensional Flux-Corrected Transport Algorithms for Fluids*, Journal of Computational Physics, 31 (1979), pp. 335-362.